Lecture Notes in Computer Science   12690

More information about this subseries at http://www.springer.com/series/7407

Ying Tan · Yuhui Shi (Eds.)

Advances in Swarm Intelligence

12th International Conference, ICSI 2021
Qingdao, China, July 17–21, 2021
Proceedings, Part II

 Springer

Editors
Ying Tan ⓘ
Peking University
Beijing, China

Yuhui Shi
Southern University of Science
and Technology
Shenzhen, China

ISSN 0302-9743 ISSN 1611-3349 (electronic)
Lecture Notes in Computer Science
ISBN 978-3-030-78810-0 ISBN 978-3-030-78811-7 (eBook)
https://doi.org/10.1007/978-3-030-78811-7

LNCS Sublibrary: SL1 – Theoretical Computer Science and General Issues

This Springer imprint is published by the registered company Springer Nature Switzerland AG
The registered company address is: Gewerbestrasse 11, 6330 Cham, Switzerland

Preface

This book and its companion volume, comprising LNCS volumes 12689 and 12690, constitute the proceedings of The Twelfth International International Conference on Swarm Intelligence (ICSI 2021) held during July 17–21, 2021, in Qingdao, China, both on-site and online.

The theme of ICSI 2021 was "Serving Life with Swarm Intelligence." The conference provided an excellent opportunity for academics and practitioners to present and discuss the latest scientific results and methods, innovative ideas, and advantages in theories, technologies, and applications in swarm intelligence. The technical program covered a number of aspects of swarm intelligence and its related areas. ICSI 2021 was the twelfth international gathering for academics and researchers working on most aspects of swarm intelligence, following successful events in Serbia (ICSI 2020, virtually), Chiang Mai (ICSI 2019), Shanghai (ICSI 2018), Fukuoka (ICSI 2017), Bali (ICSI 2016), Beijing (ICSI-CCI 2015), Hefei (ICSI 2014), Harbin (ICSI 2013), Shenzhen (ICSI 2012), Chongqing (ICSI 2011), and Beijing (ICSI 2010), which provided a high-level academic forum for participants to disseminate their new research findings and discuss emerging areas of research. ICSI 2021 also created a stimulating environment for participants to interact and exchange information on future challenges and opportunities in the field of swarm intelligence research.

Due to the ongoing COVID-19 pandemic, ICSI 2021 provided opportunities for both online and offline presentations. On the one hand, ICSI 2021 was held normally in Qingdao, China, but on the other hand, the ICSI 2021 technical team provided the ability for authors who were subject to restrictions on overseas travel to present their work through an interactive online platform or video replay. The presentations by accepted authors were made available to all registered attendees on-site and online.

The host city of ICSI 2021, Qingdao (also spelled Tsingtao), is a major sub-provincial city in the eastern Shandong province, China. Located on the western shore of the Yellow Sea, Qingdao is a major nodal city on the 21st Century Maritime Silk Road arm of the Belt and Road Initiative that connects East Asia with Europe, and has the highest GDP of any city in the province. It had jurisdiction over seven districts and three county-level cities till 2019, and as of 2014 had a population of 9,046,200 with an urban population of 6,188,100. Lying across the Shandong Peninsula and looking out to the Yellow Sea to its south, Qingdao borders the prefectural cities of Yantai to the northeast, Weifang to the west, and Rizhao to the southwest.

ICSI 2021 received 177 submissions and invited submissions from about 392 authors in 32 countries and regions (Algeria, Australia, Bangladesh, Belgium, Brazil, Bulgaria, Canada, China, Colombia, India, Italy, Japan, Jordan, Mexico, Nigeria, Peru, Portugal, Romania, Russia, Saudi Arabia, Serbia, Slovakia, South Africa, Spain, Sweden, Taiwan (China), Thailand, Turkey, United Arab Emirates, UK, USA, and Vietnam) across 6 continents (Asia, Europe, North America, South America, Africa, and Oceania). Each submission was reviewed by at least 2 reviewers, and had on average 2.5 reviewers. Based on rigorous reviews by the Program Committee members and additional reviewers, 104 high-quality papers were selected for publication in this proceedings, an acceptance rate of 58.76%. The papers are organized into 16 cohesive sections covering major topics of swarm intelligence research and its development and applications.

On behalf of the Organizing Committee of ICSI 2021, we would like to express our sincere thanks to the International Association of Swarm and Evolutionary Intelligence (IASEI), which is the premier international scholarly society devoted to advancing the theories, algorithms, real-world applications, and developments of swarm intelligence and evolutionary intelligence. We would also like to thank Peking University, Southern University of Science and Technology, and Ocean University of China for their co-sponsorships, and the Computational Intelligence Laboratory of Peking University and IEEE Beijing Chapter for their technical co-sponsorships, as well as our supporters including the International Neural Network Society, World Federation on Soft Computing, International Journal of Intelligence Systems, MDPI's journals Electronics and Mathematics, Beijing Xinghui Hi-Tech Co., and Springer Nature.

We would also like to thank the members of the Advisory Committee for their guidance, the members of the Program Committee and additional reviewers for reviewing the papers, and the members of the Publication Committee for checking the accepted papers in a short period of time. We are particularly grateful to Springer for publishing the proceedings in the prestigious series of Lecture Notes in Computer Science. Moreover, we wish to express our heartfelt appreciation to the plenary speakers, session chairs, and student helpers. In addition, there are still many more colleagues, associates, friends, and supporters who helped us in immeasurable ways; we express our sincere gratitude to them all. Last but not the least, we would like to thank all the speakers, authors, and participants for their great contributions that made ICSI 2021 successful and all the hard work worthwhile.

May 2021   Ying Tan
                                                                           Yuhui Shi

Organization

General Co-chairs

Ying Tan — Peking University, China
Russell C. Eberhart — IUPUI, USA

Program Committee Chair

Yuhui Shi — Southern University of Science and Technology, China

Advisory Committee Chairs

Xingui He — Peking University, China
Gary G. Yen — Oklahoma State University, USA

Technical Committee Co-chairs

Haibo He — University of Rhode Island, USA
Kay Chen Tan — City University of Hong Kong, China
Nikola Kasabov — Auckland University of Technology, New Zealand
Ponnuthurai Nagaratnam Suganthan — Nanyang Technological University, Singapore
Xiaodong Li — RMIT University, Australia
Hideyuki Takagi — Kyushu University, Japan
M. Middendorf — University of Leipzig, Germany
Mengjie Zhang — Victoria University of Wellington, New Zealand
Qirong Tang — Tongji University, China
Milan Tuba — Singidunum University, Serbia

Plenary Session Co-chairs

Andreas Engelbrecht — University of Pretoria, South Africa
Chaoming Luo — University of Mississippi, USA

Invited Session Co-chairs

Andres Iglesias — University of Cantabria, Spain
Haibin Duan — Beihang University, China

Special Sessions Chairs

Ben Niu   Shenzhen University, China
Yan Pei   University of Aizu, Japan

Tutorial Co-chairs

Junqi Zhang   Tongji University, China
Shi Cheng   Shanxi Normal University, China

Publications Co-chairs

Swagatam Das   Indian Statistical Institute, India
Radu-Emil Precup   Politehnica University of Timisoara, Romania

Publicity Co-chairs

Yew-Soon Ong   Nanyang Technological University, Singapore
Carlos Coello   CINVESTAV-IPN, Mexico
Yaochu Jin   University of Surrey, UK
Rossi Kamal   GERIOT, Bangladesh
Dongbin Zhao   Institute of Automation, China

Finance and Registration Chairs

Andreas Janecek   University of Vienna, Austria
Suicheng Gu   Google, USA

Local Arrangement Chair

Gai-Ge Wang   Ocean University of China, China

Conference Secretariat

Renlong Chen   Peking University, China

Program Committee

Ashik Ahmed   Islamic University of Technology, Bangladesh
Rafael Alcala   University of Granada, Spain
Abdelmalek Amine   Tahar Moulay University of Saida, Algeria
Sabri Arik   Istanbul University, Turkey
Carmelo J. A. Bastos Filho University of Pernambuco, Brazil
Sandeep Bhongade   G.S. Institute of Technology, India
Sujin Bureerat   Khon Kaen University, Thailand
Bin Cao   Tsinghua University, China

Zhongzhi Shi	Institute of Computing Technology, China
Joao Soares	GECAD, Portugal
Xiaoyan Sun	China University of Mining and Technology, China
Yifei Sun	Shaanxi Normal University, China
Ying Tan	Peking University, China
Qirong Tang	Tongji University, China
Mladen Veinović	Singidunum University, Serbia
Cong Wang	Northeastern University, China
Dujuan Wang	Sichuan University, China
Guoyin Wang	Chongqing University of Posts and Telecommunications, China
Rui Wang	National University of Defense Technology, China
Yong Wang	Central South University, China
Yuping Wang	Xidian University, China
Ka-Chun Wong	City University of Hong Kong, China
Shunren Xia	Zhejiang University, China
Ning Xiong	Mälardalen University, Sweden
Benlian Xu	Changsu Institute of Technology, China
Peng-Yeng Yin	National Chi Nan University, Taiwan, China
Jun Yu	Niigata University, Japan
Saúl Zapotecas Martínez	UAM Cuajimalpa, Mexico
Jie Zhang	Newcastle University, UK
Junqi Zhang	Tongji University, China
Tao Zhang	Tianjin University, China
Xingyi Zhang	Huazhong University of Science and Technology, China
Xinchao Zhao	Beijing University of Posts and Telecommunications, China
Yujun Zheng	Zhejiang University of Technology, China
Zexuan Zhu	Shenzhen University, China
Miodrag Zivkovic	Singidunum University, Serbia
Xingquan Zuo	Beijing University of Posts and Telecommunications, China

Additional Reviewers

Bao, Lin	Márquez Grajales, Aldo
Chen, Yang	Ramos, Sérgio
Cortez, Ricardo	Rivera Lopez, Rafael
Gu, Lingchen	Rodríguez de la Cruz, Juan Antonio
Han, Yanyang	Vargas Hakim, Gustavo Adolfo
Hu, Yao	Yu, Luyue
Lezama, Fernando	Zhou, Tianwei
Liu, Xiaoxi	

Contents – Part II

UAV Cooperation and Control

Machine Learning

Data Mining

Other Applications

Contents – Part I

Particle Swarm Optimization

Fireworks Algorithms

Brain Storm Optimization Algorithm

Bacterial Foraging Optimization Algorithm

DNA Computing Methods

Multi-objective Optimization

A Multi-objective Evolutionary Algorithm
Based on Second-Order Differential Operator

Ruizhi Wan, Yinnan Chen, and Xinchao Zhao[✉]

School of Science, Beijing University of Posts and Telecommunications, Beijing 100876, China
zhaoxc@bupt.edu.cn

Abstract. Differential evolution (DE) is a swarm intelligence algorithm based on population, which has been used to solve multi-objective optimization problems (MOP). The distribution and convergence of the non-dominated solutions set are often the key indicators to evaluate the merits of MOP algorithms. In this paper, we propose a decomposition multi-objective evolutionary algorithm based on second-order differential operator (MOEA/D-SODE). By selecting the commonly used ZDT, DTLZ and IMOP benchmark functions, comparing with the existing differential evolution MOEA/D-DE, the solutions set obtained by MOEA/D-SODE algorithm has better convergence and distribution. The experimental results verify the effectiveness of MOEA/D-SODE algorithm, which provides a new and effective method for MOP.

Keywords: Differential Evolution · Multi-objective Optimization · Second-order Differential Operator · SODE

1 Introduction

Differential Evolution was proposed by Price et al. in 1995 [1], which is a population evolution algorithm through cooperation and competition among individuals. By virtue of the unique differential evolutionary operator, DE has good robustness and strong convergence. Because it is simple and excellent global optima search ability, DE was introduced to solve multi-objective optimization problems by Robic et al. [2] in 2005. Combined with non-dominated sorting genetic algorithm-II (NSGA-II) [3], DE is to solve the single-objective optimization problem. In 2009, Zhang et al. combined the MOEA/D [4] with DE and proposed the MOEA/D-DE [5] algorithm, which extended the method of solving MOP problems. The new MOEA/D algorithm based on adaptive differential evolution proposed by Chen et al. [6] not only can avoid falling into local optimum but also ensure diversity. Geng et al. [7] proposed MOEA/ D-IDE algorithm which has a better performance in speed and accuracy of convergence. Wu et al. [8] proposed a self-adaptive differential evolution with random neighborhood-based strategy and generalized opposition-based learning, which was proved to have high searching accuracy, fast convergence speed and strong robustness. The wavelet basis function

© Springer Nature Switzerland AG 2021
Y. Tan and Y. Shi (Eds.): ICSI 2021, LNCS 12690, pp. 3–12, 2021.
https://doi.org/10.1007/978-3-030-78811-7_1

and normal distribution are used to control parameters by Deng et al. [9], proposed a MOEA/D-WMSD which has greatly improved the convergence and distribution, and can effectively solve the multi-objective optimization problems.

In this paper, we proposed a second-order differential algorithm (SODE), which changes the usual differential vector producing in the DE into the second-order differential vector. Its effect is verified when solving the multi-objective problems. The second section presents concepts of MOP problems. The third and fourth section introduces SODE and the main framework of MOEA/D-SODE. The fifth section provides experimental results and analysis. The sixth section is conclusion.

2 MOP Problem

According to the duality principle of optimization theory, the optimization maximization problems can be solved by the optimization minimization problems. Without loss of generality, suppose the MOP with n decision variables, m objective functions, and $p + q$ constraint conditions, the MOP can be defined as

$$
\begin{cases}
\min F(x) = (f_1(x), f_2(x), \ldots, f_m(x))^T \\
s.t. \, g_i(x) \geq 0, i = 1, 2, \ldots, p \\
h_j(x) = 0, j = 1, 2, \ldots, q \\
x^L \leq x \leq x^U
\end{cases}
\tag{1}
$$

Where $x = (x_1, x_2, \ldots, x_n)^T \in X$ is a n-dimensional decision vector, X is called n-dimensional decision space; $x^L = (x_1^L, x_2^L, \ldots, x_n^L)$, $x^U = (x_1^U, x_2^U, \ldots, x_n^U)$, x_i^L and x_i^U are the lower and upper boundary of x_i separately. $g_i(x)$ is inequality constraint condition. $h_j(x)$ is equality constraint condition. Constraint functions $g_i(x)$ and $h_j(x)$ decide the feasible domain of vector x.

3 Second-Order Differential Evolution

SODE consists of four stages: initialization, mutation, crossover and selection. Except for mutation stage, the other parts of SODE are the same as original differential evolution.

At the mutation operation stage, the mutation operation DE/rand/1 is used in the model strategy. In order to effectively utilize the direction information and search state of the current population, the second order differential vector mechanism based on the classical mutation strategy is shown in Eqs. (2)–(7).

$$
d^G = x_{r_1}^G - x_{r_2}^G
\tag{2}
$$

$$
d_1^G = x_{r_3}^G - x_{r_4}^G
\tag{3}
$$

$$d_1^G = x_{best}^G - x_{r_4}^G \tag{4}$$

$$d_2^G = x_{r_5}^G - x_{r_6}^G \tag{5}$$

$$d_r^G = d^G + \lambda(d_1^G - d_2^G) \tag{6}$$

where $r_1, r_2, r_3, r_4, r_5, r_6$ are six randomly selected integers in [1, NP]. According to the reference [10], λ is set to 0.1. Equations (3, 4) are two ways of generating d_1^G. x_{best}^G is the best vector of the G-th generation.

$d_1^G - d_2^G$ is a second-order differential vector in Eq. (6). Two ways of generating variation vector v_i^G are as Eq. (7), (8):

$$v_i^G = x_{r_7}^G + F \cdot d_r^G \tag{7}$$

$$v_i^G = x_{best}^G + F \cdot d_r^G \tag{8}$$

F is a scale parameter which is set to 0.5. r_7 is a randomly selected integer in [1, NP].

4   MOEA/D-SODE Algorithm

4.1   General Framework of MOEA/D-SODE

Replacing the MOEA/D genetic operator by the SODE operator, and using the polynomial mutation, the MOEA/D-SODE based on Chebyshev decomposition mechanism is obtained. To maintain the population diversity, the times that a new solution is allowed to replace old solutions are restricted to the number n_r. The pseudo code of the algorithm is shown in Algorithm 1.

Algorithm 1 General Framework of MOEA/D-SODE

Input: N (population size), $\lambda^1, \lambda^2, \cdots, \lambda^N$ (a set of weight vectors)

Output: $\{x^1, \cdots, x^N\}$ (approximation to the Pareto Set),

$(F(x^1), \cdots, F(x^N))$ (approximation to the Pareto Front)

1) Calculate the Euclidean distances between any two weight vectors and then figure out the closest weight vectors $B(i)$ ($i = 1, \cdots, N$) for each weight vector. Set $B(i) = \{i_1, \cdots, i_T\}$.

2) $x^1, \cdots, x^N \leftarrow RandomInitialize(N)$.

Set $FV^i = F(x^i)$.

3) $z = (z_1, \cdots, z_m) \leftarrow z_j = \min_{1 \le i \le N} f_j(x^i)$

4) **While** termination criterion not fulfilled **do**

5) **For** $i = 1, \cdots, N$

6) $rand \leftarrow$ Generate a uniformly random number from $[0, 1]$

7) then

8) $P = \begin{cases} B(i) & \text{if } rand < \delta \\ \{1, \cdots, N\} & \text{otherwise} \end{cases}$, δ is a control parameter.

9) $y \leftarrow SODE(x^{r_1}, x^{r_1}, \cdots, x^{r_5}, x^{best})$, $r_1, r_2, \cdots, r_5$ are 5 indexes randomly selected

from P, x_{best}^G is randomly selected from the non-dominant solutions of $\{x^1, \cdots, x^N\}$.

10) **For** $j = 1, \cdots, m$ if $z_j > f_j(y)$

then $z_j \leftarrow f_j(y)$ **end For**

11) Set $c=0$ then

① If $c=n_r$ or P is empty, go to step 4)

else $j \leftarrow$ randomly picked index from P

end If

12) ② If $g(y \mid \lambda^j, z) \le g(x^j \mid \lambda^j, z)$ ($g(x \mid \lambda, z^*) = \max_{1 \le i \le m} \{\lambda_i \mid f_i(x) - z_i^* \mid\}$) then set

$x^j = y$, $FV^j = F(y)$, and $c=c+1$

end If

13) ③ Remove j from P and go to ①

14) **end For**

15) **end while**

16) **return** $\{x^1, \cdots, x^N\}$ and $F(x^1), \cdots, F(x^N)$

4.2 A SODE-Best Second-Order Differential Operator

According to experimental tests, the algorithm using SODE-best strategy combined with Eqs. (2), (4), (5), (6), (8) has better performance. The best individuals x_{best}^G are randomly selected from the non-dominant solutions of each generation. When the differential vector d^G is generated by Eq. (2), if it exists one dominating another one between the two randomly selected individuals, the dominant solution is used to subtract the non-dominant solution. The pseudo code based on a SODE-best second-order differential operator with polynomial mutation is shown in Algorithm 2:

Algorithm 2 SODE($x^1, x^2, \cdots, x^5, x^{best}$)

Input: $x^1, x^2, \cdots, x^5, x^{best}$

Output: y

1) **If** x^1 is dominated by x^2 then $d = x^2 - x^1$
2) **else** $d = x^1 - x^2$ **end if**
3) $d_1 = x^{best} - x^3$
4) $d_2 = x^4 - x^5$
5) $d_r = d + \lambda * (d_1 - d_2)$
6) $\bar{y} = x^{best} + F * d_r$ (λ, F are two scale factors)

7) $y_k = \begin{cases} \bar{y}_k + \sigma_k \times (b_k - a_k) & \text{with probability } p \\ \bar{y}_k & \text{with probability } 1-p \end{cases}$

8) $\sigma_k = \begin{cases} (2 \times rand)^{\frac{1}{\eta+1}} - 1 & \text{if } rand < 0.5 \\ 1 - (2 - 2 \times rand)^{\frac{1}{\eta+1}} & \text{otherwise} \end{cases}$

9) $y = (y_1, y_2, \cdots, y_n)$, $\bar{y} = (\bar{y}_1, \bar{y}_1, \cdots, \bar{y}_n)$. *rand* is a uniformly random number which generates from [0, 1], η and p are two control parameters, a_k and b_k are the lower and upper bounds of the k-th decision variable respectively.

10) **return** $y = (y_1, y_2, \cdots, y_n)$

4.3 The Flow Chart of MOEA/D-SODE

In summary, the flow chart of MOEA/D-SODE is shown in Fig. 1.

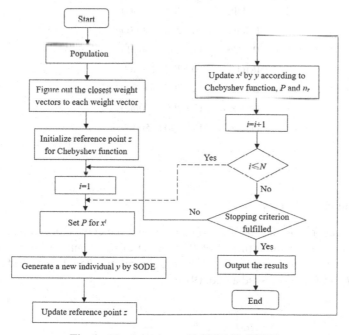

Fig. 1. The flow chart of MOEA/D-SODE

5 Experimental Results and Analysis

To verify the performance of MOEA/D-SODE, we select ZDT, DTLZ and IMOP benchmark functions [12] which include ZDT1, ZDT2, ZDT6, DTLZ2, DTLZ5, DTLZ6, IMOP2, IMOP5, IMOP6, IMOP8. In these 10 functions, ZDT1, ZDT2, ZDT6, IMOP2 are 2-objective functions, others are 3-objective functions. In addition, IMOP5, IMOP6, IMOP8 have complicated Pareto fronts.

5.1 Experimental Environment and Parameter Setting

The parameters are the decision variables D, population size N and maximum function evaluations which are shown in Table 1. According to [11], SODE with scale factor $F = 0.5$ and $\lambda = 0.1$ has the best performance in single-objective problems. Simple tests show taking the same setting would have the best performance correspondingly. Other parameters are crossover probability $CR = 0.5$, $\eta = 20$ and $p = 1/n$ in polynomial mutation, neighborhood size $T = 20$, $\delta = 0.9$, $n_r = 2$. Each algorithm runs independently 20 times on each test function.

Table 1. Parameters for different test functions

Functions	D	N	Evaluations
ZDT1	30	100	50000
ZDT2	30	100	50000
ZDT6	10	100	50000
IMOP2	10	200	50000
IMOP5	10	200	50000
IMOP6	10	300	50000
IMOP8	10	300	100000
DTLZ2	12	200	50000
DTLZ5	12	200	50000
DTLZ6	12	200	50000

5.2 Performance Metrics

The inverted generation distance (IGD) can measure the convergence and distribution of the algorithm at the same time. Let P^* be a set of points uniformly distributed on the PF, P is a set of points approximated by the algorithm. Then the inverted generation distance of P^* to P is expressed as Eq. (9):

$$IGD(P^*, P) = \frac{\sum\limits_{v \in P^*} d(v, P)}{|P^*|} \tag{9}$$

where m is the number of targets, $d(v, P)$ represents the minimum of the *Euclidean* distances the point v to each point in P. P^* is a sufficiently wide and uniform set selected from the PF, $|P^*|$ represents the number of solutions in the P^*. And the smaller value of the IGD (P^*, P) shows that the Pareto optimal solution set has better convergence and diversity.

Hypervolume (HV) index is the volume of the region in the target space formed by the non-dominant solution set and the reference point obtained by the algorithm. It is expressed as Eq. (10):

$$HV = \delta(\bigcup_{i=1}^{|S|} v_i) \tag{10}$$

where δ represents Lebesgue measure used to measure volume. The $|S|$ represents the number of solutions in the non-dominant solution set. v_i represents the volume of the hypercube formed by the reference point and the i-th solution of the solution set. The larger value of HV, the better diversity of the Pareto optimal solution set and the better the comprehensive performance of the algorithm.

5.3 Analysis of Experimental Results

To verify the effectiveness of MOEA/D-SODE, well-known MOEA/D-DE [5] is selected for the comparison algorithm. MOEA/D-DE uses DE operator in the mutation and crossover stage, and two algorithms are the same for other parts. The parameter values are the same in [5]. By running the MOEA/D-DE and MOEA/D-SODE, we compute the average results of IGD and HV metrics and show them in Table 2 and Table 3 respectively. In Tables, M represents the objective number of the functions, the values in parentheses represent standard deviations. The better average values are highlighted with a gray background. The results of $\pm/\approx$ indicate that the results of MOEA/D-SODE are significantly better than/ significantly worse than/ no significant difference with those of MOEA/D-DE under Wilcoxon test with 5% significance level.

We can see from Table 2 that in all 10 test functions, the IGD mean values obtained by MOEA/D-DE are all greater than MOEA/D-SODE, and the numbers of $+/-/\approx$ obtained are 6/0/4 respectively. It shows that MOEA/D-SODE performs significantly better than MOEA/D-DE on these 10 test functions under the IGD metric.

As is shown in Table 3, for all 10 functions, MOEA/D-DE obtains IGD mean values of 4 greater than MOEA/D-SODE, 6 less than MOEA/D-SODE, and the numbers of $+/-/\approx$ are 3/5/2 respectively. It shows that the performance of MOEA/D-SODE on these 10 functions is slightly better than those of MOEA/D-DE under the HV metric.

All the IGD values of MOEA/D-SODE are better in Fig. 2. For the figures of IMOP2, MOEA/D-SODE successfully detects more solutions on the boundary of PF. For the figures of IMOP5, there are more solutions in one "circle" out of eight "circles" of true PF detected by MOEA/D-SODE, and the solutions form a semi-circle approximated to the true PF. For the figures of IMOP6, the solutions of MOEA/D-SODE form a chessboard-shaped figure with straighter "lines".

Table 2. IGD metric values obtained by two algorithms on 10 functions

Problem	M	MOEA/D-DE	MOEA/D-SODE
ZDT1	2	4.8305e−3 (7.80e−4)≈	4.7693e−3 (5.36e−4)
ZDT2		4.2733e−3 (2.79e−4) +	3.9225e−3 (8.36e−5)
ZDT6		3.1071e−3 (1.12e−5)+	3.1049e−3 (1.81e−6)
IMOP2		3.4405e−2 (2.40e−3)+	9.5405e−3 (2.22e−3)
IMOP5	3	5.9432e−2 (3.27e−3)≈	5.8929e−2 (5.03e−3)
IMOP6		2.6318e−2 (4.78e−4)+	2.6047e−2 (5.31e−4)
IMOP8		1.3215e−1 (5.32e−3)+	1.2096e−1 (6.82e−3)
DTLZ2		4.9077e−2 (3.49e−4)≈	4.9006e−2 (4.15e−4)
DTLZ5		7.2774e−3 (5.51e−5)≈	7.2627e−3 (7.61e−5)
DTLZ6		7.3721e−3 (1.60e−5)+	7.3625e−3 (1.30e−5)

Table 3. HV metric values obtained by two algorithms on 10 functions

Problem	M	MOEA/D-DE	MOEA/D-SODE
ZDT1	2	7.1758e−1 (1.45e−3)≈	7.1761e−1 (1.07e−3)
ZDT2		4.4286e−1 (8.92e−4)−	4.4414e−1 (4.27e−4)
ZDT6		3.8885e−1 (1.88e−4)≈	3.8889e−1 (1.50e−5)
IMOP2		2.3036e−1 (1.52e−4)−	2.3110e−1 (5.95e−5)
IMOP5	3	5.0697e−1 (8.79e−3)+	5.0401e−1 (1.18e−2)
IMOP6		5.3536e−1 (5.21e−4)+	5.3393e−1 (1.43e−3)
IMOP8		4.8280e−1 (5.64e−3)≈	4.8459e−1 (4.82−3)
DTLZ2		5.5278e−1 (1.10e−3)+	5.5150e−1 (1.32e−3)
DTLZ5		1.9840e−1 (3.43e−5)≈	1.9832e−1 (6.73e−5)
DTLZ6		1.9864e−1 (1.01e−5)≈	1.9866e−1 (9.13e−6)

Synthesizing Table 2, Table 3 and Fig. 2, the test functions with two and three objectives, regular and irregular PFs are solved using MOEA/D-SODE. The obtained approximately optimal solution sets of MOEA/D-SODE are better, with better distribution and coverage, indicating that the proposed MOEA/D-SODE algorithm has better optimization effect and distribution when solving these test functions. It is mainly due to the fact that the SODE operator of MOEA/D-SODE has advantages in population diversity maintenance and search ability enhancement. Therefore, the second-order differential evolution has more advantages than the original differential evolution in MOEA/D. SODE is used to improve the global search performance, population diversity and distribution of solution sets.

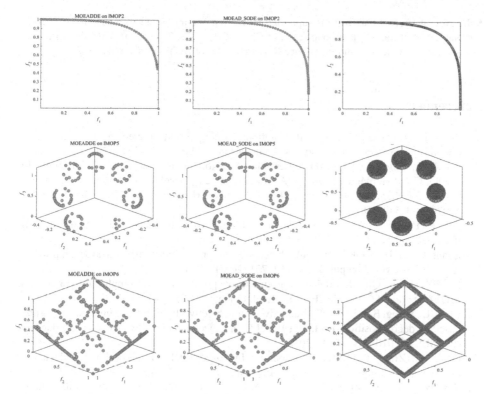

Fig. 2. PF comparison of IMOP2, IMOP5, IMOP6 between MOEA/D-DE and MOEA/D-SODE and corresponding true PFs. For each line, the left graph is the graph of MOEA/D-DE, the middle is the MOEA/D-SODE, and the right is the true PF.

6 Conclusion

Focusing on MOPs, we proposed a MOEA/D algorithm based on second-order differential strategy in this paper. We mainly change the mutation and crossover stage from genetic operator to second-order differential operator in MOEA/D. Through 10 benchmark functions, we compare MOEA/D-SODE with the MOEA/D-DE algorithm based on differential strategy. The results show that the optimal solution set obtained by MOEA/D-SODE has better convergence and distribution performance. It obviously can be seen from Fig. 2 that MOEA/D-SODE has less difference with the true PF especially on some functions with complicated PFs. Overall, SODE operator can utilize more and deep beneficial information from more individuals, which is helpful to accelerate the convergence of the algorithm and obtain better solutions.

For future research, we have the following two research directions: 1) it is meaningful to extend the application of SODE operator to the common many-objective optimization efficiency; 2) it is worthwhile applying SODE strategy to vehicle path optimization, transportation problem optimization, portfolio optimization etc.

Acknowledgement. This work is supported by Beijing Natural Science Foundation (1202020) and National Natural Science Foundation of China (61973042, 71772060). We will express our awfully thanks to the Swarm Intelligence Research Team of BeiYou University.

References

1. Storn, R., Price, K.: Differential evolution – a simple and efficient heuristic for global optimization over continuous spaces. J. Global Optim. **11**(4), 341–359 (1997)
2. Robič, T., Filipič, B.: DEMO: differential evolution for multiobjective optimization. In: Coello, C.A., Hernández Aguirre, A., Zitzler, E. (eds.) Evolutionary multi-criterion optimization. LNCS, vol. 3410, pp. 520–533. Springer, Heidelberg (2005). https://doi.org/10.1007/978-3-540-31880-4_36
3. Deb, K., Pratap, A., Agarwal, S.: A fast and elitist multiobjective genetic algorithm: NSGA-II. IEEE Trans. Evol. **6**(2), 182–197 (2002)
4. Zhang, Q., Li, H.: MOEA/D: a multiobjective evolutionary algorithm based on decomposition. IEEE Trans. Evol. Comput. **11**(6), 712–731 (2007)
5. Li, H., Zhang, Q.: Multiobjective optimization problems with complicated pareto sets, MOEA/D and NSGA-II. IEEE Trans. Evol. Comput. **13**(2), 284–302 (2009)
6. Chen, L., Wang, B., Liu, W.: Self-adaptive multi-objective differential evolutionary algorithm based on decomposition. In: International Conference on Computer Science and Education (ICCSE), pp. 610–616. IEEE (2016).
7. Geng, H., Zhou, L., Ding, Y., Zhou, S.: Improved MOEA/D algorithm based on new differential evolution model. Comput. Eng. Appl. **55**(8), 138–146 (2019)
8. Wu, W., Guo, X., Zhou, S., Gao, L.: Self-adaptive differential evolution with random neighborhood-based strategy and generalized opposition-based learning. Systems Engineering and Electronics (2020)
9. Deng, W., Cai, X., Zhou, Y., Zhao, H., Xu, J.: A novel decomposition multi-objective evolutionary algorithm based on differential evolution model with multi-strategy. Control and decision (2021)
10. Zhao, X., Xu, G., Liu, D.: Second-order DE algorithm. CAAI Trans. Intell. Technol. **2**(2), 80–92 (2017)
11. Zhao, X., Liu, J., Hao, J., Chen, J., Zuo, X.: Second order differential evolution for constrained optimization. In: Tan, Y., Shi, Y., Niu, B. (eds.) Advances in Swarm Intelligence. LNCS, vol. 11655, pp. 384–394. Springer, Cham (2019). https://doi.org/10.1007/978-3-030-26369-0_36
12. Tian, Y., Cheng, R., Zhang, X.: PlatEMO: A MATLAB platform for evolutionary multiobjective optimization [educational forum]. IEEE Comput. Intell. Mag. **12**(4), 73–87 (2017)

An Improved Evolutionary Multi-objective Optimization Algorithm Based on Multi-population and Dynamic Neighborhood

Shuai Zhao, Xuying Kang, and Qingjian Ni[✉]

School of Computer Science and Engineering, Southeast University, Nanjing, China
nqj@seu.edu.cn

Abstract. In the field of meteorology, atmospheric duct has important implications for the transmission of electromagnetic wave. When the electromagnetic wave signal is received by the signal receiving antenna of the global navigation satellite system (GNSS), the propagation loss and phase delay of the electromagnetic wave in the actual propagation process can be recorded, and the predicted values of the atmospheric dust parameters can be obtained through the inversion process. Atmospheric duct inversion problem can be modeled as a multi-objective optimization problem. Based on the classic MOEA/D algorithm, this paper designs an evolutionary multi-objective optimization algorithm for a single GNSS received signal, which introduces multiple population strategy and dynamic neighborhood mechanism. This paper also compares and analyzes the proposed algorithm with the classical evolutionary optimization algorithms through experiments. The experimental results show that the algorithm has higher accuracy and can better solve the atmospheric duct inversion problem.

Keywords: Atmospheric duct inversion · Multi-objective optimization · Multi-population strategy · Dynamic neighborhood

1 Introduction

The electromagnetic wave propagation in the atmospheric boundary layer is affected by atmospheric refraction. In some conditions, the curvature of the path can be more than the bending degree of the surface of the earth and some electromagnetic wave trapped in a thin layer of the atmosphere and has a certain thickness, this phenomenon is called the atmospheric duct. Some key parameters of the atmospheric refractive index model for reverse deduction and calculation, are called atmospheric duct inversion problems. This problem can be modeled as a multi-objective optimization problem.

In the research of multi-objective optimization problem solving, MOEA/D is a commonly used algorithm. In order to improve the search performance

© Springer Nature Switzerland AG 2021
Y. Tan and Y. Shi (Eds.): ICSI 2021, LNCS 12690, pp. 13–22, 2021.
https://doi.org/10.1007/978-3-030-78811-7_2

of MOEA/D, Zhao et al. [1] proposed an improved algorithm ENS-MOEA/D, which can adaptively change the neighborhood size of each individual according to the current population convergence state. Janssens et al. [2] proposed a method to analyze the influence of parameters and instance characteristics on the quality of the Pareto front. Yang et al. [3] proposed MOEA/D/D-DIDS, which used the dual selection strategy to enhance the convergence speed and global exploration ability of the population. Mohamed et al. [4] proposed an equalization optimizer algorithm, which increased the diversity among population members. Soheyl et al. [5] proposed a multi-objective stochastic fractal search (MOSFS) algorithm, which can accurately approach the real Pareto optimal front. Alejandro et al. [6] proposed a fuzzy adaptive multi-objective evolutionary algorithm (FAMES), which uses an intelligent operator controller to dynamically select the most potential mutation operators in different stages of search. Sun et al. [7] proposed an improved algorithm that adaptively learns the manifold structure of the Pareto optimal set by using the clustering method.

For the atmospheric duct problem, the multi-objective evolutionary optimization algorithm has not been widely used at present. Zheng et al. [8] used the real radar sea clutter data to invert atmospheric duct parameters by using genetic algorithm. Cristian et al. [9] applied the multi-objective evolutionary optimization algorithm to guide the task planning of unmanned aerial vehicle (UAV). Mai et al. [10] proposed an improved algorithm (EN-MOEA/ACD-NS) based on hybrid decomposition and applied it to GNSS signal joint retrieval of atmospheric air ducts. Zhang et al. [11] proposed a multi-objective method based on machine learning and meta-heuristic algorithm to optimize the concrete mix ratio, and successfully obtained the Pareto front of the double-objective mixed optimization problem of high-performance concrete and the three-objective mixed optimization problem of plastic concrete.

In this paper, a multi-population strategy and a dynamic neighborhood mechanism are proposed for atmospheric dust. An evolutionary multi-objective optimization algorithm for a single GNSS received signal is designed and implemented. The comparison and analysis between the algorithm and the classical evolutionary optimization algorithm show that the algorithm performs better in atmospheric dust.

The structure of this paper is as follows. Section 2 introduces the model of atmospheric dust and some methods. Section 3 introduces the proposed algorithm. Section 4 introduces the related experiments and analysis. Section 5 summarizes the main work of this paper and points out the further research direction.

2   Problem Statement and Related Methods

2.1   Problem Statement

The parameter model of atmospheric refractive index is shown in Eq. 1, including atmospheric profile parameters c_1, h_1, c_2, h_2, signal transmission frequency F_{send}

and receiving antenna height $H_{receive}$.

$$M(z) = M_0 + \begin{cases} c_1 z, & z < h_1 \\ c_1 h_1 + c_2(z - h_1), & h_1 < z < h_1 + h_2 \\ c_1 h_1 + c_2 h_2 + 0.118z, & z > h_1 + h_2 \end{cases} \quad (1)$$

Where M_0 is the sea level refractive index, h_1 is the starting height of the absorption layer, h_2 is the thickness of the reflective layer, c_1 is the slope when height is less than h_1, c_2 is the slope between absorbing layer and reflecting layer, it's between h_1 and h_2. When the height is greater than $h_1 + h_2$, the slope is 0.118. The atmospheric duct problem is based on the electromagnetic wave propagation model to invert the actual atmospheric profile parameters. In order to determine the atmospheric profile parameters, signal transmitting frequency and receiving antenna height, which are six parameters, the propagation loss f_1 and phase delay f_2 of the electromagnetic wave in the simulated atmosphere can be taken as the objective function, as shown in Eq. 2 and Eq. 3.

$$f_1 = f_1(c_1, h_1, c_2, h_2, F_{send}, H_{receive}) \quad (2)$$

$$f_2 = f_2(c_1, h_1, c_2, h_2, F_{send}, H_{receive}) \quad (3)$$

2.2 Related Methods

Decomposition method is a classic strategy for solving multi-objective optimization problems. MOEA/D first tried to apply this method to the evolutionary multi-objective optimization algorithm. MOEA/D divides the multi-objective optimization problem into a set of scalar optimization problems, and optimizes this set of scalar optimization problems simultaneously. Each subproblem only needs to combine the information provided by its neighboring subproblems to complete the optimization calculation, which makes MOEA/D have lower complexity in each iteration of the population compared to NAGS-II and MOGLS methods. There are mainly three ways of multi-objective decomposition: Weighted Sum (WS), Tchebycheff (TE) and Penalty-based Boundary Intersection (PBI).

3 Proposed Method

3.1 Framework of the Proposed Method

In this paper, a multi-population strategy and dynamic neighborhood mechanism combined with MOEA/D are proposed. The framework of the proposed method is as follows. Let λ be the weight vector of uniform distribution. The approximate Pareto Front problem can be transformed into N scalar optimization problems, and the form of the i-th scalar optimization problem is shown in Eq. 4.

$$maximize \ g^{te}(x|\lambda, z^*) = max_{1 \leq i \leq m}\{\lambda_i|f_i(x) - z_i^*\} \quad (4)$$

Where $\lambda_j = (\lambda_1^j, ..., \lambda_m^j)^T$, the MOEA/D algorithm will optimize N scalar optimization problems simultaneously. The neighbors of the i-th weight vector λ_i are defined as the set of several weight vectors closest to λ_i in $\{\lambda_1, \lambda_2, ..., \lambda_N\}$. In each iteration of the population, the MOEA/D algorithm using Tchebycheff decomposition will maintain the following variables: the population consisting of N points $x^1, x^2, ..., x^N$ in the decision space; The adaptive value points $FV^1, FV^2, ..., FV^N, FV^i = F(X^i) \; \forall \; i \in [1, 2, ..., N]$ in the target space corresponding to the N population individuals, $z = (z_1, z_2, ...z_m)^T$ where z_i represents the optimal value currently found on the i-th target f_i; A set of non-dominated solutions that are stored during the search process is denoised as EP. N is the number of subproblems, and T is the number of neighbors of each weight vector.

3.2   Multi-population Strategy

In the iterative process of multi-objective optimization algorithm, in order to reasonably allocate search resources, MOEA/D-M2M multi-population improvement strategy was proposed. The core idea of MOEA/D-M2M is to periodically reset the search subinterval of each subproblem by calculating the current population distribution in the target space, so that the search resources are not wasted in some areas with less potential for exploitation. Without loss of generality, we assume that all the functions of the target function are positive. MOEA/D-M2M requires K unit direction vectors $v_1, v_2, ..., v_K$ to divide the target space R_+^m into K subspace $\Omega_1, \Omega_2, ...\Omega_K$, where Ω_k is shown in Equation 5.

$$\Omega_k = \{u \in R_+^m | \langle u, v^k \rangle \leq \langle u, v^j \rangle, \forall j \in [1, 2, ..., K]\} \qquad (5)$$

Where $\langle u, v^k \rangle$ represents the acute angle between the vector u and the vector v^k. The vector u belongs to the region Ω_k only if the acute angle between the vector u and the vector v^k is less than or equal to the acute angle $\langle u, v^k \rangle$ in any other unit direction. According to the above division of the target space, we divide the problem into K multi-objective optimization subproblems with constraints. The form of the K-th subproblem is shown in Eq. 6.

$$minimize \; F(x) = \{f_1(x), f_2(x), ..., f_m(x) \; subject \; to \; F(x) \in \Omega_k\} \qquad (6)$$

MOEA/D-M2M maintains K subpopulations $P_1, P_2, ..., P_k$, K multi-objective optimization subproblems were jointly optimized and solved in a cooperative way. The set of all K subpopulations together constituted the population of MOEA/D-M2M, as shown in Eq. 7.

$$P = \bigcup_{i=1} KP_k \qquad (7)$$

Where the number of individuals contained in population P is denoted as N, and the number of individuals contained in subpopulation P_K is denoted as N_K. In each iteration, population will be updated in the following way:

1. Generate a new offspring population Q, size of N.
2. The population Q is divided into K subpopulation $Q_1, Q_2, ..., Q_K$, such that Q_k contains all the individuals belonging to subspace ω_k in the population Q.
3. For k = 1,.. ,K, select N_k individuals from the union of Q_k and P_k to replace the original P_k.

In the MOEA/D-M2M framework, the size of N_k determines how the search resources are allocated in different multi-objective optimization subproblems, that is to say, $\frac{N_k}{N}$ determines the proportion of the search resources allocated in the k-th multi-objective optimization subproblem.

3.3   Dynamic Neighborhood

In order to improve the effectiveness of the evolutionary multi-objective optimization algorithm, this paper proposed a dynamic neighborhood mechanism which is shown in Eq. 8 and Eq. 9. The neighborhood size refers to the extent to which each individual in the population learns from its neighbors. When the neighborhood is large, it is beneficial to enhance the global search ability of the population. When the neighborhood is small, it is beneficial to enhance the local search ability of the population.

$$ p_{k,G} = \frac{S_{k,G}}{\sum_{i=1}^{k} S_{k,G}} \tag{8} $$

$$ S_{k,G} = \frac{\sum_{g=G-LP}^{G-1} SC_{count\ k,G}}{\sum_{g=G-LP}^{G-1} SC_{all_{k,G}}} + \delta, k = 1, 2, ...K, G > LP \tag{9} $$

Where $S_{k,G}$ represents the probability of improvement compared with the previous generation among the descendants generated in the previous LP (Learning Period) generation when the neighborhood size is k and the current iteration algebra is G. Where $SC_{all_{k,G}}$ is the total number of all newborn individuals generated in the previous LP generation, and $SC'_{count\ k,G}$ represents the number of newborn individuals improved in the objective function compared with the previous generation. δ is a small constant that makes $S_{k,G}$ greater than 0.

The dynamic neighborhood mechanism dynamically improves the quality of the next generation population by limiting the selection range of individual mates. Within the limited range, how to select the best individual for hybridization and generation of the next generation has a significant impact on the final convergence effect. When looking for the best matching individual x_j of the current individual x_i, Eq. 10 holds.

$$ x_j = \arg_{x_k} \min\left(g\left(x_k \mid \lambda_i, z^*\right)\right), \text{dis}\left(x_k, x_i\right) \le K, \text{ang}\left(x_k - Z^*, F_i\right) < 0.5\theta \tag{10} $$

The formula for calculating the adaptive value of $g\left(x_k \mid \lambda_i, z^*\right)$ in the i-th subproblem under a specific decomposition mode is given here. $\text{dis}\left(x_k, x_i\right)$ represents that the distance between individual x_k and individual x_i is less than or equal to K, that is, x_k is the neighbor individual of x_i. ang $(x_k - Z^*, F_i)$ represents the angle between the vector formed by the connection between x_k and Z^* and the direction vector of the i-th subproblem is less than 0.5θ. Here Z^* is the position reference point. The size of angle θ has an important influence on the selection of learning objects. When θ is large, there are many individuals in the neighborhood of x_i who can compete for selection; when θ is small, there is no chance to compete for matching with x_i like other individuals outside the neighborhood. The dynamic adjustment strategy of angle θ is shown in Eq. 11 and Eq. 12.

$$\theta_i^{g+1} = \begin{cases} \max\left\{\theta_i^t - \theta_i^{\min}, \theta_i^{\min}\right\}, & \text{if angr } (x_i) > 1 \\ \theta_i^t, & \text{if angr } (x_i) = 1 \\ \min\left\{\theta_i^t + \theta_i^{\min}, \theta_i^{\max}\right\}, & \text{if angr } (x_i) < 1 \end{cases} \quad (11)$$

$$\text{angr}_i = \frac{ang_i}{\sum_{j=1}^{N} ang_j} \quad (12)$$

Where, θ_i^t represents the corresponding θ value of the i-th subproblem in the t generation, and θ_i^{min} and θ_i^{max} are the maximum and minimum values of ang_i, respectively. $ang_i = ang(x_i - z^*, F_i)$ is the included angle between the vector formed by the line between x_i and Z^* and the direction vector of the i-th subproblem. $angr_i$ is the normalized value of ang_i.

4   Experiment and Analysis

4.1   Settings

The experimental data are collected on the basis of different combinations of F_{send} and $H_{receive}$, and the propagation loss and phase delay of electromagnetic wave are simulated by atmospheric duct forward calculus. The receiving height is set within the range of $S_H = \{20, 50, 100, 150, 400\}$ and the transmitting frequency is set within the range of $S_F = \{1200, 1300, 1400, 4500\}$.

In this experiment, MOEA/D-ACD, MOEA/D-ACD/NS and MOEA/D-ACD/NS-THETA algorithms were used for comparison. Tchebycheff decomposition is used for the other algorithms, except for the PBI decomposition for MOEA/D-ACD/NS-THETA algorithm. The population size was set at 150, the number of iterations was 700, and the maximum number of evaluations was 105000. The actual values of the four parameters are $c_1 = 100.00, h_1 = 300.00, c_2 = -0.02$, and $h2 = -0.20$.

4.2   Results and Analysis

Table 1, Table 2 and Table 3 show the experimental operation results of MOEA/ D-ACD, MOEA/D-ACD/NS and MOEA/D-ACD/NS-THETA algorithms for atmospheric duct inversion problems respectively. Each row in the table represents the final inversion results of the algorithm for target parameters c_1, h_1, c_2 and h_2.

Table 1. Experimental results of MOEA/D-ACD applied to atmospheric duct inversion at different transmitting frequencies and receiving heights.

ID	F_{send}	$h_{receive}$	c_1	h_1	c_2	h_2
1	1200	20	100.39180	301.88290	−0.02008	−0.20119
2	1200	50	99.56011	299.28010	−0.02007	−0.19863
3	1200	100	99.28210	300.44480	−0.02011	−0.19892
4	1200	150	99.46810	300.01060	−0.01984	−0.20046
5	1200	400	99.27532	298.62560	−0.02003	−0.20053
6	1300	20	99.56736	302.09340	−0.01994	−0.19852
7	1300	50	99.90076	302.11360	−0.02016	−0.19965
8	1300	100	100.48170	298.71830	−0.01986	−0.19916
9	1300	150	100.52230	300.35260	−0.01991	−0.20055
10	1300	400	99.55754	299.09950	−0.01997	−0.19894
11	1400	20	99.66474	299.53230	−0.01988	−0.19963
12	1400	50	100.78590	298.57780	−0.01990	−0.19944
13	1400	100	99.59832	299.88380	−0.02003	−0.19968
14	1400	150	100.48080	298.10430	−0.01990	−0.19891
15	1400	400	99.76721	299.66430	−0.01998	−0.19936
16	1500	20	100.38800	301.23780	−0.02004	−0.20023
17	1500	50	100.73020	300.35030	−0.01989	−0.20072
18	1500	100	100.19570	300.42410	−0.01985	−0.20133
19	1500	150	100.00080	300.10360	−0.02013	−0.19871
20	1500	400	100.61500	299.70720	−0.01991	−0.20112

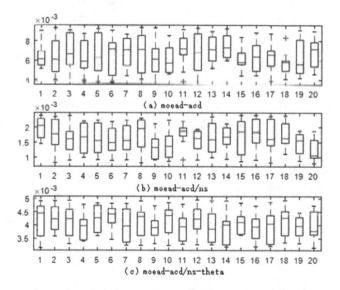

Fig. 1. Results of the three algorithms at different transmitting frequencies and receiving heights.

Table 2. Experimental results of MOEA/D-ACD/NS applied to atmospheric duct inversion at different transmitting frequencies and receiving heights.

ID	F_{send}	$H_{receive}$	c_1	h_1	c_2	h_2
1	1200	20	99.80346	300.27250	−0.02001	−0.19978
2	1200	50	99.97466	299.92390	−0.02004	−0.20036
3	1200	100	99.83644	300.11280	−0.02002	−0.19973
4	1200	150	100.14290	300.55630	−0.02000	−0.20022
5	1200	400	99.89048	300.04420	−0.01998	−0.20012
6	1300	20	99.98449	300.16720	−0.01997	−0.20027
7	1300	50	100.08630	300.09330	−0.02001	−0.19969
8	1300	100	99.95722	299.61480	−0.01999	−0.19990
9	1300	150	99.93118	300.36360	−0.01996	−0.19962
10	1300	400	99.85081	299.67870	−0.02004	−0.19991
11	1400	20	99.84432	299.88900	−0.01997	−0.19996
12	1400	50	99.94760	299.65000	−0.02000	−0.19964
13	1400	100	99.84961	299.96490	−0.01997	−0.20037
14	1400	150	100.07670	300.57480	−0.02002	−0.20029
15	1400	400	100.07410	300.49140	−0.01999	−0.19968
16	1500	20	99.87737	300.30530	−0.02001	−0.20007
17	1500	50	99.86229	300.38280	−0.01999	−0.19981
18	1500	100	100.12200	299.48070	−0.01996	−0.20000
19	1500	150	100.10210	300.29090	−0.01997	−0.20027
20	1500	400	99.98292	300.14170	−0.01997	−0.19973

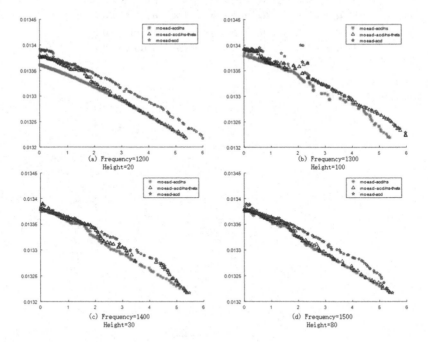

(a) Frequency=1200 Height=20

(b) Frequency=1300 Height=100

(c) Frequency=1400 Height=30

(d) Frequency=1500 Height=80

Fig. 2. Results of the three algorithms at different transmitting frequencies and receiving heights.

Table 3. Experimental results of MOEA/D-ACD/NS-THETA applied to atmospheric duct inversion at different transmitting frequencies and receiving heights.

ID	F_{send}	$H_{receive}$	c_1	h_1	c_2	h_2
1	1200	20	100.35590	300.22280	−0.01999	−0.19936
2	1200	50	99.58138	301.04060	−0.02002	−0.20083
3	1200	100	100.22150	299.06800	−0.02006	−0.19981
4	1200	150	99.77903	299.52520	−0.02002	−0.20017
5	1200	400	99.89757	300.29650	−0.02006	−0.20056
6	1300	20	99.63517	299.52260	−0.01995	−0.20048
7	1300	50	100.21630	300.52060	−0.01994	−0.19941
8	1300	100	99.81403	299.48530	−0.02000	−0.20031
9	1300	150	100.29870	300.83780	−0.01999	−0.20043
10	1300	400	100.16250	299.28090	−0.02001	−0.20021
11	1400	20	100.03470	301.32760	−0.01995	−0.19914
12	1400	50	99.76130	300.07710	−0.02008	−0.19954
13	1400	100	100.09180	300.96440	−0.01991	−0.19923
14	1400	150	99.91856	298.65090	−0.01999	−0.20053
15	1400	400	99.74736	299.52970	0.02007	−0.19955
16	1500	20	100.22370	300.11690	−0.02003	−0.19940
17	1500	50	100.04730	301.235480	−0.01993	−0.20026
18	1500	100	100.04180	299.58600	−0.01998	−0.19947
19	1500	150	100.22130	298.98900	−0.01994	−0.20085
20	1500	400	99.92299	300.62480	−0.01995	−0.20024

Figure 1 shows the experimental results of running MOEA/D-ACD, MOEA/D-ACD/NS and MOEA/D-ACD/NS-THETA algorithms for 10 times under 20 transmission frequencies and antenna height settings. It can be seen the average convergence accuracy of the MOEA/D-ACD algorithm is poorer than others. The average convergence accuracy of MOEA/D-ACD/NS algorithm is the best. The operating stability and convergence accuracy of MOEA/D-ACD/NS-THETA algorithm is between MOEA/D-ACD and MOEA/D-ACD/NS.

Figure 2 shows the convergence of MOEA/D-ACD, MOEA/D-ACD/NS and MOEA/D-ACD/NS-THETA algorithms at four given transmit frequencies and receive heights. It can be seen from Fig. 2(b) that the final convergence result of MOEA/D-ACD algorithm has a large number of solutions in some regions that deviate from the frontier where most of the population set is located. It can be seen that the convergence effect of MOEA/D-ACD/NS algorithm is the best among the three algorithms.

5   Conclusion

In this paper, the atmospheric duct inversion problem was modeled as a multi-objective optimization problem. Taking the phase delay and propagation loss of GNSS as the optimization objectives, a multi-population strategy was proposed based on the MOEA/D algorithm, and a dynamic neighborhood mechanism was introduced to design and implement an evolutionary multi-objective optimization algorithm for a single GNSS received signal. Relevant experimental results show that the proposed algorithm has better solving accuracy, and the uniformity and diversity of the population search process are enhanced.

Acknowledgements. This paper is supported by National Key R&D Program of China (2018YFB1004300).

References

1. Zhao, S.Z., Suganthan, P.N., Zhang, Q.: Decomposition-based multiobjective evolutionary algorithm with an ensemble of neighborhood sizes. IEEE Trans. Evol. Comput. **16**(3), 442–446 (2012)
2. Yang, Yu., Huang, M., Wang, Z.-Y., Zhu, Q.-B.: Dual-information-based evolution and dual-selection strategy in evolutionary multiobjective optimization. Soft Comput. **24**(5), 3193–3221 (2019). https://doi.org/10.1007/s00500-019-04081-5
3. Janssens, J., Sörensen, K., Janssens, G.K.: Studying the influence of algorithmic parameters and instance characteristics on the performance of a multiobjective algorithm using the promethee method. Cybern. Syst. **50**(5), 444–464 (2019)
4. Abdel-Basset, M., Mohamed, R., Abouhawwash, M.: Balanced multi-objective optimization algorithm using improvement based reference points approach. Swarm Evol. Comput. **60**, 100791 (2021)
5. Khalilpourazari, S., Naderi, B., Khalilpourazary, S.: Multi-objective stochastic fractal search: a powerful algorithm for solving complex multi-objective optimization problems. Soft Comput. **24**(4), 3037–3066 (2020)
6. Santiago, A., Dorronsoro, B., Nebro, A.J., Durillo, J.J., Castillo, O., Fraire, H.J.: A novel multi-objective evolutionary algorithm with fuzzy logic based adaptive selection of operators: Fame. Inf. Sci. **471**, 233–251 (2019)
7. Jianyong Sun, H., Zhang, A.Z., Zhang, Q., Zhang, K.: A new learning-based adaptive multi-objective evolutionary algorithm. Swarm Evol. Comput. **44**, 304–319 (2019)
8. Zheng, S., Han-Xian, F.: Inversion for atmosphere duct parameters using real radar sea clutter. Chin. Phys. B **21**(2), 029301 (2012)
9. Atencia, C.R., Del Ser, J., Camacho, D.: Weighted strategies to guide a multi-objective evolutionary algorithm for multi-UAV mission planning. Swarm Evol. Comput. **44**, 480–495 (2019)
10. Mai, Y., et al.: Using the decomposition-based multi-objective evolutionary algorithm with adaptive neighborhood sizes and dynamic constraint strategies to retrieve atmospheric ducts. Sensors **20**(8), 2230 (2020)
11. Zhang, J., Huang, Y., Wang, Y., Ma, G.: Multi-objective optimization of concrete mixture proportions using machine learning and metaheuristic algorithms. Constr. Build. Mater. **253**, 119208 (2020)

A Multiobjective Memetic Algorithm for Multiobjective Unconstrained Binary Quadratic Programming Problem

Ying Zhou[1(✉)], Lingjing Kong[1], Lijun Yan[1], Shaopeng Liu[2], and Jiaming Hong[3]

[1] School of Computer Sciences, Shenzhen Institute of Information Technology,
Shenzhen, China
{zhou_y,konglj,yanlj}@sziit.edu.cn
[2] Department of Computer Science, Guangdong Polytechnic Normal University,
Guangzhou, China
[3] School of Medical Information Engineering, Guangzhou University of Chinese Medicine,
Guangzhou, China

Abstract. This study introduces a multiobjective memetic algorithm for multiobjective unconstrained binary quadratic programming problem (mUBQP). It integrates multiobjective evolutionary algorithm based on decomposition and tabu search to search an approximate Pareto front with good convergence and diversity. To further enhance the search ability, uniform generation is introduced to generate different uniform weight vectors for decomposition in every generation. The proposed algorithm is tested on 50 mUBQP instances. Experimental results show the effectiveness of the proposed algorithm in solving mUBQP.

Keywords: Mutiobjective optimization · Multiobjective memetic algorithm · Multiobjective unconstrained binary quadratic programming problem

1 Introduction

Unconstrained binary quadratic programming problem (UBQP) is one of the most widely studied combinatorial optimization. Many real-world problems, such as finance, project selection, cluster analysis, economic analysis, traffic management, computer aided design and so on [1], can be formulated as UBQP. Moreover, many graph problems can be converted to UBQP including max-2sat problem, max-cut problem, maximum clique problem and number partitioning problem [1]. UBQP has NP-hard complexity [2]. Exact methods are only applicable to small or moderate-scale problem instances. On the other hand, metaheuristic algorithms have been shown to be effective in solving hard combinatorial optimization problems in reasonable computation time [3]. Therefore, metaheuristic approaches have been successfully applied to UBQP [4–8].

Supported by the Natural Science Foundation of Guangdong Province of China (2018A0303 130055, 2018A030310664, 2019A1515012048) and the Opening Project of Guangdong Key Laboratory of Big Data Analysis and Processing (202001).

Y. Tan and Y. Shi (Eds.): ICSI 2021, LNCS 12690, pp. 23–33, 2021.
https://doi.org/10.1007/978-3-030-78811-7_3

The original UBQP problem is a single-objective problem. In [9], a multiobjective version of UBQP (denoted as mUBQP) is proposed. It can express many multiobjective scenarios, including multiobjective knapsack problem and multiobjective max-cut problem [9]. Due to the structure of mUBQP, improving one objective function may deteriorate the others. Therefore, multiobjective fashion is necessary for mUBQP to obtain a set of solutions that represent the tradeoffs between the objectives. Recently, several multiobjective algorithms have been applied to mUBQP [9–14]. To the best of our knowledge, DMTS [12] is the most effective method for solving mUBQP at present. However, it is a local search based algorithm that cannot balance the convergence and diversity well.

This study proposes a multiobjective memetic algorithm, termed MOMA, to better solve mUBQP. The proposed MOMA integrates multiobjective evolutionary algorithm based on decomposition (MOEA/D) [15] and tabu search (TS) [12] to search an approximate Pareto front with good convergence and diversity. MOEA/D is a popular multiobjective evolutionary algorithm (MOEA) framework which decomposes mUBQP into multiple subproblems by scalarizing function. TS, one of the best performing metaheuristics in solving single-objective UBQP, is designed to solve the subproblems. MOEA/D is mainly utilized for global search, while tabu search focuses on local search. To further improve the search ability of MOMA, the uniform generation [12] is adopted to generate different uniform weight vectors for decomposition in each generation. As a consequence, the proposed MOMA can achieve good balance between convergence and diversity and can be expected to obtain high-quality solutions.

To assess the performance of MOMA, two algorithms are adopted as its competitors. First, to validate the effectiveness of the uniform generation, a variant of MOMA which uses the simplex-lattice design to generate weight vectors is proposed as a competitor. Second, DMTS [12] is adopted as another competitor. All algorithms are tested on 50 mUBQP instances [9]. Experimental results shows the effectiveness of the proposed MOMA in solving mUBQP.

The contributions of this study are twofold. First, a novel multiobjective memetic algorithm is proposed for mUBQP. Second, a comprehensive experimental comparison of MOMA with the competitors is provided. MOMA can be seen as a new benchmark algorithm for mUBQP, which can be used for comparison by later researchers.

The remaining sections are organized as follows. Section 2 presents background, including related works on multiobjective optimization and the formulation of mUBQP. In Sect. 3, MOMA is proposed. Experimental results are presented in Sect. 4. Finally, conclusion and future works are given in Sect. 5.

## 2	Background

### 2.1	Multiobjective Optimization

A multiobjective optimization problem (MOP) can be formulated as follows:

$$\max \; F(x) = \{f_1(x), \dots, f_m(x)\} \tag{1}$$

subject to $x \in \Omega$, where Ω is the decision variable space. $F : \Omega \to R^m$ consists of m objective functions that often conflict with each other. Let $x, y \in R^m$, x dominates

y if and only if $f_i(\boldsymbol{x}) \geq f_i(\boldsymbol{y})$ for every $i \in \{1, \ldots, m\}$, and $f_j(\boldsymbol{x}) > f_j(\boldsymbol{y})$ for at least one $j \in \{1, \ldots, m\}$. A solution $\boldsymbol{x}^*$ is Pareto optimal if there is no solution $\boldsymbol{x} \in \Omega$ dominates $\boldsymbol{x}^*$, The corresponding $F(\boldsymbol{x}^*)$ is called a Pareto optimal objective vector. The set contains all Pareto optimal solutions is called *Pareto set*, and the set contains all Pareto optimal objective vectors is called *Pareto front*. Multiobjective algorithms aim at finding a set of nondominated solutions with good convergence and diversity, i.e., solutions should be close to and well-distributed over the whole Pareto front.

Over the years, many metaheuristics have been applied to solve MOPs. Most of the popular MOEAs, such as NSGA-II [16] and MOEA/D [15], are mainly proposed for multiobjective continuous problems. Beside MOEAs, multiobjective local search algorithms, such as Pareto local search [17] and multi-directional local search [18], are promising approaches to solve multiobjective combinatorial problems. The main advantage of local search is that problem-specific knowledge can be directly utilized to guide the search toward the Pareto front. Local search can be integrated into MOEAs to form multiobjective memetic algorithms. This kind of algorithms achieves good performance on multiobjective combinatorial optimization.

2.2 Formulation of mUBQP

mUQBP is an MOP and can defined as follows [9]:

$$\max\ f_k(\boldsymbol{x}) = \sum_{i=1}^{n} \sum_{j=1}^{n} q_{ij}^k x_i x_j, \ k \in \{1, \ldots, m\} \tag{2}$$

subject to $x_i \in \{0, 1\}$, $i \in \{1, \ldots, n\}$. $F(\boldsymbol{x}) = (f_1(\boldsymbol{x}), \ldots, f_m(\boldsymbol{x}))$ is a vector with $m \geq 2$ objective functions to be maximized, and $Q^k = (q_{ij}^k)$ are m matrices of size $n \times n$, $k \in \{1, \ldots, m\}$.

3 Proposed Algorithm: MOMA

3.1 Framework of MOMA

The general framework of MOMA is presented in Algorithm 1. It decomposes mUBQP into N_W scalar optimization subproblems and solves the subproblems simultaneously. First, a weight vector set containing N_W uniform weight vectors are generated (Line 1), and a population containing N_W solutions is initialized (Line 5). The weighted sum approach is used for decomposition. Each subproblem is related to a weight vector and a solution. Given a solution $\boldsymbol{x}$ and a weight vector $\boldsymbol{w} = (w_1, \ldots, w_m)$, the corresponding single-objective subproblem is defined as:

$$\max\ g^{ws}(\boldsymbol{x}|\boldsymbol{w}) = \sum_{k=1}^{m} w_k f_k(\boldsymbol{x}) \tag{3}$$

The neighborhood of subproblems is defined based on the Euclidean distances among their weight vectors (Line 4). Each subproblem is optimized using information

---

Algorithm 1. MOMA for mUQBP

Input: the number of objectives m, a positive number H, the number of weight vectors N_W and the maximum size of the archive N_A

Output: An external archive A

1: $W \leftarrow$ *Uniform generation*(m, H, N_W) // W is a set of N_W uniform weight vectors.
2: $A \leftarrow \emptyset$
3: **for** $i = 1$ to N_W **do**
4:    Compute the Euclidean distance between each pair of weight vectors and get the N_B closest vectors. Set the neighborhood $B(i) = \{i_1, \ldots, i_{N_B}\}$.
5:    $x^i \leftarrow$ *Initialize solution*$()$
6: **end for**
7: **while** stopping criterion is not met **do**
8:    **for** $i = 1$ to N_W **do**
9:       Choose b_1, b_2 randomly from $B(i)$
10:       $x \leftarrow$ *Crossover*(x^{b_1}, x^{b_2})
11:       $x' \leftarrow TS(x, W^i)$
12:       $A \leftarrow$ *Update archive* (x', A, N_A)
13:       **for** each $j \in B(i)$ **do**
14:          **if** $g^{ws}(x'|W^j) \geq g^{ws}(x^j|W^j)$ **then**
15:             $x^j \leftarrow x'$
16:          **end if**
17:       **end for**
18:    **end for**
19:    $W^* \leftarrow$ *Uniform generation*(H, N_W)
20:    $W \leftarrow$ *Update weight vectors*(W, W^*)
21: **end while**
22: return A

---

of its neighboring subproblems. Specifically, for a subproblem, two neighboring subproblems are randomly selected as parents (Line 9), and an offspring solution is generated by crossover operator (Line 10). Then, TS is called to further improve the offspring solution (Line 11). The archive and the population are updated by the improved solution (Lines 12–17). When a generation is finished, the weight vector set is updated by uniform generation (Lines 19–20). Repeating this process until the stopping criterion is met. Finally, the archive is returned as the approximate Pareto front.

Details of MOMA are presented as follows.

3.2 Population Initialization and Stopping Criterion

Population Initialization. The population contains N_W solutions to be initialized. Each solution is created randomly, i.e., for a solution $x = \{x_1, \ldots, x_n\}$, x_i ($i \in \{1, \ldots, n\}$) is randomly set to 0 or 1. Then, these solutions are evaluated according to Eq. (2).

Stopping Criterion. The stopping criterion of MOMA is set to the maximum running time. For fair comparison, all competitors follow the same stopping criterion.

3.3 Uniform Generation

The original simplex-lattice design [15] in MOEA/D has the following problems: 1) The number of weight vectors is restricted, because it cannot be specified with any arbitrary number. 2) The fix weight vectors may not lead to a well-distributed approximate

Pareto front [12]. Therefore, in order to enhance the diversity, uniform generation [12] is adopted to generate a different uniform weight vector set in each generation.

First, simplex-lattice design is applied to obtain the $\{m, H\}$ simplex-lattice. Each component of a weight vector is in $\{0/H, 1/H, \ldots, H/H\}$ and the sum of all components is equal to 1. Then, another set of weight vectors is generated randomly. Each component of a weight vector is a random real number in $[0,1]$ and the sum of all components is equal to 1. These two sets are combined to form an initial weight vector set W. To maintain a diverse set that contains exactly N_W weight vectors, some weight vectors are eliminated from W according to the Euclidean distances among the weight vectors. More details can be found in [12].

3.4 Crossover Operator and Tabu Search

Crossover Operator. Crossover operator creates a single offspring solution of a subproblem. First, two solutions of the neighboring subproblems are randomly selected as the parent solutions. Then, an offspring solution is generated with uniform crossover [9]. Specifically, common variables of both parents are directly assigned to the offspring solution, while the remaining ones are assigned randomly.

Tabu Search. The offspring solution of a subproblem is further improved by TS [12]. TS uses 1-flip move neighborhood (i.e., x_i is changed to $1 - x_i$) to create candidate solutions. At each search step, a non-tabu variable is flipped if it lead to the maximum weighted sum value. A tabu tenure T is assigned to a flipped variable, thus it cannot be flipped in the next T moves. However, a tabu move can still be performed if it can lead to a better solution than the current best solution. In our implementation, the maximum search steps $depth$ is adopted as the stopping criterion of TS.

3.5 Archive and Weight Vector Updating

Archive Updating. Since the true Pareto front of mUBQP may be extremely large, the bounded archive in [12] is adopted to maintain a set of well-distributed nondominated solutions as the approximate Pareto front. Specifically, the archive A accept a new solution if it is not dominated of equal to any solutions in A. If A contains more than N_A solutions, a solution will be discarded to ensure that A contains exactly N_A solutions. Each solution A^i in A is associated with a value $M(A^i)$ calculated as follows:

$$M(A^i) = \min\{d(A^i, A^j) | \forall i \in \{1, \ldots, N_A\} \text{ and } i \neq j\} \quad (4)$$

where $d(A^i, A^j)$ is the Euclidean distance between A^i and A^j. The solution with the lowest M value will be removed from A.

Weight Vector Updating. When a generation is finished, uniform generation is called to create N_W new uniform weight vectors for the next generation. Each weight vector should be assigned to a subproblem. In two successive generations, the ith previous

---

Algorithm 2. Greedy Heuristic for Weight Vector Updating

Input: previous used weight vector set W and current generated weight vector set by uniform generation W^*
Output: an updated weight vector set W
1: Mark all W^{*i} in W^* unvisited
2: **for** $i = 1$ to N_W **do**
3:     $k \leftarrow \arg\min\{d(W^i, W^{*j})|\ \forall j \in \{1, \cdots \leq N_W\}$ and W^{*j} is unvisited$\}$
4:     $W^i \leftarrow W^{*k}$
5:     Mark W^{*k} visited
6: **end for**
7: return W

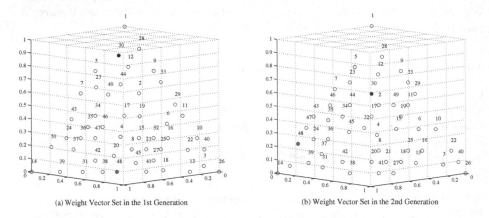

(a) Weight Vector Set in the 1st Generation   (b) Weight Vector Set in the 2nd Generation

Fig. 1. Weight vector set in (a) the 1st generation (b) the 2nd generation for a three-objective problem. Each set contains 50 weight vectors generated by uniform generation. The circle marked with number i denotes the weight vector related to the ith subproblem. It can be observed that except for the 30th and 48th weight vectors (marked with black and red respectively), the weight vectors of other subproblems are close to each other in two successive generations.

subproblem should be similar to the ith current subproblem, because two similar subproblems may have similar optimal solutions. Therefore, the previous weight vector and new weight vector of the ith subproblem should be close to each other. To this end, a simple greedy heuristic is proposed for weight vector updating, which is described in Algorithm 2. Let W be the weight vector set in the previous generation and W^* be the new weight vector set generated by uniform generation. First, all weight vectors in W^* are marked as unvisited (Line 1). Then, the weight vector W^i is set to the nearest unvisited weight vector W^{*k} in W^* (Lines 3–4). Finally, W^{*k} is marked as visited. Repeating this process until all weight vectors in W are updated. This strategy guarantees that each weight vector in W^* is assigned only once, and most of the updated weight vectors are close to the previous ones of the subproblems. Figure 1 illustrates two weight vector sets in two successive generations for a three-objective problem.

4   Computational Experiments

In this section, experiments are carried out on a server[1] to assess the performance of the proposed MOMA. Two algorithms are adopted as the competitors of MOMA. 50 benchmark instances introduced in [9] are used to evaluate the performances of the competitors. The sizes of instances n are in $\{1000,2000,3000,4000,5000\}$. All algorithms are implemented in C++ and run 30 independent times on each instance.

4.1   Experimental Settings and Performance Measures

Experimental Settings. The parameter settings in MOMA are described as follows. For uniform generation, H is set to 99 and 13 for $m = 2$ and $m = 3$, respectively, and N_W is set to 100. For TS, the tabu tenure T is set to $\sqrt{n}$, and the maximum search steps $depth$ is set to $0.1n$. The archive size N_A is set to 500. These parameter settings are directly adopted from [12] and have shown to be effective for mUBQP. To better balance the convergence and diversity, the neighborhood size N_B is set to 5. The maximum running time is set to $10^{-3}nm$ minutes as in [9].

Performance Measures. This study utilizes hypervolume (HV) [19] to measure the performances of the competitors. HV is the area which is dominated by the approximate Pareto front and dominates a reference point z^*. It is a widely used indicator due to its Pareto compliance property [20]. To compute HV, first all objective values are multiplied by -1 to convert them to minimization, then these values are normalized into the range of $[0, 1]$. The reference points are set to $z^* = (1.1, 1.1)$ and $z^* = (1.1, 1.1, 1.1)$ for the instances with $m = 2$ and $m = 3$, respectively. Higher HV value means better performance of an algorithm in terms of convergence and diversity.

In addition, the Wilcoxon signed-rank test [21] at 5% significance level is carried out to further show the differences between MOMA and the competitors.

4.2   Competitors

To assess the performance of the proposed MOMA, two algorithms are adopted as its competitors. First, to confirm the effectiveness of uniform generation in MOMA, a variant of MOMA (denoted as MOMA-S) is proposed to be compared with MOMA. In MOMA-S, the original simplex-lattice design is used to generate weight vectors at the beginning of the algorithm, and these weight vectors are always used in the following search process. The parameter settings of MOMA-S are the same as in MOMA. Second, DMTS [12], the most effective algorithm for mUBQP, is also adopted as a competitor of MOMA. Similar to MOMA, DMTS is a decomposition-based algorithm which first decomposes mUBQP into multiple subproblems, and then solves these subproblems by TS. The weight vectors in DMTS are also generated by uniform generation. DMTS adopts the original parameter settings in [12].

The main differences between MOMA and the competitors are as follows.

---

[1] Eight CPUs with Intel Xeon (Cascade Lake) Platinum 8269CY at 2.5GHz, 16.0 GB of RAM.

- MOMA vs. MOMA-S: MOMA utilizes different weight vectors generated by uniform generation in each generation, while MOMA-S utilizes fixed weight vectors generated by simplex-lattice design through the whole search process.
- MOMA vs. DMTS: MOMA is a memetic algorithm which integrates an evolutionary algorithm for global search and TS for local search, while DMTS is a local search based algorithm.

Table 1. Average values of HV of DMTS, MOMA-S and MOMA on 50 mUBQP instances.

n	ρ	$m=2$			$m=3$		
		DMTS	MOMA-S	MOMA	DMTS	MOMA-S	MOMA
1000	−0.5	0.9041−	**0.9059**	**0.9059**	**0.7975+**	0.7125−	0.779
	−0.2	0.972−	**0.9751**	**0.9751**	0.8125−	0.8077−	**0.8166**
	0	1.0008−	1.0046	**1.0047**	0.7959−	0.7982−	**0.8011**
	0.2	1.0103−	1.015−	**1.0152**	0.8431−	**0.8484+**	0.8475
	0.5	1.0268−	1.0344−	**1.0349**	0.8926−	**0.8998+**	0.8971
2000	−0.5	0.9432−	**0.946**	**0.946**	**0.7828+**	0.6882−	0.7488
	−0.2	0.9768−	**0.9808+**	0.9807	0.7952−	0.7851−	**0.7994**
	0	0.9889−	**0.9934**	**0.9934**	0.7991−	0.7993−	**0.8056**
	0.2	1.0165−	1.0217−	**1.022**	0.8274−	0.8331−	**0.8349**
	0.5	1.0343−	1.0421−	**1.0428**	0.8723−	**0.8843+**	0.8829
3000	−0.5	0.9382−	0.941−	**0.9411**	**0.7741+**	0.6796−	0.7383
	−0.2	0.9702−	0.9742−	**0.9745**	0.795−	0.7854−	**0.8009**
	0	0.9941−	0.9987−	**0.9991**	0.7828−	0.782−	**0.7909**
	0.2	1.0041−	1.0095−	**1.0098**	0.8086−	0.8149−	**0.8184**
	0.5	1.0164−	1.0242−	**1.0246**	0.8621−	**0.875+**	0.8746
4000	−0.5	0.9391−	**0.9418+**	0.9416	**0.7789+**	0.6814−	0.7206
	−0.2	0.973−	**0.9765+**	0.9763	**0.781**	0.7675−	0.7805
	0	0.9905−	**0.9951+**	0.9949	0.7676−	0.7649−	**0.774**
	0.2	0.9991−	**1.0053**	1.0053	0.808−	0.8136−	**0.8174**
	0.5	1.0104−	**1.0202**	1.0202	0.8672−	0.881−	**0.8815**
5000	−0.5	0.9383−	**0.9413+**	0.94	**0.7799+**	0.6821−	0.7099
	−0.2	0.9722−	**0.9764+**	0.976	**0.7716+**	0.757−	0.7679
	0	0.9932−	0.9984	**0.9985**	0.7712−	0.7704−	**0.7773**
	0.2	1.0022−	1.0086−	**1.0088**	0.7999−	0.8066−	**0.8115**
	0.5	1.0184−	1.0285−	**1.0288**	0.8535−	0.8728−	**0.875**
$w/t/l$		25/0/0	11/8/6		18/1/6	21/0/4	

4.3 Comparing MOMA with the Competitors

Table 1 shows the average values of HV of DMTS, MOMA-S and MOMA on 50 mUBQP instances. The column marked with "ρ" indicates objective correlation of an instance. The value marked with "-" or "+" indicates the competitor is significantly worse or better than that of MOMA according to the result of the Wilcoxon signed-rank test, respectively. The values highlighted in **boldface** are the best average values. The last row "$w/t/l$" indicates that MOMA performs better than, similar to, and worse than the competitor on w, t and l instances, respectively. From the results in Table 1, some observations can be extracted as follows.

- MOMA vs. MOMA-S: MOMA significantly outperforms MOMA-S on most instances. Specifically, for 50 mUBQP instances, MOMA performs better than, similar to, and worse than MOMA-S on 32, 8, 10 instances, respectively. Furthermore, MOMA performs much better than MOMA-S on the instances with $m = 3$. To further show the difference between two algorithms, Fig. 2 (a) shows the nondominated solutions with the highest HV value obtained by two algorithms on the instance with $n = 5000$, $m = 3$ and $\rho = -0.5$. It is clear that the solutions obtained by MOMA spread more uniformly than those obtained by MOMA-S. Due to uniform generation which generates different uniform weight vectors in each generation, MOMA can obtain better approximate Pareto front in terms of diversity. Therefore, the effectiveness of uniform generation is confirmed.

- MOMA vs. DMTS: MOMA also significantly outperforms DMTS on most instances. Specifically, for 50 mUBQP instances, MOMA performs better than, similar to, and worse than DMTS on 43, 1, 6 instances, respectively. To further show the difference between to algorithms, Fig. 2 (b) shows the solutions with the highest HV value obtained by two algorithms on the instance with $n = 5000$, $m = 2$ and $\rho = 0.5$. We can find that the nondominated solutions obtained by MOMA can approximate the Pareto front better in terms of convergence and diversity. Since MOMA is a memetic algorithm, global search (realized by MOEA/D) and local search (realized by TS) can be well-balanced in MOMA. In contrast, DMTS is a local search based algorithm that cannot balance the convergence and diversity well. Therefore, MOMA can obtain better results than DMTS.

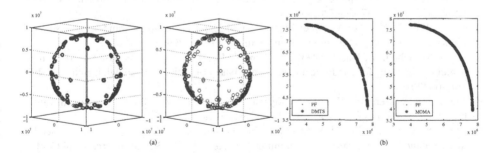

Fig. 2. Plots of nondominated solutions with the highest HV value obtained by (a) MOMA-S (left) and MOMA (right) on the instance with $n = 5000$, $m = 3$ and $\rho = -0.5$. (b) DMTS (left) and MOMA (right) on the instance with $n = 5000$, $m = 2$ and $\rho = 0.5$. Since the true Pareto front is unknown, the nondominated solutions found by all algorithms are seen as the Pareto front.

In summary, MOMA achieves the best performance in terms of convergence and diversity, which highlights the effectiveness of MOMA in solving mUBQP.

5   Conclusion and Future Work

This study has proposed a multiobjective memetic algorithm, MOMA, to solve mUBQP. In MOMA, MOEA/D is utilized for global search, while TS using problem-specific information is utilized for local search. To enhance the search ability, uniform generation is adopted to generate different uniform weight vectors for decomposition in each generation. Experimental results show the effectiveness of MOMA in solving mUBQP.

In the future, this work can be extended in the following directions. First, the proposed MOMA can be extended to solve many-objective UBQP which involves more than three objective functions. Second, the proposed MOMA can be applied to other multiobjective combinatorial optimization problems.

References

1. Kochenberger, G., et al.: The unconstrained binary quadratic programming problem: a survey. J. Comb. Optim. **28**(1), 58–81 (2014). https://doi.org/10.1007/s10878-014-9734-0
2. Garey, M.R., Johnson, D.S.: Computers and Intractability: A Guide to the Theory of NP-Completeness. Freeman, New York (1979)
3. Blum, C., Roli, A.: Metaheuristics in combinatorial optimization: overview and conceptual comparison. ACM Comput. Surv. **35**(3), 268–308 (2003)
4. Anacleto, E.A., Meneses, C.N., Ravelo, S.V.: Closed-form formulas for evaluating r-flip moves to the unconstrained binary quadratic programming problem. Comput. Oper. Res. **113**, 104774 (2020)
5. Glover, F., Hao, J.K.: f-Flip strategies for unconstrained binary quadratic programming. Ann. Oper. Res. **238**(1), 651–657 (2016)
6. Shi, J., Zhang, Q., Derbel, B., Liefooghe, A.: A parallel tabu search for the unconstrained binary quadratic programming problem. In: 2017 IEEE Congress on Evolutionary Computation (CEC), pp. 557–564 (2017)
7. Gu, S., Hao, T., Yao, H.: A pointer network based deep learning algorithm for unconstrained binary quadratic programming problem. Neurocomputing **390**, 1–11 (2020)
8. Chen, M., Chen, Y., Du, Y., Wei, L., Chen, Y.: Heuristic algorithms based on deep reinforcement learning for quadratic unconstrained binary optimization. Knowl. Based Syst. **207**, 106366 (2020)
9. Liefooghe, A., Verel, S., Hao, J.K.: A hybrid metaheuristic for multiobjective unconstrained binary quadratic programming. Appl. Soft Comput. **16**, 10–19 (2014)
10. Zhou, Y., Wang, J., Yin, J.: A directional-biased tabu search algorithm for multi-objective unconstrained binary quadratic programming problem. In: 2013 Sixth International Conference on Advanced Computational Intelligence, pp. 281–286. IEEE (2013)
11. Liefooghe, A., Verel, S., Paquete, L., Hao, J.-K.: Experiments on local search for Bi-objective unconstrained binary quadratic programming. In: Gaspar-Cunha, A., Henggeler Antunes, C., Coello, C.C. (eds.) EMO 2015, Part I. LNCS, vol. 9018, pp. 171–186. Springer, Cham (2015). https://doi.org/10.1007/978-3-319-15934-8_12

12. Zhou, Y., Wang, J., Wu, Z., Wu, K.: A multi-objective tabu search algorithm based on decomposition for multi-objective unconstrained binary quadratic programming problem. Knowl. Based Syst. **141**, 18–30 (2018)

13. Zangari, M., Pozo, A., Santana, R., Mendiburu, A.: A decomposition-based binary ACO algorithm for the multiobjective UBQP. Neurocomputing **246**, 58–68 (2017)

14. Zhou, Y., Kong, L., Wu, Z., Liu, S., Cai, Y., Liu, Y.: Ensemble of multi-objective metaheuristic algorithms for multi-objective unconstrained binary quadratic programming problem. Appl. Soft Comput. **81**, 105485 (2019)

15. Zhang, Q., Li, H.: MOEA/D: a multi-objective evolutionary algorithm based on decomposition. IEEE Trans. Evol. Comput. **11**(6), 712–731 (2007)

16. Deb, K., Pratap, A., Agarwal, S., Meyarivan, T.: A fast and elitist multiobjective genetic algorithm: NSGA-II. IEEE Trans. Evol. Comput. **6**(2), 182–197 (2002)

17. Paquete, L., Schiavinotto, T., Stützle, T.: On local optima in multiobjective combinatorial problems. Ann. Oper. Res. **156**(1), 83–97 (2007)

18. Tricoire, F.: Multi-directional local search. Comput. Oper. Res. **39**(12), 3089–3101 (2012)

19. Zitzler, E., Laumanns, M., Thiele, L.: SPEA2: Improving the strength Pareto evolutionary algorithm. In: Giannakoglou, K., Tsahalis, D.T., Periaux, J., Papailiou, K.D., Fogarty, T. (eds.) Evolutionary Methods for Design, Optimization and Control with Applications to Industrial Problems, pp. 95–100. CIMNE, Barcelona (2002)

20. Shang, K., Ishibuchi, H.: A new hypervolume-based evolutionary algorithm for many-objective optimization. IEEE Trans. Evol. Comput. **24**(5), 839–852 (2020)

21. Wilcoxon, F.: Individual comparisons by ranking methods. Biome. Bull. **1**(6), 80–83 (1945)

A Hybrid Algorithm for Multi-objective Permutation Flow Shop Scheduling Problem with Setup Times

Cuiyu Wang, Shuting Wang, and Xinyu Li[✉]

School of Mechanical Science and Engineering, Huazhong University of Science and Technology, Wuhan, China
lixinyu@mail.hust.edu.cn

Abstract. This paper studies the multi-objective permutation flow shop scheduling problem (PFSP) with setup times. Firstly, the mathematical model of multi-objective PFSP with setup time is established, then based on the theory of Pareto, Genetic algorithm and Variable Neighborhood Search, a new hybrid algorithm is proposed, named as Multiple Objective Hybrid Genetic algorithm (MOHGA). Finally, a set of benchmark instances with different scales are used to evaluate the performance of MOHGA. Experimental results show that the MOHGA obtains some solutions better than those previously reported in the literature, which reveals that the proposed MOHGA is an effective approach for the optimization of multi-objective PFSP with setup time.

Keywords: Permutation flow shop scheduling · Setup times · Multi-objective

1 Introduction

Scheduling problem is sorting the operation in certain constraints and according to the sequence, distributing time and resource to them. Effective scheduling scheme can improve productivity, resource utilization and reduce costs and inventory. This paper studies the multi-objective permutation flow shop scheduling problem (PFSP) [1], most studies of PFSP assume that the job processing time includes the setup time, or even ignore it [2], which simplify the analysis of the problem [3]. However, this assumption limits the application of research on PFSP and setup time has a great influence in practical scheduling [4]. Therefore, considering the setup times separately is necessary.

For PFSP, we consider two or more objectives in production process, such as the makespan, total completion time, average flow time, total tardiness and machine loading [5]. Multi-objective optimization means to optimize each sub-goal at the same time, however, these sub-goals may conflict with each other [6]. For example, improving the performance of a target may lead to the deterioration of other properties. Different with single-objective optimization problem, result of multi-objective optimization is not an optimal solution, but a set of Pareto optimal solutions [7].

Multi-objective optimization algorithm can be divided into two categories, one is the traditional methods, another is evolutionary algorithms. Traditional methods include

© Springer Nature Switzerland AG 2021
Y. Tan and Y. Shi (Eds.): ICSI 2021, LNCS 12690, pp. 34–44, 2021.
https://doi.org/10.1007/978-3-030-78811-7_4

weighted sum method, goal programming, min-max approach, analytic hierarchy process and so on. In most cases, these methods often provide only one pareto optimal solution. Multi-objective Optimization Evolutionary Algorithm (MOEA) can obtain a group solutions one time [8]. In 1985, David Schaffer presented the first multi-objective evolutionary algorithm: Vector Evaluated Genetic Algorithm, which has important guiding significance for the evolutionary algorithms. Since then, a large number of excellent evolutionary algorithms [9, 10] had been reported.

Based on the above situations, we study the multi-objective PFSP with setup time, and design a new Multiple Objective Hybrid Genetic algorithm (MOHGA). Based on the theory of pareto [11, 12], the proposed MOHGA combines the advantages of Genetic Algorithm (GA) and Variable Neighborhood Search (VNS). In the MOHGA, the GA guides the algorithm to explore promising areas and provides better initial solutions for VNS. The VNS intensively searches on promising areas for finding the optimal solution.

The rest of this paper is organized as follows: the problem is described in Sect. 2. Section 3 presents the proposed Multiple Objective Hybrid Genetic Algorithm (MOHGA). Results on computational experiments are reported in Sect. 4. Finally, Sect. 5 concludes the paper.

2   Problem Description

The multi-objective permutation flow shop problem with setup times considered in this study can be described as follows. There are a set of n independent jobs $N = \{1, 2, \ldots, n\}$ to be processed on m machines $M = \{1, 2, \ldots, m\}$. All jobs have the same process route and the same machining sequence on each machine which is able to process only a job one time. Setup time usually includes the time for model change, additional machine cleaning, conversion of tool and so on. In this paper, setup time is based on the machine to which the job is assigned and the processing sequence. In the PFSP, it is not enough to take just one scheduling goal into account. Not merely minimizing the makespan, we usually use the total flow time as the other objective.

3   Proposed Hybrid Algorithm for Multi-objective PFSP with Setup Times

Genetic algorithm (GA) has some advantages, including the global near-optimal, fast, easy to implement and so on, but its local search ability is poor. While the VNS is an effective local search method [13]. Based on the theory of Pareto and combining the advantages of these two algorithms, the MOHGA is proposed with both of good local and global search capabilities. In MOHGA, the GA guides the algorithm to explore promising areas and jump out from local optimal and provides better initial solutions for VNS. The VNS intensively searches promising areas for finding the optimal solution.

3.1   Encoding and Decoding of Chromosome

In this paper, we encode the chromosome by using the natural numbers and put the job as genes on chromosome. For example, if there are five jobs, the chromosome { 1, 5, 3,

4, 2} indicates the jobs' machining order is $1 \rightarrow 5 \rightarrow 3 \rightarrow 4 \rightarrow 2$. Each chromosome represents a solution for the problem.

The decoding way of PFSP with setup time is shown as in Fig. 1 and 2, in which s and t respectively represent setup time and processing time. In Fig. 1, we consider the s and t as a whole. In Fig. 2, we take s and t as two independent units. In traditional PFSP, the setup time is included in the machining time. Comparing Fig. 1 with 2, we can find that the makespan of PFSP could be shorten after treating the setup time as an independent unit, and for total flow time, the same effects can be observed. As the preparation activities for next production processes can deal with the last machining step at the same time, considering the setup time and machining time separately has obvious significance.

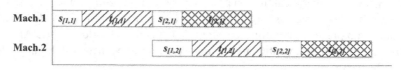

Fig. 1. Gantt chart that the setup time is included in machining time

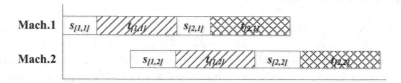

Fig. 2. Gantt chart that the setup time is treated as an independent unit

3.2 Initial Population Generation

In order to improve searching efficiency, we adopt random generation and heuristic algorithms to generate initial populations and introduce external file set to save a certain proportion of the best individuals. For FPSP, NEH [14] is one of the best heuristic methods. In this paper, part of the population is generated by NEH and another is obtained through random initialization method. Basic idea of NEH is that when sort the jobs, a job has longer machining time enjoys greater priority. The basic steps of NEH algorithm are as follows:

Step 1: Calculate each jobs' total processing time on all machines, then in descending order of total machining time arrange each job to obtain initial sort $\pi 0$;

Step 2: Take $\pi 0$'s first two genes, change the two jobs' sequence, then calculate the makspan respectively, and choose the sequence with less makespan as sort $\pi 1$;

Step 3: Take $k = k + 1$ ($k = 2, 3,..., n$), put the job k into all previously possible positions, then calculate the makespan. Put the job k into the place where make the objective function minimum to obtain the sort πk;

Step 4: If $k < n + 1$, return to *Step* 3, otherwise terminates the algorithm, output the optimal scheduling.

3.3 Pareto Sorting

We adopt the exclusive method to construct Pareto optimal solutions which is simple and clear to understand. The implementation steps are as follows:

Step 1: In initial state, non-dominated solutions set (NDSet) is set to empty;

Step 2: An individual X is selected from population randomly and put into NDSet;

Step 3: Take another individual Y from the population, and compare it with all individuals in NDSet, then delete the individuals that is dominated by Y. If there isn't individual dominated by Y, the comparison is over, and add Y to the remaining NDSet collection, then take another individual from the population, repeat the above steps.

To improve the searching efficiency of the algorithm, this paper uses the external file to retain non-dominated solutions in the process of evolution, and in order to ensure the NDSet with good dispersion, we set an upper limit M on the number of individuals in external file. If the Pareto optimal solutions in this stage don't reach limit M, all Pareto optimal solutions are stored into the external file. If the Pareto optimal solutions number exceeds the upper limit of M, M individuals selected from the NDSet are stored into the external file. The crowding distance [15] is introduced in algorithm to guarantee the multiplicity of the non-dominated solutions. Crowding distance is the sum of difference between individual and its adjacent individual on each sub-target numerical value, denoted by $L[i]_{distance}$. An individual with large crowding distance denotes the solution space density around it is small, and the individual has greater probability to participate in reproduction and evolution. After calculated the crowding distance of all non-dominated in NDSet, we use the tournament method to select M individuals with larger crowding distance and store them in external file set.

3.4 Selection and Crossover Operator

When selecting individuals for crossing operation, we choose two different cross patterns depending on probability. One pattern is based on the tournament selection method which selects two individuals from current population. Another way is selecting an individual from the external file randomly, and the other is selected based on tournament method. At the beginning of the algorithm, we adopt the second cross pattern to accelerate the convergence speed. With the increasing of iterations, we use the first cross pattern to avoid premature convergence and enhance the diversity of the population.

Large number of experimental results showed that some crossover operators can't produce offspring better than parent individuals. The reason is that crossover operations have destroyed the best mode on the later stage, which reduces the searching efficiency. This paper uses two-point crossover method, which can inherit the destruction of middle and top genes of the parents, preserves the relative position between the genes and best mode of chromosomes as far as possible.

Firstly, two crossover points are selected randomly and save the parents' top genes to offspring individuals respectively, then put the remaining parent genes into the corresponding offspring according to the previous gene sequence, which will reduce destruction of the existing mode.

3.5 Mutation Operator

In this paper, the encoding is based on the workpiece. For this method, we adopt the inverse mutation, which enable the gene cluster has greater change and with better ability to escape from local optima, and can also expand the search space of algorithm. The inverse operation is as follows:

Firstly, two mutation points will be generated randomly. Then the genes between two points are arranged with the opposite order. The process of mutation is shown as in Fig. 3 (Gene 6 and 13 are mutation points.)

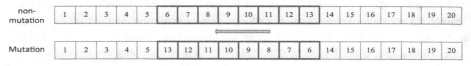

Fig. 3. Inverse mutation

3.6 Neighborhood Structure Design

Variable neighborhood search (VNS) algorithm is mainly changing variable neighborhood structure N_k systematically during the search process which can expand the search scope to improve the ability of local search. A local optimal solution can be achieved. Then change structure N_k again based on this local optimal solution to expand the search scope and find another local optimal solution. The validity of local search generally depends on the definition of neighborhood, search strategy, initial solutions and other factors.

A lot of literature has improved the VNS has been applied to a variety of combinatorial optimization problems in different areas. However, the neighborhood structures they used are combination of two simple neighborhood operations which namely insert operation and swap operation. In addition to these structures, a new inverse operation is introduced in this paper.

Inverse operation: As shown in Fig. 3, two mutation points will be generated randomly. Then the genes between two points are arranged in opposite order.

Insert operation: a gene is selected randomly, then insert the gene into a random position which is different from the original.

Swap operation: two different genes are selected randomly then exchange their positions in the chromosome.

3.7 Flowchart of Proposed MOHGA

The flowchart of proposed MOHGA is shown in Fig. 4.

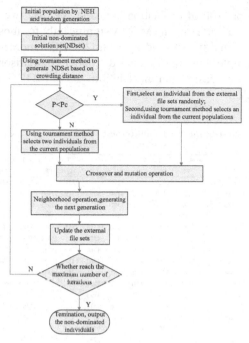

Fig. 4. Flow chart of MOHGA

4 Experimental Results and Analysis

In order to verify the effectiveness of proposed MOHGA in solving PFSP with setup time, the computational experiments are implemented on various test instances. More specifically, we compare the results with MOGA. 20 different test instances (Ta001–Ta020) with different scales provided in Vallada et al. [16] designed by Taillard are selected. And we calculate each case ten times, select the best results as the final solution sets to avoid the random error to some extent. Processing time comes from Taillard [17] typical problems, setup time is 10% of the machining time which is denoted as SSD-10. For example, if the processing time is uniformly distributed in [1, 99], the setup time is uniformly distributed in [1, 9].

For multi-objective problem, we couldn't compare the Pareto sets obtained by different algorithms directly, so we can't intuitively determine which algorithm can acquire better solutions. Therefore, in this paper we use the number of NDSet to evaluate the performance of multi-objective algorithm.

The two algorithms are coded in C++, implemented in a PC with CPU Intel(R) Core (TM) i5 2.27 GHZ. After extensive tests, the algorithm's parameters are set as follows: population size $P = 100$, crossover probability $Pc = 0.8$, mutation probability $Pm = 0.01$, maximum number of iterations $G = 100$.

For the small scales of instances (Ta001–Ta010), the results obtained by MOHGA and MOGA are summarized in Table 1 and Table 2 respectively. In these tables, MS represent maximum completion time (Makespan), TF represent the total flow time.

Table 3 gives the comparison of evaluation indicator NPS by two algorithms, which indicates the MOHGA outperforms the MOGA significantly for most test instances. To demonstrate the solution quality of hybrid algorithm MOHGA more intuitive, take Ta005 (scale 20*5) as an example, the non-dominated frontier comparison chart is shown in Fig. 5, from which we can easily find the solution sets of MOHGA can dominate the solution sets of MOGA.

Table 1. Non-dominated solution sets of MOHGA (scale 20*5)

Ta001		Ta002		Ta003		Ta004		Ta005		Ta006		Ta007		Ta008		Ta009		Ta010	
MS	TF	MS	TF	MS	TF	MS	TF	MS	TF	MS	TF	MS	TF	MS	TF	MS	TF	MS	TF
1378	15013	1469	16872	1173	14799	1406	17155	1329	15257	1297	15940	1337	16121	1306	15290	1345	16320	1203	14883
1399	14994	1470	16825	1176	14786	1410	16897	1330	15219	1304	15937	1341	15997	1307	15207	1347	16185	1205	14820
		1472	16610	1177	14785	1412	16888	1333	14965	1307	15420	1347	15515	1315	15202	1350	15991	1208	14792
		1473	16582	1179	14739	1428	16878	1339	14946	1311	15413	1351	15549	1321	15165	1351	15555	1210	14707
		1476	16574	1180	14694	1430	16837	1346	14895	1312	14584	1353	15515	1327	15147	1358	15510	1213	14679
		1482	16567	1181	14690	1432	16698	1351	14888	1314	14574	1355	15492	1331	15137	1359	15446	1216	14674
		1484	16560	1185	14627	1440	16679	1352	14843	1315	14560	1406	15485	1332	15060	1364	15363	1217	14638
		1491	16365	1190	14607	1441	16665	1357	14822	1328	14507	1431	15478	1348	15045	1365	15362	1218	14620
		1492	16338	1191	14508	1449	16620	1363	14806	1332	14486	1440	15467	1349	15018	1369	15341	1221	14600
		1495	16331	1198	14411	1470	16557	1376	14781	1341	14294			1350	14973	1371	15291	1222	14598
				1228	14323	1480	16545	1379	14778	1342	14133			1351	15202	1386	15289	1223	14536
				1240	14306			1384	14765	1346	14111			1382	14909	1442	15272	1228	14201
				1251	14265			1401	14711	1348	14079			1403	14838	1447	15254	1268	14112
				1264	14242			1405	14697	1350	13963			1416	14821				
				1265	14211			1411	14644	1355	13931			1426	14778				
				1377	14200			1430	14597										

Table 2. Non-dominated solution sets of MOGA (scale 20*5)

Ta001		Ta002		Ta003		Ta004		Ta005		Ta006		Ta007		Ta008		Ta009		Ta010	
MS	TF	MS	TF	MS	TF	MS	TF	MS	TF	MS	TF	MS	TF	MS	TF	MS	TF	MS	TF
1378	15147	1469	17022	1172	14928	1408	17410	1333	15031	1297	15944	1346	15297	1303	16354	1347	16421	1205	14927
1399	15002	1472	16639	1178	14922	1410	16834	1339	15012	1307	15560	1354	14695	1309	15207	1351	15797	1206	14925
		1474	16629	1180	14804	1430	16822	1347	14969	1312	14677	1361	14656	1314	15143	1354	15749	1208	14871
		1491	16460	1190	14503	1432	16588	1350	14958	1323	14598	1375	14632	1316	15196	1358	15741	1209	14826
		1493	16440	1191	14410	1434	16549	1351	14923	1332	14577	1378	14592	1321	15191	1359	15685	1210	14781
		1502	16421	1193	14389	1453	16510	1352	14921	1341	14311	1384	14570	1322	15187	1362	15624	1211	14775
		1508	16406	1196	14355	1457	16444	1354	14897	1346	14366			1327	15146	1367	15623	1213	14706
		1511	16390	1251	14336	1468	16434	1357	14876	1348	14207			1351	15143	1390	15613	1217	14699
				1258	14293			1376	14819	1350	14187			1367	15125	1418	15549	1221	14625
				1260	14272			1384	14810	1355	14145			1426	15099			1222	14590
								1397	14786	1357	13971							1223	14563
								1405	14727									1228	14305
								1411	14674									1237	14264
								1430	14627									1274	14249
								1460	14461									1288	14115
								1489	14441										

Table 3. Evaluation indicator of NPS (scale 20 × 5)

	Ta001	Ta002	Ta003	Ta004	Ta005	Ta006	Ta007	Ta008	Ta009	Ta010
MOHGA	2	10	16	11	16	15	9	15	13	13
MOGA	2	8	10	8	16	11	6	10	9	15

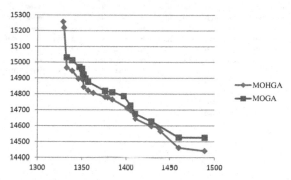

Fig. 5. Non-dominated frontier comparison chart (Ta005)

For the large scales of instances (Ta011–Ta020), the results obtained by MOHGA and MOGA are summarized in Table 4 and Table 5 respectively. Table 6 gives the comparison of evaluation indicator NPS by two algorithms, from which we can observe that the MOHGA outperforms the MOGA significantly for most test instances. For the large instance, we take Ta013 (20*10) and Ta041 (50*10) as test instances and the non-dominated frontiers are shown in Fig. 6, 7 respectively. From the figures, we can find the solution sets of MOHGA can dominate the solution sets of MOGA, which is shown that the highly effectiveness of the proposed MOHGA for solving the permutation flow shop scheduling problem with setup times.

Table 4. Non-dominated solution sets of MOGA (large-scale)

\multicolumn{2}{}{Ta011}		Ta013		Ta015		Ta017		Ta019		Ta021		Ta031		Ta041		Ta051		Ta061	
MS	TF	MS	TF	MS	TF	MS	TF	MS	TF	MS	TF	MS	TF	MS	TF	MS	TF	MS	TF
1684	23056	1604	22185	1537	19950	1590	20036	1731	22478	2453	35966	2967	73837	3367	100884	4289	142897	5994	284410
1686	22902	1607	22175	1552	19934	1594	19934	1732	22381	2460	35867	2968	73614	3370	100767	4305	142868	5995	284343
1687	22626	1608	22142	1555	19635	1595	19920	1742	22254	2462	35832	2970	73564	3373	100587	4313	142809	5998	284286
1747	22338	1610	22109	1590	19611	1599	19700	1744	22241	2467	35738	2978	72767	3377	100219	4318	142790	6052	284272
1752	22323	1615	22066	1596	19600	1605	19695	1750	21991	2473	35672	2979	72321	3383	99799	4323	142762	6053	284084
1791	22245	1623	22062	1597	19589	1609	19664	1752	21958	2484	35664	2984	72284	3423	99796	4330	142691	6106	284059
1796	22235	1628	21871	1611	19583	1611	19661	1788	21902	2485	35659	2990	71889	3443	99440	4332	142666		
1797	22174	1632	21413	1614	19540	1615	19464	1798	21856	2488	35590	3000	71676	3449	99356	4335	142650		
1802	22160	1638	21291	1619	19508	1622	19439			2491	35571	3024	71352	3450	99332	4345	142564		
1810	22101	1649	21280	1642	19490	1626	19389			2492	35402	3028	71298	3453	99273	4347	142538		
1811	22073	1654	21258	1653	19476	1641	19370			2495	35379	3049	71123			4357	142534		
1813	22053	1663	21226			1645	19321			2499	35278					4360	142517		
		1674	21180			1676	19244			2503	35260					4375	142508		
		1714	21179			1691	19226			2504	35171					4381	142452		
		1735	21157			1720	19210			2514	35117					4382	142402		
		1787	20960							2580	35113					4405	142392		

Table 5. Non-dominated solution sets of MOHGA (large-scale)

Ta011		Ta013		Ta015		Ta017		Ta019		Ta021		Ta031		Ta041		Ta051		Ta061	
MS	TF	MS	TF	MS	TF	MS	TF	MS	TF	MS	TF	MS	TF	MS	TF	MS	TF	MS	TF
1684	23030	1591	21868	1510	20180	1587	19849	1716	22492	2422	37679	2967	73913	3341	102487	4273	143409	5984	284884
1686	22884	1596	21835	1511	20087	1597	19824	1722	22469	2427	37240	2968	72851	3346	102445	4277	143299	5986	284689
1687	22866	1604	21810	1526	20076	1605	19821	1724	22378	2442	37204	2979	71646	3348	102438	4281	143279	5988	284209
1688	22848	1609	21720	1529	20055	1607	19799	1726	21720	2448	37166	2989	71382	3350	102310	4282	143230	5994	284499
1690	22829	1615	21597	1530	19882	1611	19791	1732	21623	2458	37154	2990	71178	3356	102255	4288	143183	5995	284343
1696	22767	1630	21591	1537	19785	1617	19739	1756	21612	2459	37058	2997	71128	3360	102244	4289	142439	6019	283601
1700	22699	1631	21532	1556	19763	1618	19711	1759	21598	2472	36998	3000	71076	3361	100906	4291	142361	6028	283594
1726	22689	1633	21213	1559	19714	1625	19657	1767	21596	2474	36970	3009	71072	3362	100790	4298	142106	6048	283582
1738	22654	1637	21133	1578	19661	1626	19621	1768	21565	2478	36935	3017	71062	3363	100467	4303	141770		
1745	22598	1646	21122	1582	19642	1628	19617	1772	21515	2481	35602	3021	71025	3364	100450	4305	141562		
1746	22571	1647	21110	1584	19587	1629	19584	1787	21469	2516	35491	3029	71023	3368	100268	4309	141454		
1765	22569	1651	21064	1588	19570	1633	19524	1866	21445	2520	35441	3076	71021	3410	99852	4320	141275		
1766	22546	1702	21014			1634	19511			2521	35288			3412	99546	4323	140927		
1781	22539	1735	20972			1684	19494			2548	35157			3439	98560	4337	140897		
1801	22512	1787	20940			1685	19457									4341	140753		
1812	22455	1813	20908			1691	19418									4478	140739		
1813	22451					1699	19407												
1817	22376																		
1824	22345																		
1829	22338																		
1837	22321																		

Table 6. Evaluation indicator NPS (large-scale)

	Ta011	Ta013	Ta015	Ta017	Ta019	Ta021	Ta031	Ta041	Ta051	Ta061
MOHGA	21	16	12	11	12	14	12	14	16	8
MOGA	12	16	11	15	8	16	11	10	16	6

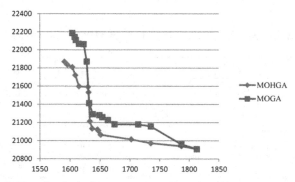

Fig. 6. Non-dominated frontier comparison chart (Ta013: scale 20*10)

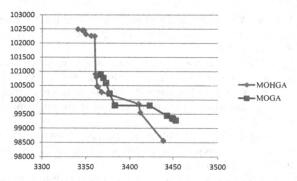

Fig. 7. Non-dominated frontier comparison chart (Ta041: scale 50*10)

5   Conclusion and Future Work

For multi-objective PFSP with setup time, a new hybrid algorithm MOHGA is designed. The proposed method shows the good searching ability, the main reason is the use of external file set to retain non-inferior solutions generated in the process of evolution, while by using tournament method to select individuals with large crowding distance from the Pareto sets and put them into external file sets to guarantee external file sets with better dispersion. Combined with variable neighborhood search, the local search capability of the algorithm is enhanced. Finally, we use the benchmarks to test the algorithm and compare the solutions with MOGA, the effectiveness of the MOHGA is verified by the experimental results.

This research can be extended into several applications. Firstly, Flow shop scheduling problem can be extended into permutation flow shop and flow shop with buffer and flexibility, etc. Secondly, we should study the flow shop model in a dynamic environment (such as machine failure, emergency orders) which can reflect the actual production well. Thirdly, solving large-scale problems usually take a long time, thus the searching efficiency of the algorithm need more in-depth research.

Acknowledgment. This research work is supported by the National Key R&D Program of China under Grant No. 2018AAA0101700, and the Program for HUST Academic Frontier Youth Team under Grant No. 2017QYTD04.

References

1. Li, Y., Wang, C., Gao, L., Song, Y., Li, X.: An improved simulated annealing algorithm based on residual network for permutation flow shop scheduling. Complex Intell. Syst., 1–11 (2020). https://doi.org/10.1007/s40747-020-00205-9
2. Cheng, E., Gupta, J., Wang, G.: A review of flow shop scheduling research with setup times. Prod. Oper. Manag. **9**(3), 262–282 (2000)
3. Gui, L., Gao, L., Li, X.: Anomalies in special permutation flow shop scheduling problems. Chin. J. Mech. Eng. **33**(1), 1–7 (2020). https://doi.org/10.1186/s10033-020-00462-2

4. Lee, J., Yu, J., Lee, D.: A tabu search algorithm for unrelated parallel machine scheduling with sequence-and machine-dependent setups: minimizing total tardiness. Int. J. Adv. Manuf. Technol. **69**, 2081–2089 (2013)
5. Yenisey, M., Yagmahan, B.: Multi-objective permutation flow shop scheduling problem: literature review, classification and current trends. Omega **45**, 119–135 (2014)
6. Wang, G., Gao, L., Li, X., Li, P., Tasgetiren, M.: Energy-efficient distributed permutation flow shop scheduling problem using a multi-objective whale swarm algorithm. Swarm Evol. Comput. **57**, 100716 (2020)
7. Deb, S., Tian, Z., Fong, S., Tang, R., Wong, R., Dey, N.: Solving permutation flow-shop scheduling problem by rhinoceros search algorithm. Soft. Comput. **22**(18), 6025–6034 (2018). https://doi.org/10.1007/s00500-018-3075-3
8. Xu, G., Luo, K., Jing, G., Yu, X., Ruan, X., Song, J.: On convergence analysis of multi-objective particle swarm optimization algorithm. Eur. J. Oper. Res. **286**, 32–38 (2020)
9. Yagmahan, B., Yenisey, M.: Ant colony optimization for multi-objective flow shop scheduling problem. Comput. Ind. Eng. **54**, 411–420 (2008)
10. Frosolini, M., Braglia, M., Zammori, F.: A modified harmony search algorithm for the multi-objective flowshop scheduling problem with due dates. Int. J. Prod. Res. **49**(20), 5957–5985 (2011)
11. Murata, T., Ishibuchi, H., Tanaka, H.: Multi-objective genetic algorithm and its applications to flow shop scheduling. Comput. Ind. Eng. **4**(30), 957–968 (1996)
12. Michele, C., Gerardo, M., Ruben, R.: Multi-objective sequence dependent setup times permutation flowshop: a new algorithm and a comprehensive study. Eur. J. Oper. Res. **227**(2), 301–313 (2013)
13. Pierre, H., Nenad, M.: Variable neighbourhood search: methods and applications. Ann. Oper. Res. **175**, 367–407 (2010)
14. Nawaz, M., Enscore, E., Ham, I.: A heuristic algorithm for the m-machine, n-job flow shop sequencing problem. Omega **11**, 91–95 (1983)
15. Deb, K., Samir, A., Amrit, P., Meyarivan, T.: A fast elitist non-dominated sorting genetic algorithm for multi-objective optimization: NSGA-II KanGAL Report 200001. Indian Institute of Technology, Kanpur, India (2000)
16. Vallada, E., Ruiz, R., Maroto, C.: Synthetic and real benchmarks for complex flow-shop problems. Technical Report, Universidad Polytechnic de Valencia, Valencia, Espana, Grupo de Investigation Operativa GIO (2003)
17. Taillard, E.: Benchmarks for basic scheduling. Eur. J. Oper. Res. **64**(2), 278–285 (1993)

Dynamic Multi-objective Optimization via Sliding Time Window and Parallel Computing

Qinqin Fan[1]([✉]), Yihao Wang[1], Okan K. Ersoy[2], Ning Li[3], and Zhenzhong Chu[4]

[1] Logistics Research Center, Shanghai Maritime University, Shanghai 201306, China
[2] School of Electronic and Computer Engineering, Purdue University, West Lafayette, IN 47906, USA
[3] Key Laboratory of System Control and Information Processing, Ministry of Education of China, Shanghai Jiao Tong University, Shanghai 200240, China
[4] Logistics Engineering College, Shanghai Maritime University, Shanghai 201306, China

Abstract. Tracking changing Pareto front (PF) in the objective space and Pareto set (PS) in the decision space is an important task in dynamic multi-objective optimization (DMO). Similarly, maintaining population diversity and reusing previous evolutionary information are useful to explore promising regions and to find high-quality solutions quickly in time-varying environments. To this end, a sliding time window based on parallel computing (STW-PC) is introduced in the present study. In the STW-PC, obtained time-sequence solution sets aim to preserve the diversity and facilitate a fast convergence since problems in successive time/environments are usually related. The parallel computing method is also employed to reduce the computational time. The STW-PC is incorporated into a multi-objective evolutionary algorithm and is compared with two competitors on 12 dynamic multi-objective optimization problems. The results show that the STW-PC can both improve the tracking performance of the selected algorithm in different degrees of changes, and significantly reduce the calculation time compared with transfer learning.

Keywords: Evolutionary computations · Dynamic multi-objective optimization · Sliding time window · Parallel computing · High-performing computing

1 Introduction

Dynamic multi-objective optimization problems (DMOPs) have been commonly found in various fields [1–6]. Compared with static optimization problems, their objective functions, constraints, decision space sizes or other parameters may change with time or under different environments. Therefore, two important tasks in the dynamic optimization are: (1) detect the change of optimization environments, and (2) find high-quality solutions quickly in dynamic or uncertain environments. For the former, how to effectively identify changes of models/environments is important. It should be noted that

Y. Tan and Y. Shi (Eds.): ICSI 2021, LNCS 12690, pp. 45–57, 2021.
https://doi.org/10.1007/978-3-030-78811-7_5

the change detection is not the focus in the current study. For the latter, the population diversity and the convergence speed are two important performance indicators for dynamic multi-objective evolutionary algorithms (DMOEAs). To improve the population diversity, diversity maintenance/improvement and multi-population [7] are two main approaches in the DMO. Moreover, memory-based and prediction-based methods are useful to promote convergence. Essentially, they belong to information reuse or knowledge transfer.

As stated above, the tracking performance of DMOEAs is impacted by population diversity and convergence speed, and thus these issues should be focused upon.

(1) Although various methods have been proposed to improve/maintain population diversity, the number of maintainable individuals is usually not enough in previous studies due to the limit of population/archive size [8]. Additionally, the current model may have a great correlation with recent "local" models on most actual dynamic optimization problems, but not with old "local" models. In other words, solutions of a dynamic optimization problem in recent successive environments are often relevant. Therefore, how to utilize the recent evolutionary information is an important step to find high-quality solutions quickly under dynamic environments.

(2) Prediction-based methods are promising for solving DMOPs, but most of them need to satisfy the independent identical distribution (IID) hypothesis [7]. Therefore, alleviating the limitation on the IID hypothesis is necessary. Additionally, their run time may be unacceptable in some cases, especially when the time complexity is high.

To solve the above-mentioned issues, a sliding time window based on parallel computing (STW-PC) is proposed to solve DMOPs in the current study. In the STW-PC, the sliding time window [9], which is an effective approach in predictive control, is used to save successive evolutionary information (i.e., trust regions [10]) with a random initial population to assist selected algorithms in improving their tracking performance in continuously changing environment. Intuitively, it can both improve the population diversity and speed up the convergence. Moreover, the parallel computing method is utilized to reduce its computational time. To demonstrate the effectiveness of the STW-PC, it is compared with transfer-learning-based MOEA and another MOEA on 12 DMOPs [11]. The results show that the STW-PC not only assists other algorithm in improving tracking performance, but also significantly reduces the computational time compared with the transfer learning method.

The remaining sections of this paper are organized as follows: the DMO, the performance metric, and the sliding time window are introduced in Sect. 2. Section 3 reviews related studies on DMOEAs. The STW-PC is presented in Sect. 4. Experimental results and analyses are reported in Sect. 5. Section 6 discusses conclusions and describes future studies.

2 Background

2.1 Dynamic Multi-objective Optimization

A dynamic multi-objective optimization problem can be defined as follows:

$$\min_{\mathbf{x} \in \Omega} \mathbf{F}(\mathbf{x}, t) = \langle f_1(\mathbf{x}, t), f_2(\mathbf{x}, t), \dots, f_m(\mathbf{x}, t) \rangle, \tag{1}$$

$$x_j \in \left(x_j^{\text{low}}, x_j^{\text{high}} \right), \ j = 1, 2, \dots, D,$$

where $\mathbf{x}$ denotes a decision vector in the feasible search space $\Omega \subset R^D$; and t ($t = 1, 2, \dots,$ T) denotes a time/environment variable; x_j^{low} and x_j^{high} are the lower and upper bounds of x_i, respectively; D denotes the dimensionality of the DMOP and m represents the number of objectives. Compared with a static multi-objective optimization problem, the DMOP consists of m time-varying objective functions under different time or environments.

Definition 1 (Dynamic dominance relation): for each time step or environment, A vector $u \in R^m$ is said to dominate another vector $v \in R^m$, which is denoted as $u \succ v$, if $\forall n \in \{1, 2, \dots, m\}$, $u_n < v_n$ and $u \neq v$.

Definition 2 (Dynamic Pareto optimal set): for each time step or environment, if there are no other solutions can be Pareto optimal to $x^* \in R^D$, x^* is called the Pareto optimal solution. The set of all Pareto optimal solutions at t is called the Pareto set (PS), denoted as X_t^*. The PS of the DMOP can be denoted as $X^* = X_1^* \bigcup X_2^* \dots \bigcup X_T^*$.

Definition 3 (Dynamic Pareto front): for each time step or environment, the Pareto front (PF) can be defined as $PF_t = \{F(x_t^*) | x_t^* \in X_t^*\}$. Like Definition 2, the PF of the DMOP can be denoted as $PF = PF_1 \bigcup PF_2 \dots \bigcup PF_T$. Similar to a static MOP, the main target of the DMO is to find a PF with good diversity and convergence at t.

2.2 Performance Metric

To evaluate the performance of DMOEAs, various performance metrics have been proposed [7, 12]. Overall, they adopt same performance indicators used in the static multi-objective optimization under each time step or environment. However, DMOPs are time-varying, thus using a cumulative performance is a common way to assess the overall performance of DMOEAs during all time steps or environments. Two IGD variants utilized in [7, 13] are introduced in this section.

An IGD variant (MIGD) is defined as follows:

$$MIGD\left(A, PF^*, C\right) = \frac{1}{|T|} \sum_{t \in T} \frac{\sqrt{\sum_{v \in PF_t^*} d(v, A_t)^2}}{|PF_t^*|}, \tag{2}$$

where A_t and PF_t^* denote the obtained PF approximation and a set of uniformly distributed points along the true PF at t, respectively; $d(v, A_t)$ is the minimum Euclidean distance between v and points in A_t; $|PF_t^*|$ represents the number of non-dominated individuals in PF_t^*; C denotes a parameter setting in the DMOP, and $|T|$ represents the number of time steps in T. A smaller value of the MIGD means that a DMOEA can achieve better PF approximations during all time steps.

2.3 Sliding Time Window

The STW, which divides the entire time window into $\triangle$ parts (called time slot), is an effective yet simple approach to solve dynamic optimization problems [14–18]. However, it has been utilized in the DMO rarely. In the STW, the size of the time window is the most important factor. If it is too large, a good result may be obtained but more computational resources are consumed. On the contrary, a small one reduces the computational budget, but it may significantly influence the performance of the STW. Therefore, its size should be set based on computational resources and final solution precision. We assume that the time window is $W(t, \triangle)$, $(1 \leq \triangle \leq T)$, which includes $\triangle$ time slots.

An example of the STW is illustrated in Fig. 1. Suppose $\triangle = 3$. When $t = t_0$, $W(t_0, \triangle) = \Phi$. If $t = t_1$, $W(t_1, \triangle)$ is equal to the first time slot. Like the above process, different time slots will be in W at different t. The number of time slots in W is equal to $\triangle$. It is observed from Fig. 1 that W is employed to store recent information or "local" models. That is why the STW can be utilized to deal with a large number of dynamic problems, especially when successive time steps/environments are relevant.

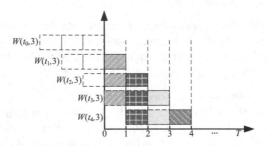

Fig. 1. Changes of the time window in the sliding time window when t is from t_0 to t_4 ($\triangle = 3$).

3 Related Work

Because objective functions or parameters in DMOPs will change over time or in different environments, their PFs or PSs, or both may be varying.

To improve the population diversity, researchers have proposed many advanced DMOEAs to solve DMOPs. Diversity-maintain/improvement-based and multi-population-based approaches are two main methods. For example, Deb et al. [19] proposed two dynamic NSGA-II variants to solve dynamic problems, in which randomly generated individuals and mutated individuals are used to replace some individuals in the current population. Yang [20] used memory-based and elitism-based immigrant methods to improve the search capability of the GA in a nonstationary environment. In this algorithm, the random immigrant approach is mainly used to increase the population diversity and the global exploration ability. However, Jiang et al. [21] pointed out that a population diversity improvement/maintain strategy may not be useful for solving complex dynamic optimization problems since it cannot provide available evolutionary information from past experiences. Besides improving the exploration capability of search engines, a multi-population strategy is an effective approach to improve/maintain the population diversity in changing environments. For example, Branke et al. [22] used two populations to balance the exploration and exploitation capabilities. Different from the above studies, Liu et al. [23] used a co-evolutionary method to share information among different populations, and a similarity detection approach to detect the environmental changing. The results show that their proposed algorithm is a promising method to solve DMOPs.

To speed up convergence, memory-based and prediction-based methods have been proposed. In [24], an archive is used to save best individuals. If the environmental change is detected, some individuals in the current population would be replaced by historical individuals stored in the archive. Similar to the above work, a dynamic competitive-cooperation co-evolutionary algorithm (dCOEA) [25] is proposed to solve DMOPs. In the dCOEA, a temporal memory is utilized to store historical evolutionary information. Stroud [26] proposed a Kalman-extended GA (KGA) method to solve dynamic optimization problems, in which information provided by the Kalman formulation is employed to determine which operator should be adopted. In [27], the Kalman filter is used to learn previous evolutionary information and then produce useful candidate solutions. The experimental results show that the algorithm can help to improve the exploitation capability of evolutionary algorithms in a dynamic environment. Unlike previous studies, Zhou et al. [28] introduced a population prediction method to divide a Pareto set into two parts, and obtain a predicted center point and an estimated manifold. Therefore, this method can provide a good predicted population under different environments. Because the quality of an initial population is important for DMOEAs to find a high-quality solution set, a transfer learning approach [7] is utilized to generate an initial population at each time step. Their results confirm that their proposed algorithm can effectively reuse previous population information and assist MOEAs in tracking time-varying PFs or PSs or both. Ruan et al. [29] stated that an original transfer learning may not be an effective method to solve DMOPs in some cases. Therefore, novel strategies and kernel functions are studied. Recently, Wang et al. [30] proposed an ensemble learning based prediction strategy to enhance the prediction precision and improve the tracking performance of algorithms.

4 Sliding Time Window Based on Parallel Computing

To solve DMOPs effectively, maintaining the diversity and reusing historical information are important in DMOEAs. The STW is an effective approach to solve problems in an uncertain or dynamic environment, such as the predictive control and the modeling of time-varying systems. Therefore, the STW-PC is proposed in the present study. Its basic framework is shown in Algorithm 1.

At each time step t, the first step is to randomly generate an initial population in Ω, i.e. P_t (**line 2**), which is mainly used to improve the exploration capability of selected algorithms and provide additional evolutionary information. Subsequently, Δ-1 obtained solution sets in time steps $[t-\Delta + 1, t-1]$ (denoted as $P_{t-\Delta+1}, P_{t-\Delta+2}, \ldots, P_{t-1}$) and P_t stored in W are used as Δ initial populations at t. Meanwhile, a DMOEA/MOEA is employed to simultaneously find Δ PF approximations and PSs via the parallel computing technique (**line 3**). After finding all approximate PFs and PSs, they are merged to obtain final approximate PF and PS using various multi-objective selection methods at t (**line 4**). The flowchart of the STW-PC is shown in Fig. 2.

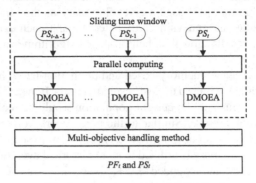

Fig. 2. Flowchart of the STW-PC.

Algorithm 1 STW-PC

Input:	W: time window ;
	$\triangle$: number of time slots in W ;
	T: number of environment changes;
1:	**for** t =1 to T **do**
2:	Generate an initial population P_t in Ω;
	Based on $\triangle$ initial populations in $W(t, \Delta) = \{P_{t-\Delta+1}, P_{t-\Delta+2}, \cdots, P_t\}$, a selected
3:	DMOEA/MOEA is used to find $\triangle$ approximate PFs at the same time via parallel computing
	technique;
4:	Obtain $P_t = Selection(P_{t-\Delta+1} \cup P_{t-\Delta+2} \cdots \cup P_t)$ and $PF_t = \{F(x_t^*)|x_t^* \in P_t\}$;
5:	**end for**
Output:	$PF = \{PF_1, PF_2 \cdots, PF_T\}$ and $PS = \{P_1, P_2 \cdots, P_T\}$.

5   Experimental Results and Analyses

To demonstrate the effectiveness of the proposed STW-PC, a regularity model-based multi-objective estimation of distribution algorithm (RM-MEDA) [31], and a transfer-learning-based algorithm are selected in the current study. Moreover, the STW-PC is incorporated into the RM-MEDA (named as STW-PC*), and 12 bi- and tri-objective DMOPs proposed in IEEE CEC2015 are used. All compared algorithms are coded in Matlab and run on a Windows 10 operating system (64 bit).

The parameter configurations of 12 DMOPs are shown in Table 1. n_t, τ_T and τ_t denote the severity of change, the maximum number of generations, and the frequency of change, respectively. It is observed from Table 1 that each DMOP changes 20 times (i.e., $\frac{\tau_T}{\tau_t}$) during the whole time T. For all compared algorithms, the population size is set to 200, and the maximum numbers of generations are set to $20 \times \tau_t + 50$ for transfer-learning-based algorithm and $20 \times \tau_t + 200$ for the STW, respectively. The main reason is that, unlike the transfer learning, no useful information can be provided by the STW in the first time step. Therefore, giving more computational budgets can improve the performance of the STW. However, from the perspective of computational time, the run time of the STW will not increase significantly, that is demonstrated in Table 1. Additionally, each compared algorithm is run for 20 independent times on all functions.

Table 1. Parameter configurations on 12 DMOPs

	n_t	τ_t	τ_T
C1	10	50	1000
C2	1	50	1000
C3	20	50	1000

5.1 Test the STW-PC Using MIGD

The mean MIGD values of all compared algorithms and the ratios of their improvement (ROI) are presented in Table 2. Note that the results of compared algorithms shown in Table 2 are directly taken from Ref. [7], except for the STW-PC*. The best mean value is highlighted in bold; [] denotes the value of ROI.

For the RM-MEDA, as shown in Table 2, the STW-PC* outperforms the RM-MEDA and the Tr-RM-MEDA on 35 and 31 cases, respectively. Also, it has a high success ratio to help the RM-MEDA solve DMOPs. Compared with the transfer learning method, the STW-PC is a more competitive approach to assist the RM-MEDA in adapting continuously changing environments. At the same time, the STW-PC* produces much better results (i.e., the ROI value is greater than 50%) on 13 and 6 cases when compared with the RM-MEDA and the Tr-RM-MEDA. Additionally, the Tr-RM-MEDA surpasses the STW-PC* on only one function HE7, in which PSs are the same during the whole time steps. The results presented in Table 2 indicate that the proposed algorithm slightly perform worse than the Tr-RM-MEDA on HE7 with three parameter configurations.

Based on the above comparisons, it can be concluded that the STW-PC is able to help other MOEAs/DMOEAs enhance their performances to track either changing PFs or changing PSs or both in continuously changing environments on different types of DMOPs.

5.2 Experimental Analysis

Impact of $\triangle$. Intuitively, a larger $\triangle$ value saves more evolutionary information in W, thus it provides more trust regions to guide the population evolution in a new optimization environment. To analyze the impact of $\triangle$, it was selected from the set $\{3, 4, 5, 6\}$, and the RM-MEDA and 12 DMOPs were used in experiments. Moreover, all parameter settings are the same as in the above experiments except for the $\triangle$ value. Note that the denominator of the ROI value is the MIGD calculated by $\triangle = 3$ in the following experiments.

Table 2. MIGD values of RM-MEDA, Tr-RM-MEDA, and STW-PC*

	C1			C2			C3		
	RM-MEDA	Tr-RM-MEDA	STW-PC*	RM-MEDA	Tr-RM-MEDA	STW-PC*	RM-MEDA	Tr-RM-MEDA	STW-PC*
FDA4	6.84E-02 [30.4%]	5.27E-02 [9.7%]	**4.76E-02**	6.93E-02 [35.24%]	5.01E-02 [10.6%]	**4.48E-02**	6.87E-02 [32.2%]	5.29E-02 [11.9%]	**4.66E-02**
FDA5	1.86E-01 [63.3%]	7.93E-02 [14.0%]	**6.82E-02**	2.68E-01 [67.1%]	**8.18E-02 [-7.8%]**	8.82E-02	1.66E-01 [60.7%]	7.52E-02 [13.2%]	**6.53E-02**
FDA5$_{iso}$	6.59E-02 [36.9%]	6.61E-02 [37.1%]	**4.16E-02**	6.11E-02 [37.3%]	6.12E-02 [37.4%]	**3.83E-02**	6.53E-02 [37.2%]	6.54E-02 [37.3%]	**4.10E-02**
FDA5$_{dec}$	6.39E-01 [87.1%]	4.10E-01 [80.0%]	**8.25E-02**	3.80E-01 [68.2%]	1.31E-01 [7.6%]	**1.21E-01**	6.74E-01 [87.9%]	3.66E-01 [77.8%]	**8.14E-02**
DIMP2	4.66E+00 [20.4%]	4.82E+00 [23.0%]	**3.71E+00**	5.01E+00 [21.8%]	4.96E+00 [21.0%]	**3.92E+00**	4.88E+00 [19.5%]	5.10E+00 [22.9%]	**3.93E+00**
DMOP2	4.10E-03 [43.9%]	2.70E-03 [14.8%]	**2.30E-03**	1.84E+01 [2.2%]	1.87E+01 [3.7%]	**1.80E+01**	3.40E-03 [38.2%]	3.80E-03 [44.7%]	**2.10E-03**
DMOP2$_{iso}$	1.90E-03 [56.4%]	1.90E-03 [56.4%]	**8.29E-04**	**1.10E-01 [0.0%]**	**1.10E-01 [0.0%]**	**1.10E-01**	1.90E-03 [56.3%]	1.90E-03 [56.3%]	**8.30E-04**
DMOP2$_{dec}$	1.10E-01 [90.0%]	4.41E-02 [75.1%]	**1.10E-02**	2.30E-01 [16.1%]	2.21E-01 [12.7%]	**1.93E-01**	1.07E-01 [90.0%]	5.40E-02 [80.2%]	**1.07E-02**
DMOP3	3.00E-03 [30.0%]	3.00E-03 [30.0%]	**2.10E-03**	1.83E+01 [1.6%]	1.85E+01 [2.7%]	**1.80E+01**	3.30E-03 [42.4%]	2.40E-03 [20.8%]	**1.90E-03**
HE2	8.91E-01 [93.0%]	1.07E-01 [41.9%]	**6.22E-02**	6.97E-01 [91.5%]	7.26E-02 [18.0%]	**5.95E-02**	8.94E-01 [93.1%]	1.08E-01 [43.1%]	**6.15E-02**
HE7	4.40E-02 [4.6%]	**3.51E-02 [-19.7%]**	4.20E-02	3.91E-02 [11.5%]	**3.33E-02 [-3.9%]**	3.46E-02	4.36E-02 [5.3%]	**3.37E-02 [-22.6%]**	4.13E-02
HE9	2.64E-01 [11.0%]	2.46E-01 [4.5%]	**2.35E-01**	2.36E-01 [11.0%]	2.18E-01 [3.7%]	**2.10E-01**	2.61E-01 [10.0%]	2.47E-01 [4.9%]	**2.35E-01**

ROI values of all cases with three parameter configurations are illustrated in Fig. 3. From Fig. 3 (a)–(e), it is seen that a large $\triangle$ value is able to improve the tracking performance of the RM-MEDA. As shown in Fig. 3 (f), although a large $\triangle$ value can help the RM-MEDA adapt to changing environments when the parameter configuration is C1, $\triangle = 6$ does not produce better results than $\triangle = 5$. This is because the RM-MEDA can find high-quality solutions on DMOP2 with C1. Moreover, Fig. 3 (f) indicates the effectiveness of $\triangle$ is limited on environments C2–C3. The main reason may be that these two environments are hard for the RM-MEDA to search good results under limited computational resources, i.e., the maximum number of generations set to 50 is not enough for the RM-MEDA in each time step. It can be observed from Fig. 3 (g) that a large $\triangle$ value can enhance the tracking performance of the RM-MEDA when parameter configurations are C1 and C3. However, it cannot assist the RM-MEDA in improving the tracking performance when the parameter configuration is C2. This is because the STW-PC assumes that "local" models are related in successive time steps. If the severity of change is high, the performance of the STW–PC may reduce. In other words, historical individuals may not be able to provide valuable evolutionary information when environment is changing dramatically. Figures 3 (g) and (i) illustrate that a large $\triangle$ value does not improve the tracking performance of the RM-MEDA significantly. Like Fig. 3(g), the tracking performance of the proposed algorithm may be limited when the severity of environmental change is high such as C2. For HE1, HE7, and HE9, they belong to Type III, i.e., PSs are the same in different time steps. From Fig. 3 (j), we can see that a large $\triangle$ value may slightly reduce the performance of the STW-PC because a large $\triangle$ value will increase the number of individuals, which may influence the selection of individuals in PF. Additionally, it can be seen from Figs. 3 (k) and (l) that the performance of the STW–PC cannot be significantly influenced by values of $\triangle$ because the STW–PC can provide good initial population in each time step.

Based on the above observations and comparisons, the following conclusions are obtained:

(1) The STW–PC is an effective approach to solve DMOPs, especially when problems are related in successive time steps or environments.
(2) If the environment is changing dramatically, like other existing methods, the performance of the STW-PC will be reduced. Therefore, its performance is influenced by selected MOEAs/DMOEAs.

Computational Time of STW-PC* and Tr-RM-MEDA. In the STW-PC, obtained solution sets in W will be used as initial populations. However, enough time cannot be given before the next time step or environment change. Therefore, the computational time of the STW-PC is investigated on FDA4 with C1–C3 in this experiment. Additionally, the computational time of the Tr-RM-MEDA is also given. All experiments were ran on a laptop computer with Windows 10 operating system (64 bit) using MATLAB (R2016a). Moreover, two compared algorithms ran 20 times on each instance independently.

The mean values of run time reported in Table 3 indicate that the STW-PC requires much less run time than the transfer learning method in which the complexity of the

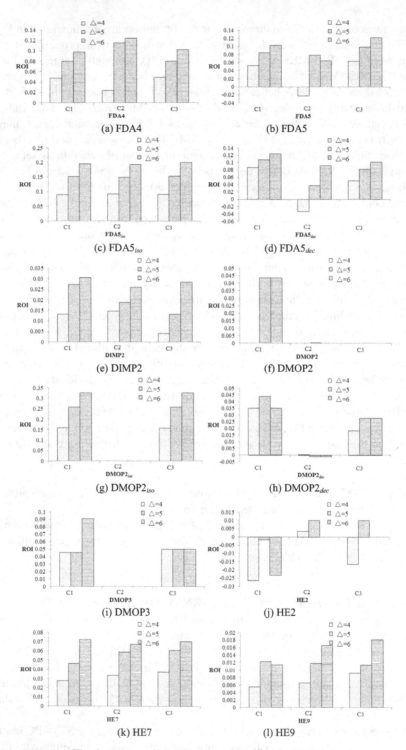

Fig. 3. Influence of different △ values on the STW-PC.

eigenvalue decomposition is high. Therefore, it is a promising method to solve DMOPs within a limit time.

Table 3. Mean run time (s) of STW-PC* and the Tr-RM-MEDA on FDA4.

FDA4	Tr-RM-MEDA	STW-PC*
C1	1.84E + 04	8.74E + 01
C2	1.93E + 04	8.65E + 01
C3	1.94E + 04	8.64E + 01

6 Conclusions and Future Work

A sliding time window based on parallel computing (STW-PC) is introduced in the current study. In the STW-PC, the STW is used to maintain the population diversity and to reuse past evolutionary information to improve the search efficiency when the time/environment changes. The results show that the STW-PC is an effective and promising approach to solve DMOPs. It is worth mentioning that the STW-PC can both significantly reduce the computational time and have better tracking performance when compared with the transfer learning method. Additionally, the STW-PC is a generic framework to solve different types of DMOPs, especially when "local" models in successive environments are relevant.

Acknowledgement. This work was partially supported by the Shanghai Science and Technology Innovation Action Plan (18550720100, 19040501600), the National Nature Science Foundation of China (No. 61603244).

References

1. Cruz, C., González, J., Pelta, D.: Optimization in dynamic environments: a survey on problems, methods and measures. Soft. Comput. **15**(7), 1427–1448 (2011)
2. Nguyen, S., Zhang, M., Johnston, M., et al.: Automatic design of scheduling policies for dynamic multi-objective job shop scheduling via cooperative coevolution genetic programming. IEEE Trans. Evol. Comput. **18**(2), 193–208 (2013)
3. Yan, X., Cai, B., Ning, B., et al.: Moving horizon optimization of dynamic trajectory planning for high-speed train operation. IEEE Trans. Intell. Transp. Syst. **17**(5), 1258–1270 (2015)
4. Yazici, A., Kirlik, G., Parlaktuna, O., et al.: A dynamic path planning approach for multirobot sensor-based coverage considering energy constraints. IEEE Trans. Cybern. **44**, 305–314 (2013)
5. Karatas, M.: A dynamic multi-objective location-allocation model for search and rescue assets. Eur. J. Oper. Res. **288**(2), 620–633 (2021)
6. Fan, Q., Wang, W., Yan, X.: Multi-objective differential evolution with performance-metric-based self-adaptive mutation operator for chemical and biochemical dynamic optimization problems. Appl. Soft Comput. **59**, 33–44 (2017)

7. Jiang, M., Huang, Z., Qiu, L., et al.: Transfer learning-based dynamic multiobjective optimization algorithms. IEEE Trans. Evol. Comput. **22**(4), 501–514 (2018)
8. Tanabe, R., Ishibuchi, H.: A review of evolutionary multi-modal multi-objective optimization. IEEE Trans. Evol. Comput. **24**(1), 193–200 (2019)
9. Yan, W., Chang, J., Shao, H.: Least square SVM regression method based on sliding time window and its simulation. J. Shanghai Jiaotong Univ. **38**(4), 524–526 (2004)
10. Fan, Q., Yan, X., Zhang, Y., et al.: A variable search space strategy based on sequential trust region determination technique. IEEE Trans. Cybern. **PP**, 1–3 (2019)
11. Helbig, M., Engelbrecht, A.: Benchmark functions for CEC 2015 special session and competition on dynamic multi-objective optimization. Technical report (2015)
12. Zhang, Q., Yang, S., Jiang, S., et al.: Novel prediction strategies for dynamic multi-objective optimization. IEEE Trans. Evol. Comput. **24**(2), 260–274 (2019)
13. Muruganantham, A., Tan, K., Vadakkepat, P.: Evolutionary dynamic multiobjective optimization via Kalman filter prediction. IEEE Trans. Cybern. **46**(12), 2862–2873 (2015)
14. Akrida, E., Mertzios, G., Spirakis, P., et al.: Temporal vertex cover with a sliding time window. J. Comput. Syst. Sci. **107**, 108–123 (2020)
15. Ma, C., Li, W., Cao, J., et al.: Adaptive sliding window based activity recognition for assisted livings. Inf. Fusion **53**, 55–65 (2020)
16. Ferland, J., Fortin, L.: Vehicles scheduling with sliding time windows. Eur. J. Oper. Res. **38**(2), 213–226 (1989)
17. Kiviniemi, V., Vire, T., Remes, J., et al.: A sliding time-window ICA reveals spatial variability of the default mode network in time. Brain Connectivity **1**(4), 339–347 (2011)
18. Fagerholt, K.: Ship scheduling with soft time windows: an optimisation based approach. Eur. J. Oper. Res. **131**(3), 559–571 (2001)
19. Deb, K., Rao N., U., Karthik, S.: Dynamic multi-objective optimization and decision-making using modified NSGA-II: a case study on hydro-thermal power scheduling. In: Obayashi, S., Deb, K., Poloni, C., Hiroyasu, T., Murata, T. (eds.) EMO 2007. LNCS, vol. 4403, pp. 803–817. Springer, Heidelberg (2007). https://doi.org/10.1007/978-3-540-70928-2_60
20. Yang, S.: Genetic algorithms with memory-and elitism-based immigrants in dynamic environments. Evol. Comput. **16**(3), 385–416 (2008)
21. Jiang, S., Kaiser, M., Wan, S., et al.: An empirical study of dynamic triobjective optimisation problems. In: Proceedings of IEEE International Conference on Evolutionary Computation, pp. 1–8 (2018)
22. Branke, J., Kaußler, T., Smidt, C., et al.: A multi-population approach to dynamic optimization problems. In: Parmee, I.C. (ed.) Evolutionary Design and Manufacture, pp. 299–307: Springer, London (2000). https://doi.org/10.1007/978-1-4471-0519-0_24
23. Liu, R., Li, J., Mu, C., et al.: A coevolutionary technique based on multi-swarm particle swarm optimization for dynamic multi-objective optimization. Eur. J. Oper. Res. **261**(3), 1028–1051 (2017)
24. Branke, S.: Memory enhanced evolutionary algorithms for changing optimization problems. In: Proceedings of the 1999 Conference on Evolutionary Computation, pp. 1875–1882 (1999)
25. Goh, C., Tan, K.: A competitive-cooperative coevolutionary paradigm for dynamic multiobjective optimization. IEEE Trans. Evol. Comput. **13**(1), 103–127 (2008)
26. Stroud, P.: Kalman-extended genetic algorithm for search in nonstationary environments with noisy fitness evaluations. IEEE Trans. Evol. Comput. **5**(1), 66–77 (2001)
27. Rossi, C., Abderrahim, M., Díaz, J.: Tracking moving optima using Kalman-based predictions. Evol. Comput. **16**(1), 1–30 (2008)
28. Zhou, A., Jin, Y., Zhang, Q.: A population prediction strategy for evolutionary dynamic multiobjective optimization. IEEE Trans. Cybern. **44**(1), 40–53 (2013)

29. Ruan, G., Minku, L., Menzel, S., et al.: When and how to transfer knowledge in dynamic multi-objective optimization. In: Proceedings of IEEE International Conference on Evolutionary Computation, pp. 2034–2041 (2019)
30. Wang, F., Li, Y., Liao, F., Yan, H.: An ensemble learning based prediction strategy for dynamic multi-objective optimization. Appl. Soft Comput. **96**, 106592 (2020)
31. Zhang, Q., Zhou, A., Jin, Y.: RM-MEDA: a regularity model-based multiobjective estimation of distribution algorithm. IEEE Trans. Evol. Comput. **12**(1), 41–63 (2008)

A New Evolutionary Approach to Multiparty Multiobjective Optimization

Zeneng She[1], Wenjian Luo[1(⊠)], Yatong Chang[1], Xin Lin[2], and Ying Tan[3]

[1] School of Computer Science and Technology, Harbin Institute of Technology, Shenzhen 518055, Guangdong, China
{20s151103,20s151150}@stu.hit.edu.cn, luowenjian@hit.edu.cn
[2] School of Computer Science and Technology, University of Science and Technology of China, Hefei 230027, Anhui, China
iskcal@mail.ustc.edu.cn
[3] Key Laboratory of Machine Perception (MOE), and Department of Machine Intelligence, School of Electronics Engineering and Computer Science, Peking University, Beijing 100871, China
ytan@pku.edu.cn

Abstract. Multiparty multiobjective optimization problems (MPM OPs) are a type of multiobjective optimization problems (MOPs), where multiple decision makers are involved, different decision makers have different objectives to optimize, and at least one decision maker has more than one objective. Although evolutionary multiobjective optimization has been studied for many years in the evolutionary computation field, evolutionary multiparty multiobjective optimization has not been paid much attention. To address the MPMOPs, the algorithm based on a multiobjective evolutionary algorithm is proposed in this paper, where the non-dominated levels from multiple parties are regarded as multiple objectives to sort the candidates in the population. Experiments on the benchmark that have common Pareto optimal solutions are conducted in this paper, and experimental results demonstrate that the proposed algorithm has a competitive performance.

Keywords: Multiobjective optimization · Evolutionary computation · Multiparty multiobjective optimization

1 Introduction

In the real world, there are a lot of optimization problems which have more than one objective, and these objectives are conflicted with each other. This type of optimization problems is called multiobjective optimization problems (MOPs)

This study is supported by the National Key Research and Development Program of China (No. 2020YFB2104003) and the National Natural Science Foundation of China (No. 61573327).

[1, 4, 11] for two or three objectives and many-objective optimization problems (MaOPs) [6] for more than three objectives.

Multiobjective evolutionary algorithms (MOEAs) have been studied for many years, such as NSGA-II [2], MOEA/D [16], SPEA2 [20] and Two-Arch2 [14]. MOEAs could find a set of optimal solutions in a run, which attracts many researchers to design efficient evolutionary algorithms for solving MOPs [12, 13, 19].

However, there are cases that there are multiple decision makers (DMs) and each decision maker only pays attention to some of all the objectives of MOPs. If MOEAs only search for the optimal solutions on some certain objectives for a DM, it may lead to deterioration of other objectives concerned by other DMs. If existing MOEAs are directly used to solve the MOP including all the objectives from all parties, they may result in too many Pareto optimal solutions for the MOP, but not Pareto optimal solutions for each DM's objectives.

Multiparty multiobjective optimization problems (MPMOPs) are used to express the above situation. An MPMOP often has multiple parties, and at least one party has more than one objective. MPMOPs are viewed as a subfield of multiobjective optimization problems in the viewpoints of different DMs, respectively. Although MPMOPs are regarded as multiple MOPs, they are quite different in most circumstances so that MPMOPs cannot be directly solved by the original MOEAs. That is because each DM does not consider all objectives in MPMOPs, but a few objectives which he/she cares.

A common method to solve MPMOPs is that the final optimal solutions are obtained by the complex negotiation of a third party among the Pareto optimal solutions from each party [7, 8, 15]. Recently, in [9], Liu et al. first proposed a method without negotiations to address a special class of MPMOPs with the common Pareto optimal solutions, called OptMPNDS. MPMOPs with the common Pareto optimal solutions means that there exits at least one solution that is Pareto optimal for all parties. In other words, the intersection of Pareto optimal solutions of MOPs in viewpoint of multiple DMs is not empty. OptMPNDS defines the dominance relation of solutions based on the corresponding Pareto optimal levels of multiple parties, where the levels are obtained according to the non-dominated sorting in NSGA-II [2]. In each generation, after sorting individuals by objectives preferred by each party, the common solutions with the same level of all parties are assigned to the same rank. Then, the rank of the rest individuals are set to the maximum level obtained by all parties. However, it cannot perfectly handle the situation that the individuals have the same maximum level with different other levels.

In this paper, we propose an improved evolutionary algorithm to solve the MPMOPs, called OptMPNDS2. Similar to OptMPNDS, OptMPNDS2 is also based on NSGA-II. However, OptMPNDS2 overcomes the shortcoming mentioned above. Experiments on the benchmark in [9] are conducted in this paper. From the experimental results, it can be seen that OptMPNDS2 has a more powerful performance on some MPMOPs.

The rest of this paper is organized as follows. Section 2 gives the related work. Section 3 clearly explains the proposed algorithm and Sect. 4 describes the performance metrics and shows the experiment results. Finally, we make a brief conclusion in Sect. 5.

2   Related Work

2.1   Multiparty Multiobjective Optimization Problems (MPMOPs)

An MPMOP is a particular class of multiobjective optimization problems (MOPs). MOPs are a type of optimization problems which have at least two objectives. Because MPMOPs are based on MOPs, the related concepts of MOPs are described first. Here, for convenience, an MOP is defined as a minimization problem [5]:

$$Minimize \quad F(x) = (f_1(x), f_2(x), \ldots, f_m(x)),$$

$$Subject \quad to \begin{cases} h_i(x) = 0, & i = 1, \ldots, n_p \\ g_j(x) \leq 0, & j = 1, \ldots, n_q \\ x \in [x_{min}, x_{max}]^d \end{cases}, \quad (1)$$

where f_i denotes the i-th objective function and F, combined with m objectives, denotes the vector function of objectives which should be minimized. And $h_i(x) = 0$ represents the i-th equality constraint, of which total number is n_p; $g_j(x) \leq 0$ represents the j-th inequality constraint, of which total number is n_q. x, a d-dimensional vector, stands for the decision variables that have the lower bounds x_{min} and upper bounds x_{max} in each dimension.

Dominance is a relation about decision vectors [19]. Given two decision vectors x and y, under the condition that all objectives satisfy $f(x) \leq f(y)$, if there exists one objective f_i satisfying $f_i(x) < f_i(y)$, it is said that x Pareto dominates y, denoting as $x \prec y$ [10]. Pareto optimal set (PS) is a set of Pareto optimal solutions. A decision vector x belongs to PS if and only if no solution dominates x. Pareto optimal front (PF) is a set of objectives of decision vectors in Pareto optimal set, which is formally defined as PF = $\{f = (f_1(x), f_2(x), \cdots, f_n(x)) | x \in$ PS$\}$.

There are multiple DMs in an MPMOP, where each DM focuses on different objectives and at least one DM has at least two objectives. Different from Formula (1), where $f_i(x)$ denotes one objective of the solution x, MPMOPs consider that $f_i(x)$ is a vector function, which represents all objectives of one party. Specifically,

$$f_i(x) = (f_{i1}(x), f_{i2}(x), \ldots, f_{ij_i}(x)),$$

and j_i denotes the number of objectives of the i-th party. Then m becomes the number of the parties.

As shown in [9], during evolution, to compare two individuals in MPMOPs, the comparison of all objectives together is not well performed. This is because one DM only focusses on the objectives concerned by himself. Therefore, the individuals are assigned the max levels among the Pareto optimal levels obtained by all parties to compare with each other in [9].

2.2 NSGA-II

Non-dominated sorting genetic algorithm II (NSGA-II) [2] is a popular evolutionary algorithm to solve MOPs. There are two core strategies in NSGA-II, i.e., the fast non-dominated sorting and the calculation of crowding distance. The fast non-dominated sorting adopts the Pareto dominance relation to rapidly achieve the Pareto levels of all the individuals in the evolutionary population. The crowding distance describes the distribution of the individuals. If the crowding distance is large, it means that the individual locates at a region with a few individuals; while it is small, the individual is in a dense region.

In each generation of NSGA-II, the process is shown as follows. First, the crossover and mutation operators are adopted to generate offspring. Next, the parent and offspring are put together to sort the Pareto levels. Then, the crowding distances of the individuals in the same level are obtained. Finally, based on the Pareto levels and crowding distances, the population of next generation is selected from the parent population and the offspring population.

3   The Proposed Algorithm

In this section, the algorithm based on NSGA-II is proposed to solve MPMOPs. Different from traditional MOPs, in MPMOPs, we should maximize the profits of each party. That is, we should try to approach to the Pareto front of each party. In order to achieve this goal, we should optimize the objectives from each party simultaneously, but group them according to each party.

The core issue is how to evaluate a given individual. For a DM, the Pareto optimal level number L_i of the party i could be regarded as an objective value for each individual. Therefore, for total m parties, we have the following multiobjective problem:

$$Minimize \quad L = (L_1, L_2, \ldots, L_m), \tag{2}$$

where m represents the number of the parties, L_i of the party i could be obtained by non-dominated sorting function in NSGA-II. It should be noted that, when calculating L_i of the party i, only the objectives of the party i are considered. Based on Formula (2), we can sort the individuals in evolutionary population according to the standard Pareto dominance, and then the corresponding algorithm could be designed.

The pseudocodes OptMPNDS2 are described in Algorithm 1, which are described as follows.

(1) The population P_0 is initialized with population size N.
(2) The number of generation t and offspring Q_t are set as 0 and $\emptyset$, respectively.
(3) From steps 3 to 15, in the loop, the population P_t and its offspring Q_t are gathered into R_t and function MPNDS2 is applied to sort these individuals into different ranks $\mathcal{F}$. Then, sort individuals in the same rank according to crowding distance. Next the parameters t and P_t are updated for next generation. Subsequently, N individuals are picked from the best to the worst and stored in the population of the next generation P_t. Finally, the offspring Q_t is generated from P_t by both crossover and mutation operators, and the loop is repeated until the termination condition is satisfied.
(4) Return the final solutions MPS.

Algorithm 1. OptMPNDS2

Require: N (the population size),
  $F = (F_1(x), F_2(x), \ldots, F_m(x))$ (the objective function)
Ensure: MPS (the multiparty Pareto optimal solutions)
 1: Initialize population P_0 with size N ;
 2: $t = 0, Q_t = \emptyset$;
 3: **while** The termination is not satisfied **do**
 4:     $R_t = P_t \cup Q_t$;
 5:     $\mathcal{F} = \text{MPNDS2}(N, R_t, F)$;
 6:     Sort $\mathcal{F}$ by crowding distance for each rank ;
 7:     $t = t + 1, P_t = \emptyset, i = 1$;
 8:     **while** $|P_t \cup \mathcal{F}_i| \leq N$ **do**
 9:         $P_t = P_t \cup \mathcal{F}_i$;
10:         $i = i + 1$;
11:     **end while**
12:     $P_t = P_t \cup \mathcal{F}_i(1 : N - |P_t|)$;
13:     Create the offspring Q_t of P_t ;
14: **end while**
15: $MPS = $ multiparty Pareto optimal solutions in P_t ;

OptMPNDS2 is similar to NSGA-II except the function MPNDS2(.), and the pseudocodes of MPNDS2(.) are given in Algorithm 2.

As depicted in Algorithm 2, the key point of OptMPNDS2 is to redefine a dominance relation among individuals in the objective space $L = (L_1, L_2, \ldots, L_m)$. From the perspective of a party, the individuals are sorted by F_i (i.e., the objective function of the party i). For individuals in R_t, the Pareto level of the party i is calculated as the new "objective" of the party and stored in $\mathcal{L}(:, i)$. After all m parties are sorted, perform the non-dominated sorting with the objectives $\mathcal{L}$ again. Here, the non-dominated sorting function $NonDominatedSorting$ adopted in our algorithm is the same as that in NSGA-II [18].

Algorithm 2. MPNDS2

Require: $N, R_t, F = (F_1(x), F_2(x), \ldots, F_m(x))$
Ensure: $\mathcal{F}$
 1: $\mathcal{L} = \emptyset$;
 2: **for** $i \in \{1, \cdots, M\}$ **do**
 3:    $\mathcal{L}(:, i) = NonDominatedSorting(R_t, F_i)$;
 4: **end for**
 5: $\mathcal{F} = NonDominatedSorting(R_t, \mathcal{L})$;

Although both OptMPNDS2 and OptMPNDS sort individuals in terms of each party first and then sort individuals according to the levels by parties, the detail behaviors of these two algorithms are different. In OptMPNDS, the maximum levels in all parties are used to sort the individuals, while OptMPNDS2 performs the non-dominated sorting again according to the non-dominated level numbers of the individuals.

Here, we use an example to explain the difference between OptMPNDS2 and OptMPNDS. Suppose there are two individuals x and y in a bi-party multiobjective optimization, and their non-dominated level numbers are (1, 3) and (2, 3), respectively.

(1) In OptMPNDS, x and y have the same rank because both the maximum non-dominated levels are 3.
(2) In OptMPNDS2, x dominates y. For the first party, their non-dominated levels are 1 and 2, respectively. For the second party, their non-dominated levels are the same. Therefore, according to Formula (2), x dominates y.

4 Experiments

4.1 Parameter Settings

In [9], 11 MPMOP test problems with the common Pareto optimal solutions are given, and two algorithms, i.e., OptMPNDS and OptAll, are used to solve MPMOPs. OptAll just views the problems as MOPs and performs the NSGA-II to obtain solutions.

To compare with these two algorithms, OptMPNDS2 uses the same parameters as [9]. All three algorithms use the simulated binary crossover (SBX) and polynomial mutation [3], of which distribution indexes are set as 20. The rates of crossover and mutation are set to 1.0 and $1/d$, respectively, where d represents the dimension of the individuals. For each test problem, all algorithms are run 30 times to obtain the average results. In each run, the population size is set to 100 and the maximum fitness evaluations is set to $1000 * d * m$, where d and m represent the dimension of MPMOPs and the number of parties, respectively.

Inverted generational distance (IGD) [17] is an indicator to evaluate the solution quality, which measures both the convergence and uniformity of solutions. To adapt metric for MPMOPs, the work [9] slightly modifies the related concept about the distance between an individual v and a PF (denoted by P) shown as follows:

$$d(v, P) = \min_{s \in S} \sum_{i=1}^{m} \sqrt{(v_{i1} - s_{i1})^2 + \cdots + (v_{ij_i} - s_{ij_i})^2}, \tag{3}$$

where v_{ij}, as well as s_{ij}, means the j-th objective values of the i-th party, respectively. There are m parties and the i-th party has j_i objectives.

Based on Formula (3), IGD is calculated as follows:

$$IGD(P^*, P) = \frac{\sum_{v \in P^*} d(v, P)}{|P^*|}, \tag{4}$$

Where P^* represents the true PF and P represents the PF that algorithms obtained.

For IGD, the smaller value means the better performance of the algorithm, since it measures the distance between the true PF and PF obtained by the algorithm.

Considering the final population returned by an algorithm could contain some dominated solutions, the solution number (SN) [9], which represents the number of non-dominated solutions in the final population, is also used to evaluate the performance of the algorithms. The algorithm with a larger SN has a better performance.

4.2　Results

Tables 1, 2 and 3 show the IGD of the three algorithms, i.e., OptMPNDS2, OptMPNDS, OptAll. The problems from MPMOP1 to MPMOP11 are used in experiments, and the dimensions are set to 10, 30 and 50. There are the mean and standard deviation values for each problem in each row. And the best results of the same problem are labeled in the bold font. The sign "—" means that the algorithm does not obtain any solution in at least one run. And nbr denotes the number of problems for which the algorithm obtains the best results.

As tables depicted, OptMPNDS2 performs better than OptAll and is comparable with OptMPNDS. In Table 1, the number of the best IGD results of OptMPNDS2 reaches 4. In Table 2, OptMPNDS2 performs the best on 8 problems of 11. The performance of OptMPNDS2 is better than OptMPNDS in Table 3, where OptMPNDS2 wins 7 problems of 11.

Table 1. Mean and standard deviation of IGD for MPMOPs with 10 dimensions.

Problems	IGD		
	OptMPNDS2	OptMPNDS	OptAll
MPMOP1	**1.6345e-05 ± 6.4522e-06**	2.7894e-05 ± 1.8206e-05	—
MPMOP2	**6.7130e-05 ± 3.2436e-05**	2.7364e-04 ± 1.0637e-03	6.5179e-02 ± 7.5853e-02
MPMOP3	3.1678e-02 ± 1.5827e-02	**2.5420e-02 ± 9.2706e-03**	3.6974e-02 ± 1.1352e-02
MPMOP4	**5.6223e-02 ± 5.2352e-03**	5.7572e-02 ± 7.0402e-03	4.7859e-01 ± 8.3774e-02
MPMOP5	6.6392e-02 ± 1.1805e-02	**6.2492e-02 ± 1.1398e-02**	1.5786e-01 ± 3.3413e-02
MPMOP6	2.0289e-02 ± 3.1342e-03	**1.9709e-02 ± 2.4359e-03**	7.6749e-01 ± 2.0057e-01
MPMOP7	4.9085e-06 ± 4.0617e-06	**4.5667e-06 ± 4.1010e-06**	—
MPMOP8	**2.3038e-05 ± 3.1462e-05**	1.2262e-02 ± 4.9124e-02	7.2167e-01 ± 2.6127e-01
MPMOP9	8.3663e-02 ± 1.1029e-02	**8.2055e-02 ± 1.1601e-02**	6.1508e-01 ± 9.2426e-02
MPMOP10	5.7807e-02 ± 5.9171e-03	**5.7619e-02 ± 6.3695e-03**	3.2451e-01 ± 8.2224e-02
MPMOP11	1.8539e-02 ± 1.1730e-03	**1.8343e-02 ± 9.0481e-04**	1.5923e+00 ± 5.8241e-01
nbr	4	**7**	0

Table 2. Mean and standard deviation of IGD for MPMOPs with 30 dimensions.

Problems	IGD		
	OptMPNDS2	OptMPNDS	OptAll
MPMOP1	**1.5425e-05 ± 5.3205e-06**	1.6166e-05 ± 5.9224e-06	8.2886e-03 ± 2.8462e-03
MPMOP2	3.0351e-02 ± 6.3732e-02	**2.4442e-03 ± 1.3105e-02**	5.6115e-02 ± 5.8392e-02
MPMOP3	**1.9912e-01 ± 1.2983e-01**	2.4422e-01 ± 1.5082e-01	2.1359e-01 ± 6.1893e-02
MPMOP4	**5.2178e-02 ± 9.0699e-03**	5.3974e-02 ± 9.7020e-03	1.4187e+00 ± 7.8597e-01
MPMOP5	4.1281e-02 ± 5.8203e-03	4.1739e-02 ± 4.2176e-03	3.8524e-01 ± 8.7829e-02
MPMOP6	1.5643e-02 ± 7.1395e-04	**1.5277e-02 ± 9.1020e-04**	2.1889e+00 ± 1.5142e+00
MPMOP7	5.4601e-06 ± 1.7764e-06	**5.1282e-06 ± 2.9116e-06**	—
MPMOP8	**1.1977e-02 ± 4.9290e-02**	1.7650e-01 ± 1.4797e-01	1.4400e-01 ± 1.1676e-01
MPMOP9	**7.6085e-02 ± 1.3514e-02**	7.8395e-02 ± 9.6008e-03	2.1323e+00 ± 9.0104e-01
MPMOP10	**3.3202e-02 ± 8.8673e-03**	4.3365e+00 ± 2.7402e+00	6.4002e-01 ± 2.2718e-01
MPMOP11	**1.7454e-02 ± 7.3541e-04**	1.7898e-02 ± 7.7425e-04	3.0805e+00 ± 2.6421e+00
nbr	**8**	3	0

In terms of SN, from Tables 4, 5 and 6, it can be observed that OptMPNDS2 performs slightly better than OptMPNDS, and these two both are better than OptAll. All the numbers of the best SN results obtained by OptMPNDS2 for all dimensions reach 9. The numbers of the best SN results obtained by OptMPNDS are 7, 7 and 6, respectively. For OptAll, its *nbr* values for the SN metric is always 0.

In summary, OptMPNDS2 obtains more solutions, and the solutions are closer to the true PF for MPMOPs with higher dimensions.

Table 3. Mean and standard deviation of IGD for MPMOPs with 50 dimensions.

Problems	IGD		
	OptMPNDS2	OptMPNDS	OptAll
MPMOP1	**1.6188e-05 ± 5.3270e-06**	1.8747e-05 ± 6.0160e-06	6.3026e-03 ± 1.1210e-03
MPMOP2	**2.1572e-02 ± 3.3461e-02**	2.7957e-02 ± 4.8711e-02	7.6744e-02 ± 7.2050e-02
MPMOP3	5.3982e-01 ± 1.9816e-01	5.1823e-01 ± 2.0427e-01	**4.2675e-01 ± 1.0779e-01**
MPMOP4	**5.2182e-02 ± 7.6690e-03**	5.4883e-02 ± 9.0317e-03	2.0794e+00 ± 1.8092e+00
MPMOP5	**3.1668e-02 ± 2.4659e-03**	3.2240e-02 ± 2.9652e-03	4.9932e-01 ± 1.5816e-01
MPMOP6	1.4525e-02 ± 9.1710e-04	**1.4232e-02 ± 6.2776e-04**	2.6863e+00 ± 2.5453e+00
MPMOP7	8.4879e-06 ± 3.3024e-06	**7.3476e-06 ± 2.4889e-06**	—
MPMOP8	**9.2591e-02 ± 1.1908e-01**	2.4295e-01 ± 1.3750e-01	2.8748e-01 ± 1.6227e-01
MPMOP9	8.0242e-02 ± 1.3628e-02	**7.3962e-02 ± 1.0786e-02**	3.9340e+00 ± 2.5023e+00
MPMOP10	**2.5628e-02 ± 1.3457e-03**	1.1150e+01 ± 1.2323e+00	8.2081e-01 ± 3.0717e-01
MPMOP11	**1.7872e-02 ± 8.9646e-04**	1.7885e-02 ± 9.2098e-04	5.0912e+00 ± 6.7435e+00
nbr	**7**	3	1

Table 4. Mean and standard deviation of SN for MPMOPs with 10 dimensions.

Problems	SN		
	OptMPNDS2	OptMPNDS	OptAll
MPMOP1	**99.10 ± 2.09**	92.53 ± 11.14	0.03 ± 0.18
MPMOP2	**91.33 ± 6.94**	89.27 ± 7.14	6.30 ± 1.37
MPMOP3	99.87 ± 0.43	**100.00 ± 0.00**	30.20 ± 1.52
MPMOP4	**100.00 ± 0.00**	**100.00 ± 0.00**	19.70 ± 4.25
MPMOP5	99.97 ± 0.18	**100.00 ± 0.00**	29.90 ± 3.74
MPMOP6	**100.00 ± 0.00**	**100.00 ± 0.00**	6.60 ± 1.87
MPMOP7	**99.67 ± 0.80**	99.53 ± 1.25	0.00 ± 0.00
MPMOP8	**97.37 ± 1.81**	85.67 ± 8.04	2.60 ± 1.16
MPMOP9	**100.00 ± 0.00**	**100.00 ± 0.00**	21.80 ± 4.12
MPMOP10	**100.00 ± 0.00**	**100.00 ± 0.00**	10.00 ± 2.03
MPMOP11	**100.00 ± 0.00**	**100.00 ± 0.00**	4.97 ± 2.33
nbr	**9**	7	0

Table 5. Mean and standard deviation of SN for MPMOPs with 30 dimensions.

Problems	SN		
	OptMPNDS2	OptMPNDS	OptAll
MPMOP1	99.33 ± 1.47	**99.47 ± 1.50**	2.00 ± 0.00
MPMOP2	**96.13 ± 4.34**	94.00 ± 5.68	5.47 ± 1.14
MPMOP3	94.57 ± 7.99	**94.87 ± 8.76**	25.67 ± 5.45
MPMOP4	**100.00 ± 0.00**	**100.00 ± 0.00**	14.87 ± 9.87
MPMOP5	**100.00 ± 0.00**	**100.00 ± 0.00**	5.73 ± 1.72
MPMOP6	**100.00 ± 0.00**	**100.00 ± 0.00**	3.30 ± 1.09
MPMOP7	**99.87 ± 0.43**	99.80 ± 0.76	1.10 ± 0.66
MPMOP8	**99.57 ± 0.77**	98.47 ± 3.25	5.20 ± 1.10
MPMOP9	**100.00 ± 0.00**	**100.00 ± 0.00**	12.83 ± 3.87
MPMOP10	**100.00 ± 0.00**	37.30 ± 33.27	2.83 ± 0.79
MPMOP11	**100.00 ± 0.00**	**100.00 ± 0.00**	2.63 ± 1.22
nbr	**9**	7	0

Table 6. Mean and standard deviation of SN for MPMOPs with 50 dimensions.

Problems	SN		
	OptMPNDS2	OptMPNDS	OptAll
MPMOP1	**99.90 ± 0.55**	99.80 ± 0.66	2.00 ± 0.00
MPMOP2	**97.20 ± 3.70**	96.87 ± 4.34	5.17 ± 1.02
MPMOP3	84.27 ± 18.57	**87.70 ± 14.24**	23.53 ± 5.20
MPMOP4	**100.00 ± 0.00**	**100.00 ± 0.00**	19.33 ± 16.69
MPMOP5	**100.00 ± 0.00**	**100.00 ± 0.00**	3.47 ± 1.01
MPMOP6	**100.00 ± 0.00**	**100.00 ± 0.00**	3.43 ± 1.79
MPMOP7	**99.97 ± 0.18**	99.33 ± 1.63	1.33 ± 0.61
MPMOP8	**99.43 ± 1.04**	98.47 ± 3.52	4.27 ± 0.87
MPMOP9	99.97 ± 0.18	**100.00 ± 0.00**	15.23 ± 6.18
MPMOP10	**100.00 ± 0.00**	21.27 ± 7.54	2.13 ± 0.63
MPMOP11	**100.00 ± 0.00**	**100.00 ± 0.00**	3.17 ± 1.32
nbr	**9**	6	0

5   Conclusion

In this paper, we propose an evolutionary algorithm called OptMPNDS2 to solve MPMOPs. A new dominance relation of individuals in MPMOPs is defined to handle the non-dominated sorting, where the Pareto optimal level number of each party is regarded as an objective value of the individual. In the experiments, OptMPNDS2 is compared with OptMPNDS as well as OptAll. Experimental results show that the overall performance of the proposed OptMPNDS2 is better. In the future, we will establish a benchmark of the MPMOPs which have no common Pareto optimal solutions, and the benchmark will be used to evaluate the proposed algorithm.

References

1. Coello, C.C.: Evolutionary multi-objective optimization: a historical view of the field. IEEE Comput. Intell. Mag. **1**(1), 28–36 (2006)
2. Deb, K., Pratap, A., Agarwal, S., Meyarivan, T.: A fast and elitist multiobjective genetic algorithm: NSGA-II. IEEE Trans. Evol. Comput. **6**(2), 182–197 (2002)
3. Deb, K., Sindhya, K., Okabe, T.: Self-adaptive simulated binary crossover for real-parameter optimization. In: Proceedings of the 9th Annual Conference on Genetic and Evolutionary Computation, pp. 1187–1194 (2007)
4. Fonseca, C.M., Fleming, P.J.: An overview of evolutionary algorithms in multiobjective optimization. Evol. Comput. **3**(1), 1–16 (1995)
5. Geng, H., Zhang, M., Huang, L., Wang, X.: Infeasible elitists and stochastic ranking selection in constrained evolutionary multi-objective optimization. In: Wang, T.-D., et al. (eds.) SEAL 2006. LNCS, vol. 4247, pp. 336–344. Springer, Heidelberg (2006). https://doi.org/10.1007/11903697_43
6. Ishibuchi, H., Tsukamoto, N., Nojima, Y.: Evolutionary many-objective optimization: a short review. In: 2008 IEEE Congress on Evolutionary Computation (IEEE World Congress on Computational Intelligence), pp. 2419–2426. IEEE (2008)
7. Lau, R.Y.: Towards genetically optimised multi-agent multi-issue negotiations. In: Proceedings of the 38th Annual Hawaii International Conference on System Sciences, pp. 35c–35c. IEEE (2005)
8. Lau, R.Y., Tang, M., Wong, O., Milliner, S.W., Chen, Y.P.P.: An evolutionary learning approach for adaptive negotiation agents. Int. J. Intell. Syst. **21**(1), 41–72 (2006)
9. Liu, W., Luo, W., Lin, X., Li, M., Yang, S.: Evolutionary approach to multiparty multiobjective optimization problems with common pareto optimal solutions. In: Proceedings of 2020 IEEE Congress on Evolutionary Computation (CEC), pp. 1–9. IEEE (2020)
10. Miettinen, K.: Nonlinear Multiobjective Optimization, vol. 12. Springer Science & Business Media, Berlin (2012)
11. Tamaki, H., Kita, H., Kobayashi, S.: Multi-objective optimization by genetic algorithms: a review. In: Proceedings of IEEE International Conference on Evolutionary Computation, pp. 517–522. IEEE (1996)
12. Tian, Y., Cheng, R., Zhang, X., Jin, Y.: PlatEMO: a MATLAB platform for evolutionary multi-objective optimization [educational forum]. IEEE Comput. Intell. Mag. **12**(4), 73–87 (2017)
13. Trivedi, A., Srinivasan, D., Sanyal, K., Ghosh, A.: A survey of multiobjective evolutionary algorithms based on decomposition. IEEE Trans. Evol. Comput. **21**(3), 440–462 (2016)
14. Wang, H., Jiao, L., Yao, X.: Two_Arch2: an improved two-archive algorithm for many-objective optimization. IEEE Trans. Evol. Comput. **19**(4), 524–541 (2014)
15. Zhang, C., Wang, G., Peng, Y., Tang, G., Liang, G.: A negotiation-based multi-objective, multi-party decision-making model for inter-basin water transfer scheme optimization. Water Resour. Manag. **26**(14), 4029–4038 (2012)

16. Zhang, Q., Li, H.: MOEA/D: a multiobjective evolutionary algorithm based on decomposition. IEEE Trans. Evol. Comput. **11**(6), 712–731 (2007)
17. Zhang, Q., Zhou, A., Zhao, S., Suganthan, P.N., Liu, W., Tiwari, S.: Multiobjective optimization test instances for the CEC 2009 special session and competition (2008)
18. Zhang, X., Tian, Y., Cheng, R., Jin, Y.: An efficient approach to nondominated sorting for evolutionary multiobjective optimization. IEEE Trans. Evol. Comput. **19**(2), 201–213 (2014)
19. Zhou, A., Qu, B.Y., Li, H., Zhao, S.Z., Suganthan, P.N., Zhang, Q.: Multiobjective evolutionary algorithms: a survey of the state of the art. Swarm Evol. Comput. **1**(1), 32–49 (2011)
20. Zitzler, E., Laumanns, M., Thiele, L.: SPEA2: Improving the strength pareto evolutionary algorithm. TIK-report **103** (2001)

Swarm Robotics and Multi-agent System

Immune System Algorithms to Environmental Exploration of Robot Navigation and Mapping

Elakiya Jayaraman[1], Tingjun Lei[1], Shahram Rahimi[2], Shi Cheng[3(✉)], and Chaomin Luo[2(✉)]

[1] Department of Electrical and Computer Engineering,
Mississippi State University, Mississippi State, MS 39762, USA
[2] Department of Computer Science and Engineering,
Mississippi State University, Mississippi State, MS 39762, USA
`Chaomin.Luo@ece.msstate.edu`
[3] School of Computer Science, Shaanxi Normal University, Xi'an 710119, China
`cheng@snnu.edu.cn`

Abstract. In real world applications such as rescue robots, service robots, mobile mining robots, and mine searching robots, an autonomous mobile robot needs to reach multiple targets with the shortest path. This paper proposes an Immune System algorithm (ISA) for real-time map building and path planning for multi-target applications. Once a global route is planned by the ISA, a foraging-enabled trail is created to guide the robot to the multiple targets. A histogram-based local navigation algorithm is used to navigate the robot along a collision-free global route. The proposed ISA models aim to generate a path while a mobile robot explores through terrain with map building in unknown environments. In this paper, we explore the ISA algorithm with simulation studies to demonstrate the capability of the proposed ISA in achieving a global route with minimized overall distance. Simulation studies demonstrate that the real-time concurrent mapping and multi-target navigation of an autonomous robot is successfully performed under unknown environments.

Keywords: Immune system algorithm · ISA · Multi-target navigation · Autonomous navigation · Path planning

1 Introduction

Autonomous robots are robots that perform tasks without human interference for any corrective measures. Autonomous robots accomplish tasks by using their sensors to gather information about their surroundings. There are many applications for autonomous robots in the real world; these include but not limited to service robots, agricultural robots, mining robots, mine searching robots, and rescue robots. In all the above applications, the fundamental requirement for the autonomous operation of these robots includes path planning and mapping. Path planning and mapping of autonomous robots is a crucial issue that needs addressing before developing a completely autonomous exploring robot.

© Springer Nature Switzerland AG 2021
Y. Tan and Y. Shi (Eds.): ICSI 2021, LNCS 12690, pp. 73–84, 2021.
https://doi.org/10.1007/978-3-030-78811-7_7

There has been considerable research in this area due to its real-world applications. Multi-target motion planning is very crucial for an autonomous robot. The multi-target path planning seeks a collision-free route to visit a sequence of targets with the minimized total distance under unknown environments. Let us consider a case in Fig. 1. There are nine points that must be visited by a robot. Let us consider Spot 1 as the starting point and Spot 9 as its final destination. There are multiple paths that could be traversed to reach Spot 9 but only a few passes through Spot 2–8 also some of the spots are longer than the others. The first stage would be to determine the best order to reach Spot 9 with Spots 2–8 in its path. The main optimization criteria are the length of the path and the cost of the travel. The second stage is focused on path planning; this stage is to ensure collision-free traversal of the global route based on the map information. Then, the robot moves to the next waypoint/target according to the global path. It may encounter some new obstacles, which is an unexpected obstacle on the map. The algorithm replans the necessary changes to avoid the obstacles and achieve the target based on the input sensor data. In this stage, there are multiple concurrent sub-steps for robot motion planning and map building. Robot motion can be controlled with a servo motor with a motion controller which obtains its input from the path planner. Also, feedback from the sensor array, which may be a combination of LIDAR, RADAR, SONAR, Optical Camera would be useful in navigation, obstacle detection, and map building.

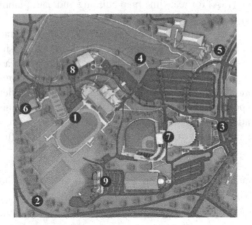

Fig. 1. Aerial cartoon of a multi-target navigation.

Many approaches have been proposed, which achieve reliable autonomous robot motion planning [1–7, 14, 15]. Gu *et al.* [1] proposed a motion planning model based on the elastic band method where tunability and stability are focused on automated passenger vehicle navigation. Raja *et al.* [2] proposed an algorithm-based Genetic Algorithm (GA) by introducing a potential field method for motion planning for a 6-wheel rover. Davies *et al.* [3] developed a GA path planner to guide an autonomous robot to reach specified multiple targets, free of collision. The main downside for this algorithm is given that artificial waypoints may need to be added to avoid deadlock scenarios due to local minima, which is computationally expensive. Luo and Yang [4, 5] developed a bio-inspired neural network model that performs both mapping and path planning.

Luo *et al.* [6] proposed a biologically inspired neural network approach to real-time collision-free motion planning of multiple mobile robots in unknown environments. However, multi-target navigation studies have not been performed much for intelligent robot systems. Faigl and Macak [7] developed a self-organizing map-based method with an artificial potential field navigation function to generate an optimal path of a mobile robot to visit multiple targets. Gopalakrishnan and Ramakrishnan [11] implemented a multi-target navigation model of an autonomous robot modeled as a point robot without map building capability. Santos *et al.* [9] implemented a Max-Min Ant system algorithm based on how ants navigate the terrain with higher efficiency in comparison with the GA and A* algorithm but it lacks map building capabilities. Serres *et al.* [10] proposed optic flow-based motion planning which imitates navigation of insects in the presence of obstacles. However, this effort lacks multi-target and map building capabilities.

In this paper, a real-time concurrent multi-target motion planning and map building approach of an autonomous robot in completely unknown environments is proposed. The global route is generated by Immune System Algorithm (ISA) from the provided multiple targets/waypoints. The sensory information obtained by onboard sensors mounted on the robot is utilized for its real-time concurrent multi-target navigation and map building in unknown sceneries. The rest of this paper is organized as follows. The ISA global path planner is proposed in Sect. 2, which first obtains the multi-target access sequence from ISA, and then obtains the collision-free trajectory with obstacle avoidance from ISA. In Sect. 3, the real-time concurrent map building and collision-free multi-target navigation of the autonomous robot are introduced. Afterward, the simulation and comparison studies of multi-target navigation and map building of the autonomous robot in various unknown environments are described in Sect. 4. Finally, several important properties of the proposed approach are concluded in Sect. 5.

2   Immune System Algorithm Based Multi-target Navigation

Immune System Algorithm (ISA) is a swarm-based bio inspired algorithm based on the human immune system. A human body is often subjected to harm due to various pathogens such as bacteria or viruses. Thanks to the highly developed and effective immune system, most of these pathogens are targeted and destroyed, keeping our bodies healthy. The different stages taken by our immune system to protect our bodies can be described as follows. When pathogens attack our bodies, the antigen proteins on the surface of the pathogen provoke an immune system response. If this is the first attack by this pathogen the immune system mass produces antibodies tailored to detect this pathogen and destroys them this is known as a primary response. During the initial response, the produced antibody is propagated by cloning during which somatic hyper-mutation occurs. Thus, produced antibody may have a better affinity to the pathogen (optimization) and hence better detect the pathogen differentiating from other cells and destroying it with T-cells. As the process of propagation and mutation continues some B cells differentiate memory B-cells with a high affinity toward the antigen of the attacking pathogen. The memory B-cells considerably im-prove immune system response time on future pathogen infection as it is optimized to detect, seek, and destroy this particular pathogen. The immune system inspired ISA follows the same iterative approach in opti-mizing the objective function because of this ISA can easily be used to solve different types of optimization problems often encountered in engineering, such as multi-target navigation problems.

2.1 Multi-target Access Order

Algorithm 1: Calculate distance between targets

N = Number of targets

$W = W_{i=1}^{N}, D = D_{i,j=1}^{N}$

$D \leftarrow$ Distance(W,N)

Require $N > 1$

Initialize: $i = 1; j = 1$

while $(i \leq N)$ **do**

    while $(j \leq N)$ **do**

        $Distance, D_{ij} = \sqrt{(X_i - X_j)^2 + ((Y_i - Y_j))^2}$

        $j = j + 1$

    end

  end

Return D

Fig. 2. Objective of the multi-target access order. Given the coordinates of the target point and optimize the access sequence according to the distance.

In the practical applications of multi-target navigation and mapping, multiple targets are provided by GPS coordinates. These targets are a series of waypoints and the relative cost of traveling between each target. The target is to search for the route of all waypoints that pass all waypoints at once and find the shortest overall journey. For example, in a rescue robot application, the robots start from a designated waypoint, visit each other, and then end at the initial waypoint. The traveling salesman problem (TSP) is an optimization problem that minimizes the travel distance in a limited number of cities while the travel cost between each city is known. The classic TSP is used to deal with the problem of multi-target access, in which a series of targets are visited, thereby minimizing the total planned length of the route. The goal of this TSP with respect to multi-objective navigation is to search for an ordered set of all waypoints so that autonomous robots can access it, thereby minimizing costs. For TSP, each waypoint and the distance or cost between them must be listed. Therefore, the target is a waypoint, where the GPS coordinates of the targets are in latitude and longitude.

The multi-target access order problem has practical applications in certain areas, such as the transportation planning problem that needs to be delivered and the problem of minimizing fuel costs and time. This specific problem is very similar to our multi-target path planning problem for autonomous robotic applications, especially for search and rescue robots. The pseudocode for the implemented ISA is explained in Algorithm 1, 2, and 3, as shown in Fig. 2 and Fig. 3. The objective of the algorithm is to minimize total path length P when n number of Cartesian coordinates (X_n, Y_n) of waypoints are given.

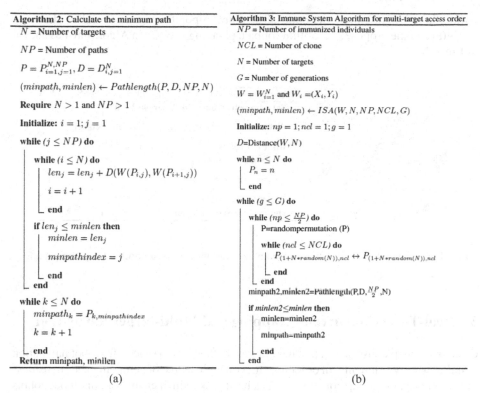

Fig. 3. Pseudo code of the (a) calculate the minimum path as fitness function (b) ISA for multi-target access order.

2.2 ISA for Path Planning

To plan the global trajectory from one waypoint to another, obstacles in the map need to be considered. We need to find a collision-free trajectory with obstacle avoidance while seeking the shortest distance to reduce energy consumption and improve efficiency. In this paper, we take the local search procedure into account for finding the shortest path in the map with ISA. The proposed method integrates the local search procedure to search and find the short and reasonable trajectory with the ISA.

The first stage of seeking the optimal path is to find a feasible solution since the unfeasible ones are not realistic. In order to improve the performance of finding the optimal path through the ISA algorithm, the path is gradually constructed from the generated random points (all in free space, outside the obstacle). On the basis of Dijkstra's algorithm to find the shortest path in a graph, the path is built from the start point S and the next path point is chosen from the N randomly generated points. The same procedure is taken place for the chosen points until the final target is reached and the edges are their connections in the map. Each point is directly connected to all other points. When its connection passes an obstacle and becomes an unfeasible solution, the distance between its nodes is set to infinite to ensure that only a feasible solution can be found as the shortest path. The benefit of the local search procedure is to filter out unfeasible connections with infinite length among the nodes. The point-to-point navigation result is shown in Fig. 4(a) and the proposed algorithm for path planning is given in Algorithm 4 as shown in Fig. 4(b).

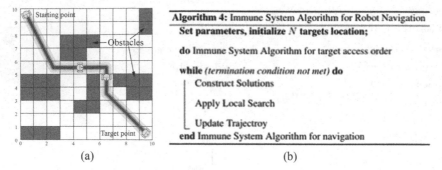

(a)   (b)

Fig. 4. (a) Illustration of ISA point-to-point navigation. (b) ISA algorithm for path planning.

3   Real-Time Concurrent Mapping and Multi-target Navigation

Concurrent map construction and navigation are the keys to successful robot navigation in unknown environments. In order to achieve a high degree of autonomy and robustness in robot navigation, map construction is a basic task, which enables autonomous robots to make decisions while avoiding obstacles. Therefore, in terms of robot navigation, when a mobile robot is walking in an unknown environment, a two-dimensional map will be constructed, which is filled with cells of equal size (marked as "occupied" or "free").

In our navigation system, it is divided into two layers, one layer is the ISA global path planner described in Sect. 2, and the other layer is the local navigator based on the histogram. Efficiency and flexibility make ISA adapt to the motion planning and map building of autonomous robots. Local navigation aims to create speed commands for autonomous mobile robots to move towards the target. A series of markers are included in the motion plan, which decomposes the global route generated by the ISA global planner into a series of fragments, making the model particularly effective for working areas that are densely filled with obstacles. Ulrich and Borenstein [12] developed a Vector Field Histogram (VFH) based robot reactive navigation method based on the histogram. In this paper, we use VFH as a local reactive navigator based on LIDAR for obstacle avoidance.

Concurrent map building and navigation are the essences of successful multi-target navigation under unknown environments. Map building is a fundamental task in order to achieve high levels of autonomy and robustness in multi-target navigation that makes it possible for autonomous robots to make decisions in positioning with obstacle avoidance. It is especially beneficial for autonomous robots to implement robust multi-target navigation in unknown terrains, given the fact that it facilitates the utilization of path planning algorithms to determine the optimal trajectory among waypoints as multiple targets. A precise estimate of the robot pose is demanded by map building so that accurate registration of the local map on the global map is capable of being carried out. The polar histogram of obstacles in the workspace with VFH is shown in Fig. 5(a). In our navigation system, the histogram-based local navigator senses obstacles through a vehicle-mounted 270° LIDAR with a radius of 2.5.

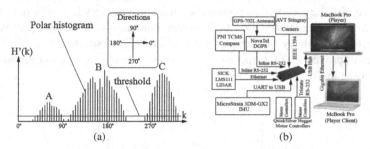

Fig. 5. (a) Polar histogram of obstacles in workspace with VFH method (redrawn from [12]). (b) The sensor configuration for multi-target navigation and mapping.

4   Simulation and Comparison Studies

In this section, simulations and comparison studies are used to verify the effectiveness and efficiency of the proposed ISA multi-target autonomous robot navigation and mapping. A 270° SICK LMS111 LIDAR is configured for the obstacle detection illustrated in Fig. 5(b). This LIDAR unit obtains data over a 270° field-of-view with 0.25° resolution, a maximum range of 20 m, and a 25 Hz scanning rate. The sensor configuration of the robot for multi-target navigation and mapping is illustrated in Fig. 5(b). The ISA based multi-target global path planning and VFH local navigator are simulated and tested in multi-target navigation with obstacle avoidance.

4.1   Comparison of the Proposed ISA Model for Path Planning with ACO

Table 1. Comparison of path length (Fig. 15 case 4 in [13])

Models	Minimum length	Mean length	Improvement from proposed model
Proposed ISA model	**35.47**	**36.17**	—
ACO [13]	38.80	40.97	11.71%
Improved ACO [13]	38.14	40.16	9.93%

The proposed ISA path planning method, as stated in Sect. 2.2, is first applied to a test scenario with populated obstacles in comparison of the test scenario identical as Fig. 15

case 4 of [13] shown in Fig. 6 in this context. The trajectories of robot motion planning are illustrated about the proposed ISA model, ACO and improved ACO, respectively, in Fig. 6. The workspace has a size of 35 × 30, which is topologically organized as a grid-based map. The starting point is (5, 3) and the target point is (23, 29). In Table 1, comparative data may be found that our proposed model is much better than the models of ACO and improved ACO, respectively, in the minimum trajectory length, and mean trajectory length. The comparison results show that the trajectory length by our proposed model is 8.58% shorter than ACO, 7.00% shorter than improved ACO, respectively. The mean trajectory length by our proposed model is 11.71% shorter than ACO, 9.93% shorter than improved ACO, respectively. The trajectory planned by our proposed ISA model is shown in Fig. 6(c).

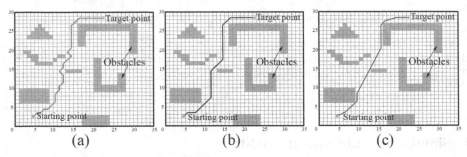

Fig. 6. Illustration of robot navigation with various models. (a) ACO model [13]; (b) improved ACO model [13]; (c) the proposed model.

4.2 Simulation of Proposed ISA Model for Multi-target Path Planning

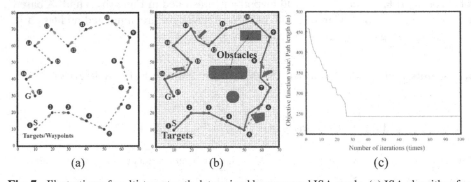

Fig. 7. Illustration of multi-target path determined by proposed ISA mode. (a) ISA algorithm for TSP applied in a 17-target workspace; (b) final trajectory of the proposed ISA model (c) the fitness value over iterations.

The proposed ISA associated with VFH local navigation is applied to the multi-target navigation application as shown in Fig. 7 and Fig. 8. Luo *et al.* [8] proposed a hybrid PSO approaches to resolve multi-target robot motion planning issues. However, their model has not considered the obstacles in the environment in global path planning which is necessary for autonomous robot navigation systems. The workspace has a size of 70 × 80, which is topologically organized as a grid-based map. There are 17 targets in the workspace and a service mobile robot is needed to reach each target. Initially, the starting point of the robot is at S (10, 10), which is also considered to be the target point 1. After executing this ISA-based TSP algorithm with only the coordinates of the 17 targets, the access order of the targets with the smallest total length of the route is obtained as shown in Fig. 7(a). The fitness value of iteration in terms of the ISA is shown in Fig. 7(c). Then in the light of the whole environment information, such as obstacles, the final *trajectory* planned by the proposed ISA model regarding the multi-target access order is shown in Fig. 7(b).

The robot is able to traverse from the initial point to plan a reasonable collision-free route to reach the final designation. The robot moves based on the planned global trajectory while it constructs the map with 270° LIDAR. The trajectory generated by the robot is shown in Fig. 8(a) whereas the map built is illustrated in Fig. 8(b) at the end of the travel of the robot.

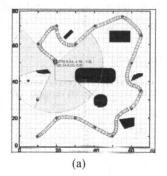
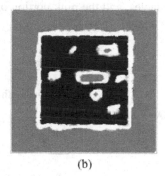

(a)   (b)

Fig. 8. Illustration of ISA based multi-targets robot navigation and mapping model (a) trajectory generated in the middle stage; (b) map built at the end of stage.

4.3 Comparison of the Proposed ISA Multi-target Path Planning Model with Genetic Algorithm

The proposed model is then applied to a test scenario with populated obstacles in comparison with the test scenario identical as Fig. 2 of [3] shown in Fig. 9(a) in this context. The workspace has a size of 20 × 30, which is topologically organized as a grid-based map. Initially, the starting point is located at S (0, 0). The trajectory planned when using our proposed ISA model is illustrated in Fig. 9(b). The generated trajectory length completed by the robot is listed in Table 2 in comparison with Davies's model [3]. The shorter trajectory is generated by our ISA model illustrated in Fig. 9(b). It is observed that our model outperforms over theirs in terms of the trajectory length.

With VFH-based local navigator, the built map from the initial position S (0, 0) and the final trajectory planned is illustrated in Fig. 9(c), and final map built exactly when

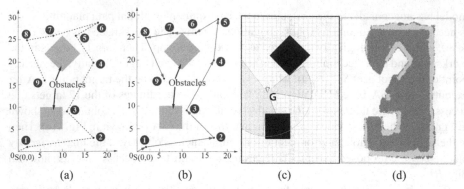

(a)   (b)   (c)   (d)

Fig. 9. Comparison studies with Davies's model [3]. (a) planned trajectory by Davies's model (redrawn from [3]); (b) planned trajectory by our ISA model; (c) illustration of robot trajectory generated at the end; (d) illustration of map built at the end.

the robot reaches the final target are illustrated in Fig. 9(d). The green fields indicate detected obstacles, and the pink portion of the image represents explored zones by the 270° LIDAR scans.

Table 2. Comparison of path length (Fig. 2 in [3])

Models	Minimum length	Improvement from proposed model
Proposed ISA model	**74.202**	—
GA [3]	77.798	4.6%

To demonstrate its efficiency in determining the path length. Figure 10 shows a plot of iterations required to reach an ideal solution in comparison with other orders of algorithms like log-linear and exponential algorithms. The ISA is run with a varying number of points and the results show the algorithm performs better than a quadratic algorithm $O(n^2)$ but worse than Log-linear algorithm $O(nlog(n))$ which is much better than the brute force method of solving which is an $O(n!)$ algorithm.

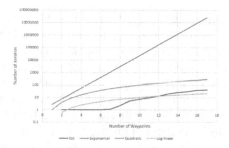

Fig. 10. Approximate estimation of the order of the algorithm with different number of targets.

5   Conclusion

The ISA based path planning algorithm was developed in this paper. ISA is first utilized to solve multi-target access order by solving the traveling salesman problem, then ISA is applied with the map information to obtain a collision-free global path. A LIDAR-based local navigator algorithm has been implemented for local navigation and obstacle avoidance. In addition to the multi-target ISA based navigation, grid-based map representations are imposed for real-time autonomous robot navigation. Simulation and comparison studies have demonstrated the effectiveness of the proposed real-time multi-target ISA approach of an autonomous mobile robot.

References

1. Gu, T., Atwood, J., Dong, C., Dolan, J.M., Lee, J.W.: Tunable and stable real-time trajectory planning for urban autonomous driving. In: 2015 IEEE/RSJ International Conference on Intelligent Robots and Systems (IROS), pp. 250–256 (2015)
2. Raja, R., Dutta, A., Venkatesh, K.S.: New potential field method for rough terrain path planning using genetic algorithm for a 6-wheel rover. Robot. Auton. Syst. **72**, 295–306 (2015)
3. Davies, T., Jnifene, A.: Multiple waypoint path planning for a mobile robot using genetic algorithms. In: 2006 IEEE International Conference on Computational Intelligence for Measurement Systems and Applications, pp. 21–26 (2006)
4. Luo, C., Yang, S.X.: A bioinspired neural network for real-time concurrent map building and complete coverage robot navigation in unknown environments. IEEE Trans. Neural Networks **19**(7), 1279–1298 (2008)
5. Yang, S.X., Luo, C.: A neural network approach to complete coverage path planning. IEEE Trans. Syst. Man Cybern. Part B (Cybern.) **34**(1), 718–724 (2004)
6. Luo, C., Yang, S.X., Li, X., Meng, M.Q.-H.: Neural-dynamics-driven complete area coverage navigation through cooperation of multiple mobile robots. IEEE Trans. Ind. Electron. **64**(1), 750–760 (2016)
7. Falgl, J., Macak, J.: Multi goal path planning using self-organizing map with navigation functions. In: European Symposium on Artificial Neural Networks (ESANN 2011), Computational Intelligence and Machine Learning, pp. 41–46 (2011)
8. Luo, C., Zhu, A., Mo, H., Zhao, W.: Planning optimal trajectory for histogram-enabled mapping and navigation by an efficient PSO algorithm. In: 2016 12th World Congress on Intelligent Control and Automation (WCICA), pp. 1099–1104 (2016)
9. Santos, V.D.C., Osório, F.S., Toledo, C.F., Otero, F.E., Johnson, C.G.: Exploratory path planning using the Max-min ant system algorithm. In: 2016 IEEE Congress on Evolutionary Computation (CEC), pp. 4229–4235 (2016)
10. Serres, J.R., Ruffier, F.: Optic flow-based collision-free strategies: from insects to robots. Arthropod Struct. Dev. **46**(5), 703–717 (2017)
11. Ramakrishnan, S., Dagli, C.H., Gopalakrishnan, K.: Optimal path planning of mobile robot with multiple targets using ant colony optimization. In: Smart Systems Engineering, pp. 25–30 (2006)
12. Ulrich, I., Borenstein, J.: VFH+: reliable obstacle avoidance for fast mobile robots. In: 1998 IEEE International Conference on Robotics and Automation (Cat. No. 98CH36146), vol. 2, pp. 1572–1577 (1998)
13. Hsu, C.C., Wang, W.Y., Chien, Y.H., Hou, R.Y.: FPGA implementation of improved ant colony optimization algorithm based on pheromone diffusion mechanism for path planning. J. Mar. Sci. Technol. **26**(2), 170–179 (2018)

14. Lei, T., Luo, C., Ball, J.E., Rahimi, S.: A graph-based ant-like approach to optimal path planning. In: 2020 IEEE Congress on Evolutionary Computation (CEC), pp. 1–6. IEEE (2020)
15. Lei, T., Luo, C., Jan, G., Fung, K.: Variable speed robot navigation by an ACO approach. In: Tan, Y., Shi, Y., Niu, B. (eds.) ICSI 2019. LNCS, vol. 11655, pp. 232–242. Springer, Cham (2019). https://doi.org/10.1007/978-3-030-26369-0_22

Primitive Shape Recognition Based on Local Point Cloud for Object Grasp

Qirong Tang[✉], Lou Zhong, Zheng Zhou, Wenfeng Zhu, and Zhugang Chu

Laboratory of Robotics and Multibody System, School of Mechanical Engineering, Tongji University, Shanghai 201804, People's Republic of China

Abstract. Object recognition and grasping are important means of interaction between robot and environment, and also two of the main tasks of robot. Due to the rich information provided by depth sensors, it has paved the way for the object recognition. The geometric information is more conducive to the primitive recognition of the object, and the primitive shape information is used as the input information for the robot to grasp. This study proposes a primitive shape recognition method using local point cloud. First, 900 sets of point cloud data including three primitive shapes was created. Then the PointNet network using the point cloud data to recognize the primitive shape of the objects was trained. Experiments in simulation and physical world shows our recognition method can effectively recognize the primitive shape of the object.

Keywords: Primitive shape · Object recognition · Point cloud · PointNet

1 Introduction

Robot is widely used in all aspects of production and life because it can replace human heavy labor, realize production automation and keep human safety. Manipulator grasping is one of the important means for robot to interact with the outside world. It is usually divided into three aspects: perception, planning and control. Model-based control methods have always dominated the field of manipulator grasping, such as model predictive control [1, 2] and force control. With the emergence of machine learning and neural networks, data-driven methods [3–5] provide new ways for robot grasping.

The modeling process of model-based methods is complex and the generalization performance of grasping unknown objects does not work well. Meanwhile, the data-driven methods are more robust, but they require enormous data for train. To avoid the disadvantages, many researches are based on the grasping of the primitive shapes.

Grasping based on the primitive shape is an approach from another point of view. In this method, the objects are not accurately modeled. The object is sampled into the primitive shape with prior knowledge, and the grasping posture

© Springer Nature Switzerland AG 2021
Y. Tan and Y. Shi (Eds.): ICSI 2021, LNCS 12690, pp. 85–91, 2021.
https://doi.org/10.1007/978-3-030-78811-7_8

is selected from a small number of grasp candidates. There is no need to do a lot of searches, and alleviate problems that require enormous data. At present, there are two kinds of shape-based grasping methods: selecting the grasping posture according to the predefined grasping postures [6,7] or sorting the grasping quality according to the known grasping modes [8].

The first of grasping with primitive is to complete the target shape recognition. What's more, the development of the depth sensors have paved the way for 3D shape recognition due to the additional 3D information. The deep learning provides an efficient and accurate method for primitive shape recognition. These methods are mainly by extracting RGB-D image [9–11] or point cloud [12–14] features.

This study mainly investigates the 3D shape recognition of the objects' point cloud using PointNet. This study is aiming at

- point cloud dataset of primitive shape creation,
- a primitive shape recognition method using local point cloud.

This paper is organized in the following manner: Sect. 2 introduces the method of establishing primitive shape point cloud dataset and the PointNet network. The effectiveness of the proposed methods is verified through a set of experiments in simulation and physical world and the results are shown in Sect. 3. Section 4 concludes the work.

2   Method of Primitive Recognition

2.1   Primitive Shape Dataset Creation

First, a small dataset is prepared with respect to PointNet structure and the dataset contains point clouds solid objects from three categories: sphere, cylinder, cuboid. The depth image was created using a Kinect V2 in V-REP. These objects are uniform in color and texture information. The size of the objects are different for each. For example, the height-diameter ratio of a cylinder is different. In the V-REP simulation platform, the object and the Kinect sensor are put at different angles and the primitive shape model is rotated around the axis. Therefore, for each shape, depth images were taken from different angles and distances by Kinect sensor. Each primitive shape includes 300 samples depth images.

In order to remove the background depth information, the background subtraction method is used. As shown in Fig. 1, background depth images are subtracted from original images to retain the object depth information. If the final information is less than 0, it is retained. The retained information adds the background information to restoring the original depth of the object. So, the ground is black expect for the objects. Then the depth information is converted to point clouds based on the intrinsics, as in Eq. (1).

$$
ZP_{uv} = \begin{bmatrix} u \\ v \\ z \end{bmatrix} = \begin{bmatrix} f_x & 0 & c_x \\ 0 & f_y & c_y \\ 0 & 0 & 1 \end{bmatrix} P = KP, \tag{1}
$$

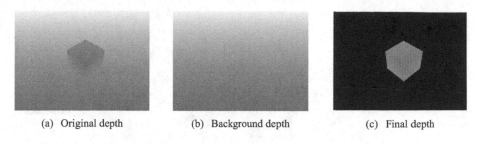

(a) Original depth   (b) Background depth   (c) Final depth

Fig. 1. Depth image subtraction

where P_{uv} represents the position (u, v) in the pixel coordinate system, z is the depth in the position (u, v) in the pixel coordinate, f_x, f_y is the focal length and c_x, c_y is the offset, K is the intrinsics, P is the camera coordinate system.

The 1024 point cloud represents the object after sampling under the point clouds. Then, the point cloud is normalized, the center of point clouds is translated to the origin of coordinates, as in Eq (2), and the size of point cloud is scaled to the unit sphere, as in Eq. (3),

$$P = P - \bar{P}, \tag{2}$$

$$P = P/max\{P\}, \tag{3}$$

where the $\bar{P}$ is the center of point clouds and $max\{P\}$ is the max distance of the points to origin.

2.2 Primitive Shape Recognition Using PointNet

The proposed approach hypothesizes that common objects can be divided into three categories: cylinder, sphere and cuboid. This part of recognition objects in terms of shape features using local point clouds extracted from depth image. Since the point clouds has the permutation invariance and rigid transformation robustness, it is necessary to pay attention to these two properties when performing point clouds feature recognition.

The state-of-the-art of deep neural networks are specifically designed to handle the irregularity of point clouds. This approach was proposed by PointNet [12]. The PointNet provides a unified architecture for object classification. Since the point cloud has the permutation invariance and rigid transformation robustness, it is necessary to pay attention to these two properties when performing point cloud feature recognition. The PointNet solves the problems by rotating transformation and constructing a symmetric function. The point clouds is the input of the PointNet, The number of outputs, N, corresponds to the numbers of class.

As shown in Fig. 2, the network structure rotates the input point cloud, and then uses the multi-layer perceptron to arise the three-dimensional point cloud to 1024-dimensional. The maximum pooling layer solves the problem of

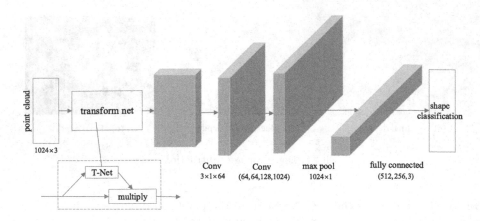

Fig. 2. Modified PointNet model

permutation invariance and extracts the most important points in the point cloud. Finally, shape classification is performed through multiple fully connected layers.

In this study, the PointNet was used to recognize the primitive shape features of the object based on local point clouds. To simplify the model, only the incoming point clouds are spatially transformed, and the rest are consistent with the PointNet.

3 Evaluation

3.1 Model Train

The input data of the PointNet is the three-dimensional point clouds, and the output is the label of the category of the object. If the output label is consistent with the shape label of the object, the shape recognition is considered correct. Those 900 local point cloud data sets created by the V-REP simulation platform are divided into training sets and test sets, which are used for network training and testing, respectively. In the training process, random point cloud interference is used to increase the diversity of training data. After training, objects recognition performance experiment are carried out in simulation and in the actual environment.

3.2 Results of Simulation Experiment

In simulation, the 3D model of the object was downloaded from the YCB model library for verifying the accuracy of the recognition. The objects are similar to the features of sphere, cylinder, and cuboid, such as cups, tennis balls, oranges, etc. Compared with standard primitive shape, the surface texture of the objects is more complicated. Meanwhile, each primitive shape contains three objects.

The method of obtaining the target point cloud is consistent with the method of the primitive 3D model point cloud. The test point cloud data contains 10 sets of point cloud data for each objects and includes 9 types of objects, totally 90 sets of point cloud data (sphere: 30, cuboid: 30 and cylinder: 30) are considered. The point cloud data only uses the trained network model for shape recognition. The PointNet extracts the features of the point cloud to recognize the primitive shape of objects. In the simulation, the primitive shape recognition results are shown in Table 1.

Table 1. The accuracy of shape recognition in simulation (%)

Objects	Can1	Can2	Cup	Cookie box	Candy box	Block	Golf	Tennis	Orange
Recognition accuracy rate of each object	100	100	80	90	100	90	100	100	100
Recognition accuracy rate of each shape	93.3			93.3			100		
Average	96								

3.3   Results of Physical Experiment

Preprocessing of Actual Depth Image. In this part, Kinect is used to obtain the 3D point cloud of the objects in the actual world to verify the model. Considering the noise of depth images in the actual environment, it is necessary to preprocess the image of the depth sensor to better recover the point cloud and reduce the interference of environmental noise.

(a) Depth image before filtering   (b) Depth image after filtering

Fig. 3. Depth image median filter

It consists of two parts: target region extraction of depth image and filtering the scene containing objects using median filter. As shown in Fig. 3, it is a bottle filtered before and after image. In Fig. 3(a), the black area is the noise of the depth image. And the Fig. 3(b) is a better depth image after median filtering.

Results in Real World. The other processes are the same which the process in simulation. 15 sets of point cloud data for each type of object were acquired by Kinect. Thus, there are a total of 135 sets of data. The primitive shape recognition results in physical world are shown in Table 2.

Table 2. The accuracy of shape recognition in physical world (%)

Objects	Bottle1	Drinkbottle	Bottle2	Box1	Box2	Rubik'scube	Golf	Tennis	Orange
Recognition accuracy rate of each object	60	100	86.7	93.3	100	93.3	86.7	100	100
Recognition accuracy rate of each shape	82.2			95.6			95.6		
Average	91.1								

Comparing Table 1 and Table 2, the accuracy of the primitive three dimensional shapes recognition of the target in the simulation is higher than that in the actual environment. In the physical world, the recognition accuracy is lower due to the noise of depth image and the point cloud of the object. Pleasantly, primitive shape recognition has both acceptable performance both in simulation and physical world. And the recognition accuracy of ball objects are better than the other two shapes.

4   Conclusions

A method of recognizing the object primitive shape using local point cloud is proposed by this study. First of all, a point cloud dataset with three primitive shapes is established via converting depth images into point cloud. Then, Point-Net is used to train the data set. The recognition accuracy rate for objects of different sizes is up to 96% in simulation. In the actual environment, the accuracy rate of object shape recognition is about 91.1%.

In the future research, the authors would like to apply the primitive shape recognition for grasp process. The shape information of the object is used as the input information of the robot controller, which may contribute to a better grasping performance of the robot.

Acknowledgements. This work is supported by the projects of National Natural Science Foundation of China (No. 61873192; No. 61603277; No. 61733001), the Quick Support Project (No. 61403110321), and Innovative Project (No. 20-163-00-TS-009-125-01). Meanwhile, this work is also partially supported by the Fundamental Research Funds for the Central Universities and the Youth 1000 program project. It is also partially sponsored by International Joint Project Between Shanghai of China and Baden-Württemberg of Germany (No. 19510711100) within Shanghai Science and Technology

Innovation Plan, as well as the projects supported by China Academy of Space Technology and Launch Vehicle Technology. All these supports are highly appreciated.

References

1. Hogan, F., Grau, E., Rodriguez, A.: Reactive planar manipulation with convex hybrid MPC. In: Proceedings of IEEE International Conference on Robotics and Automation, pp. 247–253, Brisbane, Australia (2018)
2. Righetti, L., et al.: An autonomous manipulation system based on force control and optimization. Auton. Rob. **36**(1–2), 11–30 (2014)
3. Liang, H., et al.: PointNetGPD: detecting grasp configurations from point sets. In: Proceedings of IEEE International Conference on Robotics and Automation, pp. 3629–3635, Montreal, Canada (2019)
4. Mahle, J., et al.: Dex-Net 2.0: deep learning to plan robust grasps with synthetic point clouds and analytic grasp metrics. In: Proceedings of Robotics: Science and Systems, pp. 1–12, Massachusetts, USA (2017)
5. Mahler, J., et al.: Learning ambidextrous robot grasping policies. Sci. Rob. **4**(26), 1–12 (2019)
6. Tsai, J., Lin, P.: A low-computation object grasping method by using primitive shapes and in-hand proximity sensing. In: IEEE International Conference on Advanced Intelligent Mechatronics, pp. 497–502, Munich, Germany (2017)
7. Beltran-Hernandez, C., Petit, D., Ramirez-Alpizar, I., Harada, K.: Learning to grasp with primitive shaped object policies. In: IEEE/SICE International Symposium on System Integration, pp. 468–473, Paris, France (2019)
8. Lin, Y., Tang, C., Chu, F., Vela, P.: Using synthetic data and deep networks to recognize primitive shapes for object grasping. In: IEEE International Conference on Robotics and Automation, pp. 10494–10501, Paris, France (2020)
9. Eitel, A., Springenberg, J., Spinello, L., Riedmiller, M., Burgard, W.: Multimodal deep learning for robust RGB-D object recognition. In: IEEE/RSJ International Conference on Intelligent Robots and Systems, pp. 681–687, Hamburg, Germany (2015)
10. Carlucci, F., Russo, P., Caputo. B.: $(DE)^2CO$: deep depth colorization. IEEE Rob. Autom. Lett. **3**(3), 2386–2393 (2018)
11. Loghmani, M., Planamente, M., Caputo, B., Vincze, M.: Recurrent convolutional fusion for RGB-D object recognition. IEEE Rob. Autom. Lett. **4**(3), 2878–2885 (2019)
12. Charles, R., Su, H., Kaichun, M., Guibas, L.: PointNet: deep learning on point sets for 3D classification and segmentation. In: 34th IEEE Conference on Computer Vision and Pattern Recognition, pp. 77–85, HI, USA(2017)
13. Charles, R., Su, H., Kaichun, M., Guibas, L.: PointNet++. deep hierarchical feature learning on point sets in a metric space. In: 31st Annual Conference on Neural Information Processing System, pp. 5100–5109, Long Beach, CA, USA (2018)
14. Wu, W., Qi, Z., Fu, L.: PointConv: deep convolutional networks on 3D point clouds. In: 35th IEEE Conference on Computer Vision and Pattern Recognition, pp. 9613–9622, Long Beach, CA, USA (2019)

Odometry During Object Transport: A Study with Swarm of Physical Robots

Muhanad H. M. Alkilabi[1,3](✉) ⓘ, Timoteo Carletti[2] ⓘ, and Elio Tuci[1] ⓘ

[1] Department of Computer Science, University of Namur, Namur, Belgium
{muhanad-hayder-mohammed.mohammed,elio.tuci}@unamur.be
[2] Department of Mathematics, University of Namur, Namur, Belgium
timoteo.carletti@unamur.be
[3] Department of Computer Science, University of Kerbala, Kerbala, Iraq

Abstract. Object transport by a single robot or by a swarm of robots can be considered a very challenging scenario for odometry since wheel slippage caused by pushing forces exerted on static objects and/or by relatively frequent collisions with other robots (for the cooperative transport case) tend to undermine the precision of the position and orientation estimates. This paper describes two sets of experiments aimed at evaluating the effectiveness of different sensory apparatuses in order to support odometry in autonomous robots engaged in object transport scenarios. In the first set of experiments, a single robot has to track its position while randomly moving in a flat arena with and without an object physically attached to its chassis. In the second set of experiments, a member of a swarm of physical robots is required to track its position while collaborating with the group-mates to the collective transport of a heavy object. In both sets, odometry is performed with either wheel encoders or with an optic-flow sensor. In the second set of experiments, both methods are evaluated with and without gyroscope corrections for angular displacements. The results indicate that odometry based on optic-flow sensors is more precise than the classic odometry based on wheel encoders. In particular, this research suggests that by using an appropriate sensory apparatus (i.e., an optic-flow sensor with gyroscope corrections), odometry can be achieved even in extreme odometry conditions such as those of cooperative object transport scenarios.

Keywords: Odometry · Cooperative transport · Swarm robotics · Optic-flow sensor

1 Introduction

In order to navigate complex environments successfully, autonomous robots may required to keep track of their position with respect to local landmarks or with respect to a given frame of reference. Compared to single robot systems, multi-robot systems can exploit the collectivity to improve the effectiveness of localisation algorithms. In

Dr. Alkilabi thanks the Research Institute naXys, University of Namur, Belgium, for the financial support received during the writing of this paper.

© Springer Nature Switzerland AG 2021
Y. Tan and Y. Shi (Eds.): ICSI 2021, LNCS 12690, pp. 92–101, 2021.
https://doi.org/10.1007/978-3-030-78811-7_9

particular, in distributed localisation algorithms, estimation errors can be reduced by integrating sensor data collected from different agents placed in different positions. The eventual heterogeneity of the system can be exploited to fuse information from different types of sensors [5]. The time required for localisation using dynamically built maps (SLAM) can be significantly reduced if different robots explore different parts of the environment simultaneously and subsequently share the information gathered. This process is also referred to as cooperative or collaborative localisation.

Cooperative localisation is achieved by a group of robots exchanging their relative or absolute positions while in close vicinity. In this context, most work on localisation has focused on the question of how to reduce the odometry error using a cooperative team of robots. One way to correct this odometry error is to use apriori defined landmarks which can be detected by the robot's sensors. Unfortunately, detecting and recognising landmarks is a difficult task in general, especially when the environment is much larger than the sensing range of the robot. Alternatively, some studies propose the use of *Mobile Landmarks* based on cooperative localisation [9]. In such an approach a team of robots is divided into two groups. One group remains stationary and act as landmarks while the other group navigates the environment. The two groups interchange their roles until a target position is reached. The odometry is calculated based on triangular measurements of the distances and the angles between the robots. However, relying on stationary robots will reduce the effectiveness of the group, which are obliged to dedicate a significant amount of their resources to generate landmarks instead of using them to carry out important tasks.

In swarm robotics systems [4], the generally limited sensory and computational capabilities of the robots make it hard to implement some of the odometry based methods discussed above. To overcome these limitations, several studies rely on approaches that can implement odometry in small size platforms. For example, in [8], the author proposes an odometry method for autonomous micro-robots based on the optical properties of a motor-wheel transmission. Such an approach can provide relative localisation for robots where the robot size and its computational resources are limited. In [12], the author illustrates the ODOCLUST algorithm for an aggregation task. The robots alternate between random motion and odometry guided homing based on the last contact point with other robots to achieve aggregation. The results indicate that the method is robust in coping with odometry errors. In [6], the authors study a swarm robotics foraging scenario where e-puck robots randomly initialised in a bounded arena are required to find a food source and a nest area and then to move back and forth between the two target locations. The authors discuss a navigation and localisation algorithm called *Social Odometry*. Social Odometry is based on peer-to-peer local communication which leads to a self-organised path selection without the need for any central unit.

In this paper, we illustrate and compare the performances of an alternative localisation method for e-puck robots based on the use of an optic-flow sensor and gyroscope for odometry. This method is compared with the classic wheel encoders approach, and it is evaluated in various difference scenarios, such as i) single e-puck randomly moving in an empty arena; ii) single e-puck randomly moving in an empty arena with an object attached to its chassis; iii) swarms of e-pucks transporting an object that has to be pushed by all the members of the group to be transported. In this scenario, one robot of the swarm is required to track its position. Object transport is a very challenging scenario for odometry since wheel slippage caused by pushing forces exerted on static objects, and by relatively frequent collisions with other robots during the transport, tend to undermine the precision of position and orientation estimates [11]. The results of our study show that the odometry method based on the use of the developed optic-flow sensor which integrated into the e-puck robotics platform combined with gyroscope corrections for angular displacements can significantly reduce the disruptive effects of slippage in both the single robots and the swarm of robots types of scenario.

2 Odometry Using Single Optic-Flow Sensors

This section describes a method to estimate the robot's changes in position based on the readings of a single optic-flow sensor mounted underneath the e-puck chassis. Optic-flow sensor has been used to improve the coordination among the robots of a swarm required to cooperatively transport heavy objects (as in [1,2]). The sensor is an optical camera that accurately computes the pixel displacement along x and y components of the robot's translational movements (see Fig. 1a). The readings of the sensor are delivered in counts format (i.e., cx counts in the x-direction and cy counts in the y-direction of the robot translational movement). For a detailed description of the characteristics of this sensor see the supplementary material at https://drive.google.com/file/d/1tZ4kO3PvnSm5EFjQHPdjjWJAP5Qksn55/view?usp=sharing.

The odometry method with the optic-flow sensor is based on the assumption that the robot travels only in the direction orthogonal to the wheel axis. If the kinematic constraints of a differential drive robot are violated due to the external forces (causing the robot to move in the direction of the wheels axis; for the e-puck robot, this is the lateral axis), the method cannot accurately function. In such circumstances, a single optic-flow sensor interprets the lateral motion as a rotation along the robot's centre of mass, rather than a translational movement.

This method, originally proposed in [10], is based on the principle of the relative velocity of a point on a rigid body. According to the kinematics of a rigid body that moves in a two-dimensional space, during translation motion, all points on the body move with the same velocity. In contrast, during rotational motion, all points which lay outside the body's axis of rotation have a different linear velocity. This is due to the fact that their velocity depends on the point's distance from the centre of rotation. All points lying on the rotational axis have zero linear velocity.

Let's assume that this rigid body corresponds to the e-puck body. Point o, whose linear velocity is $\vec{V_o}$, is at the robot centre of mass, while point s, whose linear velocity is $\vec{V_s}$, is at a certain distance from o not on the robot's transversal axis. This means that point o has zero linear velocity only when the robot rotates on its transversal axis. $\vec{\omega}$ is the angular velocity vector of the robot and $\vec{r}$ is the vector from the position of s to o (see Fig. 1b). The cross product $\vec{\omega} \times \vec{r}$ refers to the relative velocity of point o with respect to point s. Providing that the s local frame of reference is aligned with the robot's local frame of reference, the lateral (x) and longitudinal (y) component of the velocity vector of point o, that is $V_{o,x}$ and $V_{o,y}$, can be computed as follows:

$$\vec{V_o} = \vec{V_s} + \vec{\omega} \times \vec{r}; \qquad V_{o,x} = V_{s,x} - \omega.r_y; \qquad V_{o,y} = V_{s,y} + \omega.r_x; \qquad (1)$$

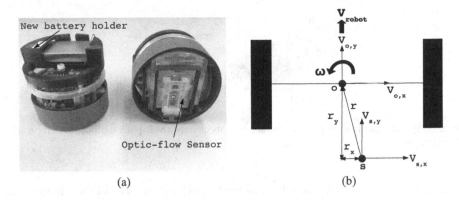

Fig. 1. (a) The optic-flow sensor mounted on the e-puck chassis. (b) Graphs indicating the velocities component of the single optic-flow sensor method to estimate the robot position.

with r_x and r_y being the lateral and longitudinal components of $\vec{r}$ within the robot local frame of reference (see Fig. 1b). Given the kinematics constraints of a differential drive robot like the e-puck, $V_{o,x} = 0$. Given that the linear velocity of the robot is always in the direction orthogonal to the wheels' axis, the linear velocity of point o does not have any lateral component. That is, $V_o = V_{o,y}$. Thus, the angular velocity $\vec{\omega}$ of the robot and the linear velocity $\vec{V_o}$ of point o can be computed as follows:

$$\omega = \frac{V_{s,x}}{r_y}; \qquad V_o = V_{s,y} + \frac{r_x}{r_y}V_{s,x}; \qquad (2)$$

Assuming the optic-flow sensor is place at s, the velocity at this point can be measured as follows:

$$V_{s,x} = \frac{cx}{R.\Delta t}; \qquad V_{s,y} = \frac{cy}{R.\Delta t}; \qquad (3)$$

where cx and cy are the number of counts returned by the optic-flow sensor for the lateral and longitudinal displacement of point s, and R is the sensor resolution. By sampling the lateral and longitudinal component of the velocity vector of point s using the sensor sampling interval $\Delta t = 50$ ms, it is possible to compute the displacement of point o in the lateral and longitudinal directions.

The change in the robot orientation $\Delta\theta$ and the change in the robot linear displacement Δp (i.e., translational movements) can be computed as follows:

$$\Delta\theta = \frac{cx}{R.r_y}; \qquad \Delta p = \frac{cy}{R} + \frac{r_x.cx}{R.r_y}; \qquad (4)$$

Finally, the relative robot position and orientation at time t can be computed as follows:

$$\theta_t = \theta_{t-1} + \Delta\theta; \qquad X_t = X_{t-1} + \Delta p\cos(\theta_t); \qquad Y_t = Y_{t-1} + \Delta p\sin(\theta_t); \quad (5)$$

Odometry using incremental wheel encoders is a common technique to estimate robot changes in position and orientation based on a method called *forward kinematics*

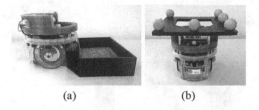

(a)   (b)

Fig. 2. (a) E-puck robots with an object attached to its chassis. (b) E-puck with markers for tracking its motion.

which relates the velocity of the robot centre of mass to the velocity of the left and right wheels. As this method is well understood, this section does not describe how these equations are formulated. For a detailed description of the mathematical derivation of forward kinematics equations, see [3] and supplementary material.

3   Results

This section illustrates the results of two sets of experiments aimed at evaluating the effectiveness of different sensory apparatus in order to support odometry in autonomous robots engaged in object transport scenarios. In the first set of experiments (referred to as single-robot condition), a single physical e-puck robot is required to localise itself while randomly exploring a bounded arena both with and without an object physically attached to its chassis (see Fig. 2a). In the second set of experiments (referred to as multi-robot condition), a single physical e-puck robot is required to localise itself while transporting with other robots a heavy object in an arbitrarily chosen direction. Both sets of experiments aimed at comparing the effectiveness of an optic-flow sensor based odometry system with a wheel encoders based odometry system.

The results of the single-robot condition test provide evidence of the accuracy of the optic-flow and wheel-encoder based odometry during a relatively simple transport scenario where the object is physically attached to the robot. The dynamics of transport can generate wheel slippage. However, lateral displacements of the robot chassis are unlikely to occur. For both the optic-flow sensor based odometry and the wheel encoders based odometry, the accuracy of the localisation process is compared with a scenario in which the robot moves randomly in a bounded arena without transporting the object. The results of the multi-robot condition provide evidence of the accuracy of the optic-flow and wheel-encoder based odometry in an object transport scenarios where both wheel slippage and lateral displacement of the robot chassis are likely to happen at any stage of the transport. Comparisons between the odometry performances in single-robot and multi-robot conditions give a quantitative estimate of the disruptive effects on odometry of the phenomena induced by robot-robot and robot-object collisions. In the multi-robot condition, it is shown that the accuracy of the optic-flow and the wheel encoders based odometry can be improved by using the gyroscope sensor for estimating changes in orientation.

In both conditions, the position estimates generated by the robot are compared with the ground truth measured by a Vicon motion capture system. The Vicon system comprises 7 cameras covering the square robot arena (220×220 cm) with a position accuracy of ± 1 mm and orientation accuracy of $\pm 1°$. During these tests, a 3D printed square (10×10 cm) with 6 markers is attached on top of the robot as shown in Fig. 2b. The aim of this structure is to obtain a better Vicon's accuracy while tracking the position and orientation of the robot during transport.

During all tests, after each update cycle, the robot communicates the estimated change in position Δp and orientation $\Delta \theta$ as calculated by the optic-flow sensor (see Eq. 4 in Sect. 2), and by the wheel encoders to an external computer via Bluetooth. In the multi-robot condition, the robot also communicates the estimated change in orientation as computed by the gyroscope. Along with this information, the external computer records the absolute position of the robot as recorded by the Vicon system and it timestamps the flux of data. The robot trajectories with respect to the Vicon frame of reference are reconstructed offline by the external computer by converting the sequence of local changes in position and orientation into a coherent set of "global" or Vicon based coordinates.

3.1 The Single-Robot Condition

In this condition, a single e-puck robot undergoes 80 trials evaluation. At the beginning of each trial, the robot is placed in the middle of a flat bounded square arena (220×220 cm). During the trial, which last 180 s, the movement of the robot is characterised by an isotropic random walk, with a fixed step length (3 s, at 3.2 cm/s), and turning angles chosen from a wrapped Cauchy probability distribution characterised by the following PDF [7]:

$$ f_w(\theta, \mu, \rho) = \frac{1}{2\pi} \frac{1 - \rho^2}{1 + \rho^2 - 2\rho \cos(\theta - \mu)}, \quad 0 < \rho < 1, \tag{6} $$

where $\mu = 0$ is the average value of the distribution, and ρ determines the distribution skewness. For $\rho = 0$ the distribution becomes uniform and provides no correlation between consecutive movements, while for $\rho = 1$ a Dirac distribution is obtained, corresponding to straight-line motion. The robot undergoes 40 trials without object, and 40 trials with an object of 60 g attached to its chassis, as shown in Fig. 2a. For both these sets of tests (with and without object), half of the trials (i.e., 20) are with $\rho = 0.3$, and half with $\rho = 0.9$ (video recording of these tests are available at https://www.youtube.com/embed/HbMRv_nuvm4).

The results of the single-robot test condition are shown in Fig. 3, where white boxes refer the Euclidean distances (in millimetre) between the robot positions at the end of each trial as recorded by the motion caption system and the positions estimated by the optic-flow sensor, for trial without object (see white boxes in correspondence of x-axis label "no-obj"), and for trial with the object attached to the robot chassis (see white boxes in correspondence of x-axis label "obj"). Grey boxes refer to the same distances recorded by the motion caption system and the positions estimated by the wheel encoders, for trials without object (see grey boxes in correspondence of x-axis

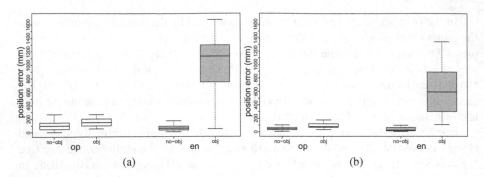

Fig. 3. Graphs showing the position errors for a random walk based on wrapped Cauchy probability distribution with (a) $\rho = 0.3$; (b) $\rho = 0.9$. Position errors correspond to the Euclidean distances (in millimetre) between the final robot's positions recorded by the Vicon (i.e., ground truth) and the estimates of the final robot's positions generated by the optic-flow sensor (white boxes above the "op" x-axis labels) and by the wheels encoder (grey box above the "en" labels). Also on the x-axis, the "no-obj" labels refer to runs without any object attached to the robot chassis, while the "obj" labels refers to runs with an object attached to the robot chassis. Each box is made of 20 points (trials), where each point indicates the position error in a single trial. Boxes represent the inter-quartile range of the data, while horizontal bars inside the boxes mark the median value. The whiskers extend to the most extreme data points within 1.5 times the inter-quartile range from the box.

label "no-obj"), and for trials with the object attached to the robot chassis (see grey boxes in correspondence of x-axis label "obj"). We can clearly notice that in both graphs (differing for the type of random walk, see also Figure caption for details), for those trials with no object attached to the robot ("no-obj" boxes in Fig. 3) the median of the position error is less than 10 cm for both the optic-flow sensor estimates, and for the wheels encoder estimates. In those trials where the robot is physically attached to an object, the errors corresponding to the optic flow final position estimates are only slightly higher than the errors recorded in trials with no object (see white boxes in Fig. 3). The errors corresponding to the wheel encoder estimates in trials with the object are instead several centimetres higher than the errors registered in the trials without the object (see grey boxes in Fig. 3). In trials where the robot has an object attached to it, the median of the position errors corresponding to the estimates generated using the wheels encoders are significantly higher than the errors corresponding to the position estimates generated using the optic-flow sensor (Wilcoxon rank-sum test, $p < 0.001$).

To summarise, without an object attached to the robot chassis, the optic-flow sensor and the wheel encoder generate position estimates quite similar to each other with the median of the position error around 10 cm computed on repeated random walks of 180 s. With an object of 60 g attached to the robot chassis, the optic-flow sensor is a better means to generate position estimates than the wheel encoder. The wheel slippage events disrupt only the wheels encoder estimates.

3.2   The Multi-robot Condition

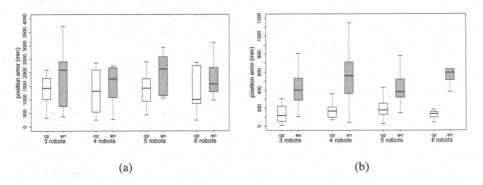

(a)   (b)

Fig. 4. Graphs showing the position errors referring to the Euclidean distances (in millimetre) between the final *R-robot*'s positions recorded by the Vicon (i.e., ground truth) and the final *R-robot*'s positions estimates generated by: (a) the optic-flow sensor (white boxes above x-axis labels "op") and the wheels encoder (grey boxes, above x-axis labels "en"); (b) the optic-flow sensor with gyroscope corrections (white boxes above x-axis labels "op"), and the wheels encoder with gyroscope corrections (grey boxes above x-axis labels "en"). Tests are run for homogeneous groups of 3, 4, 5, and 6 physical e-puck robots, in 10 trials per group. Each point in the box refers to the position error in a single trial.

This section describes the results of a further series of experiments in which homogeneous groups of 3, 4, 5 and 6 robots are required to push a cuboid object as far as possible from its initial position in an arbitrary direction. The robots are controlled by dynamic neural networks shaped using evolutionary robotics approach. In each group, there is a robot, referred to as *R-robot*, whose position is estimated using both the readings of its optic-flow sensor and its wheel encoders. The accuracy of the optic-flow and the wheel encoder based odometry of the *R-robot* are evaluated with respect to the ground truth generated by the Vicon. Each group is evaluated for 10 trials. In each of these 10 trials i) the *R-robot* is involved in the transport for the entire duration of the trial; ii) the transport is successful (i.e., the object is transported from its starting position for at least a distance of 1 m. For more details, see video recording of some trials of this test at https://www.youtube.com/embed/6khzx79OrYc).

In Fig. 4a the white boxes refer to the errors corresponding to the position estimates generated with the optic-flows sensor, and the grey boxes refer to the errors corresponding to the position estimates generated with wheel encoders. We can notice that the errors for both types of position estimates are relatively high. Also that errors corresponding to estimates generated by the wheel encoders are slightly higher than the error corresponding to position estimates generated by the optic flow sensor. Generally, wheels slippage occurring at any time while the *R-robot* pushes a static element of the transport scenario—either the cuboid object or another static robot—severely effects the precision of the position estimates generated by wheel encoders. The optic-flow sensor is relatively immune from the consequences of wheel slippage. Unfortunately, in the collective object transport task, wheel slippage events are associated with rather frequent lateral sliding of the robot's body (i.e., robot movements in the direction perpendicular to the wheel direction of rotation), which is a consequence of pushing forces mainly exerted by other robots. Lateral sliding events disrupt the precision of the optic-flow estimates relative to changes of the *R-robot*'s orientation more than they affect the precision of estimates relative to the change in position.

In order to overcome the disruptive effects on odometry associated with the object transport scenario, a further set of tests, were run in which the changes in orientation estimates for both the optic-flow sensor and the wheel encoders are generated using the gyroscope mounted in the *R-robot*. This means that the optic-flow sensor and the wheel encoders are used only to estimate the linear displacements in the two-dimensional navigation space. The results of this set of tests are shown in Fig. 4b. We can notice that for both types of final position estimate and for all groups, the errors are much smaller. Moreover, the errors corresponding to the position estimates generated by the optic-flow sensor with gyroscope corrections (see white boxes in Fig. 4b) are significantly lower (Wilcoxon rank-sum test, $p < 0.001$) than the errors corresponding to position estimates generated by the wheel encoders with gyroscope (see white boxes in Fig. 4b). We conclude that in this cooperative object transport scenario, the optic-flow sensor with gyroscope correction results in a better means for odometry than the wheel encoder with gyroscope correction. The optic-flow generates fairly accurate position estimation compared to the wheel encoder. The latter method generates position estimates characterised by trajectory lengths longer than those of the actual trajectory (see the trajectory graphs described in supplementary material at https://drive.google.com/file/d/1tZ4kO3PvnSm5EFjQHPdjjWJAP5Qksn55/view?usp=sharing).

4   Conclusions

This study has illustrated the results of a series of experiments aimed at testing the effectiveness of the optic-flow sensor as a means to perform odometry. In the first test, a single robot moves in an isotropic random walk with and without an object physically attached to its chassis. The results of this test indicate that, the optic-flow and the wheel encoders based odometry are fairly accurate. However, when the object is attached to the robot, the optic-flow sensor outperformed the wheel encoders due to its robustness to the disruption of the wheel slippage caused by the transport.

In the second test, a multi-robot system engaged in a cooperative transport scenario were evaluated. In this test, the estimate of the linear displacement generated by the optic-flow sensor and the wheel encoder methods have been combined with the estimates of the angular displacement generated by the gyroscope. The results indicate that the optic-flow sensor generates fairly accurate position estimation compared to the wheels encoders. The latter generates position estimates characterised by trajectory lengths longer than the robot's actual trajectories. In addition to the importance of optic-flow sensor in the context of odometry, previous research has shown that in cooperative object transport scenarios, the optic-flow sensor allows a swarm of robots to effectively coordinate their actions and to align individual contributions to transport without the necessity to feel the forces applied to the object (see [2]). The sensory information generated by the optic-flow sensors allows the robots to recover from deadlocks and to coordinate their forces in order to avoid working against each other. This study suggests that in order to accurately perform odometry with the optic-flow sensor and gyroscope correction for angular displacement, the best position of the optic-flow sensor is at the robot's centre of mass since, positioned in this way, the optic-flow sensor records no lateral displacements when the robot rotates on the spot. This improves the accuracy of the method by accounting for the lateral displacements in the odometry calculation.

In conclusion, cooperative object transport scenarios are particularly challenging for odometry since frequent wheel slippage and lateral displacement severely disrupt the precision of both the position and orientation estimates. Yet, the results of this research suggest that by using an appropriate sensory apparatus (e.g., optic-flow and gyroscope sensor), odometry can be achieved even in extreme odometry conditions such as those of cooperative object transport scenarios. In future, we plan to develop a cooperative transports scenario in which swarms of robots deliver objects to a target location known by one robot of the group. In this scenario, the robots of a swarm can improve each other localisation estimates by exchanging calculated positions information in order to move the object to a target location.

References

1. Mohammed Alkilabi, M.H., Narayan, A., Lu, C., Tuci, E.: Evolving group transport strategies for e-puck robots: moving objects towards a target area. In: Groß, R., et al. (eds.) Distributed Autonomous Robotic Systems. SPAR, vol. 6, pp. 503–516. Springer, Cham (2018). https://doi.org/10.1007/978-3-319-73008-0_35
2. Alkilabi, M.H.M., Narayan, A., Tuci, E.: Cooperative object transport with a swarm of e-puck robots: robustness and scalability of evolved collective strategies. Swarm Intell. **11**(3), 185–209 (2017). https://doi.org/10.1007/s11721-017-0135-8
3. Borenstein, J., Feng, L.: Umbmark: A method for measuring, comparing, and correcting dead-reckoning errors in mobile robots. Technical Report UMMEAM-94-22 (1994)
4. Öchsner, A.: Guest editorial. J. Phase Equilib. Diffus. **26**(5), 406 (2005). https://doi.org/10.1007/s11669-005-0025-4
5. Fox, D., Burgard, W., Kruppa, H., Thrun, S.: Collaborative multi-robot localization. In: Burgard, W., Cremers, A.B., Cristaller, T. (eds.) KI 1999. LNCS (LNAI), vol. 1701, pp. 255–266. Springer, Heidelberg (1999). https://doi.org/10.1007/3-540-48238-5_21
6. Gutiérrez, A., Campo, A., Monasterio-Huelin, F., Magdalena, L., Dorigo, M.: Collective decision-making based on social odometry. Neural Comput. Appl. **19**(6), 807–823 (2010)
7. Kato, S., Jones, M.: An extended family of circular distributions related to wrapped Cauchy distributions via Brownian motion. Bernoulli **19**(1), 154–171 (2013)
8. Kernbach, S.: Encoder-free odometric system for autonomous microrobots. Mechatronics **22**(6), 870–880 (2012)
9. Kurazume, R., Nagata, S., Hirose, S.: Cooperative positioning with multiple robots. In: Robotics and Automation, 1994. Proceedings, 1994 IEEE International Conference on, pp. 1250–1257. IEEE (1994)
10. Lee, S., Song, J.B.: Robust mobile robot localization using optical flow sensors and encoders. In: Robotics and Automation, 2004. Proceedings, ICRA2004. 2004 IEEE International Conference on, vol. 1, pp. 1039–1044. IEEE (2004)
11. Tuci, E., Alkilabi, M., Akanyety, O.: Cooperative object transport in multi-robot systems: a review of the state-of-the-art. Front. Robot. AI **5**, 1–15 (2018)
12. Vardy, A.: Aggregation in robot swarms using odometry. Artif. Life Robot. **21**(4), 443–450 (2016). https://doi.org/10.1007/s10015-016-0333-2

Active Disturbance Rejection Control
of Underwater Manipulator

Qirong Tang[✉], Daopeng Jin, Yang Hong, Jinyuan Guo, and Jiang Li

Laboratory of Robotics and Multibody System, School of Mechanical Engineering,
Tongji University, Shanghai 201804, People's Republic of China

Abstract. In this study, a method of active disturbance rejection controller (ADRC) is presented for 2-DOF underwater manipulator. The ADRC basically does not rely on the accurate mathematical model of the object and can decouple the model. This method can eliminate the influence of model errors, time-varying parameters and external interference on the control effect. Firstly, the manipulators is divided into two subsystems. For each joint subsystem, the hydrodynamic force, coupling term between joints and unknown environment disturbances are considered as the total disturbance. Subsequently, an extended state observer (ESO) is designed to estimate and compensate the total disturbance. Moreover, in order to improve the disturbance observation effect of the extended state observer, the inertia matrix of the manipulator system is used to decouple the static part. Finally, the effectiveness of ADRC is verified by simulation and it is demonstrated that ADRC's control effect outperforms PD and CSMC in either accuracy, dynamic characteristics or robustness.

Keywords: Underwater manipulator · Active disturbance rejection controller · Trajectory tracking

1 Introduction

Underwater manipulators are usually working in a complex hydraulic environment so it is subject to additional fluid viscous resistance, additional mass forces and some unknown water flow disturbances [1], which make the model of underwater manipulator more complicated than general manipulator. The nonlinearity and uncertainty of the system, as well as the coupling of multiple external disturbances in the underwater environment have become major challenges under conventional control strategies.

The control methods of the underwater manipulator system are not limited to PID control, adaptive control, neural network control and sliding mode control [2]. In [3], a controller including model parameter estimation and multi-layer closed loop PID is designed to control a multi-DOF underwater manipulator, but it does not compensate for nonlinear disturbances. Tomei et al. in [4] proposed an adaptive robot control algorithm combining a PD controller with a dynamic

© Springer Nature Switzerland AG 2021
Y. Tan and Y. Shi (Eds.): ICSI 2021, LNCS 12690, pp. 102–110, 2021.
https://doi.org/10.1007/978-3-030-78811-7_10

compensation model. This algorithm improves the accuracy of nonlinear control, but a more accurate dynamic model of the manipulator is required. Neural network control [5] does not rely on the precise mathematical model, however, its sample data for training is a key problem. Sliding mode control is widely used for uncertain models and anti-interference. Bin Xu et al. in [6] designed an improved sliding mode controller based on fuzzy logic, but the output torque of the controller still had chattering problem.

The Active Disturbance Rejection Controller (ADRC) [7] was proposed in response to the limitations of PID control. On the basis of traditional PID control, new nonlinear dynamic structures are proposed: tracking differentiator (TD), extended state observer (ESO) [8] and nonlinear state error feedback (NLSEF) [9]. MahMoud et al. in [10] applied the active disturbance rejection controller to the trajectory tracking of a two-link manipulator and the simulation showed that it had better anti-interference performance than PID. Radoslaw in [11] used the Lyapunov analysis to prove the stability of ADRC in manipulator control and proposed model low-order estimation compensation to further improve its stability.

This study takes the advantage of the ADRC to control a 2-DOF underwater manipulator. The coupling term between joints and unknown environment disturbances are considered as the total disturbance. The extended state observer is built to estimate them and then the feedback control law is used for active compensation. Finally, it is compared with a PD controller and a continuous sliding mode controller.

The rest of the study is organized as follows: Sect. 2 presents the system overview. Section 3 recalls basic components behind the ADRC method. ADRC decoupling design of underwater manipulator is presented in Sect. 3. Simulation and results are presented in Sect. 4 and finally, Sect. 5 concludes the study.

2   System Overview of Underwater Manipulator

The system consists of a 2-DOF manipulator shown in Fig. 1. It can be simplified to a two-link system with M_1, M_2. The end of manipulator is equipped with an actuator that can perform some underwater tasks. In this study, ADRC is used to control the manipulator in order to realize the tracking control of the desired trajectory in the joint space.

The Cartesian coordinate system of underwater manipulator is shown in Fig. 1, which consists of the reference fixed coordinate system $O-x_0y_0z_0$, the joint coordinate system $O_i-x_iy_iz_i(i = 1, 2)$ and the end actuator coordinate system $O_3-x_3y_3z_3$. Here O is the origin of the reference fixed coordinate system. Assuming that each link of the manipulator is a homogeneous element, the center of the link and the center of gravity are coincident.

Based on the second type of Lagrangian energy equation, the dynamic model of underwater manipulator is given as follow,

$$\boldsymbol{\tau} + \mathbf{w}(t) = \mathbf{M}\left(\mathbf{q}\right)\ddot{\mathbf{q}} + \mathbf{C}\left(\mathbf{q}, \dot{\mathbf{q}}\right)\dot{\mathbf{q}} + \mathbf{D}\left(\mathbf{q}, \dot{\mathbf{q}}\right)\dot{\mathbf{q}} + \mathbf{G}\left(\mathbf{q}\right). \qquad (1)$$

where $\mathbf{M}(\mathbf{q}) \in \Re^{2 \times 2}$ is the inertia matrix, $\mathbf{C}(\mathbf{q}, \dot{\mathbf{q}}) \in \Re^{2 \times 2}$ is the centrifugal and Coriolis vector, $\mathbf{D}(\mathbf{q}, \dot{\mathbf{q}})$ is the hydrodynamic damping matrix formed by the terms due to fluid viscous resistance and additional mass force, and $\mathbf{G}(\mathbf{q}) \in \Re^{2 \times 1}$ is the gravity vector. $\mathbf{q}$, $\dot{\mathbf{q}}$ and $\ddot{\mathbf{q}}$ are joint position, joint velocity and joint acceleration vectors, respectively. Here $\tau \in \Re^{2 \times 1}$ is a joint input torque vector and $\mathbf{w}(t)$ is an unknown disturbance caused by complex fluid flow.

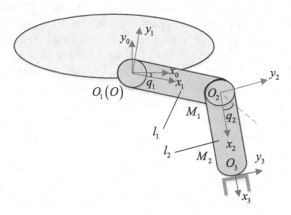

Fig. 1. A 2-DOF underwater maniplator model

3   ADRC Decoupling Design for Underwater Manipulator

The controlled object in this case is a 2-DOF underwater manipulator, which is a Multi Input Multi Output (MIMO) second-order system describedly Eq. (1). Each joint can be independently regarded as a Single Input Single Output (SISO) second-order system. The coupling relationship between the joints can be regarded as part of the total disturbance. Firstly, dynamic model (1) can be described as

$$\ddot{\mathbf{q}} = \begin{bmatrix} q_1 & q_2 \end{bmatrix}^T = \underbrace{\mathbf{M}^{-1}(\mathbf{w}(t) - \mathbf{C}\dot{\mathbf{q}} - \mathbf{D}\dot{\mathbf{q}} - \mathbf{G})}_{f(\mathbf{q}, \dot{\mathbf{q}}, \mathbf{w}(t), t)} + \underbrace{\mathbf{M}^{-1}}_{b} \begin{bmatrix} \tau_1 & \tau_2 \end{bmatrix}^T, \qquad (2)$$

where $\mathbf{w}(t)$ is the external fluid flow disturbance, $f(\mathbf{q}, \dot{\mathbf{q}}, \mathbf{w}(t), t)$ is the unknown time-varying part of the system, including the system uncertainty, internal disturbances caused by the coupling between joints and external fluid flow disturbance. If each joint of the underwater manipulator is treated as an SISO second-order independent system, $\mathbf{M}^{-1}(\mathbf{w}(t) - \mathbf{C}\dot{\mathbf{q}} - \mathbf{D}\dot{\mathbf{q}} - \mathbf{G})$ can be called dynamic coupling part of system while $\mathbf{M}^{-1}\begin{bmatrix} \tau_1 & \tau_2 \end{bmatrix}^T$ can be the static coupling part of the manipulator. Using inertia matrix $\mathbf{M}$ to define the virtual control input

$[U_1\ U_2] = \mathbf{M}^{-1}[\tau_1\ \tau_2]^T$, the static coupling part can be decoupled, and the state space of the i-th $(i = 1, 2)$ joint can be obtained

$$\begin{cases} \ddot{q}_i = f(\mathbf{q}, \dot{\mathbf{q}}, w_i(t), t) + U_i \\ y_i = q_i \end{cases} \tag{3}$$

For the i-th joint, state x_1 is the joint angle value q_i and state x_2 is the joint angle velocity value $\dot{q}_i$. Here U_i is i-th virtual driving torque of joint i. $w_i(t)$ is the external disturbance, f is defined as the total disturbance of the system, which is a nonlinear function. It contains all internal disturbance, external disturbance and nonlinear terms. In order to estimate and compensate the f term, f is defined as an extended state x_3 of the system in ADRC, as

$$x_3 = f(\mathbf{q}, \dot{\mathbf{q}}, w_i(t), t). \tag{4}$$

Aiming at the dual-joint underwater manipulator, this study designs ADRC separately for the two joints. An ADRC decoupling scheme for the 2-DOF underwater manipulator takes the form as shown in Fig. 2, and several used concepts of ADRC decoupling scheme are shown as follows.

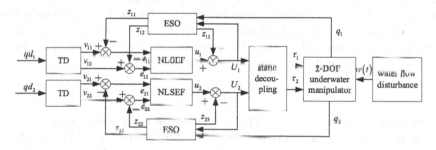

Fig. 2. An ADRC decoupling scheme for the 2-DOF underwater manipulator

1) Tracking Differentiator: For a given reference i-th joint angle position signal qd_i, the TD generates a transient trajectory v_{i1} and extracts the derivative v_{i2}. The linear TD is discretized which is constructed as follows,

$$\begin{cases} FH = -r(v_{i1} - qd_i) - 2r \\ v_{i1} = v_{i1} + hv_{i2} \\ v_{i2} = v_{i2} + hFH \end{cases} \tag{5}$$

where $h = 0.01s$ is the discretization step size, r is the tracking speed factor and FH is a self-defined linear function.

2) Extended State Observer: The ESO is used to estimate system states and the total disturbances based on the output and input signals of the object. The state space expression of each joint is shown in Eq. (3). ESO is designed independently for each joint as

$$
\begin{cases}
e_i = z_{i1} - q_i \\
\dot{z}_{i1} = z_{i2} - \beta_{i1} fal(e, \alpha_{i1}, \delta_i) \\
\dot{z}_{i2} = z_{i3} - \beta_{i2} fal(e, \alpha_{i2}, \delta_i) + U_i \\
\dot{z}_{i3} = -\beta_{i3} fal(e, \alpha_{i3}, \delta_i)
\end{cases}
\tag{6}
$$

where z_{i1}, z_{i2} and z_{i3} are estimated i-th joint states correspond to q_i, $\dot{q}_i$ and total disturbance x_{i3}, respectively. Here $\beta_i (i = 1, 2, 3)$ is the observer gain coefficient and *fal* is a continuous power function with a linear segment near the origin, which is defined as

$$
fal(e, \alpha, \delta) = \begin{cases}
\dfrac{e}{(\delta^{1-\alpha})} & |e| \leq \delta \\
|e|^{\alpha} \operatorname{sgn}(e) & |e| > \delta
\end{cases}
\tag{7}
$$

where δ is the length of the linear segment, α_i is an adjustable parameter and α_1, α_2 and α_3 are usually set as 1, 0.5 and 0.25, respectively. ESO finally realizes the estimation of the state variables (including the total disturbance) of the system (6). The estimation value of joint state z_{i1}, z_{i2} is input into the state feedback control law as feedback value, while the estimated value of the total disturbance z_{i3} is used to compensate the control torque as following,

$$
U_i = u_i - z_{i3} .
\tag{8}
$$

Combining Eqs. (3), (6) and (8), each joint system is compensated by ESO into a linear integral series system and the input amplification coefficients of the two joints are both 1. After that, The virtual control torque **U** defined in Eq. (3) can be restored to joint driving torque τ of the underwater manipulator by the static decoupling law $[\tau_1 \ \tau_2]^T = \mathbf{M} [U_1 \ U_2]^T$.

3) Nonlinear State Error Feedback Control Law: The NLSEF introduces a way to integrate the nonlinear function $fhan(e_1, ce_2, r_1, h_1)$ with feedback control, which sometimes achieves dramatically better performance than linear control [9]. For i-th joint, the constructure of NLSEF is given by

$$
\begin{cases}
e_{i1} = v_{i1} - z_{i1} \\
e_{i2} = v_{i2} - z_{i2} \\
u_i = -fhan(e_{i1}, ce_{i2}, r_1, h_1)
\end{cases}
\tag{9}
$$

where e_{i1} is the joint angle position error, e_{i2} is the joint angle velocity error, c is the damping coefficient, r_1 is the control value gain, h_1 is the precision factor and u_i is the control law output. The *fhan* is fastest control synthesis function defined in [9].

After the compensation of ESO for the total disturbance and the decoupling of the static decoupling law (SDL), the final control torque input applied to the joint subsystem then becomes

$$\begin{bmatrix} \tau_1 \\ \tau_2 \end{bmatrix} = \mathbf{M} \begin{bmatrix} -fhan(e_{11}, ce_{12}, r_1, h_1) - z_{13} \\ -fhan(e_{21}, ce_{22}, r_1, h_1) - z_{23} \end{bmatrix}. \tag{10}$$

4 Simulation and Results

In order to verify the performance of the proposed method, simulation has been established in MATLAB. The simulation object is a 2-DOFs underwater manipulator system as shown in Fig. 1. The basic physical parameters are shown in Table 1, the DH parameters are shown in Table 2 and the parameters of ADRC have been adjusted, which is shown in Table 3.

Table 1. Basic physical parameters

Items	Link M_1	link M_2
$m/$kg	2.38	1.96
$l/$m	0.32	0.32
$I_{zz}/$kg$\cdot$m^2	0.0198	0.0168

Table 2. D-H parameters

i	α_{i-1}(rad)	a_{i-1}(m)	d_i(m)	θ_i(rad)
1	0	0	0	q_1
2	0	0.32	0	q_2
3	0	0.32	0	0

Table 3. ADRC parameters

joint	TD		ESO				NLSEF		
	h	r	β_1	β_2	β_3	δ	r_1	h_1	c
O_1	0.01	10	2000	1290	1200	0.03	100	0.01	2.7
O_2	0.01	10	2000	1936	780	0.03	100	0.01	1.3

According to the parameters of ADRC designed previously as in Table 3 and the basic physical parameters of the underwater manipulator as in Table 1 and in Table 2, a simulation model is established to simulate the joint trajectory tracking control compared with traditional PD and CSMC. The expected tracking trajectory is defined as

$$\begin{cases} qd_1 = \dfrac{\pi}{4} + \dfrac{\pi}{4} \sin\left(0.5t + \dfrac{\pi}{4}\right) \\ qd_2 = \dfrac{\pi}{6} \sin\left(0.8t + \dfrac{\pi}{4}\right) \end{cases}. \tag{11}$$

The entire simulation process spends 20 s. In order to verify the disturbance adaptability of ADRC, at the 10 s, disturbance applied to two joints is defined as follow,

$$\begin{cases} w_1 = 5 + 15\sin\left(0.5t - \dfrac{\pi}{4}\right) \\ w_2 = 5 + 2\sin\left(0.5t + \dfrac{\pi}{6}\right) \end{cases}. \tag{12}$$

The Fig. 3 depicts the results of signal tracking simulator. The result shows that under the condition of no external disturbance in the first 10 s, both the ADRC and CSMC can achieve better tracking results, while the traditional PD controller has a certain steady-state error. When the external disturbance shown in Eq. (12) is added at 10 s, the tracking errors of the three controllers all have a certain degree of sudden change. By contrast, with the efficacy of ESO designed in ADRC, the unknown disturbance is estimated and actively compensated, so it can eliminate the sudden error faster, showing better disturbance immunity than CSMC and PD controller. At the same time, as shown in Fig. 4, the joint input torque of the decoupled ADRC controller and the CSMC controller is compared. It can be found that the joint decoupling ADRC controller in this study hardly has chattering phenomenon, and the system has good dynamic performance.

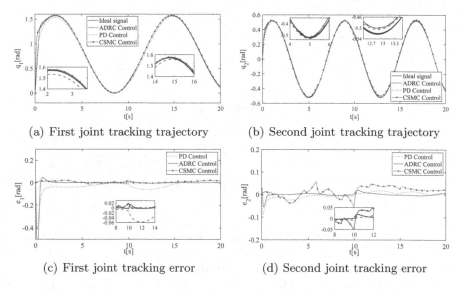

(a) First joint tracking trajectory   (b) Second joint tracking trajectory

(c) First joint tracking error   (d) Second joint tracking error

Fig. 3. Joint angle signal tracking results and disturbance rejection

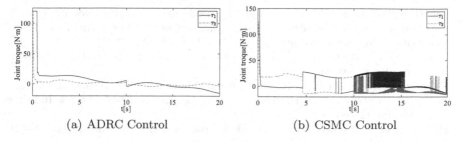

(a) ADRC Control   (b) CSMC Control

Fig. 4. Joint input torque results of ADRC Control and CSMC Control

5 Conclusion

This study presents a decoupling ADRC technique for the 2-DOF underwater manipulator system, which is subject to complex model, nonlinearity and external disturbance. The ADRC does not require an accurate data of the manipulator and defines the internal disturbance and external fluid disturbance as the total disturbance, which is estimated and compensated by ESO. In the comprehensive comparisons with traditional PD control and continuous sliding mode control (CSMC), The results illustrate the effectiveness of the proposed design and it is demonstrated that ADRC's control effect outperforms PD and CSMC in either accuracy, dynamic characteristics or robustness.

Acknowledgements. This work is supported by the projects of National Natural Science Foundation of China (No. 61873192; No. 61603277; No. 61733001), the Quick Support Project (No. 61403110321), and Innovative Project No. 20-163-00-TS-009-125-01). Meanwhile, this work is also partially supported by the Fundamental Research Funds for the Central Universities and the Youth 1000 program project. It is also partially sponsored by International Joint Project Between Shanghai of China and Baden-Wrttemberg of Germany (No. 19510711100) within Shanghai Science and Technology Innovation Plan, as well as the projects supported by China Academy of Space Technology and Launch Vehicle Technology. All these supports are highly appreciated.

References

1. Zhao, S., Yuh, J.: Experimental study on advanced underwater robot control. IEEE Trans. Rob. **21**(4), 695–704 (2005)
2. Zhang, Q., Zhang, A.: Research on coordinated motion of an autonomous underwater vehicle-manipulator system. Ocean Eng. **24**(3), 79–84 (2006)
3. Soylu, S., Buckham, B., Podhorodeski, R.: Development of a coordinated controller for underwater vehicle-manipulator systems. In: IEEE Proceeding of Oceans 2008, 15–18 September, Quebec City, Canada, pp. 1–9 (2008)
4. Tomei, P.: Adaptive PD controller for robot manipulators. IEEE Trans. Robot. Autom. **7**(4), 565–570 (1991)
5. Zhang, W., Qi, N., Yin, H.: Neuralnetwork tracking control of space robot based on sliding mode variable structure. Control Theory Appl. **28**(9), 1141–1144 (2011)

6. Xu, B., Pandian, S., Petry, F.: A sliding mode fuzzy controller for underwater vehicle-manipulator systems. In: Annual Meeting of the North American Fuzzy Information Processing Society, pp. 181–186. Detroit, USA (2005)
7. Han, J.: From PID to active disturbance rejection control. IEEE Trans. Industr. Electron. **56**(3), 900–906 (2009)
8. Han, J.: Active Disturbance Rejection Control Technique-the Technique for Estimating and Compensating the Uncertainties, 1st edn. National Defense Industry Press, Beijing (2008)
9. Han, J.: Nonlinear states error feedback control law–NLSEF. Control Decis. **10**(3), 221–225 (1995)
10. Mahmoud, A., Abdallah, Y., Fareh, R.: Tracking control of serial robot manipulator using active disturbance rejection control. In: IEEE Proceeding of Advances in Science and Engineering Technology International, pp. 1–5. Dubai, United Arab Emirates (2019)
11. Radoslaw, P., Piotr, D.: On the stability of ADRC for manipulators with modelling uncertainties. ISA Trans. **102**(1), 295–303 (2020)

Distributed Position-Force Control for Cooperative Transportation with Multiple Mobile Manipulators

Pengjie Xu[1], Jun Zheng[2]($\boxtimes$), Jingtao Zhang[1], Kun Zhang[1], Yuanzhe Cui[1], and Qirong Tang[1]($\boxtimes$)

[1] Laboratory of Robotics and Multibody System, School of Mechanical Engineering, Tongji University, Shanghai, China
[2] Research Institute of Zhejiang University, Taizhou, China
dbzj@netease.com

Abstract. This paper presents a distributed position-force control framework for multiple mobile manipulators in charge of achieving a tightly cooperative transportation task. Since the effect of each robot is different in the whole system, a three-layer control framework is designed. For the first layer, mobile bases run distributed observer which uses global states. At the second layer, the position deviation is adopted to improve the accuracy of general manipulators. Then, position control works in combination with force to ensure that the most important manipulator achieves cooperative transportation accurately and compliantly. The designed controller is extensible, which suits not only for pure transportation tasks but can also be exploited in those cases where a closed kinematic chain is generated by multi-robots manipulations. An analysis of the proposed controller is validated by simulation with three UR5 manipulators mounted on differential driven mobile bases separately.

Keywords: Multiple mobile manipulators · Distributed position-force control · Cooperative transportation

1 Introduction

Multiple and even swarm robots include a wide range of researches and applications, which can deal with tasks that are tough to be operated by an individual robot, such as multi-robots cooperative transportation, exploration and assembling, etc. The scene of cooperative transportation has recently attracted the interest of the robotics community [1]. Many industrial applications such as large-scale material moving and reorienting require multiple mobile manipulators (MMMs) to operate collectively. Hence, the design and realization of robust cooperative transportation methods for MMMs are of utmost significant. In order to accomplish described operations here that need robots to cooperate tightly. Of course, it is essential that each robot in the whole system requires to operate synchronously along desired trajectories at concordant times. This means that

© Springer Nature Switzerland AG 2021
Y. Tan and Y. Shi (Eds.): ICSI 2021, LNCS 12690, pp. 111–118, 2021.
https://doi.org/10.1007/978-3-030-78811-7_11

coordinated trajectory planning and reliable tracking control algorithms are the basics for MMMs.

A large number of studies have been reported regarding the investigation and development of algorithms for cooperative operation with multiple robots[2]. Issues such as motion planning, task compatibility and other desired payload properties have been focused, along with the problem of adaptive control. In the approaches of motion planing, related researches can be mainly classified into two aspects by considering whether there are obstacles in the workspace. Trajectory planning requires precise position and velocity information of MMMs and the object. The fundamental challenges are encountered include real-time motion, cooperative motion and collision-free [3]. In [4], a screw theoretical approach is employed for planar payload operation used multi-robots. They propose a geometric model that includes a measure for handling object, a self-defined relevance for the mobile bases and manipulators. For spatial payload transportation, the authors in [5] present an integer program to geometrically plan motion with MMMs. The control of MMMs for cooperative transportation is made difficult by the fact that each robot has access only to every robot's information while global coordination for the whole system might be required [6]. For example, a proposed controller in [7] required only local position/velocity coupling feedback by using contraction theory, which is exploited to synchronize in joint space based on Lagrangian systems. Furthermore, leader-following structures are widely used for decentralized control of multiple robotic systems. In [8], a pushing leader and redundant followers are used for transporting an planar object in simulation. A stable grasping is ensured by followers estimate leader's motion and control feedback forces [9]. Further researches study the relationships between robots and environments. Remarkably, for the distributed control scheme, the uncertainties in the knowledge of the dynamic parameters should also be considered [10].

Trajectory planning and control methods can be employed to the high-dimensional systems where the major considered points are between mobile manipulators and the operated object. These algorithms and schemes are very essential and effective in generating a feasible solution, however the internal interaction generated by task demands should be considered as well. For example, in the scene of cooperative transportation, as shown in Fig. 1, mobile manipulator 3 follows a relatively complicated trajectory and bears more payload of the carried object. The motional and mechanical characteristics of robot 3 create a need for controller to ensure successful completion of tasks.

In this study, a distributed position-force control framework is proposed to guide each mobile manipulator through different control laws synchronously. Furthermore, the mobile manipulators with less payload realize trajectory tracking through position controller, the position-force controller play a role for another one. In the simulation scenarios (Fig. 1), the performance of the proposed control framework is verified using three mobile manipulators. The transporting object with size of 1.2 m × 1 m × 0.04 m and weight 8 kg. Each mobile manipulator can be simply described by a differential driven mobile base equipped with a UR5(6 DoFs) serial arm. Gripper with load capacity 3 kg is installed on UR5's flange. The remaining of this paper is organized as follows. Section 2 addresses the kinematic and dynamic models of the mobile manipulator, and Sect. 3 explains the distributed position-force controller. The simulation is given in Sect. 4. The conclusion is drawn in Sect. 5.

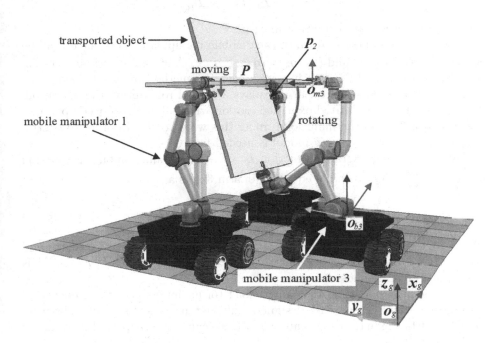

Fig. 1. Analysis scenarios for cooperative transportation.

2   System Modelling

Each mobile manipulator (Fig. 1) results in redundantly 8 DoFs. For analysis purposes, one assumes that the mobile manipulator can be split into two parts, the mobile base and manipulator arm. Hence, the kinematic models of mobile base and arm are established respectively.

The pose of objects P in word frame o_g is denoted by $\boldsymbol{p}_g = [x, y, z, \phi, \theta, \psi]^T \in \Re^6$. Here the angle vector $[\phi, \theta, \psi]$ is described with roll-pitch-yaw Euler angles. Here $\boldsymbol{p}_k = [\boldsymbol{p}_{basek}, \boldsymbol{p}_{armk}]^T$ is the state vector of each mobile manipulator, $k \in [1, 2, 3]$ in word frame o_g. Mobile base position and posture in o_g are denoted as $\boldsymbol{p}_{basek} = [x_k, y_k, \varphi_k]^T, k \in [1, 2, 3]$. Each arm is considered as an open kinematic chain with 6 DoFs joints ($\boldsymbol{p}_{armk} = [\boldsymbol{p}_1, \boldsymbol{p}_2, \cdots, \boldsymbol{p}_6]^T, k \in [1, 2, 3]$). One supposes that each gripper k grasps P_k rigidly, which is pre-planned and depended on the features of P.

The forward kinematic of $\boldsymbol{q}_{basek}$ and driving wheels $\boldsymbol{\theta}_{basek} = [\theta_{lk}, \theta_{rk}]^T$ can be described as:

$$A\dot{\boldsymbol{p}}_{basek} = 0, \tag{1}$$

where $A = [-\sin(\varphi_k), \cos(\varphi_k), -d]$, d is mobile base's rear tracks.

Using Denavit-Hartenberg(DH) convention [11] to describe the open chain manipulator, the homogeneous transform matrix of mobile arm writes as:

$$^g T_6 = {}^g T_b \, {}^b T_1 \cdots {}^5 T_6, \tag{2}$$

where $^{i-1}T_i$ is the transformation matrix.

The position and orientation of each mobile manipulator in frame o_g can be described with Eqs. (1) and (2). $q_k = [q_{basek}, q_{armk}]^T$ in joint coordinate system can also be calculated through inverse kinematics method by Eqs. (1) and (2).

In the concerned scenario, arm 3 plays a crucial role during the operation. It completes the complicated expected motion when adjusting object's posture. Besides, arm 3 bears the main load under this work scenes. In order to control arm 3 effectively, its dynamic model is needed firstly.

By constructing Lagrange function $L = T - V$ [12] and computing equation $\frac{\partial}{\partial t}\left(\frac{\partial L}{\partial P}\right) - \frac{\partial L}{\partial P} = Q$, dynamic equation of arm 3 gets as

$$M(P)\ddot{P}_g + C\left(P, \dot{P}\right)\dot{P}_g + G(P) = \tau, \tag{3}$$

where $M \in \Re^{6 \times 6}$ is the inertia matrix, $C \in \Re^{6 \times 6}$ is the matrix of centripetal and coriolis forces, $G \in \Re^{6 \times 1}$ represents the gravitational vector, $\tau \in \Re^{6 \times 1}$ is the input fore or torque.

The operating force and the control joint torque inputs need to be converted through Jacobian transpose for designed controller in the next section. Substitute Eq. (2) and Jacobian transpose into Eq. (3), one can get the expected joint torque of arm 3.

3   Distributed Position-Force Controller

The structure of the proposed controller is shown in Fig. 2, which is in terms of distributed position-force control framework. More specifically, an inner feedback of x, y and an outer one φ are used to rectify the deviations of mobile bases, the control law u_{base} is listed in Eq. (4).

$$u_{base} = k_i e_{xy} + k_p \dot{e}_{xy} + \dot{\varphi}^{ref} - k_3 e_\varphi - \eta sgn e_\varphi, \tag{4}$$

where e_{xy} and e_φ are the tracking errors for x, y and φ separately, k_i, k_p, k_3 and η are control parameters.

The control inputs are designed by PD control law in order to make up tracking error of arm 1 or arm 2. As shown by part three in Fig. 2, to achieve both trajectory tracking and joint torque controlling of arm 3, the expected torque obtained by Eq. (3) is adopted to estimate joint torque. The position-force control law with feedback errors is given in Eq. (5).

$$u_{arm3} = \tau^{ref} + k_p e_q + k_d \dot{e}_q, \tag{5}$$

where e_q is the tracking error of arm 3.

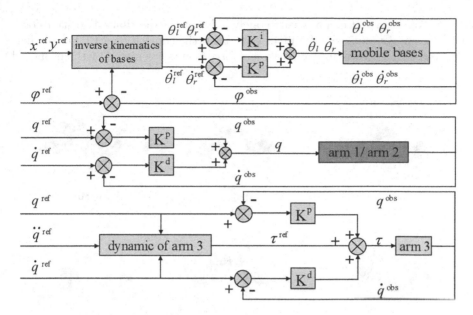

Fig. 2. Controller showing the main components of distributed scheme.

4   Simulation

In this section, a simulation for cooperative transportation is carried out to confirm the availability of the proposed distributed controller. Through MATLAB and Vrep, three mobile manipulators move the transported object 0.3 m along z_g, and then rotate it 0.3π rad around w_g. Parameters of dynamic and kinematic are listed in Table 1.

Table 1. Table of parameters

Symbol	Value	Unit	Description
r_{wheel}	0.235	m	radius of wheels
m_{wheel}	5	kg	mass of wheels
I_{wheel}	[0.018 0.018 0.07]	kg · m²	inertia of wheels
m_{base}	30	kg	mass of mobile base
I_{base}	[0.035 0.045 0.069]	kg · m²	inertia of mobile base
m_{arm}	[2.58 1.55 1.20 0.52 0.41 0.09]	kg	mass of arm

The process of simulation is shown in Fig. 3. The position and orientation data of transported object are depicted in Fig. 4. As it is shown by simulation result, three mobile manipulators successfully adjust posture of the transported object. From Fig. 4, it can be easily concluded that the distributed position-force controller enjoys much smaller jitter error than the conventional PID controller. Moreover, Fig. 5 shows the comparison of force inputs calculated by the proposed

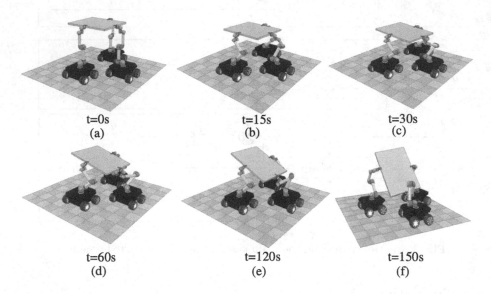

t=0s   t=15s   t=30s
(a)   (b)   (c)

t=60s   t=120s   t=150s
(d)   (e)   (f)

Fig. 3. Isometric view of the system motion.

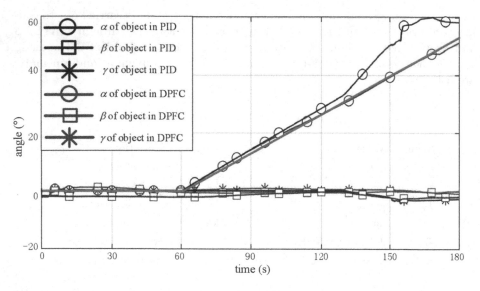

Fig. 4. Trajectory tracking results. The object's desired trajectory in alpha direction is specified with red solid line, three black solid lines with symbols are acquired by PID controller, the blues are real trajectories under the distributed position-force controller (DPFC). (Color figure online)

control scheme. It remains within the specified bounds and never reaches the
joint limits $(25 \text{ N} \cdot \text{m})$ at all times.

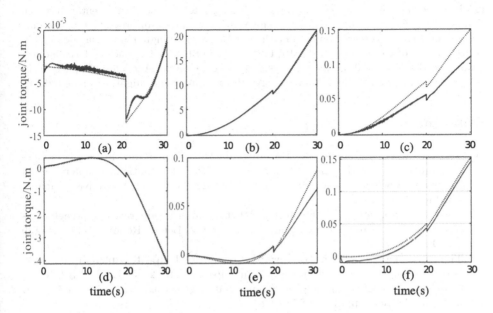

Fig. 5. Desired torque input (black dotted line) and real output (blue solid line) on
arm 3. (Color figure online)

5   Conclusion

In this investigation, a cooperative transportation by three mobile manipulators is studied. An analysis of the possible instability is given during adjustment
object process. The effects of each mobile manipulator is analysed during the process of cooperative transportation. Aiming at improving stability of the whole
system, a distributed position-force control scheme is preliminarily proposed.
For the first part, the observed states of position and orientation is devised to
estimate mobile bases. In the second part, arm 1 and arm 2 realize trajectory
tracking control in joint space based on PID method. The third part, which is
also the most important one, the dynamic of arm 3 is used to compute desired
torques, which lays a foundation for it to achieve force control. Then, based on
desired torques and feedback data of joints, position-force controller is proposed
to improve the task tracking performance and operating compliance. Comparative simulation with pure PID methods shows the effectiveness and better performance of the proposed method. It is necessary to consider internal forces of
end-effector and couplings with mobile base for dynamic model in future work.
Furthermore, a higher robust and adaptive controller should be designed, which
aims to eliminate those impacts of internal uncertainty and disturbance.

Acknowledgement. This work is supported by the projects of National Natural Science Foundation of China (No. 61873192; No. 61603277; No. 61733001), the Quick Support Project (No. 61403110321), and Innovative Project No. 20-163-00-TS-009-125-01). Meanwhile, this work is also partially supported by the Fundamental Research Funds for the Central Universities and the Youth 1000 program project. It is also partially sponsored by International Joint Project Between Shanghai of China and Baden-Württemberg of Germany (No. 19510711100) within Shanghai Science and Technology Innovation Plan, as well as the projects supported by China Academy of Space Technology and Launch Vehicle Technology. All these supports are highly appreciated.

References

1. Tallamraju R., Salunkhe D., Rajappa S.: Motion planning for multi-mobile-manipulator payload transport systems. In: 15th International Conference on Automation Science and Engineering (CASE), 22–26 August, Vancouver, Canada, pp. 1469–1474 (2019)
2. Bella, S., Belbachir, A., Belalem, G.: A hybrid air-sea cooperative approach combined with a swarm trajectory planning method. J. Behav. Robot. **11**(1), 118–139 (2020)
3. Thakar, S., Fang, L., Shah, B.: Towards time-optimal trajectory planning for pick-and-transport operation with a mobile manipulator. In: 2018 IEEE 14th International Conference on Automation Science and Engineering(CASE), 20–24 August, Munich, Germany, pp. 981–987 (2018)
4. Menon, A., Cohen, B., Likhachev, M.: Motion planning for smooth pickup of moving objects. In: 2014 IEEE International Conference on Robotics and Automation (ICRA), 20–24 August, Munich, Germany, pp. 981–987 (2018)
5. Jiao, J., Gao, Z.: Transportation by multiple mobile manipulators in unknown environments with obstacles. IEEE Syst. J. **11**(5), 2894–2904 (2017)
6. Nascimento, T.: Multi-robot nonlinear model predictive formation control: the obstacle avoidance problem. Robotica **34**(3), 307–321 (2016)
7. Navarro, B., Cherubini, A., Fonte, A.: A framework for intuitive collaboration with a mobile manipulator. In: 2017 IEEE/RSJ International Conference on Intelligent Robots and Systems (IROS), 24–28 September, Vancouver, Canada, pp. 6293–6298 (2017)
8. Han, H., Park, J.: Robot control near singularity and joint limit using a continuous task transition. Int. J. Adv. Rob. Syst. **10**(10), 346–357 (2013)
9. Caccavale, F., Chiacchio, P., Marino, A., Villani, L.: Six-DOF impedance control of dual-arm cooperative manipulators. IEEE/ASME Trans. Mechatron. **13**(5), 576–586 (2008)
10. Li, Z., Chen, W., Luo, J.: Adaptive compliant force-motion control of coordinated non-holonomic mobile manipulators interacting with unknown non-rigid environments. Neurocomputing **71**(9), 1330–1344 (2016)
11. Kebria, P., Al-Wais, S., Abdi, H.: Kinematic and dynamic modelling of UR5 manipulator. In: 2016 IEEE International Conference on Systems, Man, and Cybernetics (SMC), 9–12 October, Budapest, Hungary, pp. 452–457 (2016)
12. Donner, P., Endo, S., Buss, M.: Physically plausible wrench decomposition for multi-effector object manipulation. IEEE Trans. Rob. **34**(4), 1053–1067 (2018)

Real-Time Sea Cucumber Detection Based on YOLOv4-Tiny and Transfer Learning Using Data Augmentation

Thao NgoGia, Yinghao Li, Daopeng Jin, Jinyuan Guo, Jiang Li, and Qirong Tang[✉]

Laboratory of Robotics and Multibody System, School of Mechanical Engineering,
Tongji University, Shanghai 201804, People's Republic of China

Abstract. You Only Look Once version 4 (YOLOv4) model has an outstanding performance in object detection and recognition. However, the YOLOv4 is too complex, requiring high computing resources with a lot of training data, which is difficult in the underwater environment. YOLOv4-tiny is proposed based on YOLOv4 to simplify the network structure and reduce parameters, which makes it be suitable for developing on mobile and embedded devices. In this paper, in order to implement a real-time cultured sea cucumber detector to the autonomous underwater vehicle (AUV), YOLOv4-tiny and transfer learning are applied. The model has a good performance in speed but the accuracy is unsatisfactory while evaluated on the real-world underwater datasets. Therefore, a data augmentation method based on improved Mosaic data augmentation is further proposed to improve the quality of the training dataset. The proposed method is evaluated on the real-world sea cucumber underwater videos and has a good performance.

Keywords: Real-time underwater detection · YOLOv4-tiny · Transfer learning · Data augmentation

1 Introduction

Inshore aquaculture is an important part of the marine economy of coastal countries in East and Southeast Asia. With the continuous improvement of the local's living standards, the scale of the offshore aquaculture industry is gradually increasing and becoming a new economic growth point. To improve the economic effect of aquaculture, reduce the labor force of the diving fishermen, and protect the ecosystem, usage of automation and intelligent technology in offshore aquaculture is very important. This investigation aims to study the real-time detection algorithm and provide a method for cultured sea cucumber detection based on underwater videos in the offshore aquaculture environment.

The original YOLO (You Only Look Once) was proposed by Joseph Redmon in the Darknet framework, it is the first object detection network to combine the problem of drawing bounding boxes and identifying class labels in one end-to-end differentiable network. After the success of the first three versions of the YOLO network [1–3], the release of YOLOv4 in 2020 has obtained special attention from most people in the field

© Springer Nature Switzerland AG 2021
Y. Tan and Y. Shi (Eds.): ICSI 2021, LNCS 12690, pp. 119–128, 2021.
https://doi.org/10.1007/978-3-030-78811-7_12

[4]. The fourth generation of the YOLO family has improved again in terms of accuracy (mAP) and speed (FPS) - the two metrics are generally used to evaluate an object detection algorithm. YOLOv4-tiny model is proposed based on YOLOv4 to simplify the network structure and reduce parameters, which makes it suitable for developing on mobile and embedded devices [5]. In this study, to detect sea cucumber in an aquaculture environment based on an underwater video collected by underwater robots, a real-time detection method is proposed based on YOLOv4-tiny and transfer learning using data augmentation.

The main contributions of this paper are as follows:

• Provide a YOLOv4-tiny architecture implementation for sea cucumber dataset via transfer learning method. The YOLOv4-tiny model is firstly learned from COCO datasets with 121,408 images, 883,331 object annotations, and 80 classes, and the transfer learning method is used to transfer the learned feature map to training on our sea cucumber dataset.
• When using deep learning algorithms to detect an underwater object, the biggest problem is the low quality of datasets. To solve this problem, a data augmentation method is proposed based on improving Mosaic data augmentation and basic data augmentation methods. The proposed method is evaluated on the real-world cultured sea cucumber underwater dataset, which validates the effectiveness of the proposed methods for poor data environment detection.

2   Related Works

Convolutional Neural Network (CNN) is widely used for object detection and recognition in many fields and has achieved marvelous accomplishments. In the underwater environment, with the development of underwater vehicles (ROVs, AUVs, Gliders), the computer vision methods based on the principles of CNN have become increasingly important.

Object detection algorithms based on the convolutional neural network are mainly divided into two categories: one is target detection algorithm based on proposed region, i.e., two-stage object detector, and the other is object detection algorithm based on regression, i.e., single-stage object detector. The two-stage object detector firstly proposes various regions from the input image, and then classifies and regresses the proposed regions, so as to realize object detection. The single-stage object detector omits the step of region generation and directly integrates the process of feature extraction, object classification, and position regression into a convolutional neural network thus simplified the process into an end-to-end regression problem.

• one-stage detector: YOLO, SSD, RetinaNet, CenterNet.
• two-stage detector: R-CNN, R-FCN, Mask R-CNN, Fast R-CNN, Faster R-CNN.

In contrast, the YOLO model is the most reasonable system for real-time detection and classification of various objects [6]. Many underwater object detection methods are proposed based on YOLO architecture. Wu et al. [7] used YOLO architecture to construct an underwater feature extractor of fishes images and used the SGBM method to predict the depth map, which is used to estimate the length and width of the fishes. By accumulating the object classification results from the past frames to the current frame, Park and Kang [8] propose a method based on the YOLO algorithm to accurately classify objects and count their number in sequential video images with a high classification probability. Tansel Akgül et al. [9] using YOLO-V2, YOLO-V3, YOLO-V3 Tiny, and MobileNet-SSD networks to compare the effectiveness of these popular object detection algorithms in detecting underwater objects tasks. Wang et al. [10] proposed an object detection method based on improved YOLO network and transfer learning for the detection of different types of fish. Wu et al. [11] built a deep learning model based on YOLO v4 architectures for underwater trash detection. The training images applied several filters for noise reduction and image enhancement to improve the accuracy. The above results showed a competitive real-time performance of the detection model based on YOLO in the underwater environment.

Sufficient training data is very important for training a deep learning-based object detection model. However, with today's technology, collecting sufficient underwater data is still a difficult task. Data augmentation is a very important way to take full advantage of a small dataset that we have collected. The basic data augmentations like rotation, flipping, cropping, brightness, contrast, saturation, and hue are simple methods but very useful and have widely applied to deep learning model training, in which the size of the dataset is not sufficient. Cui et al. [12] used common data augmentation to provide more learning samples from the Gulf of Mexico sea to training a convolutional neural network based on fish detection. In order to reduce the recognition error of plant leaf, Zhang et al. [13] adopted common data augmentation methods (translation, scaling, sharpen, rotation) to suppress the over-fitting degree. Huang et al. [14] proposed three data augmentation methods to increase the robustness of Farter R-CNN to marine turbulence variations, shooting angle variations, and uneven illumination variations. Yun et al. [15] proposed a CutMix augmentation strategy: patches are cut and pasted among training images where the ground truth labels are also mixed proportionally to the area of the patches. This augmentation method has a good performance on CIFAR and ImageNet classification tasks, as well as on the ImageNet weakly-supervised localization task. When the YOLOv4 was released in early 2020, the Mosaic augmentation strategy was proposed, which combines 4 training images into one for training (instead of 2 in CutMix), allow detection of objects outside their normal context.

In this study, YOLOv4-tiny and transfer learning is used for real-time cultured sea cucumber detection. In addition, based on improved Mosaic augmentation we proposed an underwater sea cucumber dataset augmentation method and evaluated the proposed method on the real-world underwater dataset.

3   Real-Time Sea Cucumber Detection Method

3.1   Method Overview

The overall framework is shown in Fig. 1.

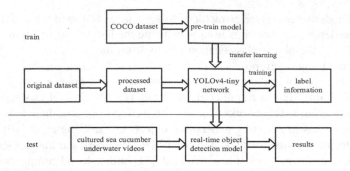

Fig. 1. Real-time sea cucumber detection method framework.

3.2 YOLOv4-Tiny Introduction

The YOLOv4 network is too complex, requiring high computing power. However, the embedded processing system carried by the underwater robot has limited computing ability. To implement a real-time underwater object detector based on an autonomous underwater vehicle (AUV), the YOLOv4-tiny model is considered to be used. YOLOv4-tiny is proposed based on YOLOv4 to simple the network structure and reduce parameters, which makes it suitable for developing on mobile and embedded devices. The speed of object detection for YOLOv4-tiny can reach 371 frames per second (FPS) using 1080Ti GPU with the accuracy that meets the demand of the real application.

Compared to the YOLOv4 network, the size of the YOLOv4-tiny model is much smaller. The YOLOv4-tiny model uses the CSPDarknet53-tiny network instead of the CSPDarknet53 network. The numbers of convolutional layers are compressed, but high accuracy is still guaranteed. In the YOLOv4-tiny network, the LeakyReLU function is used as an activation function instead of the Mish activation function that was used in YOLO4, and in order to increase object detection speed, YOLOv4-tiny used two different scales feature maps that are 13*13 and 26*26 to predict the detection results.

3.3 Training YOLOv4-Tiny with Transfer Learning

Transfer learning is the improvement of learning in a new task through the transfer of knowledge from a related task that has already been learned. Initializing with transferred features can improve generalization performance even after substantial fine-tuning on a new task, which could be a generally useful technique for improving deep neural network performance. Therefore, the transfer learning technique is the best way to custom object detection tasks. COCO dataset is a well-known object detection and recognition dataset with 121,408 images, 883,331 object annotations, and 80 classes. In this study, the YOLOv4-tiny model is firstly trained with the COCO dataset and the resulting model was saved as a pre-train detection model. While training with the sea cucumber dataset, the transfer learning method is used to transfer the learned feature from the pre-train model to train the sea cucumber detection model.

3.4 Data Augmentation

Data is very important for deep learning algorithms, however collecting data in underwater environments is not an easy task. So, the lack of data is the biggest problem while using deep learning algorithms for underwater object detection. Data augmentation is a very effective technique to solve this problem.

Data Augmentations Based on Basic Image Manipulations. The data augmentation methods based on basic image manipulations (basic data augmentations) are generally divided into two categories: geometric augmentation methods and color space augmentation methods.

The geometric augmentation methods contain flipping, scaling, cropping, rotation, translation, etc. They implement some basic geometric transformations to generate new images with different position information. Geometric transformations are very good solutions for positional biases present in the training data.

The color space augmentation methods contain brightness, contrast, saturation, hue, etc. Image data is encoded into three stacked matrices, which represent pixel values for RGB color. Lighting biases are amongst the most frequently occurring challenges to image recognition problems. Similarly to geometric augmentation, the color space augmentation methods implement basic photometric transformations to generate new images with different optical properties, solved the lighting biases problem in training data [17]. Some basic augmentation methods are shown in Fig. 2.

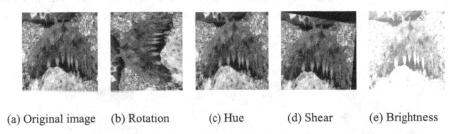

(a) Original image   (b) Rotation   (c) Hue   (d) Shear   (e) Brightness

Fig. 2. Basic data augmentation methods.

CutMix and Mosaic Data Augmentation. In the CutMix method, a portion of an image is cut-and-paste over another image. The ground truth labels are readjusted proportionally to the area of the patches. The cutout area forces the model to learn object classification with different sets of features, which helps avoid overconfidence. Since that area is replaced with another image, the amount of information in the image and the training efficiency will not be impacted significantly also.

Instead of combining two images into one for training in CutMix, the Mosaic data augmentation method combines four training images into one for training. It enhances the detection of objects outside their normal context.

Our Method. Based on the idea of Mosaic data augmentation, in this paper, we propose a data augmentation method, which effectively improves the performance of deep learning detectors in a poor data environment. Our augmentation method includes three steps, which is described as follows:

(1) Mixing basic data augmentation: In this paper, the original dataset is collected from the public internet platforms like Google or Baidu with various types of images including underwater images and on-land images. The underwater environment is complex, light is absorbed and scattered in the water, the brightness of underwater images is uneven and the color of underwater images is distorted. To make the detection model have better performance in the real-world underwater environment, some basic data augmentation methods are used to improve the original training dataset. Firstly, the geometric augmentation methods (flipping, rotation, etc.) are used to generate various object positions. Then, the color space data augmentation methods (hue, brightness, etc.) are used to generate different optical properties for the images obtained from position augmentation methods. By mixing geometric transformations with color space transformations randomly, the original dataset is extended with various image properties. An example for this step is shown in Fig. 3(a) and Fig. 3(b).

(2) Shuffle the dataset: To improve the training process of our method, before using Mosaic augmentation, the dataset obtained from step (1) is randomly shuffled.

(3) Mosaic augmentation: In most of the images in the dataset, the objects have a large scale and locate in the center of the frame. To allow the model to learn how to identify objects at a smaller scale and detect the object located in different portions of the frame, the Mosaic augmentation method is used. By combining four different types of images into one, a new training dataset is generated with enhanced features, as shown in Fig. 3(c).

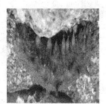 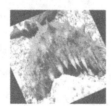 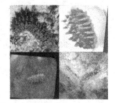

(a) Flipping and hue.   (b) Rotation and brightness.   (c) Mosaic augmentation.

Fig. 3. Mixing basic data augmentation method and Mosaic augmentation method.

4   Experiment

4.1   Experimental Setup

The original sea cucumbers training dataset that contains 234 images is collected from the public internet platforms like Google and Baidu with different sizes of images. A part of the original dataset is shown in Fig. 4(a). Two color-space transformations and three geometric transformations are used in this experiment, which including exposure (randomly between $-45°$ and $+45°$), brightness (randomly between $-45°$ and $+45°$), flipping (horizontal), rotation (randomly between $-45°$ and $+45°$), and shear ($\pm30°$ horizontal and $\pm30°$ vertical). A new dataset is generated by mixing basic data augmentation methods, which contains 1872 images (each original image is generated to 8 new images) with a part of the dataset is shown in Fig. 4(b). After shuffle the dataset obtained from the previous step, the Mosaic augmentation with the equal scales is used to generate another dataset with enhanced features. By combining four images into one, 468 images are generated with a part of the dataset is shown in Fig. 4(c).

The test dataset is collected from a cultured sea cucumber farm, filming by an autonomous underwater vehicle (AUV). The test dataset contains 100 images, which are shown in Fig. 4(d).

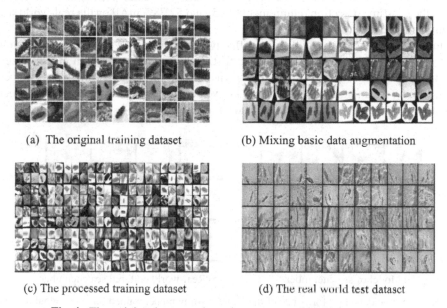

(a) The original training dataset   (b) Mixing basic data augmentation

(c) The processed training dataset   (d) The real world test dataset

Fig. 4. The training dataset and test dataset are used in this experiment.

In this experiment, the pre-train model can be reused to train our datasets on the YOLO model via transfer learning, and the Tensorflow framework is used to implement the training process. However, the YOLO algorithms were written on the Darknet framework. Therefore, before reusing the pre-train model, we need to convert it from Darknet to Tensorflow framework.

The sizes of both the original dataset and the processed dataset are very small, the model can be easily trained on a single GPU or even a single CPU. In this experiment, an NVIDIA RTX 3090 GPU is used for training model with 200 epochs, the batch size set as 8 images, and the learning rate follows the cosine schedule [18] (learning rate initially is 0.0001, it reduces slowly at the start and reduces quickly in halfway and ends up with

0.000001). The software configurations are used in this experiment include Window 10, Python 3.8, PyCharm 2020, and Tensorflow 2.4 with CUDA 11.1 and cuDNN v8.0.5.

4.2 Train Model on Original Dataset

Table 1. Experimental results of based on original dataset object detector on the real world test dataset (IoU threshold = 0.50)

	mAP (mean Average Precision)	FPS (frames per second)
YOLOv4-tiny	52.78%	41.1

As shown in Table 1, the YOLO4-tiny has the FPS (frames per second) value that meets the real-time detection requirement (>25) and 52.78% mAP on the real-world sea cucumber test dataset.

In this experiment, to validate the effectiveness of the real-time object detection model, two test videos are collected. The first one is taken by filming a model of sea cucumber on a table (on-land environment). The second one is taken from a sea cucumber farm, filming by an autonomous underwater vehicle (AUV). The results of the object detection model on two test videos are shown in Fig. 5.

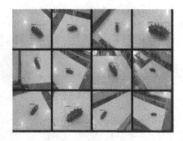
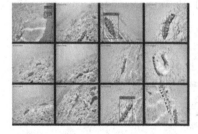

(a) Model of sea cucumber   (b) Real-world cultured sea cucumber

Fig. 5. Real-time detect performance of the YOLOv4-tiny model trained on the original dataset (score threshold = 0.65).

As shown in Fig. 5, in the on-land model of sea cucumber, the model has a pretty good performance. It can detect the model of sea cucumber at a close enough distance, but it is undetectable when the object is too far from the camera. In the real-world underwater sea cucumber video, the detection model has a very poor performance. The object in the underwater test video seems to be unfamiliar with that the model has been learned.

4.3 Train Model on Processed Dataset

As shown in Table 2, the performance of the detection model is improved while training on our processed dataset with the mAP of YOLOv4-tiny reaches 77.25% and maintains a high detection speed (41.6).

Table 2. Experimental results of based on processed dataset object detector on the real world test dataset (IoU threshold = 0.50)

	mAP (mean Average Precision)	FPS (frames per second)
YOLOv4-tiny	77.25%	41.6

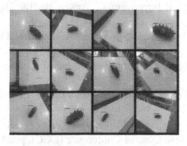 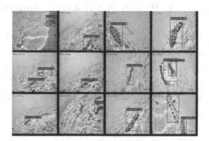

(a) Model of sea cucumber   (b) Real-world cultured sea cucumber

Fig. 6. Real-time detect performance of the YOLOv4-tiny model trained on the processed dataset (score threshold = 0.65).

As shown in Fig. 6, the YOLOv4-tiny train on the processed dataset has an excellent performance in both test videos. In the on-land sea cucumber model test, the object is detected at a farther distance. In the real-world underwater test video, almost of sea cucumber objects are detected.

5 Conclusions

In this study, a real-time sea cucumber detector based on YOLOv4-tiny and transfer learning using the data augmentation method is proposed. In this method, the YOLOv4-tiny model is selected to be the detection model. By using transfer learning, the model is effectively trained on a small custom dataset. To improve the performance of the detection model on a real-world underwater dataset, a data augmentation method is proposed to improve the quality and quantity of the original dataset. The proposed method is evaluated on a real-world cultured sea cucumber underwater dataset and achieved a competitive performance. The work in the future is to diving deeper into the underwater object detection model and data processing to find out a more effective way to improve the ability for the underwater object detection task.

Acknowledgements. This work is supported by the projects of National Natural Science Foundation of China (No.61873192; No.61603277; No.61733001), the Quick Support Project (No.61403110321), and Innovative Project No.20–163-00-TS-009-125-01). Meanwhile, this work is also partially supported by the Fundamental Research Funds for the Central Universities and the "National High Level Talent Plan" project. It is also partially sponsored by International Joint Project Between Shanghai of China and Baden-Württemberg of Germany (No. 19510711100), and project of Shanghai Key Laboratory of Spacecraft Mechanism (18DZ2272200). All these supports are highly appreciated.

References

1. Redmon, J., Divvala, S., Girshick, R., Farhadi, A.: You only look once: unified, real-time object detection. In: 2016 IEEE Conference on Computer Vision and Pattern Recognition (CVPR), pp. 779–788 (2016)
2. Redmon, J., Farhadi, A.: YOLO9000: better, faster, stronger. In: 2017 IEEE Conference on Computer Vision and Pattern Recognition (CVPR), pp. 6517–6525 (2017)
3. Redmon, J., Farhadi, A.: YOLOv3: An Incremental Improvement. ArXiv, abs/1804.02767 (2018)
4. Bochkovskiy, A., Wang, C., Liao, H.: YOLOv4: Optimal Speed and Accuracy of Object Detection. ArXiv, abs/2004.10934 (2020)
5. Jiang, Z., Zhao, L., Li, S., Jia, Y.: Real-time object detection method based on improved YOLOv4-tiny. ArXiv, abs/2011.04244 (2020)
6. Jiao, L., et al.: A Survey of deep learning-based object detection. IEEE Access **7**, 128837–128868 (2019)
7. Wu, H., He, S., Deng, Z., Kou, L., Huang, K., Suo, F., Cao, Z.: Fishery monitoring system with AUV based on YOLO and SGBM, In: 2019 Chinese Control Conference (CCC), pp. 4726–4731. Guangzhou, China (2019)
8. JinHyun, P., Changgu, K.: A study on enhancement of fish recognition using cumulative mean of YOLO network in underwater video images. J. Mar. Sci. Eng. **8**(11), 952 (2020)
9. Akgül, T., Çalik, N., Töreyın, B.: Deep Learning-Based Fish Detection in Turbid Underwater Images. In: 2020 28th Signal Processing and Communications Applications Conference (SIU), Gaziantep, Turkey (2020)
10. Wang, C., Samani, H.: Object detection using transfer learning for underwater robot. In: 2020 International Conference on Advanced Robotics and Intelligent Systems (ARIS), pp. 1–4. Taipei, Taiwan (2020)
11. Wu, C., Shih, P., Chen, L., Wang, C., Samani, H.: Towards underwater sustainability using ROV equipped with deep learning system. In: 2020 International Automatic Control Conference (CACS), pp. 1–5. Hsinchu, Taiwan (2020)
12. Suxia Cui, Y., Zhou, Y.W., Zhai, L.: Fish detection using deep learning. Appl. Comput. Intell. Soft Comput. **2020**, 1–13 (2020)
13. Zhang, C., Zhou, P., Li, C., Liu, L.: A convolutional neural network for leaves recognition using data augmentation. In: 2015 IEEE International Conference on Computer and Information Technology; Ubiquitous Computing and Communications; Dependable, Autonomic and Secure Computing; Pervasive Intelligence and Computing, pp. 2143–2150. Liverpool, UK (2015)
14. Huang, H., Zhou, H., Yang, X., Zhang, L., Qi, L., Zang, A.: Faster R-CNN for marine organisms detection and recognition using data augmentation. Neurocomputing **337**, 372–384 (2019)
15. Yun, S., Han, D., Chun, S., Oh, S., Yoo, Y., Choe, J.: CutMix: regularization strategy to train strong classifiers with localizable features. In: 2019 IEEE/CVF International Conference on Computer Vision (ICCV), pp. 6022–6031. Seoul, South Korea (2019)
16. Shorten, C., Khoshgoftaar, T.M.: A survey on image data augmentation for deep learning. J. Big Data **6**(1), 1–48 (2019)
17. Loshchilov, I., Hutter, F.: SGDR: Stochastic Gradient Descent with Warm Restarts. arXiv: Learning (2017)

Toward Swarm Robots Tracking: A Constrained Gaussian Condensation Filter Method

Shihong Duan[1,2], Hang Wu[1,2], Cheng Xu[1,2,3(✉)], and Jiawang Wan[1,2]

[1] School of Computer and Communication Engineering, University of Science and Technology Beijing, Beijing, China
xucheng@ustb.edu.cn
[2] Shunde Graduate School, University of Science and Technology Beijing, Foshan, China
[3] Beijing Key Laboratory of Knowledge Engineering for Materials Science, Beijing, China

Abstract. Real-time high-precision navigation has a wide range of applications in scenarios. In practice, the measurement models are often non-linear, and sequential Bayesian filters, such as Kalman and particle filter, suffer from the problem of accumulative errors, which cannot provide long-time high-precision services for localization. To solve the problem of arbitrary noise distribution, this paper proposes a Gaussian condensation filter to achieve high-precision localization in a non-Gaussian noise environment. To this end, we proposed an error-ellipse re-sampling-based Gaussian condensation (EER-GCF) filter, which establishes error-ellipses with different confidence probabilities and implements a re-sampling algorithm based on the sampling points' geometrical positions. Furthermore, a cooperative Gaussian condensation filter based on error-ellipse re-sampling (CEER-GCF) is proposed to enhance information fusion in the swarm robots network. This study accomplishes swarm robots tracking based on spatial-temporal constraints to enhance tracking accuracy. Experiment results show that the accuracy of EER-GCF reaches 0.80 m, while CEER-GCF achieves a localization accuracy of 0.27 m.

Keywords: Error-ellipse re-sampling · Swarm robots tracking · Spatial-temporal constraints

1 Introduction

Nowadays real-time and high-accuracy localization has been considered for many civil and military applications. In many scenarios, fusion and cooperative methods are sufficient to provide qualified accuracy positioning support for general requirements [15,20]. Zihajehzadeh et al. [20] combined the characteristics of instantaneous high-precision measurement of IMU and accumulated-error-free

© Springer Nature Switzerland AG 2021
Y. Tan and Y. Shi (Eds.): ICSI 2021, LNCS 12690, pp. 129–136, 2021.
https://doi.org/10.1007/978-3-030-78811-7_13

of TOA. However, external beacons are still required, which is not suitable for large area tracking. Xu et al. [15] also provided a reliable implementation with IMU/TOA fusion for human motion tracking. They mounted several sensing nodes onto human joints, to realize long-term and large-distance requirements. Nevertheless, it is difficult to generalize to generic swarm robots tracking applications due to its strict model requirements. Filtering methods are also widely considered to solve cooperative fusion problems. Kalman-like filters have been widely applied for many important problems. Kalman Filter [16] is the most widely adopted Bayesian estimator to minimize the variance of the estimation error. However, it strictly requires the system model to be linear and assumes the noise to be Gaussian white noise. Extended Kalman filter (EKF) [3] is considered with linearizing the nonlinear state model. However, due to the selection of linearization points and the abandonment of higher-order terms, there will be an inevitable linearization error. To solve this, You et al. [18] proposed an unscented Kalman filter (UKF) to realize UWB and IMU fusion in indoor localization of quad-rotors. It can approximate the posterior distribution based on sampling points, but the non-Gaussian noise problem remains unsolved.

Most of the existing IMU/TOA fusion literature mainly considered the ambient noise as Gaussian, such as [11,13,17]. However, in certain conditions such as suburban and urban environments, due to man-made environmental factors, different sensor measurements often show various distribution characteristics, most of which are usually non-Gaussian. TOA distance ranging is easily influenced by the multi-path and non-line of sight (NLOS) factors. Typically, ranging errors can be modeled as a Gaussian distribution in line of sight (LOS) scenarios [14]. In NLOS scenarios, ranging errors can be modeled as Gaussian distribution [17], log-normal [1], or other non-Gaussian distributions [10,12]. Besides, IMU measurement noise is often non-Gaussian and random with significant impulse characteristics [4]. α-stable distribution [8,10] and Student-t distribution [7,19] are often considered in noise modeling of inertial sensors, and it is generally believed that IMU's noise is non-Gaussian and exhibits different characteristics in various environments.

Therefore, the Gaussian assumption to some extent does not conform to the real-world noise situation. There are still challenges in realizing a fusion positioning solution for arbitrary noise distribution. Wang et al. [11] proposed the particle filter (PF) based on Monte Carlo sampling, which uses the average value of a set of weighted particles to estimate the mean and covariance of the state, and approximates the posterior distribution in the region containing the significance probability. However, in the process of re-sampling, it faces the problem of particle degradation and depletion [5]. Besides, for high-dimensional problems, high complexity is usually inevitable [9].

In this perspective, A Gaussian condensation filter method is proposed to effectively handle the arbitrary noise distribution, aiming at the non-Gaussian noise problem of swarm robots tracking. To conquer the cumulative error problem of swarm robots tracking, an error-ellipse re-sampling-based Gaussian condensation filter is proposed. Furthermore, a spatial-constrained Gaussian condensation filter is proposed to effectively improve the information fusion in swarm robots network.

2   Cooperative Gaussian Condensation Filter

In this section, we first detail the Gaussian condensation filter to handle non-Gaussian noise under non-ideal conditions. On this basis, we optimize the sampling points with the use of its geometric position and fuse spatial distance measurements between mobile robots. Then, we introduce an error-ellipse re-sampling-based Gaussian condensation filter algorithm to accomplish the temporal estimation. After that, a cooperative Gaussian condensation filter is established by considering spatial-temporal constraints.

2.1   Gaussian Condensation Filter Based on Error-Ellipse Re-Sampling

GCF [6] recursively generates the *posteriori* probability density function. The key ideas of GCF are illustrated:

1) Prediction. It is assumed that the state transition process obeys the first-order Markov model, namely $p(X_k|X_{1:k-1}) = p(X_k|X_{k-1})$. At time k, *a priori* probability distribution $p(X_k|Z_{1:k-1})$ can be calculated by integration of the product of $p(X_k|X_{k-1})$ with $p(X_{k-1}|Z_{1:k-1})$. However, the integral is extremely complicated in non-Gaussian systems. It can only be efficiently solved when the involved functions are Gaussian or sums of deltas, which are the intrinsic properties of Kalman-like and particle filters. Moreover, according to the central limit theorem, any statistical distribution could be approximated by a mixture of Gaussian, whose number of components is much smaller than that of using a mixture of deltas [11]. In this study, we consider the state equation is linear but the *posteriori* distribution is the Gaussian mixture model, namely $\hat{p}(X_{k-1}|Z_{1:k-1}) = \sum_{i=1}^{m} \alpha_i \mathcal{N}(X_{k-1}; \mu_{k-1|k-1}^{(i)}, Q_{k-1|k-1}^{(i)})$. Then the prediction step is expressed as:

$$\tilde{p}(X_k|Z_{1:k-1}) = \sum_{i=1}^{m} \alpha_i \mathcal{N}(X_k; \mu_{k|k-1}^{(i)}, Q_{k|k-1}^{(i)}) \tag{1}$$

where $\mu_{k|k-1}^{(i)} = F_k \mu_{k-1|k-1}^{(i)}$, $Q_{k|k-1}^{(i)} = F_k Q_{k-1|k-1}^{(i)} F_k^T + A$. m is the number of Gaussian kernels and A is the covariance matrix of noise σ_k.

2) Update. The a *posteriori* probability density function is updated by measurements Z_k at time k. Based on Eq. (1), the update step is expressed as:

$$\tilde{p}(X_k|Z_{1:k}) \propto \sum_{i=1}^{m} \alpha_i \mathcal{N}(X_k; \mu_{k|k-1}^{(i)}, Q_{k|k-1}^{(i)}) p(Z_k|X_k) \tag{2}$$

3) Gaussian Condensation. For general nonlinear/non-Gaussian filters, the number of sufficient statistics characterizing the true *posteriori* distribution increases without bound [2]. To avoid this situation, we intend to obtain a closed-form solution with the resort to approximate the *posteriori* distribution into a Gaussian mixture model. The following theorem could be used.

Let $p(X)$ be the probability density function of a random vector $X \epsilon R^n$ and $\lambda = (\alpha_1, \cdots, \alpha_m, \mu_1, \cdots, \mu_m, \Sigma_1, \cdots, \Sigma_m)$ be the parameters characterizing a mixture of m Gaussian distributions, namely $q(X; \lambda) = \sum_{i=1}^m \alpha_i \mathcal{N}(X; \mu_i, \Sigma_i)$. If $\Phi(\lambda)$ is the KL-divergence between $p(X)$ and $q(X; \lambda)$, namely

$$\Phi(\lambda) = D_{KL}(p(X), q(X; \lambda)) = E_p\{log\frac{p}{q}\} \tag{3}$$

then $\lambda^* = (\alpha_1^*, \cdots, \alpha_m^*, \mu_1^*, \cdots, \mu_m^*, \Sigma_1^*, \cdots, \Sigma_m^*)$ is a stationary point of Φ, where

$$q_i(X; \lambda) = \alpha_i \mathcal{N}(X; \mu_i, \Sigma_i), \qquad \alpha_i = E_p\{\frac{q_i(X; \lambda)}{q(X; \lambda)}\} \tag{4}$$

$$\mu_i = \frac{E_p\{\frac{q_i(X;\lambda)}{q(X;\lambda)}X\}}{E_p\{\frac{q_i(X;\lambda)}{q(X;\lambda)}}, \qquad \Sigma_i = \frac{E_p\{\frac{q_i(X;\lambda)}{q(X;\lambda)}(X-\mu_i)(X-\mu_i)^T\}}{E_p\{\frac{q_i(X;\lambda)}{q(X;\lambda)}} \tag{5}$$

and $E_p(\cdot)$ indicates the expectation over random vector p.

Based on the Bayesian recursive inference criterion, we can conclude that if the *priori* estimation is biased, the subsequent state estimation would also be affected. To suppress the impact of previous estimation errors, error-ellipse re-sampling performs replicating, retaining, and discarding on sampling points at different levels [11]. With this constrained re-sampling strategy, the *posteriori* probability density $\tilde{p}$ could obtained by recursively *prediction* and *update*. Considering that the sufficient statistic $\tilde{p}$ may infinitely increase in temporal series, Gaussian condensation theory is used to approximate $\tilde{p}$ as Gaussian mixture distribution $\hat{p}$. Finally, the state corresponding to the maximum of $\hat{p}$ is the expected estimation at the current moment.

2.2 Cooperative Gaussian Condensation Filter Based on Error-Ellipse Re-sampling

$\{P_{k,1}, P_{k,2}, \cdots, P_{k,M}\}$ works as the prior knowledge of further cooperative Bayesian optimization. Then, we define the joint state of ith robot and jth robot at time k as $r_k = [P_{k,i}, P_{k,j}]^T$. We introduce a new state vector $z = Tr \in \epsilon R^4$, where r represents the joint state vector of the robot, and z is a Gaussian distribution with mean u_z and covariance C_z, i.e., $z \sim \mathcal{N}(u_z, C_z)$. Therein, $u_z = Tu_r$ and $C_z = TC_rT^T$ could be obtained. After adding the distance constraint, we obtain new constraint information $c: ||\rho T^{-1}z|| = ||z_1|| \leq S$. Therefore, we only need to calculate the integral about $p(z_1|c)$. Then, the *posteriori* mean $u_{z_1|c}$ and covariance $C_{z_1|c}$ could be got by affine transformation without calculating $p(z|c)$ directly.

In order to avoid calculating complex numerical integration, we use a convex combination to approximate the conditional mean and covariance:

$$\hat{u}_{z_1|c} \simeq \sum_{i=0}^{2n} w^{(i)} z_1^{(i)}, \qquad \hat{D}_{z_1|c} \simeq \sum_{i=0}^{2n} w^{(i)} z_1^{(i)} \left(z_1^{(i)}\right)^T \tag{6}$$

where $z_1^{(i)}$ and $w^{(i)}$ denote the sampling points and weights respectively, and n is the dimensions of the state. When the probability mass β of the sampling point $z_1^{(i)}$ is within the constraint range, the approximate value remains unchanged. Otherwise, re-sample the points to ensure that the approximated average value falls within the convex boundary, thereby reducing the dispersion. Therefore, parameter β determines whether the sampling point is valid. Use the following equation to select sampling points:

$$h^{(i)} = \begin{cases} u_{z_1} & i = 0 \\ u_{z_1} + s_\beta^{1/2}[C_{z_1}^{1/2}]_i, & i = 1, \ldots, n \\ u_{z_1} - s_\beta^{1/2}[C_{z_1}^{1/2}]_i, & i = n+1, \ldots, 2n \end{cases} \tag{7}$$

where n indicates the dimension of position variable, and S_β is the confidence scale that satisfies $\Pr(s \le s_\beta) = \beta$ and $s = (z_1 - u_{z_1})^{\mathrm{T}} C_{z_1}^{-1} (z_1 - u_{z_1})$.

The deterministic sampling method will select $2n+1$ sampling points, and the sampling points that do not satisfy the constraint conditions will be orthogonally projected to the constraint boundary. The following equation is used to sample:

$$z_1^{(i)} = \begin{cases} h^{(i)} & \text{if } \left\| h^{(i)} \right\| \le S_k \\ \frac{S_k}{\|h^{(i)}\|} h^{(i)} & \text{others} \end{cases} \quad i = 0, \ldots, 2n \tag{8}$$

The weight of the sampling point is updated as

$$w^i = \begin{cases} 1 - \frac{n}{s_\beta}, & i = 0 \\ \frac{1}{2s_\beta}, & i = 1, \ldots, 2n \end{cases} \tag{9}$$

The *posteriori* mean and covariance can be obtained by the weighted average of the re-sampled sampling points $\{z_1^{(i)}\}_{i=0}^{2n}$.

Gaussian Condensation Filter based on Error-Ellipse Re-Sampling (EER GCF) firstly makes a rough estimation of the robot state according to the center point of a sampling set. Then, two error-ellipses with different scales of 3σ and σ are established, and we achieve a re-sampling algorithm by hierarchical screening. After obtaining the mutual information between swarm robots, the spatial distance constraint c is established. $\hat{u}_{r|c}$, which is closer to the real position, is obtained based on constrained Bayesian optimization. Therefore, the position estimation $\hat{u}_{r|c}$ optimized by the spatial constraint is considered to update the center point (x_p, y_p). Furthermore, the filter estimation at the next moment will obtain the spatial information gain of the previous moment. It can achieve cooperative localization based on the fusion of both temporal and spatial information.

3 Experimental Results

We carried out the location tracking experiment by MATLAB. The simulation runs on a PC with Windows 10 operating system, Intel 4-core i5 CPU, and 16 GB memory. In the experiment, the measurement noise obeys α-stable distribution.

Our proposed swarm robots tracking algorithm, i.e., Cooperative Gaussian Condensation Filter based on Error-Ellipse Re-Sampling (CEER-GCF), integrates spatial distance information to obtain high precision. In order to verify its effectiveness, the random walking experiments were repeated 100 times considering CPF [11], CGCF, and CEER-GCF. Furthermore, the statistical positioning errors of all-mentioned algorithms are compared with PCRLB under the same noise condition. The results are shown in Fig. 1, where the following conclusions could be drawn:

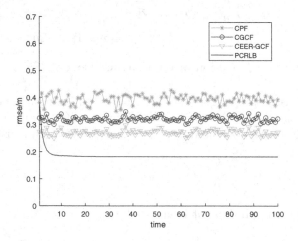

Fig. 1. RMSE of different algorithms in swarm robots tracking.

1. The CRMSE curves of all-mentioned cooperative algorithms are stable in general, which shows that the cooperative method could effectively fuse the information of multiple robots, namely cooperative algorithms have higher stability.
2. The positioning accuracy of CPF reaches 0.39 m, while that of CGCF is 0.32 m, and that of CEER-GCF reaches 0.27 m. CEER-GCF has higher accuracy, and the CRMSE curve of CEER-GCF is much closer to the cooperative PCRLB. It verifies the effectiveness of CEER-GCF in cooperative tracking.

Figure 2 shows how the number of swarm robots impacts the performance of different cooperative tracking algorithms, from which we can conclude that:

1. With the increase of the number of robots, the positioning error curve is smooth and CEER-GCF is of more accuracy. Therefore, when the larger number of robots is, the algorithm described in this paper is more effective, which is quite suitable for applications with large-scale deployment.
2. With the increase of the number of robots, the execution time of all algorithms gradually grows and that of CEER-GCF is slightly higher than other algorithms. However, it can satisfy the real-time requirements of general systems.

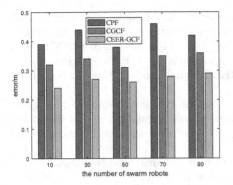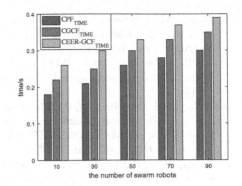

Fig. 2. The influence of the number of swarm robots on the performance of the algorithm

4 Conclusion

In this paper, EER-GCF is proposed to solve the problem of non-Gaussian noise in robot tracking. To suppress the cumulative errors of inertial robot tracking, CEER-GCF is proposed. We take the information gain in temporal series as the prior knowledge of spatial cooperative optimization and establish the distance constraint between swarm robots. Then, we obtain the state optimization based on spatial distance constraints. Finally, we realize swarm robots tracking of spatial-temporal fusion. Results verify that the CEER-GCF algorithm can achieve high-precision positioning in harsh environments.

Acknowledgment. This work is supported in part by China National Postdoctoral Program for Innovative Talents under Grant BX20190033, in part by Guangdong Basic and Applied Basic Research Foundation under Grant 2019A1515110325, in part by Project funded by China Postdoctoral Science Foundation under Grant 2020M670135, in part by Postdoctor Research Foundation of Shunde Graduate School of University of Science and Technology Beijing under Grant 2020BH001, and in part by the Fundamental Research Funds for the Central Universities under Grant 06500127.

References

1. Alsindi, N.A., Alavi, B., Pahlavan, K.: Measurement and modeling of ultrawideband toa-based ranging in indoor multipath environments. IEEE Trans. Veh. Technol. **58**, 1046–1058 (2009). https://doi.org/10.1109/TVT.2008.926071
2. Anderson, K., Iltis, R.: A distributed bearings-only tracking algorithm using reduced sufficient statistics. IEEE Trans. Aerosp. Electron. Syst. **32**, 339–349 (1996). https://doi.org/10.1109/7.481273
3. Baghdadi, A., Cavuoto, L.A., C.L.J.: Hip and trunk kinematics estimation in gait through Kalman filter using IMU data at the ankle. IEEE Sens. J. **18**(10), 4253–4260 (2018)

4. Hoang, M.L., Iacono, S.D., Paciello, V., Pietrosanto, A.: Measurement optimization for orientation tracking based on no motion no integration technique. IEEE Trans. Instrum. Meas. **70**, 1–10 (2021). https://doi.org/10.1109/TIM.2020.3035571

5. Li, T., Bolic, M., Djuric, P.M.: Resampling methods for particle filtering: classification, implementation, and strategies. IEEE Signal Process. Mag. **32**(3), 70–86 (2015)

6. Mehryar, S., Malekzadeh, P., Mazuelas, S., Spachos, P., Plataniotis, K.N., Mohammadi, A.: Belief condensation filtering for RSSI-based state estimation in indoor localization, pp. 8385–8389. IEEE, Brighton, United Kingdom (2019). https://doi.org/10.1109/ICASSP.2019.8683560

7. Salois, M.J., Balcombe, K.G.: A generalized bayesian instrumental variable approach under student t-distributed errors with application. Manchester School **83**(5), 499–522 (2015)

8. Song, A., Tong, Z., Qiu, T.: A new correntropy based tde method under α-stable distribution noise environment. J. Electron. (China) **28**, 22–26 (2011)

9. Villacrés, J.L.C., Zhao, Z., Braun, T., Li, Z.: A particle filter-based reinforcement learning approach for reliable wireless indoor positioning. IEEE J. Sel. Areas Commun. **37**, 2457–2473 (2019). https://doi.org/10.1109/JSAC.2019.2933886

10. Wang, P., He, J., Xu, L., Wu, Y., Xu, C., Zhang, X.: Characteristic modeling of toa ranging error in rotating anchor-based relative positioning. IEEE Sens. J. **17**, 7945–7953 (2017). https://doi.org/10.1109/JSEN.2017.2757700

11. Wang, X., Xu, C., Duan, D., Wan, J.: Error-ellipse-resampling-based particle filtering algorithm for target tracking. IEEE Sensors J. **20**(10), 5389–5397 (2020)

12. Xu, C., Chai, D., He, J., Zhang, X., Duan, S.: Innohar: a deep neural network for complex human activity recognition. IEEE Access, pp. 1–1 (2019)

13. Xu, C., He, J., Li, Y., Zhang, X., Zhou, X., Duan, S.: Optimal estimation and fundamental limits for target localization using IMU/TOA fusion method. IEEE Access **7**, 28124–28136 (2019). https://doi.org/10.1109/ACCESS.2019.2902127

14. Xu, C., He, J., Zhang, X., Tseng, P.H., Duan, S.: Toward near-ground localization: modeling and applications for TOA ranging error. IEEE Trans. Antennas Propag. **65**, 5658–5662 (2017). https://doi.org/10.1109/TAP.2017.2742551

15. Xu, C., He, J., Zhang, X., Yao, C., Tseng, P.H.: Geometrical kinematic modeling on human motion using method of multi-sensor fusion. Inf. Fusion **41**, 243–254 (2018). https://doi.org/10.1016/j.inffus.2017.09.014

16. Xu, C., Ji, M., Qi, Y., Zhou, X.: Mcc-ckf: a distance constrained Kalman filter method for indoor toa localization applications. Electronics **8**(5), 478 (2019). https://doi.org/10.3390/electronics8050478

17. Xu, C., Wang, X., Duan, S., Wan, J.: Spatial-temporal constrained particle filter for cooperative target tracking. J. Network Comput. Appl. **176**, 102913 (2021). https://doi.org/10.1016/j.jnca.2020.102913

18. You, W., Li, F., Liao, L., Meili, H.: Data fusion of UWB and IMU based on unscented Kalman filter for indoor localization of quadrotor UAV. IEEE Access **8**, 64971–64981 (2020)

19. Zhu, H., Leung, H., He, Z.: A variational bayesian approach to robust sensor fusion based on student-t distribution. Inf. Sci. **221**, 201–214 (2013)

20. Zihajehzadeh, S., Park, E.J.: A novel biomechanical model-aided IMU/UWB fusion for magnetometer-free lower body motion capture. IEEE Trans. Syst. Man Cybern. Syst. **47**(6), 927–938 (2017)

Adaptive Task Distribution Approach Using Threshold Behavior Tree for Robotic Swarm

Li Ma[1], Weidong Bao[1(✉)], Xiaomin Zhu[1], Meng Wu[1], Yutong Yuan[2], Ji Wang[1], and Hao Chen[3]

[1] College of Systems Engineering, National University of Defense Technology, Changsha 410073, China
{mali10,wdbao,xmzhu,wumeng15,wangji}@nudt.edu.cn
[2] Department of Electronic Engineering, Shantou University, Guangdong, China
[3] Unit 75831 of PLA, Guangdong 510000, China

Abstract. Online task distribution is a typical problem for the cooperation of robotic swarm. However, there exist several challenges such as the distributed system, large-scale swarm and limited communication. Therefore, this paper proposes an adaptive task distribution approach based on the threshold behavior tree, which solves the task distribution problem of the large-scale robotic swarm. Through observation and perception towards outside, the robot obtains information of targets and neighboring robots in the field of view. In the condition of lacking communication, each robot makes its own decision on which kind of tasks to perform according to its threshold behavior tree. Finally, the robotic swarm can achieve the expected task distribution ratio which equals to that of the targets.

Keywords: Robotic swarms · Task distribution · Multi-agent systems · Threshold behavior tree

1 Introduction

With the development of artificial intelligence, big data, 5G communication, the autonomy of robots and the cooperation between robots have been greatly developed. Robotic swarm has been applied to many fields including light show using UAV swarm [1], transportation in warehouse using UGA swarm [2], manufacturing using industrial robots swarm [3]. They replace human to perform boring and dangerous tasks, which has created great commercial value. However, with the increasing types and number of tasks [9,10], the scale of robotic swarm is also increasing. Moreover, The communication network between robots and control center is limited by environment, so the traditional centralized control principles are not able to satisfy task distribution requirements of robotic swarm.

Traditional task distribution method is mainly designed relying on assumptions of high-quality communication. Qi Zemin [4] built a cooperative task allocation model and algorithm for heterogeneous UAV swarm, which assumed that

© Springer Nature Switzerland AG 2021
Y. Tan and Y. Shi (Eds.): ICSI 2021, LNCS 12690, pp. 137–145, 2021.
https://doi.org/10.1007/978-3-030-78811-7_14

robots communicated smoothly and shared the same environmental information; Zhang Ruiwen [5] proposed a behavior mechanism for UAV considering modeling environment information and sharing route information, which assumed the UAVs shared same pheromone diagram of the whole map; Duan Haibin [6] proposed a cluster task allocation method based on the energy model of biological predation, which realized reasonable task allocation through the global information interaction between the leader and follower; Lin Chen [7] realized multi-task allocation using bidding method. These researches have solved the problem of task allocation in the robotic swarm under the condition of global information, but this kind of method needs key robots to fuse global information and many times of exchanges of information between robots, which is difficult to achieve in the communication-limited environment. The challenges mainly include:

- Limited communicaiton. In real outdoor environment, there may exist obstacles and magnetic interference, which leads to limited communication. The traditional task distribution methods are difficult to deal with these difficulties.
- Large-scale swarm. Large scale of tasks needs large quantity of robotic swarm, which brings out frequent cooperation and a lot of conflicts.

Fortunately, biological groups in nature give a good solution to the above problem. For example, the ant colony forages without global communication [8,9]. Each ant decides which kind of food to catch at the current moment relying on the information it perceives, so as to form a reasonable distribution of labor and efficient cooperation of the whole colony. Inspired by this phenomenon, we propose an adaptive task distribution model for robotic swarm, which can reach the expected task distribution without communication. The main contributions of this paper are as follows:

- For limited communication, an adaptive algorithm using threshold method is proposed. It is noted that only the perceived neighbors' task information and the surrounding target information are considered.
- For large-scale swarm, the decision-making and motion mechanism of each robot are optimized. The conflict of task distribution among robots is reduced, so that the accuracy and stability of task distribution are improved.

The rest of this paper is organized as follows: The problem description and the hypothesis are described in Sect. 2. The robotic swarm model and the adaptive approach are proposed in Sect. 3 detailly, while Sect. 4 introduces the experimental setups and analyzes the experimental results. Section 5 concludes this paper and points out the directions of the further work.

2   Problem Statement

The scenario of this paper comes from clearing dangerous targets in a fixed area, such as the accurate attack on the scattered targets in the landing of islands.

As shown in Fig. 1, there are two kinds of targets to be cleared in the environment. The 1-type target is represented by the black fork, while the 2-type target is the red fork. There are two kinds of robots in the environment, the 1-type robot (black circle) is to clear the 1-type targets, while the 2-type robot (red circle) is to clear 2-type targets. At the beginning, the same number of robots and targets are randomly placed in the area. It is worth noting that robots are of the same type, while targets have two types. A part of robotic swarm needs to change their types to achieve the expected task distribution ratio of the robotic swarm, that is, the ratio of the targets.

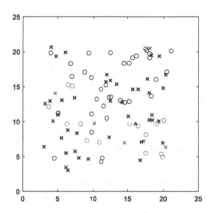

Fig. 1. The scenario of clearing dangerous targets using robotic swarm (Color figure online)

3   Model

Figure 2 shows an overview of the robotic swarm. Each robot consists of sensors, information manager module, decision maker module and motion manager module. The sensors include infrared sensor and panoramic camera. The former obtains obstacle information, and the latter obtains the target information and task information of neighbor robots within the field of vision. These information are stored in the information manager module. The information manager module updates the environment information and its own information obtained by sensors in real time, and outputs it to the decision maker module and the motion module. The decision maker module relies on the threshold behavior tree to decide whether to change its task type and which target to clear. According to the results of the decision maker module, the motion manager module chooses the motion modes of the robot. It is worth noting that the infrared sensing information from the information module should be considered in both modes to avoid collision. Specifically, obstacles consist of boundary and neighbor robots.

We define the set of the robotic swarm as $R = \{r_i | i = 1, 2, \cdots, N_r\}$, where N_r is the number of the robot in the robotic swarm. The target set is $T = \{t_j | j = 1, 2, \cdots, N_t\}$, where N_t is the number of targets in the map.

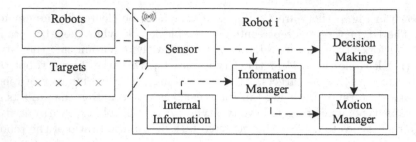

Fig. 2. The overview of the robotic swarm

3.1 Information Manager

Each robot r_i observes the environment in real time and saves the local environment information. In order to estimate the global target proportion, two lists are maintained in the information manager module of each robot: the observed target list, $target_i^D$, and the target clear list of neighboring robots, $target_i^O$. It is related to which targets are in high demand and which are being cleared by neighboring robots. If the proportion of the robot assigned to the target is close to that of the target, the target assignment is desirable. These two lists can only store a limited amount of information, so they store the latest information and delete the most outdated information. Based on these records, each robot can estimate the proportion of global targets.

3.2 Decision Maker

The design of the decision-making module is inspired by foraging ants. Decision mechanism of each robot is modeled as a threshold behavior tree, as shown in Fig. 3. If the robot has a target clearing task in progress, the current task will be kept; Otherwise, the selection probability of different tasks will be calculated. If the selection probability of the 1-type target is greater than that of the 2-type target, the robot will turn to clear the 1-type target; Otherwise, the robot will turn to clear the 2-type target. The calculation process of target selection probability is as follows. Each robot r_i ($i = 1, 2, ..., N$) is assigned the expected target demand $d_{iv}(t)$ and the corresponding response threshold $\alpha_{iv}(t)$ for the v-type target ($v = 1, 2$). Then we can obtain the target-selecting probability function $P_{iv}(t)$, which represents the probability of the robot r_i to select the v-type target.

$$P_{iv}(t) = \frac{\{d_{iv}(t)\}^n}{\{d_{iv}(t)\}^n + \{\alpha_{iv}(t)\}^n}. \tag{1}$$

The rate of v-type target in the list of target demands of robot r_i is as follows:

$$d_{iv}(t) = \frac{\sum_{l=1}^{L} target_{iv}^D(l)}{L}, \tag{2}$$

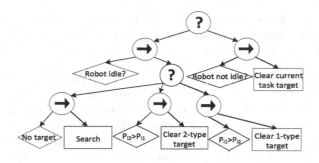

Fig. 3. The decision-making process designed by threshold behavior tree

where L is the length of the list, and t is the timepoint. Specifically, $target_{iv}^{D}(l)$ equals 1 as long as the type of the target observed is v, otherwise, it is 0.

Similarly, we define the rate of v-type target in the list of targets being disposed by neighboring robots.

$$x_{iv}(t) \approx \frac{\sum_{l=1}^{L} target_{iv}^{O}(l)}{L}, \tag{3}$$

where L is the length of the list, t is the timepoint. Specifically, $target_{iv}^{O}(l)$ equals to 1 as long as the type of the target in the list is v, otherwise, it is 0.

The robot r_i updates its response threshold with $d_{iv}(t)$ and $x_{iv}(t)$ as follows:

$$\alpha_{iv}(t+1) = \alpha_{iv}(t) - \lambda\{d_{iv}(t) - x_{iv}(t)\}, \tag{4}$$

where $\lambda > 0$ is a learning rate of the updating process of response threshold. $\alpha_{iv}(t) \in [0, 1]$. If $d_{iv}(t) > x_{iv}(t)$, $\alpha_{iv}(t)$ is decreased to encourage more participation of robots on v type target, and otherwise, $\alpha_{iv}(t)$ will be increased

Take Eq. (2) and Eq. (3) into Eq. (4), and we get Eq. (5) as follows:

$$\alpha_{iv}(t+1) = \alpha_{iv}(t) - \lambda\{\frac{\sum_{l=1}^{L} target_{iv}^{D}(l)}{L} - \frac{\sum_{l=1}^{L} target_{iv}^{O}(l)}{L}\}. \tag{5}$$

Updating the response threshold by Eq. (5), each robot tends to deal with specific targets. This specialization reduces the frequent target selection of robots in robotic swarm, resulting in the expected division of labor and improving the overall working efficiency.

3.3   Motion Manager

Under different decision results, each robot select different motion modes to ensure no interference and avoid task conflict. Therefore, two strategies are proposed. One is to use the nearest distance greedy algorithm in target selection to select the nearest target to clear; in addition, in the process of target search, the minimum density greedy algorithm is used to ensure that the robot will not collide with other robots or have target selection conflict in the process of search. Specifically, each robot has two kinds of motion mode including closing to the target and searching for the target.

- The nearest distance greedy algorithm. If the selection probability of the robot's v-type target is the largest, the robot will choose the nearest v-type target to clean up, that is, move towards the position of the target.
- The minimum density greedy algorithm. The vector from the position of each neighbor to the robot's position is calculated, and the vector sum is obtained. Then the unit vector in this direction is obtained by uniting, that is, the orientation with the minimum density of the neighbor robot.

4   Simulation Experiment

4.1   Setup

We take the parameters of E-puck2 robot as that of the simulation robot. The robot's diameter is 7 cm and its maximum speed is 15.4 cm/s. The robot is equipped with eight infrared sensors around to avoid collision. The detection distance is 6 cm, and the front infrared detection distance is 15 cm. The robot can also obtain the neighbor information through a panoramic camera, which can distinguish the type of target through image recognition. The observation range radius is about 20 cm. The related code website is at [12]. 80 Robots and 80 targets are randomly placed in the environment in the initial phase. The targets cleared by robots will be randomly regenerated and placed in the scene as a replacement. One of the two tasks of clearing 1-type and 2-type targets can be assigned to robots. Each robot keeps exploring to clear the corresponding targets. The length of the two historical queues $target_i^D$ and $target_i^O$ are set to $L = 30$, and the learning rate of the threshold is set to $\lambda = 0.02$. The initial value of the threshold for each kind of target is set to a random value between 0 to 1. Each experiment with different parameters should be run 30 times, and the average data should be counted. Task requirements are expressed as the proportion of current target quantity. All robots are initially assigned to clear black targets, and the initial values of all thresholds are randomized to ensure that each robot is not pre-determined task selection trends.

4.2   Fixed Task Demand

As shown in Fig. 4, the task requirements of black targets are set at 80%, 40% and 10%, while the red targets are set to 90%, 50% and 20% respectively. This means that 64, 32, and 8 robots should be allocated to clear black targets, 72, 40 and 16 robots to clear red targets. The progress of the robot proportions assigned in each task is shown in Fig. 4. Over time, some robots switch to different tasks, and the groups are divided into the same proportion as the expected task requirements. By changing the task of the appropriate number of robots, from black task to red task, the group makes appropriate response to the change

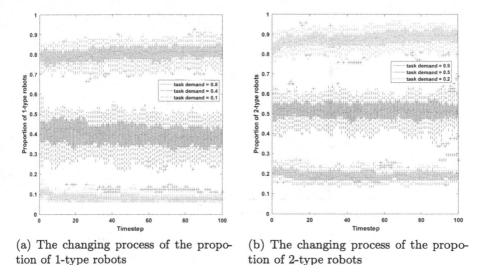

(a) The changing process of the propotion of 1-type robots

(b) The changing process of the propotion of 2-type robots

Fig. 4. The results of the fixed task demand (Color figure online)

of task demand proportion. The proportions of the two tasks initially move suddenly to the desired scale, which will stabilize near the desired scale. However, the proportion remains fluctuating near the desired level, because the individual trend is adjusted using task requirements and task supplied local information, which is estimated by the number of unfinished tasks (the number of targets to be cleared) and the number of adjacent robots performing each task. Even if the sensing range is limited, the proportion of robots assigned to each task will reach a stable level after a period of time. This trend is due to the special characteristics of the threshold method. Professional robots tend to move longer distances by maintaining the same task.

4.3 Compared with Relevant Existing Algorithms

The task requirement of black target is set to 50%. This means that 40 1-type robots should be assigned to clear black targets. The comparison method refers to the literature [13], which is a fixed threshold model. The value of the proportion of the 1-type robots is the average of 20 experimental results. It can be seen from the lines in Fig. 5 that our adaptive method makes the 1-type robot converge to 50% (the expected proportion) faster. In addition, from the real-time variance of the actual proportion and the expected proportion of labor division, we can see that the variance of the two methods show a significant downward trend over time, which proves that they can achieve convergence. But the variance of our adaptive method is always smaller than that of the comparison method, especially before 20 time steps, which is much smaller than that of the comparison method. Therefore, our method has better division performance.

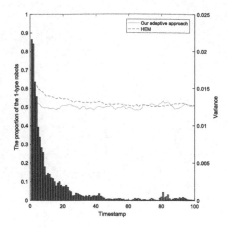

Fig. 5. The compared results of our adaptive approach and HEM [13]

5   Conclusion

This paper proposes an adaptive task distribution approach using threshold behavior tree for robotic swarm. Specifically, the updating rule of response threshold designed in the adaptive task distribution approach of robotic swarm creates professional robots for specific tasks by reducing the task threshold, which can form the desired task division by self-organization without relying on communication and central control. However, the current task environment is relatively simple. In the future, we will study how to realize task division in complex dynamic environment, and implement the approach in the actual robot cluster system.

Acknowledgments. This work is supported by National Natural Science Foundation of China (61872378, 71702186), Science Fund for Distinguished Young Scholars in Hunan Province (2018JJ1032), and Scientific Research Project of National University of Defense Technology (ZK19-03).

References

1. Waibel, M., Keays, B., Augugliaro, F.: Drone shows: Creative potential and best practices. Verity Studios (2017). http://veritystudios.com/whitepaper
2. TMTPOST (2018). China's JD.com releases world-class unmanned warehouse standards. https://www.tmtpost.com/3272511.html
3. Lei, Z., Qixin, C., Baoshun, L.: Cluster industrial robot failure diagnosis method based on outlier excavation. China (2009)
4. Zeyang, Q., Qiang, W., Yingjie, H.: Task assignment Modeling and simulation for cooperative surveillance and strike of multiple unmanned aerial vehicle. Comput. Simul. **32**(9), 142–146 (2015)

5. Ruiwen, Z., Bifeng, S., Yang, P., et al.: Modeling and simulation of UAV swarm self-organized surveillance in complex mission scenarios. Adv. Aeronaut. Sci. Eng. **11**(3), 316–325 (2020)
6. Haibin, D., Daifeng, Z., Yimin, D., et al.: Autonomous task allocation method for UAV cluster based on bio-predator energy model. China, 201911264700.0[P] (2019)
7. Chen, L.: Distributed Algorithm Research for Multi-UAV Task Assignment Problem. University of Electronic Science Technology of China, Chengdu, China (2019)
8. Labella, T.H., Dorigo, M., Deneubourg, J.-L.: Division of labor in a group of robots inspired by ants' foraging behavior. ACM Trans. Auton. Adapt. Syst. **1**(1), 4–25 (2006)
9. Feinerman, O., Pinkoviezky, I., Gelblum, A., et al.: The physics of cooperative transport in groups of ants. Nature Physics (2018)
10. Talal, H.: Gremlins are coming: DARPA enters Phase III of its UAV programme (2018). https://www.army-technology.com/features/gremlins-darpa-uav-programme/
11. Raytheon gets $29m for work on US Navy LOCUST UAV prototype. https://navaltoday.com/2018/06/28/raytheon-wins-contract-for-locus-inp/
12. https://github.com/Downloadmarktown/ATDA
13. Halasz, A., Hsieh, M.A., Berman, S., et al.: Dynamic redistribution of a swarm of robots among multiple sites. In: IEEE International Conference on Intelligent Robots and Systems, pp. 2320–2325 (2007)

Map Fusion Method Based on Image Stitching for Multi-robot SLAM

Qirong Tang$^{(\boxtimes)}$, Kun Zhang, Pengjie Xu, Jingtao Zhang, and Yuanzhe Cui

Laboratory of Robotics and Multibody System, School of Mechanical Engineering,
Tongji University, Shanghai 201804, People's Republic of China

Abstract. Compared with the single-robot SLAM, the SLAM task completed by a multi-robot system in cooperation has the advantages of more accuracy, more efficiency and more robustness. This study focuses on the map fusion problem in the multi-robot SLAM task, which is to fuse the local maps created by multiple independent robots into an integrated map. A multi-robot SLAM map fusion method based on image stitching is therefore proposed. A single robot uses lidar SLAM to build a local environment map and upload it to a central node. The central node then maps each local map from a two-dimensional occupancy grid map to a grayscale image. The SuperPoint network is used to extract the depth features from the grayscale images, and the transformation relationships between the local maps are calculated via the feature matching. The matching topology graph is used to realize the final map fusion. It carries out experimental verification in the indoor environment on three mobile robots, which were developed by our own, and the experiment proved that the method has good real-time performance and robustness. After obtaining the global map, some new robots were placed in the environment, and realized the task of multi-robot target search by using the relocalization function.

Keywords: Multi-robot system · Simultaneous localization and mapping · Map fusion · Image stitching

1 Introduction

Simultaneous localization and mapping (SLAM) technology is the main solution to the key problem in the field of intelligent robot research that how the robot can obtain the map of an unknown environment while realizing its own localization in the map simultaneously, and provide technical support for subsequent tasks.

Due to the difficulties for a single robot to complete some complicated tasks, such as limited movement capability, insufficient computing power, poor anti-interference capability and so on, a multi-robot system is demanded. In the face of a large-scale complex environment, multiple robots can be dispersed into various sections of regions of the environment. A single robot uses SLAM technology to establish a local environment map/model. Then the system fuses these local

© Springer Nature Switzerland AG 2021
Y. Tan and Y. Shi (Eds.): ICSI 2021, LNCS 12690, pp. 146–154, 2021.
https://doi.org/10.1007/978-3-030-78811-7_15

models into a larger-range environment model. Compared with the single-robot SLAM, the SLAM completed by a multi-robot system in collaboration has the advantages of more accuracy, more efficiency and more robustness. The main challenges in implementing a multi-robot SLAM system include bandwidth limitations, map fusion, asynchronous communications, coherent information integration, and data association between robots [1].

Most of the multi-robot SLAM methods are based on the corresponding single-robot SLAM algorithms. Earlier implementations are based on filtering methods, such as Extended Kalman Filter (EKF) [2,3]. These methods inherit some of the shortcomings of filter-based SLAMs. The current mainstream multi-robot SLAM algorithms are based on graph optimization or pose graph methods [4,5].

At present, most of the researches on the multi-robot SLAM problem are still in the simulation stage, where the information sensed and the communication conditions are often set to be completely ideal. This study focuses on the problem of multi-robot SLAM map fusion, proposes a new method based on image stitching [6], and conducts experimental verification in the real scene on the mobile robot platform developed by our own. With the help of ROS data transmission mechanism, our map fusion method does not depend on the specific single-robot SLAM algorithm, as long as the maps generated by the algorithm can be converted to grayscale images. It can process maps from any number of robots within the allowable range of computing power, and allows dynamic addition or removal of robots in the system.

The method closest to this work is proposed by Hörner [7]. We both draw on the principle of image stitching in computer vision. The difference is that this work uses depth features, and includes relocalization function using the global map. In addition, this work has been verified by experiments in the real scene.

The rest of this paper is organized as follows. Section 2 presents the method overview. The map fusion method based on image stitching for multi-robot SLAM is detailed in Sect. 3. The experiments are presented in Sect. 4, and conclusions are drawn in Sect. 5.

2   Method Overview

The proposed method is used to fuse 2D occupancy grid maps independently established by multiple robots, and method pipeline is already shown in Fig. 1.

Assume that there are n robots participating in the mapping task. The map built by a single robot through lidar SLAM is called a local occupancy grid map (short for local map) $Lmap_i$ $(i = 1, 2, ..., n)$, and the fused occupancy grid map is called a global map $\mathcal{M}$. The local environment maps are represented in the form of a common occupancy grid. One establishes a mapping relationship $\mathcal{F}$: $OccupancyGridMap \rightarrow GrayscaleImage$ from occupancy grid map to grayscale image. Obviously, the proposed method also supports other formats of map, as long as the map format can be mapped to a grayscale image.

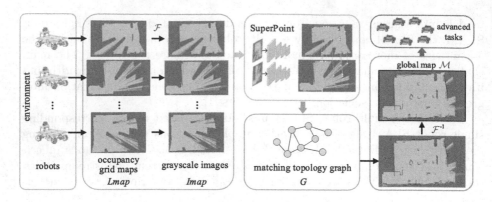

Fig. 1. Method pipeline

After $Lmap_i$ has been converted into grayscale image $Imap_i$ $(i = 1, 2, ..., n)$, one then uses the neural network SuperPoint [8] to extract feature points and calculate descriptors, which can obtain higher image matching accuracy compared with traditional ORB [9] or SIFT [10]. After feature matching, it estimates the coordinate transformation relationships between local maps, that is, solves the homography matrices. The local maps and their matching relationships form a weighted graph, which is called the matching topology graph G. The largest spanning tree Tr is established in the largest connected component H of G, and map fusion is achieved by exploring Tr. Finally, the inverse mapping $\mathcal{F}^{-1}$ needs to be used to convert the global map in the form of grayscale image back into an occupancy grid map $\mathcal{M}$.

The obtained global map can then provide support for the robot to perform other advanced tasks in the area, such as target search and path planning.

3   Multi-robot SLAM Map Fusion

3.1   Estimate the Coordinate Transformations Between Local Maps

In order to fuse the local maps, calculating the transformations between the local occupancy grid maps established by each robot is the necessity. The method follows these steps:

(1) Convert the local grid map $Lmap_i$ $(i = 1, 2, ..., n)$ built by each robot into a grayscale image $Imap_i$ $(i = 1, 2, ..., n)$. The value of each cell is within the range of [0,100]. This value represents the probability of obstacles in the cell. If the probability is unknown, it is represented by -1. The local map in this form is converted to a grayscale image, and if the value is -1, it maps to 255 in the pixel value of the grayscale image, and one can get a standard 8-bit depth grayscale image.

(2) Use the SuperPoint network to extract the feature points and calculate feature descriptors of each local grid map $Imap_i$.

(3) Feature matching is carried out for each pair of local maps afterwards. The brute force matching algorithm is used. If there are a large number of robots, the parallel hierarchical clustering algorithm will be used to speed up the matching.

(4) Solve the coordinate transformation matrix $T_{i,j}$ between each pair of local grid maps $(Imap_i, Imap_j)$. Use the RANSAC [11] method to filter the matched features, use the SVD method to solve the $T_{i,j}$, and calculate the matching confidence $c_{i,j}$ of the corresponding match

$$c_{i,j} = \frac{number\_inliers_{i,j}}{8 + 0.3 * number\_matches_{i,j}}, \tag{1}$$

where $number\_inliers_{i,j}$ is the number of inliers found in the RANSAC method, and $number\_matches_{i,j}$ is the number of matched feature points between each pair of local maps $(Imap_i, Imap_j)$.

(5) Eliminate matches with a confidence less than 1, and form a matching topology graph G. The vertices of the G are the local maps $Imap_i$ built by single robots, and the edges are $T_{i,j}$, $c_{i,j}$, $number\_inliers_{i,j}$ and $number\_matches_{i,j}$.

Each time a robot updates the local map, it will upload the updated incremental map to the central node instead of uploading the entire local map, which can effectively reduce the amount of data transmission. Then, the central node will regenerate a corresponding grayscale image.

3.2   Map Fusion Based on Matching Topology Graph

After the matching topology graph is established, if any of the local maps has strong matching relationships with the others, the coordinate system of this local map will be fixed as the world coordinate system. The coordinate transformations will be performed on the remaining maps according to the results of feature matching. The global map is thus established.

If there are a large number of robots or there are maps that need to be eliminated, such as local maps with large errors, isolated local maps and so on, the graph method will be used for map fusion [7]. It is very common that some local grid maps cannot be successfully matched. In order to cover as large environmental area as possible, the weighted maximum connected component H in the established matching topology graph G is considered, and only the local maps included by H will be fused. This study selects the coordinate system of one of the local maps as the global coordinate system in H. The maximum spanning tree Tr is established in H, and by exploring Tr, the transformations between the local maps and the global coordinate system are finally determined. As for each local map, it can obtain the final transformation result by synthesizing paired transformations along the path. After all the transformations are completed, the map fusion is realized.

3.3 Robot Relocalization

Relocalization with maps built by lidar SLAM is often more difficult than the visual relocalization problem, because the information stored in the map established by lidar SLAM is not rich enough. At present, the relocalization problem using lidar usually uses the loop detection part of Cartographer [12] or Karto-SLAM [13]. In this study, after obtaining the global map $\mathcal{M}$, the newly entering robot k uses its own lidar to build a local grid map $Rmap_k$, and uses the method described in Sect. 3.1 to perform feature matching between the local grid map $Rmap_k$ and the global map $\mathcal{M}$. The coordinate transformation between the local grid map and the global map is estimated, and then the position of the newly entering robot in the global coordinate system is obtained, so as to realize the relocalization and provide support for the robot to realize the tasks such as navigation and target search.

4 Experiments

4.1 Multi-robot Map Fusion Experiment

A real world experiment using three ground omnidirectional mobile robots as shown in Fig. 2(a) is conducted. The experimental area is an indoor environment, as shown in Fig. 2(b). The three mobile robots, each carrying a laser range scanner, are distributed in different areas to measure the environment and perceive environmental information. In the experiment, it uses remote control to make those robots move in a certain part of the field and build local maps. The map established by each robot covers only one part of the entire environment. The fused map on the central node includes the contents of all three local maps.

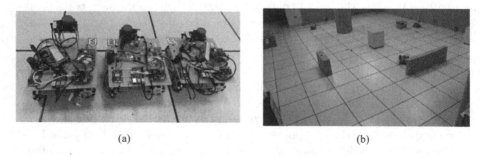

(a)   (b)

Fig. 2. (a) Mobile robot running single-robot SLAM. In order to avoid the lidar interference between each other, the lidars of the three robots are set up in different heights. (b) Experimental area.

Each robot uploads the environmental information obtained to the central node. In this experiment, the central node is only responsible for controlling the movement of the mobile robots and for performing map fusion, and single-robot

SLAM is implemented independently by mobile robots and does not depend on the central node. The central node is a laptop computer with Intel Core i7-8750H and 8G RAM in this case. The communication method between the central node and the mobile robots is WiFi, and our method is run based on ROS. The data format of the local map established by a single robot in ROS is *nav_msgs/OccupancyGrid* message, and the data format of the corresponding incremental map updates is *map_msgs/OccupancyGridUpdate* message. Common SLAM algorithms that support this format rule in ROS include Karto-SLAM [13], Gmapping [14], and so on.

The map fusion process observed in the display interface of the center node is shown in Fig. 3.

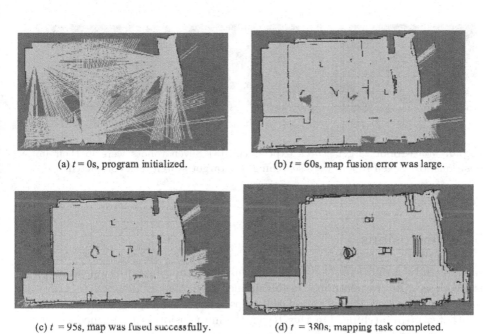

(a) t = 0s, program initialized.

(b) t = 60s, map fusion error was large.

(c) t = 95s, map was fused successfully.

(d) t = 380s, mapping task completed.

Fig. 3. Map fusion in the exploration process of three ground mobile robots.

As shown in Fig. 3, after 380 s of exploration, the three ground mobile robots finally completed the task of building a global map. On the display interface of the central node, the process of localization and mapping by the mobile robots could be observed. Initially, the fusion error between the three local maps was relatively large due to the fact that there were few feature points. After 95 s, the fusion of the mobile robot maps achieved a good effect. After 380 s, each of the three robots covered most of the entire site, the global map contained enough area information, and the task was completed.

4.2   Multi-robot Target Search Experiment

In the same field, 7 new mobile robots are placed, and the configuration of these 7 mobile robots and the lidars they carried are different from the three robots used for mapping. In this experiment, the robots themselves do not have the localization function. It only detects environmental information and uploads it to the central node. The central node uses the global map that has been obtained before to provide localization service for the robots and guide them to perform collaborative search task. Seven mobile robots use the improved PSO algorithm for target search, and the specific implementation method refers to our previous work [15]. The experiment process is shown in Fig. 4.

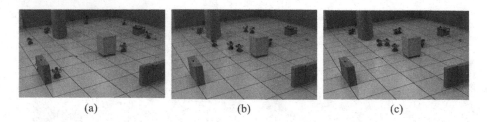

(a)   (b)   (c)

Fig. 4. Multi-robot target search experiment. (a) Seven robots were in their initial positions. (b) Seven robots were searching. (c) Target search task completed.

5   Conclusions

In this study, a multi-robot SLAM map fusion method based on image stitching is proposed. In the method, the occupancy grid map is mapped to grayscale image, and the transformation relationships between maps are estimated through depth feature matching. On this basis, the map fusion is performed using the matching topology map. This method can realize the fusion of multiple local maps consuming fewer computing resources. We have carried out experiments on three mobile robots developed by our own in an indoor environment, and the experiments have proved that the method has good real-time performance and robustness. After obtaining the global map, some new robots are used in the environment and the multi-robot target search task is realized successfully by using the relocalization function.

Future works might include using more robots to conduct experiments in real scenes and to realize autonomous robotic exploration.

Acknowledgements. This work is supported by the projects of National Natural Science Foundation of China (No.61873192; No.61603277; No.61733001), the Quick Support Project (No.61403110321), and Innovative Project (No.20-163-00-TS-009-125-01). Meanwhile, this work is also partially supported by the Fundamental Research

Funds for the Central Universities and the Youth 1000 program project. It is also partially sponsored by International Joint Project Between Shanghai of China and Baden-Württemberg of Germany (No. 19510711100) within Shanghai Science and Technology Innovation Plan, as well as the projects supported by China Academy of Space Technology and Launch Vehicle Technology. All these supports are highly appreciated. The authors also would like to thank Zhongqun Zhang for his helpful suggestions.

References

1. Lázaro, M., Paz, L., Piniés, P., Castellanos, J., Grisetti, G.: Multi-robot SLAM using condensed measurements. In: Proceedings of the 2013 IEEE/RSJ International Conference on Intelligent Robots and Systems, pp. 1069–1076. Tokyo, Japan (2013)
2. Roumeliotis, S., Bekey, G.: Distributed multirobot localization. IEEE Trans. Robot. Autom. **18**(5), 781–795 (2002)
3. Sasaokas, T., Kimoto, I., Kishimoto, Y., Takaba, K., Nakashima, H.: Multi-robot SLAM via information fusion extended Kalman filters. IFAC-PapersOnLine **49**(22), 303–308 (2016)
4. Dellaert, F., Kipp, A., Krauthausen, P.: A multifrontal QR factorization approach to distributed inference applied to multi-robot localization and mapping. In: Proceedings of the 20th National Conference on Artificial Intelligence and the Seventeenth Innovative Applications of Artificial Intelligence Conference, Vol. 3, pp. 1261–1266. Pittsburgh, USA (2005)
5. Chang, H., Lee, C., Hu, Y., Yung-Hsiang, L.: Multi-robot SLAM with topological/metric maps. In: Proceedings of the 2007 IEEE/RSJ International Conference on Intelligent Robots and Systems, pp. 1467–1472. San Diego, USA (2007)
6. Brown, M., Lowe, D.: Automatic panoramic image stitching using invariant features. Int. J. Comput. Vis. **74**(1), 59–73 (2007)
7. Hörner, J.: Map-merging for multi-robot system. Bachelor's thesis, Charles University in Prague (2016)
8. DeTone, D., Malisiewicz, T., Rabinovich, A.: SuperPoint: Self-supervised interest point detection and description. In: Proceedings of the 2018 IEEE/CVF Conference on Computer Vision and Pattern Recognition Workshops, pp. 337–349. Salt Lake City, USA (2018)
9. Rublee, E., Rabaud, V., Konolige, K., Bradski, G.: ORB: An efficient alternative to SIFT or SURF. In: Proceedings of the 2011 International Conference on Computer Vision, pp. 2564–2571. Barcelona, Spain (2011)
10. Lowe, D.: Distinctive image features from scale-invariant keypoints. Int. J. Comput. Vis. **60**(2), 91–110 (2004)
11. Fischler, M., Bolles, R.: Random sample consensus: a paradigm for model fitting with applications to image analysis and automated cartography. Commun. ACM **24**(6), 381–395 (1981)
12. Hess, W., Kohler, D., Rapp, H., Andor, D.: Real-time loop closure in 2D LIDAR SLAM. In: Proceedings of the 2016 IEEE International Conference on Robotics and Automation, pp. 1271–1278. Stockholm, Sweden (2016)
13. Konolige, K., Grisetti, G., Kümmerle, R., Burgard, W., Limketkai, B., Vincent, R.: Efficient sparse pose adjustment for 2D mapping. In: Proceedings of the 2010 IEEE/RSJ International Conference on Intelligent Robots and Systems, pp. 22–29. Taipei, China (2010)

14. Grisetti, G., Stachniss, C., Burgard, W.: Improved techniques for grid mapping with Rao-Blackwellized particle filters. IEEE Trans. Robot. **23**(1), 34–46 (2007)
15. Tang, Q., Yu, F., Xu, Z., Eberhard, P.: Swarm robots search for multiple targets. IEEE Access **8**, 92814–92826 (2020)

Robotic Brain Storm Optimization: A Multi-target Collaborative Searching Paradigm for Swarm Robotics

Jian Yang, Donghui Zhao, Xinhao Xiang, and Yuhui Shi[✉]

Department Computer Science and Engineering, Southern University of Science
and Technology (SUSTech), Shenzhen 518055, China
shiyh@sustech.edu.cn

Abstract. Swarm intelligence optimization algorithms can be adopted
in swarm robotics for target searching tasks in a 2-D or 3-D space by
treating the target signal strength as fitness values. Many current works
in the literature have achieved good performance in single-target search
problems. However, when there are multiple targets in an environment
to be searched, many swarm intelligence-based methods may converge
to specific locations prematurely, making it impossible to explore the
environment further. The Brain Storm Optimization (BSO) algorithm
imitates a group of humans in solving problems collectively. A series of
guided searches can finally obtain a relatively optimal solution for particular optimization problems. Furthermore, with a suitable clustering operation, it has better multi-modal optimization performance, i.e., it can
find multiple optima in the objective space. By matching the members
in a robotic swarm to the individuals in the algorithm under both environments and robots constraints, this paper proposes a BSO-based collaborative searching framework for swarm robotics called Robotic BSO.
The simulation results show that the proposed method can simulate the
BSO's guided search characteristics and has an excellent prospect for
multi-target searching problems for swarm robotics.

Keywords: Swarm robotics · Multi-target searching · Brain Storm
Optimization

1 Introduction

Swarm robotics simulates the emergent behaviors of social insects or animals,
such as bees, ants, birds, fish schools, wolves, and even humans [17]. Many works

This work is partially supported by the Science and Technology Innovation Committee Foundation of Shenzhen under the Grant No. JCYJ20200109141235597 and
ZDSYS201703031748284, National Science Foundation of China under grant number
61761136008, Shenzhen Peacock Plan under Grant No. KQTD2016112514355531, Program for Guangdong Introducing Innovative and Entrepreneurial Teams under grant
number 2017ZT07X386, and Special Funds for the Cultivation of Guangdong College
Students Scientific and Technological Innovation ("Climbing Program" Special Funds,
PDJH2020b0522).

© Springer Nature Switzerland AG 2021
Y. Tan and Y. Shi (Eds.): ICSI 2021, LNCS 12690, pp. 155–167, 2021.
https://doi.org/10.1007/978-3-030-78811-7_16

in the literature have shown that this kind of system can be applied to different aspects varies from space exploration to military inspection, from industrial maintenance to medical solutions [16,18]. Collaborative searching problem is one of the popular topics in this field. It intends to determine the specific target(s) location in a particular region in a collective way. It belongs to the coordinated motion and decision making of multi-robot systems. This problem strongly correlates with many practical applications, such as search and rescue applications (scream search, radiation source location, pollution source location, etc.). The characteristics of this problem are that the target number and location(s) are both unknown. Because swarm robot systems have excellent redundancy based robustness, they have more advantages than single robots or a small group of robots to perform such tasks.

The swarm intelligence optimization algorithms can be applied to target searching tasks of robotic swarms by introducing practical constraints and mapping individuals in an algorithm into robots in a swarm, including the Particle Swarm Optimization (PSO) [19], Bees Algorithm (BA) [6], Artificial Bee Colony (ABC) [1], Ant Colony Optimization (ACO) [4], Bacterial Foraging Optimization (BFO) [13], Glowworm Swarm Optimization (GSO) [9], Firefly Algorithm (FA) [11], and Grey Wolf Optimizer (GWO) [5], etc. However, when multiple targets need to be located in an environment, many current solutions imitating swarm intelligence algorithms will converge to specific positions prematurely and lose the ability to explore the searching space further. The multi-target search ability is a critical issue that needs to be solved in the collaborative searching domain of swarm robotics. What is illustrated in this paper is a new paradigm for collaborative searching of swarm robotics called Robotic Brain Storm Optimization (RBSO). It applies the Brain Storm Optimization (BSO) algorithm [12] as the source of inspiration and aims to solve the premature problem in the multi-target collaborative searching tasks.

The original BSO algorithm imitates the brainstorming process to solve optimization problems heuristically. It can guide the searching process to converge to optimal solutions over iterations. It has been verified successfully in many real-world applications [15]. Individuals in BSO are grouped and diverged in the search space. The search performance could be benefited from this inherent advantage of clustering, making it more suitable for multi-modal optimization problems with multiple peaks. By matching the members in a robotic swarm to the individuals in the algorithm, the BSO has a significant potential to be applied as a new collaboration paradigm for multi-robot systems. Essentially, the multi-target search problem in swarm robotics can be transformed into a multi-modal optimization problem in a two-dimensional or three-dimensional space [20]. The BSO's multi-modal optimization properties make it an excellent candidate to solve the collaborative multi-target searching problems. It worth to mention that there are related applications of swarm robotics that use the SI algorithms as optimization tools. For example, the Brain Storm Robotics (BSR) framework can automatically design the corresponding swarm behavior collectively [14]. Li et al. expressed swarm robotics exploration as an optimization problem and then used the BSO method to solve it [10]. There are three kind of operations in the original BSO algorithm: clustering, new individuals generation, and selection, as shown in Algorithm 1.

Algorithm 1. The BSO Procedure

1: Randomly generate n potential solutions (individuals);
2: Evaluate the generated n solutions;
3: **while** not terminated **do**
4:     **Clustering**: Cluster n individuals into m clusters by a clustering algorithm;
5:     **New individuals generation**: randomly select one or two cluster(s) to generate n new individuals;
6:     **Selection**: The fitness values of the newly generated individuals are compared with the existing individuals with the same index, the better one is kept;

The purpose of clustering in the solution space is to converge the solution to a smaller area [2]. Then, a new individual can be generated based on one or several cluster(s). In the original BSO, a probability value p_{one} is adopted to determine while a new individual will be generated by one or two cluster(s). Generating an individual from a cluster can refine the search area and improve exploitation capabilities. Conversely, individuals generated from two or more clusters may be far away from these clusters. In this case, the exploration ability is enhanced. Furthermore, the BSO uses another predefined probability value p_{center} to determine whether to generate a new solution based on the cluster center(s) or non-cluster center(s). In the one cluster generation case, the new individual from center or normal individual can control the exploitation region. While in several clusters generation case, the normal individuals could increase the population diversity of swarm. The selection strategy in BSO is to keep good solutions in all individuals. The better solution is kept by the selection strategy after each new individual generation, while clustering strategy and generation strategy add new solutions into the swarm to keep the diversity for the whole population.

The rest of the paper is organized as follows: Sect. 2 defines the collaborative multi-target searching problem and the assumptions for this paper. Section 3 introduces the RBSO framework, including operations of grouping, new position generation, task allocation, fitness evaluation, etc. The simulation results with the proposed framework are given in Sect. 4. The conclusion is reached in Sect. 5 with the forecast of our future works.

2   Problem Statement

2.1   Assumptions

Environments and Targets. For the sake of simplicity, we intend to conduct related research in this article in a two-dimensional environment. This work can also be further extended to three-dimensional environments. In this hypothetical two-dimensional environment, some simulated obstacles will be randomly distributed. During the movement, the robot must avoid these obstacles, and the robots cannot collide either. Meanwhile, a number of static targets will be distributed in the same environment outside the obstacles. The targets can broadcast non-directional beacon signals. Futhermore, it can only cover a small area around it, and the signal strength will attenuate with distance, as shown in the following formula:

$$s = \frac{1}{a\sqrt{\pi}} e^{-d/a^2} \tag{1}$$

where s is the signal strength, d is the distance to the center position of the target, a is the attenuation coefficient.

Member Robots. Simple but without generality loss, this article uses a homogenous swarm with the omnidirectional model for member robots. Furthermore, in terms of perception, we assume that every member robot is equipped with a sensor that can detect the target signal strength, i.e., it can measure the target signal strength at the current position. Also, it can perceive other robots and obstacles within a specific range. In terms of positioning, it is assumed that the member robot can obtain its position in the reference coordinate system and knows the boundary of the search area. In terms of communication, it is assumed that information can be shared with all other robots through corresponding information interaction within a time slot. In terms of storage, it is assumed that member robots have storage capacity and can record the target signal strength of the visited location.

Target Handling. For target handling, we assume that when the detected target signal is stronger than a certain threshold, which can be converted into a distance less than ϵ to the target, the target is treated as found. Before the next iteration, the robot will stay to process the target until the next iteration. After the handling, the target will no longer broadcast beacon signals to the environment, and the robot that processes the target will become available again.

2.2   Multi-target Searching Problem

Generally, based on the number of targets in a given search area, collaborative search tasks can be divided into the following two categories: single-target search and multi-target search. If the target is moving, the problem is a dynamic single-target or multi-target search problem correspondingly. This article aims to design a method of controlling a swarm of mobile robots, so that the path of the robots will contain the position of the targets as many as possible, i.e., to locate the multi-target positions. Denote m is the target number in an unknown aera, L_m is the location of target m, ϵ is the allowed tolerance vector, $P_i(t)$ is the position of robot i at time t, the multi-target collaborative searching problem of swarm robotics can be formally expressed by Eq. (2):

$$\max_{P_i} |\{m \in \mathbb{N} | L_m \pm \epsilon \in \sum_{t=0}^{T} \sum_{i=1}^{N} P_i(t)\}|$$

$$s.t. \quad P_i(t) \neq P_j(t) \quad \forall i, j \in N, i \neq j$$
$$P_i(t) \neq P_o(t) \quad \forall i \in N, o \in O$$

(2)

where t and T is the current and maximum searching time respectively, N is the population of the robotic swarm. $| \cdot |$ is the cardinal number of the set. $P_i(t) \neq P_j(t)$ represents the anti-collision between robots, and $P_i(t) \neq P_o(t)$ means the robobts in the swarm need to avoid obstacles in environments, which expressed by set O in Eq. (2). It should be noted that although we have formalized the multi-object search problem of swarm robotics above in an optimization form, it can not be solved directly by optimization mathmatically. The expression contains physical constraints, such as the relationship between the robot's moving speed and its position, collision avoidance, etc., which needs to be achieved in the physical world by robotic techniques.

3 Robotic Brain Storm Optimization

The basic idea of this paper is to map the BSO to Robotic BSO. As shown in Table 1. The target searching environments correspond to the BSO algorithm's solution space. The signal strength is treated as the fitness value at a position, and the swarm members are mapped to the individuals in the algorithm. The decision of the robotic swarm is imitating the iterative process of the BSO algorithm. Furthermore, the operations such as the clustering, new individual generation can be mapped to position grouping, new position generation, correspondingly. As modeled above, the multi-target searching problem is related to a multi-modal optimization problem.

Table 1. Mapping BSO algorithm to RBSO.

BSO	RBSO
N-D solution space	2-D or 3-D searching space
Fitness value	Signal strength
Individuals	Robots
Clustering	Grouping
New individual generation	New position generation
Multi-modal optimization	Multi-target searching

As Fig. 1 shows, the proposed method mainly includes the following steps. First, grouping a swarm of robots into several subgroups according to their fitness values in their current locations, i.e., the clustering procedure in the BSO algorithm. Secondly, according to the subgroups, a series of new positions are generated, i.e., the new solution generation procedure in the BSO algorithm. Unlike the update method in the BSO algorithm, the fitness values of the newly generated positions can not be obtained immediately. Instead, it has to be physically visited to get the fitness values by the robots. Thus we designed an additional task allocation procedure, which optimally assigns the newly generated positions to each robot in the swarm, then controls the robots move to the assigned positions.

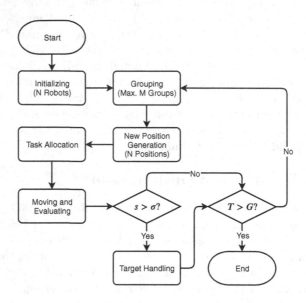

Fig. 1. Flowchart of RBSO.

Furthermore, during the movements to a generated goal point, the robot can measure the signal strength along the moving path. Therefore, different from the evaluation of a single point by the BSO algorithm, in the proposed RBSO, when evaluating the signal strength of a newly generated position, it evaluates the signal strength of all points on the path between the robot and the allocated position. If the target signal value in a position exceeds the predetermined threshold (σ) during the movements, it will be marked as a potential target position. As assumed, the robot will stay there to handle the target until the next planning procedure. After all robots in the swarm stop moving, the procedure will return to the grouping and new location generation operations until all the targets have been found or the specified search time has been reached.

3.1 Grouping

Corresponding to the clustering operation in the BSO algorithm, the purpose of grouping is to allow each robot in a swarm to be divided into several subgroups in the search space. In this article, we simply use a top-down hierarchical clustering method to complete the task of robot grouping, namely DIvisive ANAlysis Clustering (DIANA), which constructs the hierarchy in the inverse order [7]. Initially, all robots are seen as in the same group, and the largest group is split until the one of the following termination condition is met, i.e. the mean distance between elements of each group ($\bar{D}$) larger than a threshold (m_d), or the number of groups reach a pre-defined number m_g. In order to ensure that the new solution can be generated according to the BSO mechanism, here $m_g \leq 2$.

$$\bar{D} = \frac{1}{|\mathcal{A}| \cdot |\mathcal{B}|} \sum_{x \in \mathcal{A}} \sum_{y \in \mathcal{B}} d(x, y) \tag{3}$$

where $\mathcal{A}$, $\mathcal{B}$ is two groups, $d(x, y)$ is the Euclidean distance between two elements x in group $\mathcal{A}$ and y in group $\mathcal{B}$.

Algorithm 2. Grouping Precedure in RBSO

1: Inputs: N *pbest*, maximum groups m_g, maximum iterations T_g, m_d;
2: Consider all robots as a whole group;
3: Selected Group $\mathcal{G}_s$ = the whole group;
4: **while** not terminated **do**
5:     Find the most dissimilar pairs of robots (i, j) in $\mathcal{G}_s$;
6:     Assign the robots closer to i to a new group 1 and the robots closer to j to a new group 2.
7:     Split the $\mathcal{G}_s$ to group 1 and group 2.
8:     $\mathcal{G}_s$ = Find the group with the maximum internal distance in all groups
9:     **if** $|\mathcal{G}_s| \leq 2$ **then**
        return

3.2 New Position Generation

After grouping, the new positions can be generated according to the procedure of the BSO. The difference is that the class center is selected according to the fitness values in the BSO, i.e., individuals with a larger fitness value in the group will be selected as the class center. In the collaborative search task of a swarm of robots, since the targets' influence ranges are not global, all the fitness values of *pbest* of robots in a group maybe 0. Also, there is a case that more than one robots reaches the same maximum fitness value in a group. In these cases, the group center will be randomly selected among the individuals in a group.

3.3 Task Allocation

Multi-robot task allocation (MRTA) is an essential aspect of many multi-robot systems. It is a problem of determining which robots should execute which tasks to achieve overall system goals. The features and complexity of the MRTA problem depend on which requirements are under consideration [8]. In this paper, the task allocation aims to match the newly generated positions to each robot in the swarm for a new round of fitness evaluation.

In this article, the member robot can be regarded as a single-task robot (ST) since it only needs to evaluate the signal strength or handling targets at a time during the searching process. Besides, only one robot is needed for the signal strength evaluation task of a specific position, i.e., the task is a single robot task (SR). Also, the corresponding task allocation is performed after each step of the new position is generated, i.e., the instantaneous allocation (IA). Therefore, the task allocation problem is a typical ST-SR-IA allocation problem [3], which is an instance of the optimal assignment problem (OAP) from the field of combinatorial optimization. Given n robots and m tasks, each task requiring one worker. The OAP can be cast in many ways, including as an integral linear program that find mn nonnegative integers α_{ij} that minimize the cost of the system:

$$\min C = \sum_{i=1}^{n} \sum_{j=1}^{m} \alpha_{ij} c_{ij} \tag{4}$$

subject to:

$$\begin{aligned} \sum_{i=1}^{n} \alpha_{ij} = 1, \qquad 1 \leq j \leq m \\ \sum_{j=1}^{m} \alpha_{ij} = 1, \qquad 1 \leq i \leq n \end{aligned} \tag{5}$$

where C in Eq. (4) is the overall system cost, which is the weighted sum of each assignment costs. Equation (5) enforces the constraints of single-robot tasks and single-task robots. The α_{ij} are integers that must all be either 0 or 1. This problem is also known as zero-one type integer linear programming. Here the robot number is equal to the generated new locations, i.e., $|m| = |n|$. The linear assignment problem can be solved in polynomial time with algorithms such as the Hungarian algorithm [8].

3.4 Moving and Evaluating

After allocating the tasks, each robot corresponds to a newly generated position in the environment. They will move to the assigned position and evaluate the fitness values along the path. Due to obstacles and other robots in the environment, the robot needs a suitable motion planning method to move to the target point as much as possible. Here we use a relatively simple modified Bug Algorithm to achieve the corresponding motion planning [21]. When encountering obstacles, the robot will move along the edges of the obstacles. The evaluating procedure for each robot is as shown in Algorithm 3. During the process of a member robot moves towards the target, it will evaluate the target signal strength at the current position. As mentioned earlier, if the signal strength of the current position is higher than the pre-defined threshold, it will enter the target handling operations. After the target is processed, the will continue to participate in subsequent tasks. To avoid the uncertainty caused by individual robots' motion and possible failures, and to ensure the searching process, two termination conditions are set here: one is for all robots to reach the assigned positions, the other is the moving steps reach a maximum movement step m_s.

Algorithm 3. Signal Evaluating Precedure in RBSO

1: Inputs: Goal position and *pbest* for each robot, maximum moving steps m_s;
2: **for all** robot in the swarm **do**
3:     **while** not terminated **do**
4:         Move forward one step to the goal;
5:         Evaluate the target strength of current position;
6:         **if** $s > \sigma$ **then**
7:             Target Handling;

---

4   Results

Using the above method, we have obtained some preliminary results. The simulation environment is as shown in Fig. 2, where 20 member robots and 10 targets are randomly distributed in a 1000×1000 2-D environment with obstacles. The member robots are represented by the hollow circles, the targets are represented by the solid circles, and the obstacles are shown as black rectangles. The range of influence of the target signal is marked with R. The signal influence region is shown in Fig. 2(b).

The parameters configurations is shown in Table 2, where p_{one} and p_{center} are parameters in BSO algorithms for new positon generation. Other parameters such as signal attenuation coefficient a, maximum groups m_g, maximum iteration generations T_g, maximum inter-group mean distance m_d, and maximum moving steps m_s for each round of evaluation are RBSO parameters. The sampling time for each moving step is set to 0.1 s. The simulations are conducted with mobile robot toolbox in Matlab 2020a on an iMac with 3.6 GHz Intel Core i9, 40 GB DDR4 memory. The preliminary simulated results is shown in Fig. 3, where he hollow circle represents the robot, and the solid circle represents the target to be searched. When it is not found, it is marked in green and displayed in red after handling. The solid line in the figures are pathes of the member robots.

Figure 3(a) shows the initialization distribution of robots and targets, the green dots are the targets to be searched. It will be marked in red when reached. The followed Fig. 3(b)–Fig. 3(h) show the iterative search process. It can be seen

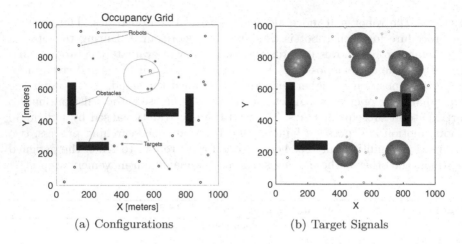

(a) Configurations   (b) Target Signals

Fig. 2. The simulation environment.

Table 2. Simulation configurations

BSO parameters		RBSO parameters				
p_{one}	p_{center}	a	m_g	T_g	m_d	m_s
0.4	0.8	10	$\frac{N}{4}$	20000	250	500

that after a certain number of iterations, 10 targets in the environment can be all found. Since the robot needs to arrive at the newly generated location or reach the targets before a stop, the paths in the graph are in polygonal lines. We further test massive times, the proposed method can find all the targets under the above configurations. The proposed method maps the clustering operation of the original BSO to the grouping operation, which produces the same utility as the original algorithm, i.e., to ensure the convergence of the search process. Besides, the operation of new location generation not only considers the members with the largest fitness value in the group but also according to the members with smaller fitness value with a certain probability, which ensures diversity of the searching process. Furthermore, since each generated position needs to be accessed by a robot, the newly added task allocation module allocates the generated positions to each robot in an optimal way without conflicts. The above results indicate the effectiveness of the proposed method, which is worthy of further development.

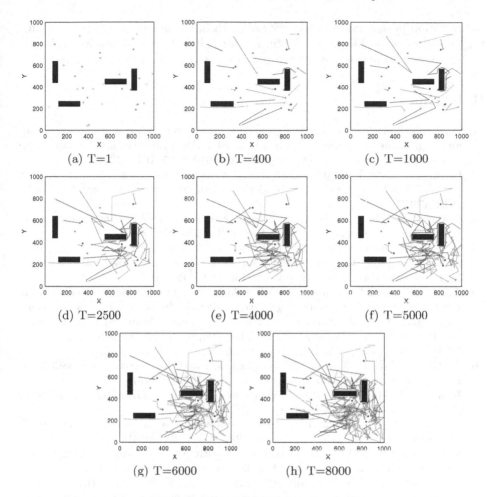

Fig. 3. The simulation results

5   Conclusion

In this paper, the Robotic Brain Storm Optimization (RBSO) for cooperative search tasks of swarm robotics is proposed. In this method, individuals in the BSO algorithm are mapped to the members in robotics swarms. Relevant constraints are introduced for the multi-target cooperative search task of swarm robot systems. The proposed method imitates the optimal searching process of the original BSO algorithm with the considerations of robot physical limits and motion constraints. In the BSO optimization algorithm, the position update of the individuals in the solution space is neither restricted by physical limitations such as movement velocity nor restricted by requirements such as collision avoidance. The proposed paradigm takes both of the above into consideration. The preliminary results show that the proposed method inherits the multi-mode optimization performance of the BSO algorithm, which can be applied in multi-target searching problems. In the future, we will further evaluate its statistical performance and perform more comparative research with other methods.

References

1. Banharnsakun, A., Achalakul, T., Batra, R.C.: Target finding and obstacle avoidance algorithm for microrobot swarms. In: 2012 IEEE International Conference on Systems, Man, and Cybernetics (SMC), pp. 1610–1615. IEEE (2012)
2. Cheng, S., et al.: A comprehensive survey of brain storm optimization algorithms. In: 2017 IEEE Congress on Evolutionary Computation (CEC), pp. 1637–1644. IEEE (2017)
3. Gerkey, B.P., Matarić, M.J.: A formal analysis and taxonomy of task allocation in multi-robot systems. Int. J. Robot. Res. **23**(9), 939–954 (2004)
4. Hoff, N.R., Sagoff, A., Wood, R.J., Nagpal, R.: Two foraging algorithms for robot swarms using only local communication. In: 2010 IEEE International Conference on Robotics and Biomimetics, pp. 123–130. IEEE (2010)
5. Jain, U., Tiwari, R., Godfrey, W.W.: Odor source localization by concatenating particle swarm optimization and grey wolf optimizer. In: Bhattacharyya, S., Chaki, N., Konar, D., Chakraborty, U.K., Singh, C.T. (eds.) Advanced Computational and Communication Paradigms. AISC, vol. 706, pp. 145–153. Springer, Singapore (2018). https://doi.org/10.1007/978-981-10-8237-5_14
6. Jevtić, A., Gazi, P., Andina, D., Jamshidi, M.: Building a swarm of robotic bees. In: 2010 World Automation Congress, pp. 1–6. IEEE (2010)
7. Kaufman, L., Rousseeuw, P.J.: Finding Groups in Data: An Introduction to Cluster Analysis, vol. 344. Wiley, Hoboken (2009)
8. Korsah, G.A., Stentz, A., Dias, M.B.: A comprehensive taxonomy for multi-robot task allocation. Int. J. Robot. Res. **32**(12), 1495–1512 (2013)
9. Krishnanand, K., Ghose, D.: Detection of multiple source locations using a glowworm metaphor with applications to collective robotics. In: Proceedings 2005 IEEE Swarm Intelligence Symposium, 2005, SIS 2005, pp. 84–91. Cham, IEEE(2005)
10. Li, G., Zhang, D., Shi, Y.: An unknown environment exploration strategy for swarm robotics based on brain storm optimization algorithm. In: 2019 IEEE Congress on Evolutionary Computation (CEC), pp. 1044–1051. IEEE (2019)
11. Palmieri, N., Marano, S.: Discrete firefly algorithm for recruiting task in a swarm of robots. In: Yang, X.-S. (ed.) Nature-Inspired Computation in Engineering. SCI, vol. 637, pp. 133–150. Springer, Cham (2016). https://doi.org/10.1007/978-3-319-30235-5_7
12. Shi, Y.: An optimization algorithm based on brainstorming process. Int. J. Swarm Intell. Res. (IJSIR) **2**(4), 35–62 (2011)
13. Yang, B., Ding, Y., Hao, K.: Target searching and trapping for swarm robots with modified bacterial foraging optimization algorithm. In: Proceeding of the 11th World Congress on Intelligent Control and Automation, pp. 1348–1353. IEEE (2014)
14. Yang, J., Shen, Y., Shi, Y.: Brain storm robotics: An automatic design framework for multi-robot systems. In: 2020 IEEE Congress on Evolutionary Computation (CEC), pp. 1–8. IEEE (2020)
15. Yang, J., Shen, Y., Shi, Y.: Visual fixation prediction with incomplete attention map based on brain storm optimization. Appl. Soft Comput. **96**, 106653 (2020)

16. Yang, J., Wang, X., Bauer, P.: Formation forming based low-complexity swarms with distributed processing for decision making and resource allocation. In: 2016 14th International Conference On Control. Automation, Robotics and Vision (ICARCV), pp. 1–6. IEEE (2016)
17. Yang, J., Wang, X., Bauer, P.: Line and v-shape formation based distributed processing for robotic swarms. Sensors **18**(8), 2543 (2018)
18. Yang, J., Wang, X., Bauer, P.: V-shaped formation control for robotic swarms constrained by field of view. Appl. Sci. **8**(11), 2120 (2018)
19. Yang, J., Wang, X., Bauer, P.: Extended PSO based collaborative searching for robotic swarms with practical constraints. IEEE Access **7**, 76328–76341 (2019)
20. Yang, J., Xiong, R., Xiang, X., Shi, Y.: Exploration enhanced RPSO for collaborative multitarget searching of robotic swarms. Complexity **2020**, (2020)
21. Zohaib, M., Pasha, S.M., Javaid, N., Iqbal, J.: IBA: intelligent bug algorithm – a novel strategy to navigate mobile robots autonomously. In: Shaikh, F.K., Chowdhry, B.S., Zeadally, S., Hussain, D.M.A., Memon, A.A., Uqaili, M.A. (eds.) IMTIC 2013. CCIS, vol. 414, pp. 291–299. Springer, Cham (2014). https://doi.org/10.1007/978-3-319-10987-9_27

Distributed Multi-agent Shepherding
with Consensus

Benjamin Campbell[1]($\boxtimes$), Heba El-Fiqi[2], Robert Hunjet[1], and Hussein Abbass[2]

[1] Defence Science and Technolgy Group, Edinburgh, SA 5111, Australia
{benjamin.campbell,robert.hunjet}@dst.defence.gov.au
[2] University of New South Wales, Canberra, ACT 2612, Australia
{heba.el-fiqi,hussein.abbass}@adfa.edu.au

Abstract. The field of swarm guidance and control can rely on intrinsic strategies such as a rule-based system within each member of the swarm or extrinsic strategies, whereby an external agent guides the swarm. In the shepherding problem, sheepdogs drive and collect a flock (swarm) of sheep, guiding them to a goal location. In the case of multiple dogs guiding the swarm, we examine how shared contextual awareness of the sheepdog agents improves the performance when solving the shepherding problem. Specifically consensus around the dynamic centre of mass of a flock is shown to improve shepherding performance.

Keywords: Shepherding · Consensus · Distributed control

1 Introduction

Communications scalability can be a challenge for swarm control. It has been seen in the past that large numbers of drones are difficult to control via communications channels directly, and indeed much research has looked at the Human Autonomy Teaming concept to increase the number of drones that can be controlled from a single ground station [9]. Shepherding offers an alternative approach, where shepherding agents (sheepdogs) round up and control a flock [15]. In the case where a flock is segregated, the sheepdogs collect the flock; when it is cohesive, they drive the flock towards the goal. Of course, in this biological system, or any robotic implementation which takes inspiration from it, the sensing range of a sheepdog is not infinite. It therefore only has partial knowledge of the flock. We argue that a shared contextual awareness which allows multiple sheepdogs to collaborate to enable a broader horizon estimate of the flock will allow for more accurate calculation of the flock state (is it fragmented or cohesive) and the selection of more effective collecting and driving points. Within [15] these points are calculated using the flock's global centre of mass (GCM) as a reference point, under the unrealistic assumption that global information about individual flock members is available. We relax this assumption, with the shared contextual awareness realised through the use of communication channels (i.e. wireless networking). As such, a lightweight and scalable approach to state

Y. Tan and Y. Shi (Eds.): ICSI 2021, LNCS 12690, pp. 168–181, 2021.
https://doi.org/10.1007/978-3-030-78811-7_17

estimation is required to minimise network congestion and provide quick convergence time. The literature on estimation provides many approaches to "track" the state of objects, ranging from Kalman filtering based approaches [1], to non-linear filtering approaches [16], with multiple efforts provided to translate these approaches to collaborative and distributed algorithms [8]. These approaches are reasonably computationally expensive and provide highly accurate estimates of position and orientation. We contend that the highly dynamic nature of the flock requires an approach which converges quickly; as such lighter-weight approaches such as average consensus [12] may suffice in certain contexts.

We investigate this concept in simulation with our main contributions being a consideration of sensing range limitations and flock dispersal on GCM estimate accuracy, a consensus based approach to flock state estimation with both cooperative (instrumented) and non-cooperative flocks, a distributed approach to multi-agent shepherding utilising local knowledge and simulations that show the consensus approaches proposed enhance shepherding performance in situations where the sheepdogs have a limited sensing horizon.

2   Relevant Previous Work

There is a wide variety of work in the literature for shepherding utilising different representations of the state of the flock. This includes the approach in Strömbom [15] which classifies the flock as collected or scattered relative to the GCM of the flock, the approach proposed by Harrison et al. [7] which represents the flock as a "deformable blob" and Zhi et al. [17] which simplifies the state space in a learning shepherding system by representing the flock as a circle. In all cases, sensing of the sheep is required. As communication constraints are an important reason for utilising shepherding for swarm control, it is safe to assume that in many cases a centralised controller with global knowledge of the flock is unlikely to be practical. Despite this, there is little work in the literature exploring the impact of the constraints imposed by local sensing on the performance of a distributed multi agent shepherding (DMAS) system. These restrictions were explored in [13] when the authors transitioned a shepherding robot trained in simulation to real systems, however, they considered only a single sheep and sheepdog, which limits the relevance of their findings when considering DMAS systems. Sensing range was also addressed in [17] which assumed that sheepdogs could not measure the position of sheep beyond a certain range but could still determine direction. In a real system, depending on the sensors used, distance as well as lack of line of sight could render sheep completely undetectable. In a DMAS system it is likely that at times some sheep will be detectable by some sheepdogs and not others. This suggests that a co-operative approach that allows the sheepdogs to build a complete picture of the flock state would be advantageous. There is a great deal of literature that focuses on tracking a target based on incomplete or unreliable information from multiple sensors. Kalman filter based approaches are widely accepted as the standard for the distributed tracking of dynamic targets [1,4,10]. There have also been alternative approaches suggested such as non linear filtering [16], and Markovian Network based approaches[6]. These approaches generally rely on robust communication networks [11] and may even need a fixed network topology [6]. In a shepherding system in a constrained and unreliable wireless network with a switching topology, a more lightweight approach that converges quickly, has minimal communication overheads and is robust to a dynamic topology may be advantageous. We identify the average consensus algorithm proposed by Olfati-Saber et al. [12] as a potential approach. In this approach each distributed node i adjusts its estimate of the value under consensus x based on the average of the difference between its own estimate x_i and the estimate of its adjacent nodes j. In a fully connected network, consensus can be achieved in as little as 2 time steps. The concept of trust can also be utilised by weighting the consensus values. The core equation for this approach is shown below.

$$u_i(t) = \sum_{v_j \in G_i} a_w(i,j)(x_j - x_i) \tag{1}$$

Where x is the value under consensus and a_w is the weighted adjacency matrix of the network G and $u_i(t) = \dot{x}_i(t)$ for continuous time or $x_i(k+1) = x_i(k) + \epsilon u_i(k)$ for discrete time with step size $\epsilon > 0$.

A potential issue with this approach is that after initialisation there is no method to introduce new sensor data. This is problematic in a dynamic system such as shepherding, where the target is continuously moving. Spanos et al. [14] proposed an extension to address this problem. By adding the gradient of the locally measured value to Eq. 1, the value under consensus can be updated based on up to date sensor data. This gives Eq. 2

$$u_i(t) = \dot{z}_i(t) + \sum_{v_j \in G_i} a_w(i,j)(x_j - x_i) \tag{2}$$

where z_i is the measured value at node i.

The remaining sections explore the utility of this approach in approximating the global centre of mass (GCM) for DMAS systems with limited sensor range.

3   Problem Definition

This paper will utilise the notation system for shepherding defined by El-Fiqi et al. [2] where Π denotes the set of sheep agents $\{\pi_1, \pi_2, ..., \pi_n\}$ and B denotes the set of sheepdog agents $\{\beta_1, \beta_2, ..., \beta_m\}$. N and M denote the cardinality of Π and B respectively. As [2] considers a centrally controlled shepherding system with global knowledge, its only concept of sensor range concerns the range at which the sheep agents sense other sheep agents, $R_\pi\pi$, and sheepdog agents, $R_\pi\beta$. This paper considers the impact of limited sensor range on the sheepdog's performance so we denote the sensor range of the sheepdogs as R_β (we do not assume a separate sensor range for β to π and β to β sensing). We also utilize the concept of communication range, which is denoted here as R_c.

Strömbom et al.'s model [15], which forms the basis of several other shepherding models [2,3], relies on the sheepdogs knowing the GCM, which is used to determine driving and collecting positions and the state of the flock (e.g. collected or fragmented). In DMAS systems (e.g. no centralised controller or global knowledge) each sheepdog develops its own estimate of the GCM, GCM_e. If the sheepdog has limited sensing range R_β, only $\Pi_\beta \subset \Pi$, are used for the local centre of Mass (LCM) calculation, where $\pi \in \Pi_\beta$, $\forall|(P_\beta - P_\pi)| <= R_\beta$ ie only sheep within the sheepdog's sensing range are used.

Intuitively in cases where a sheepdog is unable to sense all the sheep, the accuracy of GCM_e will be reduced. If the sheepdogs are able to communicate, this can be mitigated by allowing the pack to come to a consensus on the position of the GCM. Thus the central aim of this work can be expressed as; *Given sheep and sheepdogs with sensing range R_β and communication range R_c, use consensus across the agents to estimate the GCM of the flock.*

3.1 Assumptions

1. The sheepdogs can identify individual sheep.
2. The sheepdogs are homogeneous.
3. The accuracy of the position measurements of individual sheep is not dependant on distance, i.e. if a sheep is in range of a sheepdog its position can be perfectly measured.
4. The sheepdogs are able to sense all sheep in range R_B even if other sheep are in the way.
5. The sheep are homogeneous and follow the behaviour described in Strombom et al. [15]

Assumption 1 simplifies consensus as there is no need to deconflict separate tracks between sheepdogs. Consensus then is about disseminating and merging dissimilar information between the sheepdogs.

Assumptions 2 and 3 are included to simplify the question of trust when forming consensus. Due to these two assumptions, all sheep position measurements are equally accurate, and trust is either a function of the age of the individual position measurement or a function of the number of sheep positions that contributed to the LCM.

4 Proposed Solutions

We propose two potential approaches to estimate the state of the flock. Firstly, the LCM can be used as the input to the consensus algorithm described in Eq. 2, in order to estimate the Global Centre of Mass (GCM). Alternatively the sheepdogs can come to consensus on the individual sheep positions (ISP) allowing each sheepdog agent to estimate GCM using raw position data.

Another question to consider is the role of the sheep in forming a consensus. As applications of the shepherding problem are not limited to the herding of biological sheep (the sheep could for example be a swarm of drones), it is feasible to consider the case where the sheep could sense the positions of their neighbours and participate in the distributed consensus activity. For the purposes of this paper, this sheep behaviour will be referred to as active sheep, with the more traditional shepherding scenario where the sheep are simple reactive agents (taking no part in this calculation) referred to as passive sheep.

The following section describes five distributed consensus algorithms that will be investigated in this paper. The first four algorithms are based on the dynamic average consensus algorithm described by Spanos et al. [14] and estimate the GCM of the flock using the LCM approach. The fifth algorithm forms consensus on ISP. The five algorithms will be referred to as; Unweighted Passive Spanos (UPS), Weighted Passive Spanos (WPS), Unweighted Active Spanos (UAS), Weighted Active Spanos (WAS) and Individual Sheep Position (ISP).

Each time step each agent runs its algorithm once and all agents complete their iteration of the algorithm before the time step ends. We assume that all data transmissions occur instantly and concurrently. We acknowledge that to transition the methodology to real-robotic platforms, time delays and synchronisation would need to be taken into account.

4.1 Unweighted Passive Spanos

In this algorithm, each sheepdog agent β_i, shares its previous GCM estimate (GCM_e^{t-1}) with all other sheepdogs within R_c of its position, i.e. $\beta_j \in B_{iR_c}$ and measures the positions of all sheep within R_β of its position, i.e. $\pi \in \Pi_{\beta_i}$. Once all sheepdog agents have shared their estimate, each sheepdog uses Eq. 2 to estimate the GCM. Here $\dot{z}_i(t) = LCM_i^t - LCM_i^{t-1}$ and LCM_i^t is the COM_i, using only $\pi \in \Pi_{\beta_i}$, i.e. only sheep within R_B of β_i. No weightings are applied to the adjacency matrix, i.e. $\forall \beta_j \in B_{iR_c}$, $a_w(i,j) = 1$.

Algorithm 1. Unweighted Passive Spanos

Step 1: β_i receives $GCM_e^{t-1}{}_j$, $\forall \beta_j \in B_{iR_c}$
Step 2: β_i receives P_π^t, $\forall \pi \in \Pi_{\beta_i}$
Step 3: calculate the adjacency matrix a as follows,
if $\beta_j \in B_{iR_c}$, $a(i,j) = 1$, else $a(i,j) = 0$
Step 4: Calculate GCM_{ei}^t using equation 2

4.2 Weighted Passive Spanos

This algorithm is similar to UPS but has two extra steps. In addition to sharing $GCM_e^{t-1}{}_j$ each sheepdog agent β_j also shares the number of sheep w_j, that were detected at time $t - 1$ i.e. the number of sheep that contributed to the LCM_j^{t-1}. β_i then calculates the weighted adjacency matrix a_w by multiplying the adjacency matrix by the weighting vector $W_i = \{w_1, w_2, ...w_m\}$ and dividing by $\sum W_i$ in order to normalise the weighting.

Algorithm 2. Weighted Passive Spanos

Step 1: β_i receives $GCM_e^{t-1}{}_j$, $\forall \beta_j \in B_{iR_c}$
Step 2: β_i receives w_j^{t-1}, $\forall \beta_j \in B_{iR_c}$
Step 3: β_i receives P_π^t, $\forall \pi \in \Pi_{\beta_i}$
Step 4: calculate the weighting vector W_i as follows,
if $\beta_j \notin B_{iR_c}$, $w_j = 0$, $W_i = \{w_1, w_2, ...w_m\}$
Step 5: calculate the weighted adjacency matrix a_w as follows,
if $\beta_j \in B_{iR_c}$, $a_{i,j} = 1$, else $a_{i,j} = 0$
$a_w = a * W / \sum W$
Step 6: Calculate $GCM_{e\,i}^t$ using equation 2

4.3   Unweighted Active Spanos

This algorithm is identical to UPS except all agents participate in the consensus, i.e. both sheep and sheepdog agents share their GCM_e^{t-1} with all other agents $\alpha \in A_{R_c}$, and carry out step 2 and 3. Here we define A as the set of all sheep and sheepdog agents.

Algorithm 3. Unweighted Active Spanos

Step 1: α_i receives $GCM_e^{t-1}j$, $\forall \alpha_j \in A_{iR_c}$
Step 2: α_i receives P_π^t, $\forall \pi \in \Pi_{\beta_i}$
Step 3: calculate the adjacency matrix a as follows,
if $\beta_j \in B_{iR_c}$, $a(i,j) = 1$, else $a(i,j) = 0$
Step 4: Calculate $GCM_{e\,i}^t$ using equation 2

4.4   Weighted Active Spanos

As described in Subsect. 4.3 this approach is similar to its passive equivalent except all agents $\alpha \in A$ participate in the consensus.

Algorithm 4. Weighted Active Spanos

Step 1: α_i receives $GCM_e^{t-1}{}_j$, $\forall \alpha_j \in A_{iR_c}$
Step 2: α_i receives w_j^{t-1}, $\forall \alpha_j \in A_{iR_c}$
Step 3: α_i receives P_π^t, $\forall \pi \in \Pi_{\beta_i}$
Step 4: calculate the weighting vector W_i as follows,
if $\alpha_j \notin A_{iR_c}$, $w_j = 0$, $W_i = \{w_1, w_2, ...w_m\}$
Step 5: calculate the weighted adjacency matrix a_w as follows,
if $\alpha_j \in A_{iR_c}$, $a_{i,j} = 1$, else $a_{i,j} = 0$
$a_w = a * W / \sum W$
Step 6: Calculate $GCM_{e\,i}^t$ using equation 2

4.5   Individual Sheep Position

In this approach each sheepdog agent maintains κ, a 2 * N flock state matrix. The 1st column corresponds to the position of $\pi_i \in \Pi$, and the second column is a time stamp of the measurement of π_i. Each time step t, each sheepdog β_i broadcasts κ^{t-1} to all of its neighbors then senses and updates the position and time stamp of all sheep $\pi \in \Pi_{\beta_i}$. It then carries out a merge operation on the set R where κ_i^t gets the most recent position value and time stamp for each sheep in R. Here R is the set of all received κ and β_i's own κ. Once κ is updated, $\kappa_{current}$ is found, which is the list of sheep positions in κ with age $<= T_{timeout}$. Where $T_{timeout}$ is a set timeout period used to ensure stale position information is not used. The GCM is estimated using the position values contained in $\kappa_{current}$.

Algorithm 5. Individual Sheep positions

Step 1: β_i receives κ_j^{t-1}, $\forall \beta_j \in B_{iR_c}$

Step 2: β_i receives P_π^t, $\forall \pi \in \Pi_{\beta_i}$

Step 3: update κ_i^t with the most recent position information and detection timestamp for all $\pi \in \Pi$

Step 4: Find $\kappa_{current}$, the list of $P_{\pi j}$ $\forall P_\pi \in \kappa_i^t$ such that $P_{\pi j}$'s age $<= T_{timeout}$

Step 6: Calculate $GCM_{e i}^t$ using the sheep position values from $\kappa_{current}$

4.6   Distributed Multi Agent Shepherding (DMAS)

Once a GCM is calculated, the shepherding tasks of collecting and driving can occur. We propose a distributed multi-agent shepherding behaviour based on the centralised approach described in El-fiqi et al. [2].

In algorithm 6, Distributed Collecting, each timestep, the sheepdog creates a list of locally detected, scattered sheep. Here sheep are considered scattered if they are $2\sqrt{N}$ from GCM_e. This list is ordered based on proximity to the sheepdog's position. If the sheepdog is closest to the first sheep in the list (according to sensed knowledge of the other sheepdog's positions), the sheepdog will collect the sheep according to the collecting behaviour described in [15]. If it is not the closest sheepdog, it will move to the next sheep in the ordered list and repeat the process. If the sheepdog reaches the end of the list or the list is empty, the sheepdog will begin driving, i.e. begin Algorithm 7. This ensures that any sheepdogs not allocated a collecting task begin driving immediately. Only locally sensed sheep are collected to ensure only one sheep dog is assigned per scattered sheep.

Algorithm 7 utilises a driving formation based on the formation described in [2]. This is an arc formation centred on the flock GCM with radius equal to the driving distance defined in [15] and angular width $\pi/2$. The arc is bisected by a ray that extends from the goal through the GCM. The point where the ray bi-sects the arc is the central driving position in the formation, referred to as C. The other driving positions are on the arc to the left and right of C at intervals of

Algorithm 6. Distributed Collecting

Step 1: $\beta_i \leftarrow (P_\pi(t), P_\beta(t)), \forall (\pi, \beta) \in A_{\beta_i}$
Step 2: Create an ordered list of scattered sheep based on proximity
Step 3: If β_i is the closest sheepdog to the first scattered sheep, collect as per [15], else remove from the list
Step 4: Repeat step 3 until the list is empty or a collecting task is chosen

---

$\pi/2M$. During driving, each sheepdog occupies the unoccupied driving position in the formation closest to C, with occupation being defined as any sheepdog agent being within a certain threshold distance of the driving position. If there are two unoccupied driving positions equidistant from C the sheepdog will select the driving position closest to its own position. The sheepdogs move as per [15] while in driving mode. As the driving positions in the formation are dependant on the individual sheepdog's estimation of the GCM, the greater the consensus the more effectively the sheepdogs will be able to take up correct positions in the formation.

Algorithm 7. Distributed Driving

Step 1: Determine the driving positions as per the location of the goal and the estimate of the GCM
Step 2: Determine the unoccupied driving position closest to the centre of the formation, if there are two equidistant positions choose the driving position closest to own position.
Step 3: Move to the selected driving position and perform driving behaviour as per Strömbom et al. [15]

4.7 Communication Overheads

The communication overheads of the algorithms impact their scalability and applicability. We assume 802.11 packet header/CRC size (i.e. 34 bytes per packet [5]) and no acknowledgment messages. We also assume 16 bytes per (x, y) coordinate, 2 bytes for weighting information, 8 bytes per time stamp, that each message can be contained in a single packet and no packet re-transmission occurs. This gives communication overheads per time step for UPS, WPS, UAS, WAS and ISP of 50 * M, 52 * M, 50 * (M+N), 52 * (M + N) and 58 * M * N respectively.

5   Consensus Accuracy

In order to determine the efficacy of our proposed approach, we use the mean difference between the GCM and the GCM_e at each sheepdog agent as the metric of comparison.

Initially, we utilise two contrived scenarios (devoid of the act of shepherding itself) to compare and contrast consensus approaches. This allows us to investigate factors of interest such as dispersal of the flock and proximity of each sheepdog to the sheep. The scenarios are depicted in Fig. 1. For all simulations the sheep are randomly distributed in a circle with radius r_f, N = 100, M = 10 and R_c was set to ensure each sheepdog could reach its closest neighbouring sheepdog to the left and right but no further. Each algorithm was simulated 20 times for each scenario using different random seed sheep starting positions. The same seeds were used for each algorithm in order to ensure fair comparison between results. Each simulation ran for 100 time steps.

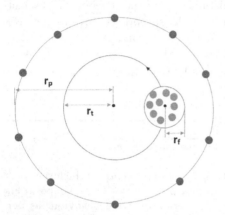

(a) Flock dispersal scenario. The sheepdogs are in a circular formation and all sheep moving in a circle.

(b) Sensor range scenario. The sheepdogs are in a linear formation and all agents moving diagonally.

Fig. 1. Scenarios used to investigate flock dispersal and sensor range. Sheep dogs are red, sheep are blue and all agent movements are indicated by black arrows. (Color figure online)

To determine impact of flock dispersal on accuracy the sheepdogs are placed in a circular formation with radius, $r_p = 50$ m. The sheepdogs are stationary. The sheep are placed in a circle with radius r_f. Their positions, relative to the circle, are fixed, but the circle moves at a constant angular velocity on a circular track with a radius $r_t = 25$ m and the same centre as the sheepdog formation. This scenario provides movement of the flock relative to the sheepdogs and consistent distance between the sheepdogs and the sheep. Flock dispersal was varied while all other factors held constant by increasing the value r_f from 1–25. R_B was set at 50 m to ensure all sheep could be sensed by at least one sheepdog at all times. The results of this experiment are shown in Fig. 2.

To investigate the impact of sensor range, the sheepdogs were placed in a linear formation at the bottom of the operating area. This change was made

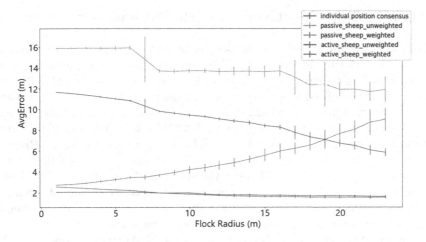

Fig. 2. Error for the various consensus algorithms with varied flock dispersal. The radius of the flock increases from r = 1–25 and the sheepdogs are in a circular formation. $R_B = 50$ m. Error bars indicate the 90% confidence interval.

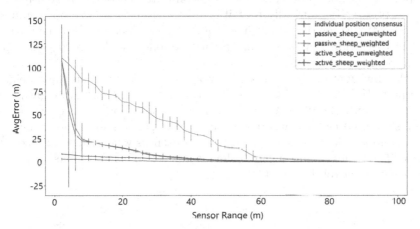

Fig. 3. Error for the various consensus algorithms with the range of the sheepdogs sensors increasing from 2–100 and the sheepdogs in a linear formation. $r_f = 25$ m. Error bars indicate the 90% confidence interval.

because when the sheepdog agents detect no sheep they utilise their own position for their LCM. At low senor ranges this causes a bias towards the centre of the pack. The flock is held in a stationary position relative to the pack and all sheepdogs and sheep move at a fixed velocity in a diagonal direction. The flock was positioned to the left of the pack to ensure any bias did not co-incidentally reduce the error. The effect of reduced sensor range was investigated by increasing R_B from 2–100 m. The dispersal of the flock is fixed at $r_f = 25$ m. The results of this experiment are shown in Fig. 3.

6   Shepherding Effectiveness with Consensus

In order to investigate the effect of consensus on DMAS the sheep were initialised randomly in a 150 by 150 m area. There were no boundaries and the sheep behaved according to Strömbom [15]. For all simulations N = 100, M = 10. The sheepdogs were initialised randomly in the bottom third of the 150 by 150 m area and the target was placed randomly in the 150 by 150 m area. The shepherding task was considered complete when the flock COM was within 5 m of the target while the flock was in the collected state. The sensor range was set at $R_B = 20$ i.e. the driving distance as per [15], for N = 100 sheep. This meant that during driving no single sheepdog could sense the entire flock. R_C was set to 75 m in order to give reasonable but not complete communication coverage of the initial area. The sheep move at 2 ms^{-1} and the sheepdogs at 3 ms^{-1}. The consensus algorithm is run every 500 ms and the sheepdogs re-asses their direction every time they update their COG estimate (ie also every 500 ms). The timeout period for the ISP (i.e. the time to live for individual sheep position information) is M timesteps. This is the amount of time that sheep position information will take to propagate from one side of a minimally connected network to the other.

It was observed empirically that if the shepherding task was not completed within 2000 timesteps (i.e. 100 s) it was unlikely to. Therefore, for the purposes of this investigation if a simulation did not complete within 2000 timesteps it was considered unsuccessful.

As the unweighted algorithms are always less accurate than their weighted counterparts (Fig. 2, 3), and there is little communication cost to weighting (see Subsect. 4.7), only the WPS, WAS and ISP consensus algorithms were simulated. We also simulated the case where sheepdogs have global knowledge of the sheep positions and the distributed case where no consensus was used. All 5 cases were simulated 500 times. Figure 4 shows the success rate of each consensus approach as a percentage of total simulations.

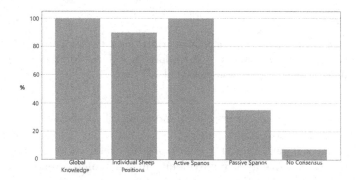

Fig. 4. Success rate of the distributed shepherding algorithm with various consensus strategies. N = 100, M = 10

7   Analysis and Discussion

Figure 2 shows the average difference between the GCM and the individual sheep-dog's consensus based GCM_e as the flock radius increases from 1–25. In these scenarios the WAS and the ISP consensus approaches had lowest error in their GCM_e independent of the width of the flock. The two unweighted algorithms become more accurate while the weighted passive sheep algorithm becomes less accurate. This is because when the flock is more collected a single sheepdog can make an accurate estimate of the centre of mass as well as the direction of the flock. For the weighted approaches this quickly propagates across the pack. As the flock width becomes larger a single sheepdog can not sense the entire flock, making the individual sheepdog's centre of gravity estimates less accurate. This results in a decrease in accuracy for the weighted algorithms. In the unweighted approaches, at low dispersal the majority of the sheepdogs have no visibility of the flock, resulting in a large error in the consensus. As the flock spreads out more dogs can sense sheep, resulting a more accurate consensus.

Figure 3 shows the results of increasing the sensor range from 2–100 m while keeping flock radius r_f constant. This figure indicates that the active sheep algorithms were least effected by limited sensor range. This was due to the sheep effectively acting as extra sensors distributed throughout the flock. This suggests that the active sheep algorithm will be particularly effective in situations where sensor range is extremely limited.

In both Figs. 2 and 3 the unweighted algorithms (UPS,UAS) were shown to have greater average errors than their weighed counterparts (WPS, WAS). As there is little communication overhead to performing the weighting (see Sect. 4.7) we see no reason to pursue the unweighted algorithms further. The consistency of these results are evident in the small error bars in Figs. 2 and 3 indicating the 90% confidence interval.

Figure 4 shows the success rate of the distributed shepherding approach described in Sect. 4.6 under various consensus algorithms. It shows that while shepherding is successful in 100% of cases with global knowledge, reducing R_B to 20 m (i.e. the driving distance) will render distributed shepherding ineffective without consensus. WAS was the most effective consensus algorithm with a success rate of 100%. This is due to the active sheep approach effectively adding 100 extra sensors, distributed throughout the flock, almost eliminating the impact of reduced sensor range (as shown in Fig. 3). The ISP approach also performed well with a success rate of approximately 90%. The PWS approach showed only marginal improvement with a success rate of 38%.

As discussed in Subsect. 4.7 WAS has a lower communication overhead than ISP and Fig. 4 shows it also has the highest success rate. If the communication cost can be met and the sheep can be instrumented, it is likely that in most cases AWS will give the best shepherding performance.

8   Conclusion and Future Work

Reduced sensor range has a significant impact on the performance of a distributed shepherding system. This impact can be mitigated using distributed consensus approaches. Five consensus algorithms were proposed with different characteristics and performance. Of these 5 the WAS approach was shown to be most effective in all circumstances, however it requires the sheep agents to be instrumented and actively participate in the consensus, it also requires a wireless network that can support N+M nodes. The Individual Sheep Position Consensus approach was also shown to work well, with a success rate of approximately 90%. Sheep participation is not required, however it has an even higher communication cost than WAS. The WPS algorithm provided a small improvement over no consensus and could be used in systems with extremely limited communication capacity. The two remaining approaches UAS and UPS were shown to be less accurate than their weighted counterparts with minimal communication overhead savings.

This study has only simulated scenarios where there are no obstacles in the environment, meaning the flock is unlikely to separate once it has been collected. Future work should investigate the performance of distributed consensus in more complex environments. This work also assumes that the sheepdog's ability to sense individual sheep is not affected by line of sight. In a real system this is unlikely to be the case. Future work should investigate the impact of line of sight based sensing on the ability to estimate the GCM and the effectiveness of distributed shepherding with consensus.

References

1. Chen, C., Zhu, S., Guan, X., Shen, X.S.: Distributed consensus estimation of wireless sensor networks. Wireless Sensor Networks. SCS, pp. 5–11. Springer, Cham (2014). https://doi.org/10.1007/978-3-319-12379-0_2
2. El-Fiqi, H., et al.: The limits of reactive shepherding approaches for swarm guidance. IEEE Access **8**, 214658–214671 (2020)
3. Elsayed, S., et al.: Path planning for shepherding a swarm in a cluttered environment using differential evolution. arXiv preprint arXiv:2008.12639 (2020)
4. Fan, S., Sun, C., Yang, C., Ye, B.: Fast distributed kalman-consensus filtering algorithm with local feedback regulation. In: 2015 IEEE International Conference on Information and Automation, pp. 2345–2350. IEEE (2015)
5. Foh, C.H., Zukerman, M.: Performance analysis of the IEEE 802.11 mac protocol. In: Proceedings of European Wireless, vol. 2 (2002)
6. Ge, X., Han, Q.L.: Consensus of multiagent systems subject to partially accessible and overlapping markovian network topologies. IEEE Trans. Cybern. **47**(8), 1807–1819 (2016)
7. Harrison, J.F., Vo, C., Lien, J.-M.: Scalable and robust shepherding via deformable shapes. In: Boulic, R., Chrysanthou, Y., Komura, T. (eds.) MIG 2010. LNCS, vol. 6459, pp. 218–229. Springer, Heidelberg (2010). https://doi.org/10.1007/978-3-642-16958-8_21

8. Henderson, J., Zamani, M., Mahony, R., Trumpf, J.: A minimum energy filter for localisation of an unmanned aerial vehicle. arXiv preprint arXiv:2009.04630 (2020)
9. Hocraffer, A., Nam, C.S.: A meta-analysis of human-system interfaces in unmanned aerial vehicle (UAV) swarm management. Appl. Ergon. **58**, 66–80 (2017)
10. Li, W., Jia, Y., Du, J.: Distributed kalman consensus filter with intermittent observations. J. Franklin Inst. **352**(9), 3764–3781 (2015)
11. Luo, X., Li, X., Li, S., Jiang, Z., Guan, X.: Flocking for multi-agent systems with optimally rigid topology based on information weighted kalman consensus filter. Int. J. Control Autom. Syst. **15**(1), 138–148 (2017)
12. Olfati-Saber, R., Murray, R.M.: Consensus problems in networks of agents with switching topology and time-delays. IEEE Trans. Autom. Control **49**(9), 1520–1533 (2004)
13. Schultz, A., Grefenstette, J.J., Adams, W.: Roboshepherd: learning a complex behavior. Robot. Manuf. Recent Trends Res. Appl. **6**, 763–768 (1996)
14. Spanos, D.P., Olfati-Saber, R., Murray, R.M.: Dynamic consensus on mobile networks. In: IFAC World Congress, pp. 1–6. Citeseer (2005)
15. Ströombom, D., et al.: Solving the shepherding problem: heuristics for herding autonomous, interacting agents. J. R. Soc. Interface **11**(100), 20140719 (2014)
16. Zamani, M., Hunjet, R.: Collaborative pose filtering using relative measurements and communications. In: 2019 12th Asian Control Conference (ASCC), pp. 919–924. IEEE (2019)
17. Zhi, J., Lien, J.M.: Learning to herd agents amongst obstacles: Training robust shepherding behaviors using deep reinforcement learning. arXiv preprint arXiv:2005.09476 (2020)

Non-singular Finite-Time Consensus Tracking Protocols for Second-Order Multi-agent Systems

Yao Zou[1,2(✉)], Wenfu Yang[1,2], Zixuan Wang[1,2], Keping Long[3], and Wei He[1,2]

[1] School of Automation and Electrical Engineering, University of Science and Technology Beijing, Beijing 100083, People's Republic of China
zouyao@ustb.edu.cn, {S20190610,g20198743}@xs.ustb.edu.cn, weihe@ieee.org
[2] Institute of Artificial Intelligence, University of Science and Technology Beijing, Beijing 100083, People's Republic of China
[3] School of Computer and Communication Engineering, University of Science and Technology Beijing, Beijing 100083, People's Republic of China
longkeping@ustb.edu.cn

Abstract. This paper surveys non-singular finite-time consensus tracking issues for second-order multi-agent systems subject to external disturbance. The consensus tracking protocol utilizes the sliding mode control methodology. Firstly, a novel non-singular sliding mode manifold is designed for each agent dynamics, and it ensures the finite-time convergence once the error trajectory reaches it. Further, two non-singular consensus protocols are developed so as to drive the error trajectory towards the assigned sliding mode manifold in finite time. However, the settling time with the first protocol scheme cannot be assessed due to its dependence on transient states. To overcome this dilemma, another protocol scenario is exploited, with which the settling time is assignable off-line using available initial information states and control parameters. Finite-time consensus tracking results are demonstrated based on Lyapunov stability theorems. Eventually, simulations verify the performance of the developed protocols.

Keywords: Consensus tracking · Multi-agent system · Sliding-model control

1 Introduction

In recent decades, cooperative control has attained considerable attention. As a representative cooperative control issue, the consensus issue, which aims at

This work has been supported in part by the National Natural Science Foundation of China under Grants 61933001, 62073028, and 62061160371; in part by the Scientific and Technological Innovation Foundation of Shunde Graduate School under Grant BK19AE014; and in part by the Beijing Top Discipline for Artificial Intelligent Science and Engineering, University of Science and Technology Beijing.

Y. Tan and Y. Shi (Eds.): ICSI 2021, LNCS 12690, pp. 182–192, 2021.
https://doi.org/10.1007/978-3-030-78811-7_18

driving the agents to reach an agreement on some quantities of interest through a local distributed protocol, has been intensively investigated [1–3].

Generally, consensus issues of multi-agent systems can be categorized into leader-following and leaderless [4]. A basic theoretical framework for distributed consensus issues with fixed and changing interaction topologies was introduced in [5]. With the raising of maneuver complexity, the leader-following consensus, which requires a class of agents to track or follow a dynamic leader [6], becomes dominant. For linear systems, a consensus protocol is proposed in [8] to drive agents tracking a leader with its control input unaccessible to its followers. [9,10], respectively, solved the problem of fault-tolerant synchronization and cooperative fault-tolerant tracking of nonlinear multi-agent systems under actuator faults.

Further, as far as the consensus protocol development is concerned, an important performance indicator is the convergence rate [11]. In this context, the finite time consensus problem, which accelerates the convergence rate, arises at the historic moment. For the finite time stability analysis, the settling time functions are associated with Lyapunov functions [12]. Based on the fundamental work undertaken in [12,13] settled the consensus issue for multi agent system formulated by single-integrator dynamic. [14] proposed two algorithms for multiple single-integrator dynamics to achieve the finite-time (average) consensus. And [15] extended it to second-order agents with external disturbance; however, they could just ensure the tracking errors of agents converging to a small region, not the origin. Besides, with the super-twisting and integral sliding mode, the protocol proposed in [16] achieved the finite-time consensus of second-order multi-agent systems with disturbance, but it required a certain upper bound of the disturbance derivative, which is commonly unavailable.

Based on the sliding mode control, this paper synthesizes two non-singular finite-time consensus tracking protocol schemes for second-order multi-agent systems suffering from external disturbance. A novel sliding mode manifold is firstly designated for each agent, on which the finite-time convergence of the tracking errors without singularity is ensured. Subsequently, two non-singular consensus protocols are developed to guarantee that the error trajectory can reach the sliding mode manifold in finite time. Compared with the first protocol scheme, with which the settling time is inestimable due to its dependence on transient states, the settling time with the second protocol scheme can be assessed beforehand with available initial information states and control parameters.

2   Preliminaries

2.1   Graph Theory Notations

A weighted graph $\mathcal{G} = \{\Pi, E, A\}$ is composed by an adjacency matrix $A = [a_{ij}] \in \mathbb{R}^n$, an edge set $E \subseteq \Pi \times \Pi$ and a non-empty finite node set $\Pi = \{\pi_1, \ldots, \pi_n\}$. For a directed graph, an edge $(\pi_i, \pi_j) \in E$ indicates that the information of node π_i can be conveyed to node π_j. Donate $N_i = \{\pi_j : (\pi_j, \pi_i) \in E\}$ as the set of all neighbours of node π_i. And the adjacency matrix A is defined such that $a_{ij} > 0$

if $(\pi_j, \pi_i) \in E$, and $a_{ij} = 0$ otherwise. The degree matrix of graph $\mathcal{G}$ is defined as $P = \text{diag}\{p_1, p_2, \ldots, p_n\}$ with $p_i = \Sigma_{j \in N_i} a_{ij}$, and the associated Laplacian matrix is defined as $L = P - A$. Further, a directed graph contains a spanning tree if one of its subgraphs forms a directed spanning tree [5].

2.2 Lemmas

Lemma 1 ([14]). *For $x_1, x_2, \ldots, x_n \geq 0$ and $\alpha \in (0, 1)$, $\sum_{i=1}^{n}(x_i)^{\alpha} \geq (\sum_{i=1}^{n} x_i)^{\alpha}$ holds.*

Lemma 2 ([12]). *Suppose that a $\mathcal{C}^1$ positive definite function $V(x)$ defined on $U \subset \mathbb{R}^n$ satisfies $\dot{V}(x) + \lambda V(x) \leq 0$ for $\lambda > 0$ and $\alpha \in (0, 1)$, then there exists a region $U_0 \subset \mathbb{R}^n$ such that any $V(x)$, which starts from U_0, can reach $V(x) \equiv 0$ in finite time. In particular, if t_r is the settling time required to achieve $V(x) \equiv 0$, then $t_r \leq \frac{V^{1-\alpha}x_0}{\lambda(1-\alpha)}$, where $V(x_0)$ is the initial value of $V(x)$.*

Lemma 3 ([17]). *Suppose that a $\mathcal{C}^1$ positive definite function $V(x)$ defined on $U \subset \mathbb{R}^n$ satisfies $\dot{V}(x) + \lambda_1 V(x) + \lambda_2 V^{\alpha}(x) \leq 0$ for $\lambda_1 > 0$, $\lambda_2 > 0$ and $\alpha \in (0, 1)$, then there exists a region $U_0 \subset \mathbb{R}^n$ such that any $V(x)$, which starts from U_0, can reach $V(x) \equiv 0$ in finite time. In this case, the settling time t_r satisfies $t_r \leq \frac{1}{\lambda_1(1-\alpha)} \ln \frac{\lambda_1 V^{1-\alpha}x_0 + \lambda_2}{\lambda_2}$.*

3 Problem Statements

Given a multi-agent group consisting of n continuous-time agents, labeled from π_1 to π_n, each of them satisfies following second-order dynamics:

$$\pi_i : \dot{x}_i = v_i, \quad \dot{v}_i = u_i + d_i, \tag{1}$$

where $d_i, u_i, v_i, x_i \in \mathbb{R}$ denote the external disturbance, control input, velocity and position of the i-th agent. $\eta_i = [x_i, v_i]^T$ serves as the information state.

Assumption 1. *The external disturbance d_i of the i-th agent is bounded, i.e., $|d_i| \leq \bar{d}_i$ with $\bar{d}_i$ as a known constant.*

Further, the reference system is described as

$$\pi_0 : \dot{x}_0 = v_0, \quad \dot{v}_0 = u_0, \tag{2}$$

where $x_0, v_0, u_0 \in \mathbb{R}$, respectively, denote the associated position, velocity and control input. Suppose that the reference information state $\eta_0 = [x_0, v_0]^T$ is only accessible to parts of agents, then a non-negative diagonal matrix $G = \text{diag}\{g_1, g_2, \ldots, g_n\}$ is introduced to represent the availability of η_0 to the agents, where $g_i > 0$ if the i-th agent has access to η_0, and otherwise $g_i = 0$.

Assumption 2. *The graph $\bar{\mathcal{G}}$ involving π_0 (treated as a virtual leader) into $\mathcal{G}$ contains a spanning tree with π_0 as its root node, i.e., $G \neq 0$.*

Assumption 3. *The control input u_0 of the reference system π_0 is unknown to each agent; however, its upper bound $\bar{u}_0 > 0$ is available to its child nodes.*

The objective is to develop a control input u_i for each agent, such that the multi-agent system (1) achieves (or reaches) the finite-time consensus tracking to the reference system (2). In other words, for any $\eta_i(0)$, there exists a $T_m > 0$ such that the settling time $\bar{t} < T_m$ and $\lim_{t \to \bar{t}} |\eta_i(t) - \eta_0(t)| = 0$, $\eta_i(t) = \eta_0(t), \forall t \geq \bar{t}$.

4   Non-singular Finite-Time Consensus Protocols

For the i-th agent, define the local neighboring position and velocity tracking errors as $x_i^e = \sum\limits_{j \in N_i} a_{ij}(x_i - x_j) + g_i(x_i - x_0)$, $v_i^e = \sum\limits_{j \in N_i} a_{ij}(v_i - v_j) + g_i(v_i - v_0)$
and their column stacks as $x^e = [x_1^e, x_2^e, \ldots, x_n^e]^T$ and $v^e = [v_1^e, v_2^e, \ldots, v_n^e]^T$. In terms of (1) and (2), the dynamics of x^e and v^e satisfy

$$\dot{x}^e = v^e, \qquad \dot{v}^e = (L + G)(u + d) - G1_n u_0, \tag{3}$$

where $u = [u_1, u_2, \ldots, u_n]^T$, $d = [d_1, d_2, \ldots, d_n]^T$ and $1_n = [1, 1, \ldots, 1]^T$.

To achieve the concerned finite-time consensus tracking, introduce a non singular sliding mode manifold s_i for the i-th agent as follows:

$$s_i = x_i^e + k_i \text{sign}(v_i^e)|v_i^e|^\alpha, \tag{4}$$

where $k_i > 0$ and $\alpha \in (1, 2)$.

Lemma 4. *Consider the error dynamics (3). If the sliding mode manifold (4) reaches $s_i \equiv 0$, the tracking errors x_i^e and v_i^e of the i-th agent converge to the origin in finite time.*

Proof. If $s_i \equiv 0$, based on (4), $s_i = x_i^e + k_i \text{sign}(v_i^e)|v_i^e|^\alpha$ is obtained. Assign a Lyapunov function candidate $L_1 = \frac{1}{2}\sum_{i=1}^{n}(x_i^e)^2$. Its derivative satisfies

$$\dot{L}_1 = -\sum_{i=1}^{n} k_i^{\frac{-1}{\alpha}}(x_i^e)^{\frac{\alpha+1}{\alpha}} \leq -k_m \left(\sum_{i=1}^{n}(x_i^e)^2\right)^{\frac{\alpha+1}{2\alpha}} = -2^{\frac{\alpha+1}{2\alpha}} k_m V^{\frac{\alpha+1}{2\alpha}}, \tag{5}$$

where Lemma 1 has been used with the fact that $\frac{1+\alpha}{2\alpha} \in (0, 1)$, and $k_m = \min\{k_i^{\frac{-1}{\alpha}}\}$. It thus follows from Lemma 2 that $L_1(x_i^e)$ reaches $L_1 \equiv 0$ in finite time. This implies that $\lim_{t \to \bar{t}_1} x_i^e(t) = 0$, where the settling time

$$\bar{t}_1 \leq T_1 = \frac{\alpha L_1^{\frac{\alpha-1}{2\alpha}}(x_i^e(0))}{2^{\frac{1-\alpha}{2\alpha}} k_m(\alpha - 1)}. \tag{6}$$

Furthermore, based on the fact that $s \equiv 0$, it also follows that $\lim_{t \to \bar{t}_1} v_i^e(t) = 0$.

Lemma 4 indicates that as long as the error trajectory of each agent can reach the sliding mode manifold s_i in finite time, the finite-time consensus tracking mission can be accomplished. Thus, the control objective is converted to developing a control input u_i for each agent to drive its error trajectory towards the sliding mode manifold s_i in finite time. For this purpose, two protocol schemes are presented as follows.

4.1   Consensus Tracking Protocol I

Differentiating the sliding mode manifold s_i in (4) along (3) yields

$$
\begin{aligned}
\dot{s}_i &= \dot{v}_i^e + \alpha k_i |v_i^e|^{\alpha-1} \dot{v}_i^e \\
&= \dot{v}_i^e + \alpha k_i |v_i^e|^{\alpha-1}\left[\left(\sum_{j\in N_i}(a_{ij}+g_i)\right)(u_i+d_i) - \sum_{j\in N_i} a_{ij}(u_j+d_j) - g_i u_0\right] \\
&= \alpha k_i |v_i^e|^{\alpha-1}\left[\frac{\text{sign}(v_i^e)}{\alpha k_i}|v_i^e|^{2-\alpha} + \left(\sum_{j\in N_i}(a_{ij}+g_i)\right)u_i - \sum_{j\in N_i} a_{ij}u_j + d_i^c - g_i u_0\right],
\end{aligned}
\tag{7}
$$

where $d_i^c = (\sum_{j\in N_i} a_{ij}+g_i)d_i - \sum_{j\in N_i} a_{ij}d_j$. It follows from Assumption 1 that $|d_i^c| \le \bar{d}_i^c = (\sum_{j\in N_i} a_{ij}+g_i)\bar{d}_i + \sum_{j\in N_i} a_{ij}\bar{d}_j$.

The consensus tracking protocol is presented in the following theorem.

Theorem 1. *Consider the error dynamics (3) with the sliding mode manifold s_i in (4). If each protocol u_i is designed as*

$$
u_i = \left(\sum_{j\in N_i} a_{ij}+g_i\right)^{-1}\left[-\tau_i s_i - \frac{\text{sign}(v_i^e)}{\alpha k_i}|v_i^e|^{2-\alpha} + \sum_{j\in N_i} a_{ij}u_j - \beta_i \text{sign}(s_i)\right],
\tag{8}
$$

where $\tau_i > 0$ and $\beta_i > g_i\bar{u}_0 + \bar{d}_i^c$, the error trajectory of each agent can reach the sliding mode manifold s_i in finite time.

Proof. Substituting the protocol (8) into (7) yields that

$$
\dot{s}_i = \alpha k_i |v_i^e|^{\alpha-1}[-\tau_i s_i - \beta_i \text{sign}(s_i) + d_i^c - g_i u_0].
\tag{9}
$$

The following discussion is divided into two cases.

Case 1. $v_i^e \equiv 0 (i = 1, 2, \ldots, n)$. In this case, assign a Lyapunov function candidate $L_2 = \frac{1}{2}\sum_{i=1}^n (s_i)^2$. Its derivative along (9) satisfies

$$
\begin{aligned}
\dot{L}_2 &= \alpha \sum_{i=1}^n k_i |v_i^e|^{\alpha-1} s_i[-\tau_i s_i - \beta_i \text{sign}(s_i) + d_i^c - g_0 u_0] \\
&\le -\tau_m \sum_{i=1}^n (s_i)^2 - \beta_m \sum_{i=1}^n |s_i| \le -2\tau_m L_2 - 2^{\frac{1}{2}}\beta_m L_2^{\frac{1}{2}},
\end{aligned}
\tag{10}
$$

where $\tau_m = \alpha \inf\{k_i \tau_i |v_i^e|^{\alpha-1}\}$ and $\beta_m = \alpha \inf\{k_i(\beta_i - \bar{d}_i^c - g_i\bar{u}_0)|v_i^e|^{\alpha-1}\}$. Thus, it follows from Lemma 4 that $L_2(s_i)$ reaches $L_2 \equiv 0$ in finite time. This implies that $\lim t \to \bar{t}_2 s_i(t) = 0$, where the settling time $\bar{t}_2$ satisfies

$$
\bar{t}_2 \le T_2 = \frac{1}{\tau_m}\ln\frac{2^{\frac{3}{2}}L_2^{\frac{1}{2}}(s_i(0)) + \beta_m}{\beta_m}.
\tag{11}
$$

Case 2. $v_i^e = 0 (i \in 1, 2, \ldots, n)$. In this case, it is necessary to show that the sliding mode manifold s_i is still an attractor. In other words, what remains is to demonstrate that x_i^e-axis is not attractive except the origin (refer to Fig. 1).

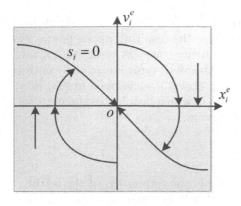

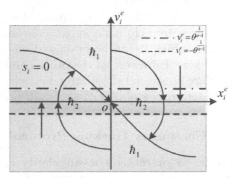

Fig. 1. Phase plot of the i-th error system (Protocol I).

Fig. 2. Phase plot of the i-th error system (Protocol II).

When $v_i^e = 0$ and $x_i^e \neq 0$, substituting the protocol (8) into (3) yields

$$\dot{v}_i^e = -\tau_i s_i - \beta_i \text{sign}(s_i) + d_i^c - g_i u_0. \tag{12}$$

In view of $\beta_i > \bar{u}_0 + \bar{d}_i^c$, then $\dot{v}_i^e > 0$ for $s_i < 0$, or $\dot{v}_i^e < 0$ for $s_i > 0$. This means that $v_i^e = 0$ is non-attractive. Thus, the error trajectory of each agent will transgress x_i^e-axis within a finite time ϵ (refer to Fig. 1). Consequently, the error trajectory of each agent can reach the sliding mode manifold s_i within a finite time $t_2' \leq T_2 + \epsilon$.

Eventually, by combining Lemma 4 and Theorem 1, it can be concluded that the protocol u_i developed in (8) achieves the consensus tracking of each agent in finite time, where the settling time $\bar{t}$ is bounded by $\bar{t} < T_1 + T_2 + \epsilon$.

It can be observed from (11) that, given initial error states, decreasing the parameters τ_m and β_m can effectively expedite the convergence of the error trajectory to the sliding mode manifold s_i. From the definitions of τ_m and β_m, they are functions with respect to the error state v_i^e, which leads to the fact that, a small $|v_i^e|$ will result in a slow convergence rate. Further, due to the dependence of the upper bound of the settling time on the transient states, it is unlikely to assess the settling time (which reflects the tracking performance) associated with initial error states and control parameters in advance.

To obviate aforementioned dilemma, an alternative consensus tracking protocol is exploited as follows:

$$u_i = \left(\sum_{j \in N_i} a_{ij} + g_i \right)^{-1} \left[u_i^c - \frac{\text{sign}(v_i^e)}{\alpha k_i} |v_i^e|^{2-\alpha} + \sum_{j \in N_i} a_{ij} u_j \right], \tag{13}$$

$$u_i^c = |v_i^e|^{1-\alpha} [-\tau_i s_i - \beta_i \text{sign}(s_i)], \tag{14}$$

where the control parameters are the same as those in Theorem 1. By following the similar analysis given in Theorem 1, the derivative of L_2 with the above protocol can be derived as $\dot{L}_2 \leq -2\tau'_m L_2 - 2\frac{1}{2}\beta'_m L_2^{\frac{1}{2}}$, where $\tau'_m = \alpha \min k_i \tau_i$ and $\beta'_m = \alpha \min k_i(\beta_i - \bar{d}_i^c - g_i \bar{u}_0)$. In this case, the resulting upper bound T_2 determined by (11) is no longer concerned with the transient error state v_i^e; in the meanwhile, it can be evaluated off-line with initial error states and control parameters. However, on the other hand, the protocol u_i developed in (13) may introduce nontrivial singularity, i.e., when $v_i^e = 0$, which invalidates the protocol and deteriorate stability of the multi-agent system.

4.2 Consensus Tracking Protocol II

In order to guarantee non-singularity and enable the estimation of the settling time independently of the transient states, another finite-time consensus tracking protocol is proposed as follows.

First, introduce a $\mathcal{C}^1$ function $\phi(z) = \begin{cases} \sin\left(\frac{\pi}{2}\frac{|z|}{\theta}\right) & if\ |z| \leq \theta \\ 1 & otherwise \end{cases}$, where θ is a positive constant. The consensus tracking protocol is presented in the following Theorem.

Theorem 2. *Consider the error dynamics (3) with the sliding mode manifold s_i in (4). If the protocol u_i is designed as*

$$u_i = \left(\sum_{j \in N_i} a_{ij} + g_i\right)^{-1}\left[u_i^c - \frac{\text{sign}(v_i^e)}{\alpha k_i}|v_i^e|^{2-\alpha} + \sum_{j \in N_i} a_{ij}u_j\right], \qquad (15)$$

$$u_i^c = \phi(|v_i^e|^{\alpha-1})|v_i^e|^{1-\alpha}[-\bar{\tau}_i s_i - \bar{\beta}_i \text{sign}(s_i)], \qquad (16)$$

where $\bar{\tau}_i > 0$ and $\bar{\beta}_i > g_i\bar{u}_0 + \bar{d}_i^c$, the error trajectory of each agent can reach the sliding mode manifold s_i in finite time.

Proof. Substituting the protocol (15) into (7) yields that

$$\dot{s}_i = \alpha k_i \phi(|v_i^e|^{\alpha-1})[-\bar{\tau}_i s_i - \bar{\beta}_i \text{sign}(s_i) + d_i^c - g_i u_0]. \qquad (17)$$

Assign a Lyapunov function candidate $L_3 = \frac{1}{2}\sum_{i=1}^{n}(s_i)^2$. Its derivative along (17) satisfies

$$\dot{L}_3 = \alpha \sum_{i=1}^{n} k_i|v_i^e|^{\alpha-1}s_i[-\bar{\tau}_i s_i - \bar{\beta}_i \text{sign}(s_i) + d_i^c - g_0 u_0]$$

$$\leq -\bar{\tau}_m \sum_{i=1}^{n}(s_i)^2 - \bar{\beta}_m \sum_{i=1}^{n}|s_i| \leq -2\bar{\tau}_m L_2 - 2^{\frac{1}{2}}\bar{\beta}_m L_2^{\frac{1}{2}}, \qquad (18)$$

where $\bar{\tau}_m = \alpha \inf\{k_i\bar{\tau}_i|v_i^e|^{\alpha-1}\}$ and $\bar{\beta}_m = \alpha \inf\{k_i(\bar{\beta}_i - \bar{d}_i^c - g_i\bar{u}_0)|v_i^e|^{\alpha-1}\}$. The subsequent discussion is divided into three cases.

Case 1. $|v_i^e| \geq \theta^{\frac{1}{\alpha-1}}$ $(i = 1, 2, \ldots, n)$. This corresponds to the fact that all the error states $[x_i^e, v_i^e]^T$ are located in area $\hbar$ (refer to Fig. 2). Within this area,

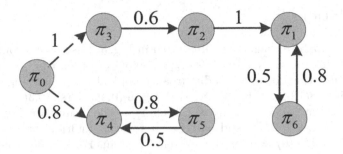

Fig. 3. A weighted graph of six agents.

$\phi(|v_i^e|^{\alpha-1}) = 1$, which implies that $\bar{\tau}_m = \{\alpha \min k_i \bar{\tau}_i\}$ and $\bar{\beta}_m = \alpha \min\{k_i(\bar{\beta}_i - \bar{d}_i^c - g_i \bar{u}_0)\}$. From Lemma 4, it follows that $L_3(s_i)$ reaches $L_3 \equiv 0$ in finite time. This implies that $\lim_{t \to \bar{t}_3} s_i(t) = 0$, where the settling time $\bar{t}_3$ satisfies

$$\bar{t}_3 \leq T_3 = \frac{1}{\bar{\tau}_m} \ln \frac{2^{\frac{3}{2}} L_2^{\frac{1}{2}}(s_i(0)) + \bar{\beta}_m}{\bar{\beta}_m}. \tag{19}$$

Case 2. $0 < |v_i^e| < \theta^{\frac{1}{\alpha-1}} (i \in \{1, 2, \dots, n\})$. This corresponds to the fact that some error state $[x_i^e, v_i^e]^T$ is situated in area $\hbar_2$ except x_i^e-axis (refer to Fig. 2). In this case, $0 < \phi(|v_i^e|^{\alpha-1}) < 1$; and due to $\bar{\tau}_m > 0$ and $\bar{\beta}_m > 0$, it follows from Lemma 4 that $s = 0$ is still an attractor.

Case 3. $v_i^e = 0 (i \subset \{1, 2, \dots, n\})$. This corresponds to the fact that some error state $[x_i^e, v_i^e]^T$ is situated in x_i^e-axis (refer to Fig. 2). In this case, it is necessary to demonstrate that x_i^e axis is not attractive except the origin.

In view of $\phi(|v_i^e|^{\alpha-1})|v_i^e|^{1-\alpha} = 1$, the protocol (15) becomes

$$u_i = \left(\sum_{j \in N_i} a_{ij} + g_i \right)^{-1} \left[-\bar{\tau}_i s_i - \bar{\beta}_i \text{sign}(s_i) + \sum_{j \in N_i} a_{ij} u_j \right], \tag{20}$$

when the error trajectory is in close proximity to x_i^e-axis. By substituting it into (3), it follows that $\dot{v}_i^e = -\bar{\tau}_i s_i - \bar{\beta}_i \text{sign}(s_i) + d_i^c - g_i u_0$.

Consider $\bar{\beta}_i > \bar{u}_0 + \bar{d}_i^c$. Then $\dot{v}_i^e > 0$ for $s_i < 0$, or $\dot{v}_i^e < 0$ for $s_i > 0$. This thus indicates that $v_i^e = 0$ is non-attractive.

By combining **Case 2** and **Case 3**, it can be concluded that, if $0 < |v_i^e| < \theta^{\frac{1}{\alpha-1}} (i \in \{1, 2, \dots, n\})$, the error trajectory of each agent will transgress area $\hbar_2$ into area $\hbar_1$ in finite time (refer to Fig. 2).

Consequently, the error trajectory of each agent can reach the sliding mode manifold s_i within a finite time $\bar{t}'_3 \leq T_3 + \epsilon(\theta)$, where $\epsilon(\theta)$ is a positive constant concerned with the parameter θ.

From Lemma 4 and Theorem 2, it follows that the protocol u_i developed in (15) achieves the consensus tracking of each agent in finite time, where the settling time $\bar{t}$ is bounded by $\bar{t} < T_1 + T_3 + \epsilon(\theta)$.

5   Simulations

A weighted graph of six agents is illustrated in Fig. 3. Designate a control input of the reference system as $u_0 = \sin(x_0)/(1 + \exp(-t))$, then $\bar{u}_0 = 1$ is obtained. Further, assume that the external disturbance applied to the multi-agent system is $d = [\sin(0.8t), 0.4\cos(0.1t), 0.8\cos(t), 1.2\sin(0.5t), 0.5\sin(0.3t), \cos(0.2t)]^T$, which implies that $\bar{d} = [1, 0.4, 0.8, 1.2, 0.5, 1]^T$. The states are initialized as $x(0) = [-5, -3, 4, 9, 4, 5]^T$ and $v(0) = 0$ for the multi-agent system, and $x_0(0) = \pi/2$ and $v_0(0) = 0$ for the reference system. For two developed protocols, the control parameters are chosen the same, that is, $\alpha = 5/3, \theta = 0.02, k_i = 0.1, \tau_i = \bar{\tau}_i = 20, \beta_i = \overline{(\beta)}_i = 4$, for $i = 1, 2, \ldots, 6$. With the selected initial information states and control parameters, it can be derived that $T_1 = 2.2564\,\text{s}$ and $T_3 = 2.2032\,\text{s}$. The simulation results are illustrated in Figs. 4 and 5.

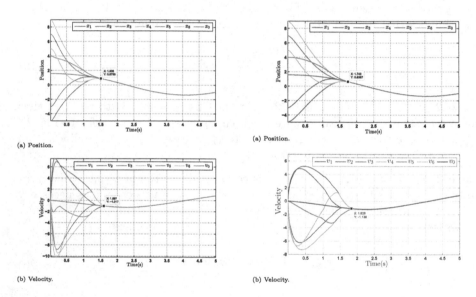

(a) Position.

(b) Velocity.

Fig. 4. Consensus tracking plots with Protocol I.

(a) Position.

(b) Velocity.

Fig. 5. Consensus tracking plots with Protocol II.

Figure 4 depicts that the consensus tracking mission with the developed protocol (8) is accomplished in finite time (about 1.508 s for the position, and about 1.597 s for the velocity) in spite of the inestimable settling time, which verifies the performance claimed in Theorem 1. Further, Fig. 5 illustrates that the exploited protocol (15) can also achieve the consensus tracking of the multi agent system in finite time (about 1.743 s for the position, and about 1.829 s for the velocity), and the corresponding settling time is below $T_1 + T_3 = 4.4596\,\text{s}$, which validates the result presented in Theorem 2.

6 Conclusions

Two non-singular protocols are put forward for second-order multi-agent systems suffering from external disturbance to achieve finite-time consensus tracking. Initially, a non-singular sliding mode manifold, which can guarantee the finite-time convergence of the error trajectory on it, is developed for each agent dynamics. Subsequently, two non-singular protocols are synthesized such that the error trajectory can reach the sliding mode manifold in finite time. As for the first protocol, the resulting settling time is imponderable due to its reliance on transient states. In contrast, the settling time with the other protocol is estimable in advance by utilizing accessible initial states and control parameters.

References

1. Ren, W.: On consensus algorithms for double-integrator dynamics. IEEE Trans. Autom. Control **53**(6), 1503–1509 (2008)
2. Ren, W.: Information consensus in multi-vehicle cooperative control. IEEE Control Syst. Mag. **27**(2), 71–82 (2007)
3. Yu, W., Chen, G., Cao, M.: Some necessary and sufficient conditions for second-order consensus in muti-agent dynamical systems. Automatica **46**(6), 1089–1095 (2010)
4. Ye, D., Zhao, X., Cao, B.: Distributed adaptive fault-tolerant consensus tracking of multi-agent systems against time-varying actuator faults. IET Control Theor. Appl. **10**(5), 554–563 (2016)
5. Ren, W., Beard, R.W.: Distributed Consensus on Multi-vehicle Cooperative Control. Springer, London (2008). https://doi.org/10.1007/978-1-84800-015-5
6. Zuo, Z.: Nonsingular fixed-time consensus tracking for second-order multiagent networks. Automatica **54**, 305–309 (2015)
7. Zhang, H., Lewis, F., Das, A.: Optimal design for synchronization of cooperative systems: state feedback, observer and output feedback. IEEE Trans. Autom. Control **56**(8), 1948–1952 (2011)
8. Li, Z., Liu, X., Ren, W., et al.: Distributed tracking control for linear multiagent systems with a leader of bounded unknown input. IEEE Trans. Autom. Control **58**(2), 518–523 (2013)
9. Shen, Q., Jiang, B., Shi, P., et al.: Cooperative adaptive fuzzy tracking control for networked unknown nonlinear multiagent systems with time-varying actuator faults. IEEE Trans. Fuzzy Syst. **22**(3), 494–504 (2013)
10. Chen, G., Ho, D.W., Li, L., et al.: Fault-tolerant output synchronisation control of multivehicle systems. IET Control Theor. Appl. **8**(8), 574–584 (2014)
11. Wang, L., Xiao, F.: Finite-time censensus problem for networks of dynamic agents. IEEE Trans. Autom. Control **55**(4), 950–955 (2010)
12. Bhat, S.P., Bernstein, D.S.: Finite-time stability of continuous autonomous systems. SIAM J. Control Optim. **38**(3), 751–766 (2000)
13. Xiao, F., Wang, L., Chen, T.: Finite-time consensus of multi-agent systems with directed and intermittent links. In: Proceedings of the 30th Chinese Control Conference, China, Yantai, pp. 6539–6543 (2011)
14. Zuo, Z., Tie, L.: A new class of finite-time nonlinear consensus protocols for multi-agent systems. Int. J. Control **87**(2), 363–370 (2014)

15. Li, S., Du, H., Lin, X.: Finite-time consensus algorithm for multi-agent systems with double-integrator dynamics. Automatica **47**(8), 1706–1712 (2011)
16. Yu, S., Long, X.: Finite-time consensus for second-order multi-agent systems with disturbances by integral sliding mode. Automatica **54**, 158–165 (2015)
17. Yu, S., Yu, X., Shirinzadeh, B., et al.: Continuous finite-time control for 290 robotic manipulators with terminal sliding mode. Automatica **41**(11), 1957–1964 (2005)

UAV Cooperation and Control

Multi-UAV Cooperative Path Planning via Mutant Pigeon Inspired Optimization with Group Learning Strategy

Yueping Yu[1], Yimin Deng[1], and Haibin Duan[1,2(✉)]

[1] School of Automation Science and Electrical Engineering, Beihang University, Beijing 100083, China
hbduan@buaa.edu.cn
[2] Peng Cheng Laboratory, Shenzhen 518000, China

Abstract. This paper proposes a mutant pigeon-inspired optimization algorithm with group learning strategy (MGLPIO), for multi-unmanned aerial vehicle(UAV) cooperative path planning. The group learning strategy is introduced in map and compass operator to reduce computation complexity and enhance the global search ability. At the same time, the triple mutations strategy is employed in landmark operator to enhance swarm diversity. What's more, in order to synchronize multi-UAV, the time stamp segmentation technique is designed to prove waypoints, which can simplify the cost function by reducing the number of independent variables. Besides, we geometric the threat sources to quantify their dangerous level. The coordination costs can guarantee collision-free flight and real-time communication. Finally, the proposed method is applied to path planning in set scenarios. The simulation results indicate that our model is feasible and effective, and the MGLPIO algorithm can have a good balance between exploration and exploitation by comparing with other four algorithms.

Keywords: Cooperative path planning · Mutant pigeon inspired optimization · Group learning strategy

1 Introduction

Compared with single unmanned aerial vehicle(UAV), using multi-UAV to perform complex tasks can not only improve the efficiency, but also improve the reliability of the tasks. Cooperative path planning is an important guarantee for the safe flight of multi-UAV in complicated battlefield environment [1].

Path planning algorithms can be divided into traditional methods and intelligent optimization algorithms. Traditional methods, such as A* algorithm, artificial potential field and Voronoi diagram, have many limitations, which are widely used for single UAV [2]. Intelligent optimization algorithms can be applied to solve multi-dimension

© Springer Nature Switzerland AG 2021
Y. Tan and Y. Shi (Eds.): ICSI 2021, LNCS 12690, pp. 195–204, 2021.
https://doi.org/10.1007/978-3-030-78811-7_19

and multi-constraint problem in the multi-UAV cooperative path planning due to its low requirements and high robustness.

Inspired by the astounding navigation competence of homing pigeons, a novel optimization algorithm called pigeon-inspired optimization (PIO) [3] was proposed by Duan et al., which has been applied extensively and has many variants. Duan and Zhao [4] proposed a dynamic discrete PIO to handle cooperative search-attack mission planning for UAVs. Qiu et al. modified a multi-objective PIO based on the hierarchical learning behavior in pigeon flocks [5].

In this paper, the time stamp segmentation technique is embedded to the multi-UAV path planning model firstly, which can ensure interoperability between multi-UAV. Then the constraints of UAVs are modeled as cost function. At last, a mutant PIO algorithm with group learning strategy is introduced to solve the model. Group learning strategy and triple mutations strategy can improve the search efficiency and enhance search precision.

2   Model of Multi-UAV Cooperative Path Planning

It is set that our num_{uav} UAVs need to cooperate to perform a military task where one enemy target needs to be attacked. On the way to the enemy target, our UAVs will pass through the enemy's military defense zone. The threat sources mainly involve mountains, enemy artilleries and radars. UAVs should escape from the threat sources, avoid internal collision, ensure internal communication and reach the enemy target at the same time. Besides the total path should be the shortest.

2.1   Model Waypoints of Single UAV

Assume that all UAVs move at the same altitude pz in the initial stage of path planning, the waypoints of UAVs mainly are considered in the horizontal plane. In order to meet the requirement of multi-UAV reaching the enemy target simultaneously, the time stamp segmentation mechanism [6] is designed. Given the desired flight time is T and the number of waypoints is D_p, then the time stamp is $\Delta t = T/(D_p - 1)$. In the global coordinate system $X_g O_g Y_g$, the position of UAV n at waypoint d is denoted as $\mathbf{P}_d^n = [px_d^n, py_d^n]^T$, the velocity of UAV n between the waypoint d and $d+1$ is denoted as $\mathbf{V}_d^n = [vx_d^n, vy_d^n]^T$, where $n = 1, 2, \ldots num_{uav}, d = 1, 2 \ldots D_p$. The $\mathbf{P}_d^n$ can be expressed as

$$\mathbf{P}_{d+1}^n = \mathbf{P}_d^n + \mathbf{V}_d^n \cdot \Delta t \tag{1}$$

Therefore, the determination of waypoints can be converted into the determination of velocity between waypoints. To increase the search efficiency of the optimal velocity, a local coordinate system $X_l O_l Y_l$ is established, which is shown in Fig. 1. The origin of the local coordinate system is the starting point of multi-UAV and the X_l axis is going from the start point $\mathbf{Start} = [Start_x, Start_y]^T$ to the enemy target $\mathbf{End} = [End_x, End_y]^T$.

Hence, the coordinate transformation between the global and the local coordinate system can be written as

$$\begin{bmatrix} p'x_d^n \\ p'y_d^n \end{bmatrix} = \begin{bmatrix} \cos\phi & \sin\phi \\ -\sin\phi & \cos\phi \end{bmatrix} \begin{bmatrix} px_d^n \\ py_d^n \end{bmatrix}$$

$$\phi = \arctan \frac{End_y - Start_y}{End_x - Start_x}$$

(2)

where $[p'x_d^n, p'y_d^n]^T$ is the position of UAV n at waypoint d in local coordinate system.

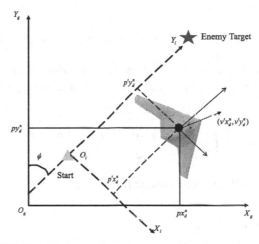

Fig. 1. UAV in global coordinate system and local coordinate system.

To reduce the number of variables to optimize, the velocity $v'x_d^n$ in the X_l direction is a constant, which is calculated by

$$v'x_d^n = \frac{\sqrt{(End_x - Start_x)^2 + (End_y - Start_y)^2}}{D_p \cdot \Delta t}$$

(3)

So only the velocity $v'y_d^n$ in the Y_l direction needs to be optimized.

2.2 Model Threat Sources

Three threat sources are considered in this paper, including mountains, enemy artilleries and radars. The dangerous degree of each threat source is measured by designing a cost function. We can regard the mountain as the trapezoidal cylinders, then the dangerous degree of the mountain fm_n^d is expressed as

$$fm_n^d = \begin{cases} \dfrac{k_m}{\|\Delta \mathbf{P}_1\|} & \text{if } pz \leq h_M \text{ and } \|\Delta \mathbf{P}_1\| \leq (1 - \dfrac{pz}{h_M}) \cdot r_M^a + \dfrac{pz}{h_M} \cdot r_M^b \\ 0 & \text{else} \end{cases}$$

(4)

where $\Delta\mathbf{P}_1 = \mathbf{P}_d^n - \mathbf{P}_M$, $\mathbf{P}_M = [px_M, py_M]^T$ is the center position of a mountain, h_M is its height, r_M^a and r_M^b are the bottom and the top radius, k_m is a constant parameter.

When modeling the threat degree of an artillery, the danger area caused by artillery is regarded as the hemisphere. Hence, the cost function of an artillery fa_n^d is presented as

$$fa_n^d = \begin{cases} 10^4 & \|\Delta\mathbf{P}_2\| \leq \dfrac{r_f}{3} \text{ and } pz \leq r_f \\ k_f * \dfrac{(r_f - \|\Delta\mathbf{p}_2\|)}{r_f} & \dfrac{r_f}{3} < \|\Delta\mathbf{P}_2\| \leq r_f \text{ and } pz \leq r_f \\ 0 & \text{else} \end{cases} \tag{5}$$

where $\|\Delta\mathbf{P}_2\| = \mathbf{P}_d^n - \mathbf{P}_A$, $\mathbf{P}_A = [px_A, py_A]^T$ is the position of an artillery, r_f is its firing radius, k_f is a constant parameter.

For the area of interference generated by a radar, it is also described by a hemisphere with its position $\mathbf{P}_R = [px_R, py_R]^T$ and the detection radius r_R, then the degree of interference fr_n^d is calculated by

$$fr_n^d = \begin{cases} (\dfrac{k_r}{\|\Delta\mathbf{p}_3\|})^2 & \|\Delta\mathbf{p}_3\| \leq r_R \text{ and } pz \leq r_R \\ 0 & \text{else} \end{cases} \tag{6}$$

where $\|\Delta\mathbf{P}_3\| = \mathbf{P}_d^n - \mathbf{P}_R$, k_r is a constant parameter.

2.3 Model Coordination Costs

It is required that the distance between UAVs should be moderate, not too close to cause collision, not too far to affect communication between UAVs. In this paper, the cost of internal collision $fc1_n^d$ exists between each UAV, and the cost of internal communication $fc2_n^d$ exists among partial UAVs.

$$fc1_n^d = \sum_{m=1}^{num_{uav}} fc1_{n\_m}^d, \quad m \neq n$$

$$fc1_{n\_m}^d = \begin{cases} 0 & \|\mathbf{P}_d^n - \mathbf{P}_d^m\| \geq r_o \\ k_{c1} & \text{else} \end{cases} \tag{7}$$

$$fc2_n^d = \begin{cases} 0 & nei_n \geq ceil(n/2) \\ k_{c2} & \text{else} \end{cases}$$

$$nei_n = \sum_{m=1}^{num_{uav}} w_n^m, \quad m \neq n \tag{8}$$

$$w_n^m = \begin{cases} 1 & \|\mathbf{P}_d^n - \mathbf{P}_d^m\| \leq r_c \\ 0 & \text{else} \end{cases}$$

where r_o and r_c are safety and communication radius, nei_n is the number of interactive neighbors of UAV n, $ceil(\cdot)$ is a rounding function, k_{c1} and k_{c2} are constant parameters.

All in all, the total cost function $fitness_n$ of an optional path of UAV n could be summarized as follows

$$fitness_n = \sum_{d=1}^{D_p} (fm_n^d + fa_n^d + fr_n^d + fc1_n^d + fc2_n^d) + fd_n$$

(9)

$$fd_n = \sum_{d=1}^{D_p} \|\mathbf{P}_{d+1}^n - \mathbf{P}_d^n\|$$

where fd_n represents the total distance of UAV n.

3 The MGLPIO Algorithm

3.1 The Basic PIO Algorithm

Map and Compass Operator. Consider N pigeons flying in a D dimensional search space. The pigeons' position and roost represent the potential solution and optimal solution respectively. At iteration t, the pigeon i's position $\mathbf{X}_i^t = [x_{i1}^t, x_{i2}^t, \ldots, x_{iD}^t]$ and velocity $\mathbf{V}_i^t = [v_{i1}^t, v_{i2}^t, \ldots v_{iD}^t]$ is updated based on the global optimal position $\mathbf{X}_{gbest}$, which is as follows:

$$\mathbf{V}_i^{t+1} = \mathbf{V}_i^t \cdot e^{-R \cdot (t+1)} + rand \cdot (\mathbf{X}_{gbest} - \mathbf{X}_i^t)$$
$$\mathbf{X}_i^{t+1} = \mathbf{X}_i^t + \mathbf{V}_i^{t+1}$$

(10)

where $0 \leq t \leq NC_{\max 1}$, $NC_{\max 1}$ is the maximum iteration of the map and compass operator. R is the map and compass factor. $rand$ represents a random number between 0 and 1.

Landmark Operator. When iteration $t > NC_{\max 1}$, pigeons gradually close to the destination. The landmark operator starts to provide navigation route for pigeons. In the landmark operator, the pigeons which are far away from the roost will be gradually abandoned, so the population size is reduced by half at per iteration, which is defined as

$$N = ceil(\frac{N}{2})$$

(11)

The rest pigeons' position is updated by familiar landmark $\mathbf{X}_{center}^t$, which is designed as

$$\mathbf{X}_i^{t+1} = \mathbf{X}_i^t + rand \cdot (\mathbf{X}_{center}^t - \mathbf{X}_i^t)$$
$$\mathbf{X}_{center}^t = \frac{\sum_{i=1}^{N} \mathbf{X}_i^t \cdot Fit(\mathbf{X}_i^t)}{\sum_{i=1}^{N} Fit(\mathbf{X}_i^t)}$$

(12)

where the weight $Fit(\mathbf{X}_i^t)$ is calculated by the following equation:

$$Fit(\mathbf{X}_i^t) = \begin{cases} fitness(\mathbf{X}_i^t), & \text{for maximization} \\ \frac{1}{fitness(\mathbf{X}_i^t)+\varepsilon} & \text{for minimization} \end{cases} \tag{13}$$

where $fitness(\mathbf{X}_i^t)$ is the cost function and the ε is an arbitrary nonzero constant. The maximum number of iterations in this stage is $NC_{\max 2}$.

3.2 Mutant PIO Algorithm with Group Learning Strategy

Since the basic PIO algorithm is easily to be trapped into the local optimal solution. The group learning strategy and the triple mutations strategy are respectively introduced in two operators to balance exploration and exploitation.

Group Learning Strategy. At first, the pigeons are randomly divided into M groups in every iteration. Then each group is treated as a big "pigeon". For this big "pigeon", the worst pigeon in this group is regarded as itself. Besides, its personal best $\mathbf{X}_{pbest}^j$ $(j = 1, 2, \ldots M)$ is represented by the best pigeon in the group [7]. Hence, the big "pigeon" is updated by learning from its personal best $\mathbf{X}_{pbest}^j$ and the global optimal position $\mathbf{X}_{gbest}$, which is stated as follows:

$$\mathbf{V}_j^{t+1} = \mathbf{V}_j^t \cdot e^{-R\cdot(t+1)} + rand_1 \cdot (\mathbf{X}_{gbest} - \mathbf{X}_j^t) + rand_2 \cdot (\mathbf{X}_{pbest}^j - \mathbf{X}_j^t)$$
$$\mathbf{X}_i^{t+1} = \mathbf{X}_i^t + \mathbf{V}_i^{t+1} \tag{14}$$

where $rand_1$ and $rand_2$ represent different random number. All in all, only updating the big "pigeon" can reduce the time cost. What's more, the swarm diversity is improved by learning from the best pigeon in each group.

Triple Mutations Strategy. In the landmark operator, the updating of pigeons depends on the weighted average position $\mathbf{X}_{center}^t$, which is easy to premature convergence. The triple mutations strategy is developed to perform more explorative search and enhance global search ability, which is showed as

$$\mathbf{X}\_\mathbf{G}_{center}^t = \mathbf{X}_{center}^t + Gaussian(\mu, \sigma) \cdot (\mathbf{X}_{\max} - \mathbf{X}_{\min}) \tag{15}$$

$$\mathbf{X}\_\mathbf{C}_{center}^t = \mathbf{X}_{center}^t + Cauchy(\mu, s) \cdot (\mathbf{X}_{\max} - \mathbf{X}_{\min}) \tag{16}$$

$$\mathbf{X}\_\mathbf{S}_{center}^t = \mathbf{X}_{center}^t + F \cdot (\mathbf{X}_m^t - \mathbf{X}_n^t) \tag{17}$$

where $\mathbf{X}\_\mathbf{G}_{center}^t$, $\mathbf{X}\_\mathbf{C}_{center}^t$ and $\mathbf{X}\_\mathbf{S}_{center}^t$ is the weighted average position $\mathbf{X}_{center}^t$ after Gaussian mutation, Cauchy mutation and scaling mutation [8], $Gaussian(\mu, \sigma)$ and $Cauchy(\mu, s)$ are the Gaussian and Cauchy distribution function, μ is the mean value and σ is the standard deviation of all pigeons' position, s is a constant parameter, F represents scaling factor, $\mathbf{X}_m^t$ and $\mathbf{X}_n^t$ are chosen randomly from all pigeons, $\mathbf{X}_{\max}$ and $\mathbf{X}_{\min}$ are the upper and lower limits of $\mathbf{X}_i^t$. After triple mutations, compare $fitness(\mathbf{X}\_\mathbf{G}_{center}^t)$, $fitness(\mathbf{X}\_\mathbf{C}_{center}^t)$ and $fitness(\mathbf{X}\_\mathbf{S}_{center}^t)$, then choose the fitter one to replace $\mathbf{X}_{center}^t$.

4 Simulation and Analysis

4.1 Path Correction and Smoothing

Since the waypoints solved by the model are discrete, the constraints at the non-waypoints cannot be satisfied. Hence, the path correction is needed to meet the actual flight requirements. If there is a waypoint whose connection between the starting point has no threat sources, this waypoint and the starting point can be directly connected. Then this waypoint is regarded as a new starting point, the remaining path are simplified as above method. In addition, given that the possible crossover among the simplified path of multi-UAV, the two UAVs change path after the point which is the closest point in front of the intersection. After correction, the path is zigzagging which don't satisfy the turning constraints of the aircraft. So B-spline curve is used to smooth the correction path.

4.2 Experiment Results and Algorithm Contrast

In order to verify the effectiveness of the model proposed in this paper, we set $num_{uav}=10$ UAVs take off from the same start point **Start** $= [-90, 60, 2]^T$ km simultaneously and they need arrive the enemy target **End** $= [90, 35, 4]^T$ km. The desired flight time T is 1 h, the number of waypoints D_p is 31, the expected altitude pz is 5 km, the safety radius r_o is 0.1 km and the communication radius r_c is 20 km. The information of threat sources is shown in Table 1. The constant parameter k_m, k_f, k_r are 200 and k_{c1}, k_{c2} are 100.

For verifying the superiority of MGLPIO algorithm, the contrast experiments with artificial bee colony (ABC) [9], particle swarm optimization (PSO) [10], mutant particle swarm optimization (MPSO), PIO algorithms are introduced. In all contrast experiments, the swarm size N is 30, and the maximum iteration is 30. The rest parameter configurations of five algorithms are set in Table 2.

Table 1. The information of threat sources.

Threat category	Threat number	Coordinates (km)	Parameters (km)
Mountains	2	$[-40, 30]^T$, $[-60, 70]^T$	$r_M^a = 16$, $r_M^b = 8$, $h_M = 6$
Artilleries	4	$[15, 30]^T$, $[70, 35]^T$ $[75, 60]^T$, $[25, 55]^T$	$r_f = 8$
Radars	2	$[20, 14]^T$, $[60, 50]^T$	$r_R = 14$, $r_R = 16$

The optimal path for UAVs by MGLPIO algorithm is shown in Fig. 2. We can draw the conclusion that multi-UAV evade all kinds of threat sources and reach the enemy target together. The dots represent the waypoint of each UAV at a given moment, which represents that the motion of each drone is synchronized.

Table 2. The parameter configurations of five algorithms.

Algorithm category	Parameters (*km*)
ABC [9]	$N_{searchbee} = N/4, \alpha = 1.2, NC_{limit} = 8$
PSO [10]	$c_1 = 1.3, c_2 = 1.3, \omega = 0.8$
MPSO	$s = 0.1, F = 0.8$
PIO [3]	$NC_{max\,1} = 25, NC_{max\,2} = 5, R = 0.2$
MGLPIO	$M = 10, s = 0.1, F = 0.8$

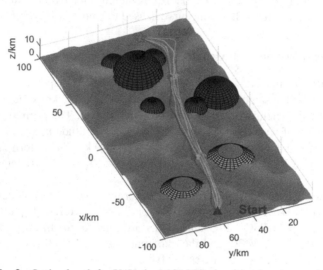

Fig. 2. Optimal path for UAVs by MGLPIO algorithm in set scenario.

To compare the performance of different optimization algorithms, we sum the cost function of all UAVs. The results are shown in Fig. 3 and Table 3.

From the Fig. 3 and the Table 3, it is obvious that the MGLPIO algorithm could find the least value and take the least cost time compared with other four algorithms. From the tenth iteration, ABC, PSO, MPSO and PIO algorithms trap into the local optimal solution, and the evolutionary curves have barely changed after this. However, the MGLPIO algorithm continues to search for the global optimal solution until the twenty-fifth iteration. The conclusion can be drawn that the MGLPIO algorithm has a good global search ability that it can plan the shortest path for multi-UAV under satisfying the constraint condition.

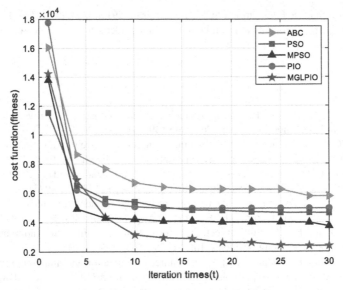

Fig. 3. Evolutionary curves of total cost function of all UAVs in five algorithms.

Table 3. The Minimum value and the cost time of total cost function of all UAVs.

Algorithm category	Minimum value	Cost time
ABC [9]	5772.2	2.0261 s
PSO [10]	4629.8	1.5622 s
MPSO	3726.7	1.7791 s
PIO [3]	4946.5	1.3348 s
MGLPIO	**2398.8**	**0.77349 s**

5   Conclusions

The multi-UAV cooperative path planning is a challenging technical problem. On the one hand, this paper introduces the time stamp segmentation mechanism to generate a serial of feasible waypoints. On the other hand, the dangerous degree of threat sources and the coordination requirements among UAVs are expressed as cost function. At the same time, the MGLPIO algorithm is employed to solve the optimal waypoints. The group learning strategy balances diversity and convergence and the triple mutations strategy prevents premature convergence. The simulation results verify that our model can provide effective path planning for multi-UAV, and the MGLPIO algorithm is superior to other algorithms with lowest cost.

Acknowledgements. This work was partially supported by Science and Technology Innovation 2030-Key Project of "New Generation Artificial Intelligence" under grant #2018AAA0102403,

National Natural Science Foundation of China under grant #91948204, #U20B2071, #U1913602 and #U19B2033, and Aeronautical Foundation of China under grant #20185851022.

References

1. Aggarwal, S., Kumar, N.: Path planning techniques for unmanned aerial vehicles: a review, solutions, and challenges. Comput. Commun **149**, 270–299 (2020)
2. Zhen, Z., Chen, Y., Wen, L., Han, B.: An intelligent cooperative mission planning scheme of UAV swarm in uncertain dynamic environment. Aerosp. Sci. Technol **100**, 1–16 (2020)
3. Duan, H., Qiu, H.: Advancements in pigeon-inspired optimization and its variants. SCIENCE CHINA Inf. Sci. **62**(7), 1 (2019). https://doi.org/10.1007/s11432-018-9752-9
4. Duan, H., Zhao, J.: Dynamic discrete pigeon-inspired optimization for multi-UAV cooperative search-attack mission planning. IEEE Trans. Aerosp. Electron. Syst **57**(1), 706–720 (2020)
5. Qiu, H., Duan, H.: A multi-objective pigeon-inspired optimization approach to UAV distributed flocking among obstacles. Inf. Sci **509**, 515–529 (2020)
6. Zhang, D., Duan, H.: Social-class pigeon-inspired optimization and time stamp segmentation for multi-UAV cooperative path planning. Neurocomputing **313**, 229–246 (2018)
7. Wang, Z., et al.: Dynamic group learning distributed particle swarm optimization for large-scale optimization and its application in cloud workflow scheduling. IEEE Trans. Cybern **50**(6), 2715–2729 (2020)
8. Mathi, D., Chinthamalla, R.: Enhanced leader adaptive velocity particle swarm optimisation based global maximum power point tracking technique for a PV string under partially shaded conditions. IET Renew. Power Gener **14**(2), 243–253 (2019)
9. Karaboga, D., Basturk, B.: A powerful and efficient algorithm for numerical function optimization: artificial bee colony (ABC) algorithm. J. Glob. Optim **39**(3), 459–471 (2007)
10. Kennedy, J., Eberhart, R.: Particle swarm optimization. In: Proceedings of the IEEE International Conference of Neural Networks, vol. 4, pp. 1942–1948. IEEE, Perth (1995)

UAV Path Planning Based on Variable Neighborhood Search Genetic Algorithm

Guo Zhang[1,2], Rui Wang[1,2(✉)], Hongtao Lei[1,2], Tao Zhang[1,2], Wenhua Li[1,2], and Yuanming Song[1,2]

[1] College of Systems Engineering, National University of Defense Technology, Changsha 410073, China
[2] Hunan Key Laboratory of Multi-Energy System Intelligent Interconnection Technology, Changsha 410073, China

Abstract. This study proposed a new genetic algorithm with variable neighbourhood search (GAVNS) for UAV path planning in three-dimensional space. First, an 0–1 integer programming mathematical model is established by inspired from the vehicle routing planning model with time window (VRPTW), and then a heuristic rule based on space vector projection is designed to quickly initialize high-quality solutions that meet constraints of upper error limit and minimum turning radius. Second, it improves mutation operator with a reselected mutation strategy, and incorporates Variable Neighborhood Search strategy based on adding and deleting route during the search process; Finally, GAVNS is compared with general Genetic Algorithm on a set of experiments. It is demonstrated that GAVNS algorithm is both effective and efficient. Moreover, the introduction of variable neighborhood search strategy enhances the local search ability of Genetic Algorithm.

Keywords: Path planning · GAVNS · 0–1 integer programming model

1 Introduction

Unmanned Aerial Vehicle (UAV) are widely applied in military fields such as battlefield reconnaissance and long-range strike due to their advantages such as easy concealment, strong maneuverability, as well as unmanned driving, and have become one of the important attack methods for unmanned and intelligent battlefields in the future [1]. In the face of complicated environmental conditions and diversified uncertain factors, advance trajectory planning is particularly important [2]. The problem of fast trajectory planning is an important subject in UAV control. The main algorithms to solve the problem include Artificial Potential Field method [3], Mixed Integer Linear Programming method [4], A* Algorithm [5], Genetic Algorithm [6], Particle Swarm Optimization [7] and Ant Colony Optimization [8], etc. The intelligent optimization algorithm is widely applied in trajectory planning problems such as Chen [9] et al. converted the UAV path optimization problem into a TSP problem and applied the Ant Colony Optimization to solve it; Literature [10] firstly used the Dubins Algorithm to generate the initial feasible path of the UAV, and then used the Genetic Algorithm to improve the path to the optimality;

© Springer Nature Switzerland AG 2021
Y. Tan and Y. Shi (Eds.): ICSI 2021, LNCS 12690, pp. 205–217, 2021.
https://doi.org/10.1007/978-3-030-78811-7_20

Peng[11] et al. proposed an effective hybrid algorithm combining genetic algorithm and variable neighborhood search together for the permutation flow shop scheduling problem. Due to the limitation of the system structure, the positioning system of the UAV cannot accurately position itself, and the accuracy is affected by the accumulated error of the built-in inertial navigation system, which may lead to mission failure once the positioning errors have accumulated to a certain level. Therefore, correcting the positioning error during the flight is an important task in the path planning of UAV.

Considering the cumulative error correction and the minimum turning radius limitation of the UAV, we establish the 0–1 integer programming model by appropriately simplifying the fast path planning problem of the UAV in three-dimensional space under the system positioning accuracy constraint and combine Variable neighborhood Search with Genetic Algorithm to solve it.

2 Model

2.1 Description and Analysis

It is assumed that the size and mass of the UAV are ignored during the flight, and it is considered as a mass point without considering the constraints such as fuel consumption, while its positioning error (i.e., horizontal and vertical error) does not have a delayed situation and can be updated synchronously in real-time with the accumulation of flight distance. Its path's constraints are as follows:

- The UAV needs real-time positioning during space flight, and its positioning error includes vertical error and horizontal error. For one meter of flight, the vertical and horizontal errors will each increase by δ. The vertical and horizontal errors should be less than θ when reaching the endpoint B. And to simplify the problem, it is assumed that the UAV can still fly according to the planned path when both vertical and horizontal errors are less than θ.
- The UAV needs to be corrected for positioning errors during the flight. There are safe locations in the flight area (called correction points) that can be used for error correction, and it will be corrected based on the type of error correction when the UAV reaches. If the vertical and horizontal errors can be corrected in time, the UAV can follow a predetermined path, passing through several correction points to correct the errors and finally reach the endpoint.
- At the start point A, the vertical and horizontal errors of the UAV are both 0.
- When the UAV at the vertical error correction point, its vertical error will be corrected 0 and the horizontal error will remain unchanged.
- When the UAV at the horizontal error correction point, its horizontal error will be corrected 0 and the vertical error will remain unchanged.
- When the UAV's vertical error is no more than α_1 and the horizontal error is no more than α_2, the vertical error correction can be performed.
- When the UAV's vertical error is no more than β_1 and the horizontal error is no more than β_2, the horizontal error correction can be performed.

- The UAV is limited by the structure and control system during the turn and cannot make the instant turn (the forward direction of the UAV cannot be changed suddenly), assuming the minimum turn radius of the UAV $R = 200$ m.

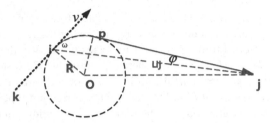

Fig. 1. Diagram of the ideal flight path of the UAV

The UAV come from i to j, the angle between i point's initial velocity $\vec{v}$ and $\vec{ij}$ is ω, at that time the trajectory is the arc ip plus line segment pj, where pj is the tangent line of circle O, p is the tangent point. It is clearly that this trajectory is the optimal flight track for the UAV (see Fig. 1).

From the theorem of the cosine of a triangle, it follows that:

$$|Oj|^2 = R^2 + L_{ij}^2 - 2RL_{ij} \sin \varpi \tag{1}$$

In addition,

$$|pj|^2 = |Oj|^2 - R^2 \tag{2}$$

$$|ip| = R(\varpi + \varphi) \tag{3}$$

Where L_{ij} denotes the Euclidean distance from point $i(x_i, y_i, z_i)$ to point $j(x_j, y_j, z_j)$.

$$L_{ij} = \sqrt{(x_i - x_j)^2 + (y_i - y_j)^2 + (z_i - z_j)^2} \tag{4}$$

Since $L_{ij} \gg R$, then $\varphi \to 0$. Thus the distance of the track between any two points $dis(i, j)$ can be obtained as follows:

$$dis(i, j) = |ip| + |pj| = R(\varpi + \varphi) + (|Oj|^2 - R^2)^{1/2}$$
$$= R\varpi + (R^2 + L_{ij}^2 - 2RL_{ij} \sin \varpi - R^2)^{1/2} = R\varpi + (L_{ij}^2 - 2RL_{ij} \sin \varpi)^{1/2} \tag{5}$$

Where $\cos \varpi = \frac{<\vec{ki}, \vec{ij}>}{|\vec{ki}||\vec{ij}|}$, then

$$dis(i, j) = R \arccos \frac{<\vec{ki}, \vec{ij}>}{|\vec{ki}||\vec{ij}|} + \left(L_{ij}^2 - 2RL_{ij} \sin \arccos \frac{<\vec{ki}, \vec{ij}>}{|\vec{ki}||\vec{ij}|}\right)^{1/2} \tag{6}$$

2.2 Mathematical Model

We refer the Vehicle Route Problem with Time Window (VRPTW) [12] and establish the following mathematical model by replacing the time window constraint with an error accumulation constraint and removing the constraint that the vehicle must pass through all customer points.

Firstly, a set of points $N = \{1, 2, ..., n\}$ is created to represent all the correction points except for the start and end of the path; The edge matrix $X = \{x_{ij}|i, j \in N\}$, where x_{ij} is the edge from i to the j, $x_{ij} = 1$ indicates the UAV will come from i to j; The distance matrix $D = \{d_{ij}|i, j \in N\}$ where d_{ij} represents the distance between the correction points i, j; φ_i, γ_i represents the horizontal error and vertical error respectively when the UAV flies to the point i, and σ_i denotes the correction type of the correction point i where $\sigma_i = 0$ indicates that i is the horizontal correction point (otherwise it is the vertical correction point). It is assumed that a feasible trajectory is represented by $M_i = \{m_0 = 0, m_1, m_2, \ldots, m_s, m_{s+1} = n + 1\}$, where $0, n + 1$ denote the point A and B respectively in the problem, as well as $m_1 \sim m_s$ is a series of the ordered correction points selected from N, then the mathematical model of the problem is as follows:

$$\min f_1 = \sum_{i=0}^{n+1} \sum_{j=0}^{n+1} x_{m_i m_j} d_{m_i m_j} \tag{7}$$

$$x_{ij} = x_{ji} , \ \forall i, j \in N \tag{8}$$

$$x_{ii} = 0, \ \forall i \in N \tag{9}$$

$$x_{ij} \odot x_{jk} = 1, \forall j \in N, \exists i, k \in N \ and \ i \neq k \tag{10}$$

$$\sum_{i=1}^{n} x_{0i} = 1, \forall i \in N \tag{11}$$

$$\sum_{i=1}^{n} x_{i(n+1)} = 1, \forall i \in N \tag{12}$$

$$\varphi_{m_i} \leq \alpha_1, \sigma_{m_i} = 1 \ and \ x_{m_i m_{s+1}} = 0 \tag{13}$$

$$\gamma_{m_i} \leq \alpha_2, \sigma_{m_i} = 1 \ and \ x_{m_i m_{s+1}} = 0 \tag{14}$$

$$\varphi_{m_i} \leq \beta_1, \sigma_{m_i} = 0 \ and \ x_{m_i m_{s+1}} = 0 \tag{15}$$

$$\gamma_{m_i} \leq \beta_2, \sigma_{m_i} = 0 \ and \ x_{m_i m_{s+1}} = 0 \tag{16}$$

$$\varphi_{m_s} + x_{m_s m_{s+1}} d_{m_i m_{s+1}} \delta < \theta \tag{17}$$

$$\gamma_{m_s} + x_{m_s m_{s+1}} d_{m_s m_{s+1}} \delta < \theta \tag{18}$$

$$|Oj| \geq R, \ \forall j \in N \tag{19}$$

$$\varphi_{m_0} = \gamma_{m_0} = 0 \tag{20}$$

Equation (7) is the optimization objective of the UAV path planning problem, indicating that the total length of the trajectory path is minimized; Eq. (8) means that the path planning is simplified to an undirected point selection and connection problem, only considering whether two points are connected to each other without the direction of the connection; Eq. (9) means that any correction point cannot be connected with itself to form a loop; Eq. (10) means that for any correction point other than the and the end point, there are only two cases where it is not connected to any point or connects to a different point; Eq. (11), (12) means that the path must contain the start and the end point; Eq. (13), (14) means that if a correction point is a horizontal correction point and not the end in the path, its horizontal and vertical errors must be less than α_1 and α_2 respectively; Similarly Eq. (15), (16) means that if a correction point is a vertical and not the end in the path, its horizontal and vertical errors must be less than β_1 and β_2 respectively; Eq. (17), (18) means that the horizontal and vertical errors of the end point must be less than θ; Eq. (19) means that the turning radius of the UAV should be greater than or equal to the minimum turning radius R; Eq. (20) means that the initial horizontal and vertical errors of the start point are set 0.

3   Variable Neighborhood Search Genetic Algorithm

In the next, we have improved the Genetic Algorithm by design the specific mutation operator and make the best individual in each generation do variable neighborhood search. It enhances the local search ability of the Genetic Algorithm, avoiding the premature convergence of the Genetic Algorithm which leads to a local optimal solution, the steps as follows.

Algorithm 1: Genetic Algorithm with Variable Neighborhood Search
Input: the number of generations $mGen$, the number of population size N
Output: bets solution

---

1   $gen \leftarrow 1$
2   $P \leftarrow$ initPopulation ();
3   **While** ($gen \leq mGen$) **do**
4       $parents \leftarrow$ Selection (P)
5       $child \leftarrow$ Crossover $(parents)$
6       $child \leftarrow$ Mutation $(child)$
7   evaluateFitness $(child)$
8       $best\_solution \leftarrow$ best_Selection $(child)$
9       $best\_solution \leftarrow$ VNS $(best\_solution)$
10      $parents \leftarrow parents | best\_solution$
11      $gen \leftarrow gen+1$
12 **End**

3.1 Chromosome Coding

0 and 1 are used to encode the trajectory path, $x = \{x_1, x_2, ..., x_n\}$ representing an individual, where $x_i = 1$ means that the correction point i is selected for position error correction (see Fig. 2). x_1, x_{n+1} is constant to 1 means that the UAV must start from the start point A to reach the end point B. The order in which the UAV passes through the correction points is determined by the size of the projection of the correction points on the track onto the spatial vector. The length of the chromosome is determined by the number of all correction points to be selected in space.

| 1 | 0 | 1 | ... | 1 | 1 | 1 |

Fig. 2. Diagram of chromosome coding

3.2 Generating the Initial Feasible Solution

The quality of the initial feasible solution has a significant impact on the convergence of speed for algorithm. To address the dilemma that the problem's constraints are too complex to generate difficultly initial feasible solutions, we design a heuristic rule based on space vector projection, which can generate quickly initial feasible solutions with high quality.

From the start point A, the correction points are selected in turn: firstly, the type of the next correction point and the set of candidate correction points are determined, and finally, a specific correction point is selected from the candidate according to the heuristic rule of space vector projection.

$$\sigma_{i+1} = \begin{cases} 0, & min(\alpha 1 - \gamma_i', \beta 1 - \gamma_i') > min(\alpha 2 - \varphi_i', \beta 2 - \varphi_i') \\ 1, & otherwise \end{cases} \tag{21}$$

$$\varphi_j = \varphi_i' + dis(i, j)\delta \tag{22}$$

$$\gamma_j = \gamma_i' + dis(i, j)\delta \tag{23}$$

The first step is to determine the next correction point type. Firstly, the remaining horizontal and vertical error margins for the current point are calculated respectively, and the type of the next correction point is determined according to Eq. (21), where $\sigma_{i+1} = 1$ indicates that the type of the next is a vertical correction point, φ_i, γ_i is the horizontal and vertical error of the i point before correction, φ_i', γ_i' is the horizontal and vertical error after correction. The error update formula is shown in Eq. (22), (23).

The second step is to determine the set of candidate correction points C_i. According to Eq. (24), a sphere of radius r is made with the current point as the center of the sphere, then all correction points within the sphere are traversed, and the distance Oj from the candidate point to the center of the sphere O is calculated by according to Eq. (1). If $|Oj| < R$ means that the candidate does not satisfy the minimum turning

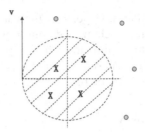

Fig. 3. Diagram of infeasible correction points

radius constraint, then it needs to be rejected. The dashed circular area is the candidate point that does not satisfy the minimum turning radius constraint (see Fig. 3), while it is a circular pipe in the three-dimensional space.

$$r_i = \frac{min(\zeta_1 - \varphi'_i, \zeta_2 - \gamma'_i)}{\delta} \tag{24}$$

$$K_{c_j} = \frac{\overrightarrow{i_{curr}c_j} \times \overrightarrow{AB}}{\left\| \overrightarrow{AB} \right\|} \tag{25}$$

$$P(c_j) = \frac{K_{c_j}}{\sum K_{c_j}} \tag{26}$$

The third step is to select a specific correction point. In order to identify a point c_j within the candidate point set C_i as the next correction point, we design a heuristic rule based on space vector projection to prioritize the points within the candidate point set. In conjunction with the greedy strategy which the closer to the end the better next point is. The spatial vector from the current point to all candidate c_j and the spatial vector from start point A to end point B are made respectively. By comparing the projections of the vectors on the vectors, if the projection is larger, it indicates that the candidate c_j is more capable of displacement relative to the end point, and the probability of being selected is greater (see Fig. 4). The specific formula is shown in Eq. (25), (26), where K_{c_j} denotes the projection of candidate correction c_j on the vector $\overrightarrow{AB}$, and $P(c_j)$ denotes the probability of c_j being selected.

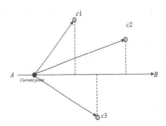

Fig. 4. Diagram of space vector projection

The fourth step, assigning the next point to the current point, it is to determine whether current point can directly reach the end point B without violating the end point error constraint Eq. (17), (18), if not, jump to the first step; if yes, current point is the last correction point of the path and then completes the construction of the initial feasible solution by adding the end point B to the path.

3.3 Cross and Mutation Strategy

After initialization of the population, the linear-rank selection is adopted to select individuals to generate a mating pool for crossover and mutation operations.

Crossover Strategy: a number *rand* in the interval [0, 1] is randomly generated, if *rand* is less than the crossover probability Pc, then individual i and $i + 1$ are selected as parents to execute single-point crossover operations (see Fig. 5).

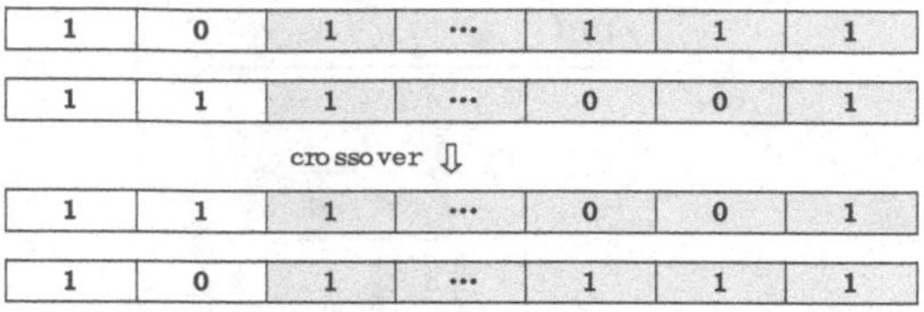

1	0	1	...	1	1	1
1	1	1	...	0	0	1

crossover ⇩

1	1	1	...	0	0	1
1	0	1	...	1	1	1

Fig. 5. Diagram of chromosome crossover

Reselected Mutation Strategy: mutation increases the diversity of the population, but the multiple constraints make the mutated individuals mostly become infeasible solutions, so we design a specific mutation strategy to increase the probability of mutated individuals satisfying the constraints, with the following steps.

A number *rand* in the interval [0, 1] is randomly generated, and if *rand* is less than the mutation probability Pm, then individual i is selected for mutation. Randomly select position1 and reselect a correction point from the hemisphere of the candidate point corresponding to position1. If it there is no correction point of the same type as position1 within the hemisphere, the inverse mutation (common mutation) is taken for the position1. If the second is available, the position1 is reversed, meanwhile the position2 corresponding to the second will become what it was before the position1 mutation, as shown in Table 1.

Table 1. Schematic table of reselected mutation strategy

	State of position1 before mutation	State of position1 after mutation	State of position2 after mutation
Reselected mutation	1	0	1
Common mutation	1	0	/
	0	1	/

3.4 Individual Evaluation

Penalizing infeasible individuals is a common method applied in genetic algorithm's constraint processing. If the crossover and mutation produce offspring individuals that do not satisfy the constraint, the evaluation function $f = f_1 + M$, where M is a large positive penalty value that makes the genetic algorithm to find the optimal solution as closely as possible.

3.5 Variable Neighborhood Search Operator

The best individual of each generation is selected for the variable neighborhood search perturbation to improve the local search ability of the genetic algorithm while ensuring the quality of the solution, the process is shown in Algorithm 2.

Algorithm 2: Variable Neighborhood Search
Input: best solution Sol
Output: new solution Sol'

1   $k \leftarrow 1$
2   Generate Neighborhood $N_1 \sim N_6$
3   **While** ($k \leq 6$) **do**
4     $Sol' \leftarrow N_k(Sol)$
5   If ($f(Sol') < f(Sol)$):
6       $k \leftarrow 1$
7   Else:
8       $Sol' \leftarrow Sol, k \leftarrow k+1$
9   **End**

Neighborhood design:

- Neighborhood N_1: for the feasible solution obtained in the current iteration step, a new feasible solution is obtained by changing one correction point in its path. For example, a possible neighborhood solution for the path $0 \rightarrow 2 \rightarrow 6 \rightarrow 3 \rightarrow 8$ in N_1 is $0 \rightarrow 2 \rightarrow 7 \rightarrow 3 \rightarrow 8$.
- Neighborhood N_2: for the feasible solution obtained at the current iteration step, a new feasible solution is obtained by changing two consecutive correction points in its path. For example, a possible neighborhood solution for the path $0 \rightarrow 2 \rightarrow 6 \rightarrow 3 \rightarrow 8$ in N_2 is $0 \rightarrow 2 \rightarrow 7 \rightarrow 4 \rightarrow 8$.
- Neighborhood N_3: for the feasible solution obtained at the current iteration step, a new feasible solution is obtained by changing three consecutive correction points in its path. For example, a possible neighborhood solution for the path $0 \rightarrow 2 \rightarrow 6 \rightarrow 3 \rightarrow 8$ in N_3 is $0 \rightarrow 5 \rightarrow 7 \rightarrow 4 \rightarrow 8$.
- Neighborhood N_4: for the feasible solution obtained in the current iteration step, a new feasible solution is obtained by deleting one correction point in its path. For example, a possible neighborhood solution for the path $0 \rightarrow 2 \rightarrow 6 \rightarrow 3 \rightarrow 8$ in N_4 is $0 \rightarrow 2 \rightarrow 3 \rightarrow 8$.

- Neighborhood N_5: for the feasible solution obtained at the current iteration step, a new feasible solution is obtained by deleting two consecutive correction points in its path. For example, a possible neighborhood solution for the path $0 \rightarrow 2 \rightarrow 6 \rightarrow 3 \rightarrow 8$ in N_5 is $0 \rightarrow 2 \rightarrow 8$.
- Neighborhood N_6: for the feasible solution obtained in the current iteration step, a new feasible solution is obtained by deleting three consecutive correction points in its path. For example, a possible neighborhood solution for the path $0 \rightarrow 2 \rightarrow 6 \rightarrow 3 \rightarrow 8$ in N_6 is $0 \rightarrow 8$.

4 Experiment

The effectiveness of the hybrid algorithm proposed in this paper is tested on an i7–9700@3.0 GHz/16G/Win7 PC. According to the 0–1 integer programming model established in the previous, the hybrid model of Variable Neighborhood Search and Genetic Algorithm is implemented based on the jMetal framework [13], where the JDK version is 1.9 and the jMetal framework version is 5.10. The parameters of the hybrid algorithm are set: population size $N = 100$, mating pool size $poolSize = 100$, the maximum number of iterations $mGen = 500$, crossover probability $Pc = 0.8$, and mutation probability $Pm = 0.25$. The test dataset has 613 correction points and the parameters for each constraint are set as follows: $\alpha_1 = 25$, $\alpha_2 = 15$, $\beta_1 = 20$, $\beta_2 = 25$, $\theta = 30$, $\delta = 0.001$.

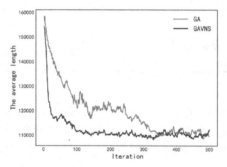

Fig. 6. Comparison of the population's average length by every iterated based on test data

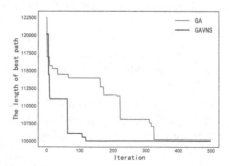

Fig. 7. Comparison of the optimal path's length by every iterated based on test data

The improved Genetic Algorithm with Variable Neighborhood Search (GAVNS) and the traditional genetic algorithm are compared in experiments on the test dataset, and the change curve of the average length for the population of each generation is shown respectively (see Fig. 6), while the length of the optimal path in the population of each generation is shown respectively (see Fig. 7). It can be seen that GAVNS starts to converge at around 100 generations, while GA is still evolving iteratively, which demonstrates that the introduction of variable neighborhood search strategy can strengthen the local search ability of the genetic algorithm.

GAVNS obtains the trajectory path after 500 iterations as $1 \rightarrow 504 \rightarrow 70 \rightarrow 507 \rightarrow 29 \rightarrow 184 \rightarrow 195 \rightarrow 451 \rightarrow 595 \rightarrow 398 \rightarrow 613$, the length is 105059.9m, passing through 9 correction points, the horizontal and vertical error of each correction point is shown in Table 2. GA also obtains the trajectory path after 500 iterations as $1 \rightarrow 504 \rightarrow 201 \rightarrow 81 \rightarrow 238 \rightarrow 283 \rightarrow 599 \rightarrow 562 \rightarrow 449 \rightarrow 486 \rightarrow 613$, the length is 105254.1m, passing through 9 correction points, the horizontal and vertical error of each correction point is shown in Table 3.This experimental comparison shows that the horizontal and vertical errors at each point of the two trajectory paths obtained by GA and GAVNS satisfy the constraint and both can complete the scheduled task, but the length of path obtained by GAVNS is better and has fewer iterations (see Fig. 7).

Table 2. GAVNS's planning path result

The no. of point	Before correction		After correction		The type of point
	Vertical error	Horizontal error	Vertical error	Horizontal error	
1	0.00	0.00	0.00	0.00	A
504	13.39	13.39	13.39	0.00	1
70	22.20	8.81	0.00	8.81	0
507	7.87	16.68	7.87	0.00	1
29	13.86	6.00	0.00	6.00	0
184	11.33	17.32	11.33	0.00	1
195	24.94	13.61	0.00	13.61	0
451	5.98	19.59	5.98	0.00	1
595	24.03	18.05	0.00	18.05	0
398	3.06	21.11	3.06	0.00	1
613	20.03	16.97	20.03	16.97	B

Table 3. GA's planning path result

The no. of point	Before correction		After correction		The type of point
	Vertical error	Horizontal error	Vertical error	Horizontal error	
1	0.00	0.00	0.00	0.00	A
504	13.39	13.39	13.39	0.00	1
201	14.25	0.87	0.00	0.87	0
81	15.75	16.61	0.00	16.61	0
238	4.63	21.24	4.63	0.00	1
283	18.31	13.68	0.00	13.68	0
599	11.11	24.79	11.11	0.00	1
562	22.06	10.96	0.00	10.96	0
449	5.15	16.11	0.00	16.11	0
486	5.73	21.84	5.73	0.00	1
613	29.30	23.57	29.30	23.57	B

5 Conclusion

In this paper, an 0–1 integer programming model is established for the path planning problem under several constraints such as the upper limit of horizontal and vertical error at the correction point as well as the minimum turning radius of the UAV. A genetic algorithm with improved population initialization and variation operator is proposed, and a two-level variable neighborhood search strategy for adding and deleting route points is introduced to strengthen the local search ability of the genetic algorithm. Finally, experiments are conducted on the above trajectory path planning problem and compare with the traditional genetic algorithm to demonstrate the stability and effectiveness of the proposed GAVNS hybrid algorithm.

The handling of unsatisfied constraint solutions is relatively straightforward, whereas the constraints on real-world optimization problems such as the fast path planning of the UAV are usually numerous and complex, improvement of the constraint handling approaches will follow so that GAVNS can be applied to more complex problems.

Acknowledgement. This work is supported in part by the National Nature Science Foundation of China under Grant 61773390 and Grant 61973310, the key project of National University of Defense Technology (ZK18–02–09), the Hunan Youth elite program (2018RS3081) and the key project ZZKY–ZX–11–04.

References

1. Aggarwal, S., Kumar, N.: Path planning techniques for unmanned aerial vehicles: a review, solutions, and challenges. Comput. Commun. **149**, 270–299 (2020)

2. Li, F., Li, Z.: Research on rapid planning of intelligent aircraft trajectory under multiple constraints. J. Phys. Conf. Ser. **1592**, 012021 (2020)
3. Chen, Y.-b., Luo, G.-c., Mei, Y.-s., Yu, J.-q., Su, X.-l.: UAV path planning using artificial potential field method updated by optimal control theory. Int. J. Syst. Sci. **47**, 1407–1420 (2016)
4. Radmanesh, M., Kumar, M.: Flight formation of UAVs in presence of moving obstacles using fast-dynamic mixed integer linear programming. Aerosp. Sci. Technol. **50**, 149–160 (2016)
5. Dong, Z., Chen, Z., Zhou, R., Zhang, R.: A hybrid approach of virtual force and A∗ search algorithm for UAV path re-planning. In: 2011 6th IEEE Conference on Industrial Electronics and Applications, pp. 1140–1145. IEEE (2011)
6. Arantes, M.d.S., Arantes, J.d.S., Toledo, C.F.M., Williams, B.C.: A hybrid multi-population genetic algorithm for UAV path planning. In: Proceedings of the Genetic and Evolutionary Computation Conference 2016, pp. 853–860 (2016)
7. Wang, Y., Bai, P., Liang, X., Wang, W., Zhang, J., Fu, Q.: Reconnaissance mission conducted by UAV swarms based on distributed PSO path planning algorithms. IEEE Access **7**, 105086–105099 (2019)
8. Li, B., Qi, X., Yu, B., Liu, L.: Trajectory planning for UAV based on improved ACO algorithm. IEEE Access **8**, 2995–3006 (2019)
9. Chen, J., Ye, F., Li, Y.: Travelling salesman problem for UAV path planning with two parallel optimization algorithms. In: 2017 Progress in Electromagnetics Research Symposium-Fall (PIERS-FALL), pp. 832–837. IEEE (2017)
10. Eun, Y., Bang, H.: Cooperative task assignment/path planning of multiple unmanned aerial vehicles using genetic algorithm. J. Aircr. **46**, 338–343 (2009)
11. Peng, K., Wen, L., Li, R., Gao, L., Li, X.: An effective hybrid algorithm for permutation flow shop scheduling problem with setup time. Procedia CIRP **72**, 1288–1292 (2018)
12. Dixit, A., Mishra, A., Shukla, A.: Vehicle routing problem with time windows using meta-heuristic algorithms: a survey. In: Yadav, N., Yadav, A., Bansal, J.C., Deep, K., Kim, J.H. (eds.) Harmony search and nature inspired optimization algorithms. AISC, vol. 741, pp. 539–546. Springer, Singapore (2019). https://doi.org/10.1007/978-981-13-0761-4_52
13. Durillo, J.J., Nebro, A.J.: jMetal: a Java framework for multi-objective optimization. Adv. Eng. Softw. **42**, 760–771 (2011)

An Improved Particle Swarm Optimization with Dual Update Strategies Collaboration Based Task Allocation

Shuang Xia, Xiangyin Zhang[(✉)], Xiuzhi Li, and Tian Zhang

Faculty of Information Technology, Beijing University of Technology, Beijing 100124, China
xy_zhang@bjut.edu.cn

Abstract. The task allocation problem is a hot topic in the field of multiple unmanned aerial vehicle (UAV). In this paper, we consider the task allocation in rescue scenarios and establish the optimization model. Then, an improved particle swarm optimization with dual update strategies collaboration (DUCPSO) is proposed to solve it. In order to simplify the solution of the problem, a real vector coding method is adopted. In addition, the ring topology update method and the mutation update method are introduced to enhance the diversity of the population, and the effective collaboration of the two strategies is realized through the adaptive strategy conversion probability. The simulation results show that the proposed algorithm can effectively obtain the optimal task allocation scheme. Compared with other algorithms, the proposed algorithm is more feasible and efficient.

Keywords: Task allocation · Particle swarm optimization · Unmanned aerial vehicle

1 Introduction

Due to the increasingly diversified mission requirements, the limited capabilities of a single unmanned aerial vehicle (UAV) cannot be satisfied, which makes the multi-UAV system gradually become the focus of researchers [1]. The task allocation of multiple UAVs is a key problem that needs to be solved, which is to assign a group of tasks to UAVs on the basis of ensuring the best overall benefits [2]. The rescue scenarios are considered in this paper, in which UAVs visit the location of survivors sequentially and provide medication and food assistance.

So far, a number of methods have been proposed to solve the task allocation problem. Traditional methods such as Hungarian algorithm (HA) [3], branch and bound (BNB) [4], and mixed integer linear programming (MILP) [5] can obtain solutions for low-dimensional problems, but the difficulty of solving the problems increases exponentially as the number of dimensions grows. The auction methods, such as the consensus-based bundle algorithm (CBBA) [6], can be distributed on each UAV to obtain a conflict-free

© Springer Nature Switzerland AG 2021
Y. Tan and Y. Shi (Eds.): ICSI 2021, LNCS 12690, pp. 218–229, 2021.
https://doi.org/10.1007/978-3-030-78811-7_21

solutions. However, its ability to handle collaborative constraints is poor. Swarm intelligence algorithms include ant colony optimization (ACO) [7], particle swarm optimization (PSO) [8], genetic algorithm (GA) [9] and so on. Due to the strong robustness, they are widely used to solve various complex problems with good performance. Therefore, they are gradually being introduced by researchers for task allocation. A discrete mapping differential evolution was proposed in [10] to deal with cooperative task assignment in a three-dimensional environment. And through a unified gene coding strategy, various allocation models can be solved in a consistent framework. Multiple algorithms were combined in [11] to implement a two-stage methodology to improve the allocation performance, in which the firefly algorithm was first used for global allocation, and then the local allocation was executed combining the advantages of quantum genetic algorithm and artificial bee colony optimization. Due to its simplicity and ease of implementation, particle swarm algorithm has become one of the most classic swarm intelligence algorithms, and many variants have been developed. Tian et al. [12] proposed a modified particle swarm optimization with chaos-based initialization and robust update mechanisms to overcome premature convergence. Chen et al. [13] introduced the differential evolution operator and proposed a dynamic multi-swarm differential learning particle swarm optimizer.

In this paper, an improved particle swarm optimization with dual update strategies collaboration (DUCPSO) is proposed for the task allocation problem in rescue scenarios. The real vector coding method is adopted, so that the continuous optimization algorithm can be applied. In order to enhance the diversity of the population, the ring topology update method is adopted. The mutation update method is introduced to make the particles in the elite population provide clearer guidance for the update of the positions. Moreover, the adaptive strategy conversion probability is adopted to balance the two update strategies. Simulation results show that the proposed DUCPSO can successfully solve the task allocation problem in rescue scenarios, which demonstrates the feasibility and high performance of the proposed algorithm.

The rest of this paper is organized as follows. The problem description of task allocation in rescue scenarios is presented in Sect. 2. In Sect. 3, the task allocation based on DUCPSO is described in detail. In Sect. 4, Simulation results and comparisons with other algorithms are provided. Finally, we conclude this paper in Sect. 5.

2 Problem Description

The problem in this paper is mainly the task allocation of UAVs in rescue scenarios. Define a set of n UAVs $U = [U_1, U_2,...,U_n]$ and m survivors $T = [T_1, T_2,...,T_m]$. In the general scenario, the number of survivors is obviously more than the number of UAVs, that is, $m > n$. For the survivors, different kinds of assistance are needed, such as medicine or food. UAVs are considered to have the same flight speed, while being able to perform both medication and food assistance tasks. Due to the different load capacity of UAVs, the number of two tasks that can be performed is different. Only when the UAV arrives at the survivor's location before the deadline can the survivor be successfully rescued. After all assistance tasks are completed, the UAVs will return to the initial position. In addition, the mission requires that one UAV can rescue multiple survivors, and each survivor can only be rescued by one UAV. The purpose of rescue is to aid all survivors at the lowest cost. The objectives of task allocation in this paper are described as follows.

$$f_1 = \frac{1}{m} \sum_{j=1}^{m} t_j \tag{1}$$

$$t_j = d(T_j)/Speed \tag{2}$$

$$f_2 = \sum_{i=1}^{n} Length_i \tag{3}$$

where f_1 and f_2 describe the average waiting time and fuel consumption, respectively. t_j is the rescue time of T_j. $d(T_j)$ is the flight distance from the initial position of the corresponding UAV to the location of T_j. $Speed$ is the flight speed of the UAV. $Length_i$ is the total flight distance of U_i.

Therefore, the optimization model is established as follows:

$$\min \ F = \omega_1 \cdot f_1 + \omega_2 \cdot f_2 + \omega_3 \cdot \delta$$

$$s.t. \begin{cases} \sum_{i=1}^{n} x_{ij} = 1, & \forall j = 1, 2, \dots, m \\ Length_i \leq L_{i,max}, & \forall i = 1, 2, \dots, n \\ \sum_{j=1}^{n} \varphi_k(x_{ij}) \leq Load_{i,k}, & \forall i = 1, 2, \dots, n, \forall k = 1, 2 \\ t_j \leq t_d(j) \ \sum_{i=1}^{n} x_{ij} = 1, & \forall j = 1, 2, \dots, m \end{cases} \tag{4}$$

where F is the fitness function, $\omega_1, \omega_2, \omega_3$ is the weight coefficient, δ is the number of constraint violations. x_{ij} is the decision variable. $x_{ij} = 1$ means that U_i rescues T_j, and $x_{ij} = 0$ means that T_j is not rescued by U_i. $L_{i,max}$ is the maximum flight distance of U_i. k is the type of assistance task. $k = 1$ means medicine assistance task, $k = 2$ means food assistance task. If U_i rescues the k-type survivor T_j, then $\varphi_k(x_{ij}) = 1$, otherwise $\varphi_k(x_{ij}) = 0$. $Load_{i,k}$ represents the maximum number of k-type tasks that U_i can complete. $t_d(j)$ is the latest rescue time of T_j.

In this paper, the initial position of the UAVs, the location of the survivors, the type of rescue, and the deadline are all known in advance. In addition, the distance between the UAV and the survivor, the survivor and the survivor is pre-stored in the distance matrix to avoid repeated calculations in the optimization process.

3   Task Allocation Based on DUCPSO

3.1   Standard PSO

PSO is an interesting intelligent optimization algorithm inspired by bird flock foraging and social interaction. The personal best positions of the particles and the global best position are used to guide the population to converge to the potential optimal area. The core of the algorithm is the update of the position and velocity of the particles, which is calculated as follows:

$$V_{i,G+1} = \omega \cdot V_{i,G} + c_1 \cdot r_1 \cdot \left(X_{\text{pbest},i} - X_{i,G}\right) + c_2 \cdot r_2 \cdot \left(X_{\text{gbest}} - X_{i,G}\right) \qquad (5)$$

$$X_{i,G+1} = X_{i,G} + V_{i,G+1} \qquad (6)$$

where $V_{i,G} = [V_{i,1}, V_{i,2}, \ldots, V_{i,D}]$ and $X_{i,G} = [X_{i,1}, X_{i,2}, \ldots, X_{i,D}]$ respectively represent the velocity and position of particle i in the G-th iteration ($i = 1, 2, \ldots, N$), N is the population size, D is the problem dimension. ω is the inertia weight, c_1 and c_2 are learning factors, r_1 and r_2 are uniformly generated random numbers in the range of $[0,1]$. $X_{\text{pbest},i}$ represents the personal best position of the particle i, X_{gbest} represents the global best position of the entire population.

3.2 Proposed DUCPSO

Encoding and Decoding Mechanism. The position of the particle corresponds to the candidate solution of the problem. Flexible coding of its structure can effectively reduce the complexity of the problem. For the rescue problem in this paper, the number of survivors is far greater than the number of UAVs, and one UAV can provide assistance to several survivors. Therefore, the rescue relationship between the UAV and the survivor, including the rescue object and the rescue sequence, is what the task allocation scheme should cover.

Real vector coding is an effective method to express this relationship. In this method, the position index is used to indicate the survivor number, and the position value corresponds to the UAV number and rescue order. The integer part of the position value represents the UAV number that rescued the survivor. The decimal part corresponds to the order in which the survivors were rescued. The smaller the value, the earlier they were rescued.

The advantage of this encoding is that continuous optimization algorithms can be adopted to solve the problem. Meanwhile, it is possible to use only one PSO population to search for solutions, not only to get the UAV number corresponding to the survivors but also to get the rescue order. In addition, the decision variable constraints are always satisfied, and do not need to be considered when calculating the fitness function.

Ring Topology Update Method. Diversity is an important factor affecting the performance of PSO. In order to search for solutions with higher performance PSO, enhancing diversity is an effective method. The global topology is adopted by the standard PSO, which easily leads to a local optimum. References prove that the ring topology is beneficial to enhance the diversity of the algorithm and make it perform better in searching for the optimal solution [15]. Therefore, the ring topology update method is adopted in this paper instead of Eq. (5) to update the particle velocity.

$$V_{i,G+1} = \omega \cdot V_{i,G} + c_1 \cdot r_1 \cdot (P_i - X_{i,G}) + c_2 \cdot r_2 \cdot (X_{\text{gbest}} - X_{i,G}) \tag{7}$$

$$P_i = \eta \cdot X_{\text{pbest},n_{i1}} + (1 - \eta) \cdot X_{pbest,n_{i2}} \tag{8}$$

$$n_{i1} = \begin{cases} i - 1, & \text{if } i > 1 \\ N, & \text{if } i = 1 \end{cases} \tag{9}$$

$$n_{i2} = \begin{cases} i + 1, & \text{if } i < 1 \\ 1, & \text{if } i = N \end{cases} \tag{10}$$

where η is a random number between 0 and 1. n_{i1} and n_{i2} are the indexes of the two particles directly connected to particle i in the ring topology.

In order to improve exploration in the early stage and improve exploitation in the later stage, the control parameters in Eq. (7) are adjusted linearly as follows:

$$\omega = \omega_{\max} - \frac{G}{G_{\max}} (\omega_{\max} - \omega_{\min}) \tag{11}$$

$$c_1 = c_{1\,\max} - \frac{G}{G_{\max}} (c_{1\,\max} - c_{1\,\min}) \tag{12}$$

$$c_2 = c_{2\,\min} - \frac{G}{G_{\max}} (c_{2\,\max} - c_{2\,\min}) \tag{13}$$

where $G_{\max}$ is the maximum number of iterations. $\omega_{\max}$ and $\omega_{\min}$ are the maximum and minimum value of ω, respectively. $c_{1\max}, c_{2\max}$ and $c_{1\min}, c_{2\min}$ are the maximum and minimum values of c_1, c_2, respectively.

Mutation Update Method. In order to further increase the diversity, a mutation method is introduced to update the position of the particles. At the beginning of each iteration, the whole population is divided into two sub-populations, including elite population (**EP**) and non-elite population (**NEP**), according to the fitness value. A random particle is selected separately from the elite population and the non-elite population to generate a difference vector, keeping its direction pointing to the particle in the elite population, which provides a clearer guide for the update of the position. The position of the particles in the mutation update method is updated as follows:

$$X_{i,G+1} = X_{i,G} + F \cdot (X_{r1} - X_{r2}) + F \cdot (X_{r3} - X_{r4}) \tag{14}$$

$$V_{i,G+1} = X_{i,G+1} - X_{i,G} \tag{15}$$

where F is the mutation factor. X_{r1} and X_{r3} are particles randomly selected from the elite population. X_{r2} and X_{r4} are particles randomly selected from the non-elite population.

Adaptive Strategy. In order to effectively combine the ring topology update method and the mutation update method, an adaptive strategy conversion method is introduced. $Pv = [Pv_1, Pv_2, ..., Pv_N]$ is the probability vector, where Pv_i is the strategy conversion probability of particle i. When selecting the update method for the position of particle i, a random number between 0 and 1 will be generated and compared with Pv_i. The ring topology update method is adopted with larger Pv_i, otherwise, the mutation update method is adopted. The strategy conversion probability of each particle will be adaptively adjusted as follows based on the fitness value of its descendants.

(1) The offspring is better: The fitness value of the generated new particle is smaller than that of the previous generation.

$$Pv_i = Pv_i + 0.1 \times (1 - Pv_i) \times \frac{G}{G_{max}} \tag{16}$$

(2) The offspring is worse: The fitness value of the generated new particle is greater than that of the previous generation.

$$Pv_i = Pv_i - 0.1 \times Pv_i \times \left(1 - \frac{G}{G_{max}}\right) \tag{17}$$

In other words, when the offspring of the particle is getting better and better, it is more inclined to update the position of the particle using the ring topology update method. The reason is that the mutation update method is more suitable for maintaining the diversity of the population in the early stage of the iteration, and the ring topology update method is more able to converge to the optimal solution in the later stage of the iteration.

DUCPSO Based Task Allocation. The DUCPSO algorithm is applied to solve the task allocation problem in rescue scenarios, and its implementation framework is shown in Algorithm 1.

Algorithm 1. DUCPSO based task allocation.

/*Environment setting*/
  Set the start points and loads of UAVs and task types, locations and deadlines
/*Input for DUCPSO*/
  Set termination conditions and the parameters ω_{max}, ω_{min}, c_{1max}, c_{1min}, c_{2max}, c_{2min}
/*Initialization*/
  Initialize the location X_i and velocities V_i of particles randomly
/*Evaluation*/
  Decode to get the task allocation scheme and calculate the fitness value by Eqs. (1)-(4)
/*Iteration computation*/
  While Termination conditions are not met **do**
    Divide the elite population and the non-elite population according to the fitness value
    Update parameters ω, c_1, c_2 by Eqs. (11)-(13)
    For $i = 1:N$ **do**
      If rand $< Pv_i$
        Update the velocity V_i and position X_i by Eqs. (6)-(10)
      Else if
        Update the velocity V_i and position X_i by Eqs. (14)-(15)
      End if
      Evaluation the task allocation scheme
      Update strategy conversion probability Pv_i by Eqs. (16)-(17)
      Update the personal best position $X_{pbest,i}$
    End for
    Update the global best position X_{gbest}
  End while
/*Output*/
  Decode and output the best task allocation scheme

4   Simulation Results

In order to verify the feasibility of applying the DUCPSO algorithm to the task allocation problem in rescue scenarios, we conducted simulation experiments based on Matlab-R2018a. Within a space range of 100km × 100km, two mission scenarios were simulated, with three UAVs rescuing ten survivors, Case 1, and four UAVs rescuing twenty survivors, Case 2. The parameters of all UAVs are set as: $Speed = 1$km/min, $L_{i,max} = 250$km. The coordinates of the UAVs and survivors are randomly generated in the mission space. In Case 1, the deadlines are randomly generated in the range of [80,180], with the corresponding information of UAVs shown in Table 1. In Case 2, the deadlines are within the range of [100,340], and the corresponding information of UAVs is shown in Table 2.

Table 1. Corresponding UAVs information for Case 1.

UAV	Coordinates(km)	$Speed$(km/min)	$L_{i,max}$(km)	Load	
				Medicine	Food
U_1	(68.37,19.49)	1	250	2	2
U_2	(57.85,19.16)	1	250	1	3
U_3	(61.71,92.06)	1	250	2	2

We compare the proposed DUCPSO with other algorithms, including standard PSO, HFPSO [16], MPBPSO [17], GDS-WOA [18], GGL-PSOD [19]. Since different algorithms have different evaluation times in each iteration, the maximum number of function evaluations is adopted as the algorithm termination condition, which is set to $FE_{max} = 20000$. Each algorithm is executed 50 times independently. The main parameters of DUCPSO are set as: $N = 10$, $\omega_{max} = 0.9$, $\omega_{min} = 0.4$, $c_{1max} = 2.5$, $c_{1min} = 0.5$, $c_{2max} = 2.5$, $c_{2min} = 0.5$, $\omega_1 = \omega_2 = 1$, $\omega_3 = 500$. The parameters of other comparison algorithms are set according to relevant references.

Table 2. Corresponding UAVs information for Case 2.

UAV	Coordinates (km)	$Speed$ (km/min)	$L_{i,max}$ (km)	Load	
				Medicine	Food
U_1	(38.73,81.45)	1	250	6	9
U_2	(51.19,85.14)	1	250	4	5
U_3	(1.01,27.71)	1	250	7	6
U_4	(47.90,34.25)	1	250	8	4

For Case 1, the optimal rescue schemes obtained by DUCPSO is shown in Fig. 1 and the convergence curve of the average fitness value is shown in Fig. 2. The statistical results are listed in Table 3, including the best, median, mean, worst and standard deviation of fitness values, with the best results marked in bold. FR represents the feasibility rate of obtaining a rescue scheme that satisfies all constraints from 50 runs. AT denotes the average time for the algorithm to run once. The optimal solutions obtained by PSO and HFPSO are the same, while the optimal solutions obtained by MPBPSO, GDS-WOA, GGL-PSOD, and DUCPSO are the same. On the basis of satisfying all constraints, it is obvious that the rescue scheme obtained by the last four algorithms is more better.

It can be seen that the results obtained by DUCPSO in Case 1 are the best among all the algorithms, which shows that the improvement of the proposed algorithm is feasible and successful. From the statistical results, the median of the fitness values obtained by the proposed algorithm is close to the best value, that is, most of the rescue solutions it finds in 50 runs can reach the optimal value, which shows that our method of improving diversity effectively makes the proposed method escape the local optimum and obtain high performance.

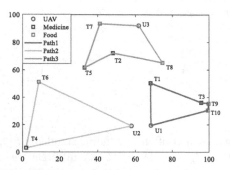

Fig. 1. Optimal result obtained by DUCPSO for Case 1.

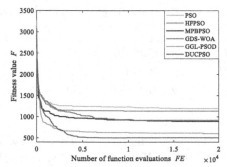

Fig. 2. Convergence curves of average fitness values for Case 1.

Table 3. The statistical results obtained by the six algorithms for Case 1.

Algorithm	Best	Mean	Median	Worst	Std	FR(%)	AT(s)
PSO	473.57	1194.54	1134.88	2761.23	558.93	32	0.88
HFPSO	473.57	1134.63	1119.70	2605.11	545.54	40	0.86
MPBPSO	**458.48**	893.34	958.11	2616.68	410.58	48	0.86
GDS-WOA	**458.48**	919.54	850.13	2131.63	429.69	5	1.35
GGL-PSOD	**458.48**	603.83	529.55	1120.07	200.02	86	0.86
DUCPSO	**458.48**	**504.15**	**462.92**	**1115.65**	**100.35**	**98**	**0.84**

For Case 2, the optimal result obtained by DUCPSO is shown in Fig. 3 and the convergence curve of the average fitness value is shown in Fig. 4. The statistical results are listed in Table 4. The performance of the proposed algorithm in larger-scale rescue mission is tested in Case 2. The statistical results show that DUCPSO and GGL-PSOD have obtained feasible rescue schemes every time in 50 plans, which is significantly

Table 4. The statistical results obtained by the six algorithms for Case 2.

Algorithm	Best	Mean	Median	Worst	Std	FR(%)	AT(s)
PSO	719.79	885.91	890.16	1064.46	83.16	68	0.95
HFPSO	716.19	875.67	887.59	1007.81	75.49	44	0.98
MPBPSO	678.96	861.61	862.92	1006.86	81.45	82	**0.94**
GDS_WOA	711.75	858.38	868.35	983.17	**74.07**	42	1.84
GGL_PSOD	540.82	667.19	664.91	**822.92**	77.77	**100**	**0.94**
DUCPSO	**509.68**	**612.30**	**577.37**	960.93	100.57	**100**	**0.94**

better than the other four algorithms. In addition, the best, median, and mean of fitness values obtained by DUCPSO are smaller than GGL-PSOD.

In the above two test cases, the proposed algorithm showed good performance. Compared with other comparison algorithms, it obtained the better rescue scheme, which proved its feasibility and effectiveness in solving the task allocation problem in rescue missions.

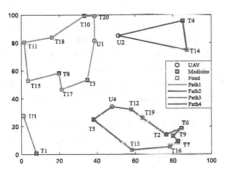

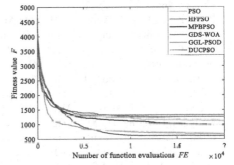

Fig. 3. Optimal results obtained by DUCPSO for Case 2.

Fig. 4. Convergence curves of average fitness values for Case 2.

5   Conclusion

In this paper, the task allocation problem in rescue scenarios is considered. An improved particle swarm optimization with dual update strategies collaboration is proposed and successfully applied to solve complex task allocation problem. The proposed algorithm adopts real vector coding to simplify the complexity of the problem. In addition, ring topology and mutation operators are introduced to enhance the diversity of the population. The simulation results show that the proposed algorithm has good performance in solving the task allocation problem, no matter in the small-scale rescue mission or the larger-scale rescue mission, and can escape the local optima. In the future research, survivors who need both medication and food assistance should be considered in order to establish an optimization model that is more suitable for practical problems.

Acknowledgment. . This work is supported by National Natural Science Foundation of China (No. 61703012, 51975011), Beijing Natural Science Foundation (No. 4182010).

References

1. Nedjah, N., Junior, L.S.: Review of methodologies and tasks in swarm robotics towards standardization. Swarm Evolut. Comput. **50**, 100565 (2019)
2. Geng, N., Meng, Q., Gong, D., Chung, P.W.H.: How good are distributed allocation algorithms for solving urban search and rescue problems? A comparative study with centralized algorithms. IEEE Trans. Autom. Sci. Eng. **16**(1), 478–485 (2019)
3. Chopra, S., Notarstefano, G., Rice, M., Egerstedt, M.: A distributed version of the Hungarian method for multirobot assignment. IEEE Trans. Rob. **33**(4), 932–947 (2017)
4. Madakat, D., Morio, J., Vanderpooten, D.: A biobjective branch and bound procedure for planning spatial missions. Aerosp. Sci. Technol. **73**, 269–277 (2018)
5. Alfa, A.S., Maharaj, B.T., Lall, S., Pal, S.: Mixed-integer programming based techniques for resource allocation in underlay cognitive radio networks: a survey. J. Commun. Netw. **18**(5), 744–761 (2016)
6. Zhao, W., Meng, Q., Chung, P.W.H.: A Heuristic distributed task allocation method for multivehicle multitask problems and its application to search and rescue scenario. IEEE Trans. Cybern. **46**(4), 902–915 (2016)
7. Zhen, Z.Y., Chen, Y., Wen, L.D., Han, B.: An intelligent cooperative mission planning scheme of UAV swarm in uncertain dynamic environment. Aerosp. Sci. Technol. **100**, 105826 (2020)
8. Li, M.C., Liu, C.B., Li, K.L., Liao, X.K., Li, K.Q.: Multi-task allocation with an optimized quantum particle swarm method. Appl. Soft Comput. **96**, 106603 (2020)
9. Wang, Z., Liu, L., Long, T., Wen, Y.L.: Multi-UAV reconnaissance task allocation for heterogeneous targets using an opposition-based genetic algorithm with double-chromosome encoding. Chin. J. Aeronaut. **31**(2), 339–350 (2018)
10. Ming, Z., Lingling, Z., Xiaohong, S., Peijun, M., Yanhang, Z.: Improved discrete mapping differential evolution for multi-unmanned aerial vehicles cooperative multi-targets assignment under unified model. Int. J. Mach. Learn. Cybern. **8**(3), 765–780 (2015). https://doi.org/10.1007/s13042-015-0364-3
11. Zitouni, F., Maamri, R., Harous, S.: FA–QABC–MRTA: a solution for solving the multi-robot task allocation problem. Intel. Serv. Robot. **12**(4), 407–418 (2019). https://doi.org/10.1007/s11370-019-00291-w
12. Tian, D.P., Shi, Z.Z.: MPSO: Modified particle swarm optimization and its applications. Swarm Evol. Comput. **41**, 49–68 (2018)
13. Chen, Y.G., Li, L.X., Peng, H.P., Xiao, J.H., Wu, Q.T.: Dynamic multi-swarm differential learning particle swarm optimizer. Swarm Evol. Comput. **39**, 209–221 (2018)
14. Kennedy, J., Eberhart, R.: Particle swarm optimization. In: Proceedings of IEEE International Conference on Neural Networks, vol. 4, pp. 1942–1948 (1995)
15. Yue, C., Qu, B., Liang, J.: A multiobjective particle swarm optimizer using ring topology for solving multimodal multiobjective problems. IEEE Trans. Evol. Comput. **22**(5), 805–817 (2018)
16. Aydilek, I.B.: A hybrid firefly and particle swarm optimization algorithm for computationally expensive numerical problems. Appl. Soft Comput. **66**, 232–249 (2018)
17. Gholami, K., Dehnavi, E.: A modified particle swarm optimization algorithm for scheduling renewable generation in a micro-grid under load uncertainty. Appl. Soft Comput. **78**, 496–514 (2019)

18. Li, Y., Han, T., Zhao, H., Gao, H.: An Adaptive whale optimization algorithm using gaussian distribution strategies and its application in heterogeneous UCAVs task allocation. IEEE Access **7**, 110138–110158 (2019)
19. Lin, A., Sun, W., Yu, H., Wu, G., Tang, H.W.: Global genetic learning particle swarm optimization with diversity enhancement by ring topology. Swarm Evol. Comput. **44**, 571–583 (2019)

Intelligent Intrusion Detection System for a Group of UAVs

Elena Basan[1]([✉]) [iD], Maria Lapina[2] [iD], Nikita Mudruk[1], and Evgeny Abramov[1]

[1] Southern Federal University, Chekov St., 2, 347922 Taganrog, Russian Federation
{ebasan,mudruk,abramoves}@sfedu.ru
[2] North-Caucasus Federal University, Pushkin St., 1, 355017 Stavropol, Russian Federation
mlapina@ncfu.ru

Abstract. Today, the creation of UAV groups is becoming a very popular and relevant task. Nevertheless, the use of UAVs is not securely, since they are vulnerable not only to attacks by an intruder, but also to environmental influences. Thus, the occurrence of anomalies must be detected in time. This work is aimed at detecting anomalies in UAV groups and determining the type of attack. To accomplish this task, the authors have developed an experimental stand emulating traffic transmission in a UAV group. The study is based on the investigation of changes in traffic transmission patterns during normal operation and under attacks. Data sets for training a neural network were collected using an developed testbed.

Keywords: UAV · Neural networks · Anomalies · Detection · Recognition · Dataset

1 Introduction

An intrusion detection system for unmanned aerial vehicles (IDS-UAV) is being developed to detect abnormal behavior or emergent events in a network by automatically analyzing behavior or analysis based on a given hypothesis and/or policies that are governed by the security rules of the network [1]. The IDS-UAV monitors system configuration, data files and/or network transmissions to check presents of attack. Attacks exploit vulnerabilities in UAV systems. Vulnerabilities may occur due to configuration of UAV networks, implementation errors, and incorrect designs and/or protocols [2]. The IDS-UAV analyzes the data flow and collects information from various UAV components during their operation.

An IDS can be deployed in two ways:

- With coordination at the base station: the ground-coordinated IDS collected information, and appropriate decisions were made based on the analyzed data;
- Autonomous: an IDS is deployed autonomously, unmanned aerial vehicles acting as nodes for IDS deployment must conduct data analysis and control other UAVs.

The key mechanisms of IDS can be classified as follows:

Y. Tan and Y. Shi (Eds.): ICSI 2021, LNCS 12690, pp. 230–240, 2021.
https://doi.org/10.1007/978-3-030-78811-7_22

– Based on the specification [4]: The IDS includes a set of relevant rules defined sue to the expected behavior of the UAV.
– Based on signatures [5]: This method aims to detect known attacks based on predefined known signatures.
– Anomaly Based [6]: Anomalous behavior is detected based on crash or illegal activity observed on the system.
– Hybrid [7]: This technique is a hybrid approach that combines two or more detection methods to provide a strong detection policy that can detect known and/or unknown attacks.

The aim of this work is to recognize the presence of an attack and differentiate different types of attacks on a group of UAVs using artificial intelligence. The idea is that UAVs, being in a group, can detect attacks when they are carried out not only on them, but also on neighboring UAVs. The impact on neighboring UAVs is usually affected on the traffic pattern. UAVs, like other cyber-physical systems, operate according to a previously defined algorithm, or have certain sets of actions. As part of their activities, UAVs exchange data and do it according to different algorithms. Thus, for the solving the problem of detecting attacks on UAVs in this work, a testbed foe emulating process of traffic transmission between UAVs was built in two versions: a distributed fully connected network and a centralized network. Attacks were carried out on every type of network. The data were normalized, and the obtained data sets were used to train the neural network to recognize the types of attacks.

2 Analysis and Research of Approaches to Detecting Attacks on UAVs

In the work of Shreekh W. and Deepak K. [8] MLP is used for the deciding the problem of binary classification (normal operation or DoS attack). The algorithm uses 80 input values to work. The training took place on the CIC IDS 2017 dataset [9]. In this work several methods are compared and the accuracy of several approaches: the RF machine learning algorithm [3], trained on the Weka1 platform. Experimental results show that the highest RF accuracy is 99.95% and the highest MLP accuracy is 98.87%. Based on these results, it is postulated that RF is more suitable for the task of detecting DoS attacks than MLP.

Abebe A. D. and Naveen C. [10] use MLP for binary classification problem (normal operation or DoS attack) and multi-class classification (normal operation, DoS attack, Probe attack, R2L or U2R attack). The algorithm uses 123 inputs to operate. The training took place on the NSLKDD dataset [11]. In this work authors compared the accuracy of several approaches: SL2; DL. Experimental results show that the highest SL2 accuracy is 97.95% and the highest DL accuracy is 99.36% for binary classification. The highest SL2 accuracy is 99.35% and the highest DL accuracy is 99.52% for multi-class classification, respectively.

Currently, most anomaly detection methods are based on a predictive model, such as an autoregressive model [12], a linear dynamic state space model [13, 14] and neural network-based regression models [15], where sensor measurements are analyzed and predicted.

The authors [16] considered an intrusion detection system based on signature analysis. The authors carried out several experiments to create a reference template describing the normal behavior of a UAV in the absence of any external influence, as well as random anomalies of behavior. Then several situations in which abnormal behavior caused by environmental conditions were simulated. Based on the collected data, a template for the normal behavior of the UAV was built. The authors considered the weighting factors based on the frequency of occurrence of a particular type of abnormal behavior. The authors have carried out several attacks that affect a certain set of physical parameters of a node. This approach demonstrates greater efficiency in detecting a malicious host than simple signature analysis, however, there are some disadvantages: necessity for constant updating and replenishment of the signature database to control data from unaccounted UAV sensors; aanalysis of changes only for the physical parameters of the node and the absence of network data analysis.

Authors of [17] decided Jamming attack detection in FANET poses new challenges due to its key differences from other peer-to-peer networks. Due to the different communication frequency range and power consumption limitations, any centralized detection system becomes unacceptable for the FANET. Therefore, the authors propose for FANET an integrated security architecture for device interference detection based on deep learning. They improve on the proposed federated learning model by using a client group prioritization technique based on Dempster-Schafer theory. The authors evaluated the mechanism on datasets from publicly available standardized RFI attack scenarios from CRAWDAD. They simulated FANET ns-3 architecture and showed that, in terms of accuracy, their proposed solution (82.01% for the CRAWDAD dataset and 89.73% for ns-3 for the simulated FANET ns-3 dataset) outperforms the traditional solution (49.11% for the CRAWDAD dataset and 65.62% for the simulated FANET ns-3 dataset).

With the increasing popularity of the use of UAVs)/drones, it is important to detect and identify the causes of failure in real time for timely recovery or forensics after an accident [18]. The cause of the accident can be either a malfunction of the sensor/actuator system, physical damage/attack, or a cyber-attack on the drone software. In [19], the authors propose new architectures based on deep convolutional neural networks and long short-term memory neural networks (CNNs and LSTMs) for detecting (using an autoencoder) and classifying drone mis operations based on real-time sensor data. The proposed architectures are capable of automatically calculating high-level functions from raw sensor data and studying the spatial and temporal dynamics of sensor data. Empirical results show that the solution is capable of detecting (with over 90% accuracy) and classifying various types of drone anomalous actions (with an accuracy of about 99% (simulation data) and up to 85% (experimental data)).

Most of the considered methods are based on processing signals or raw data, after collecting them from the device. It should be noted that such solutions require the collection of a large amount of data for training. In addition, even during normal UAV operation, occasional malfunctions may occur.

3 Development of a Simulation Model of Traffic Exchange Between UAVs

Figure 1 shows the architecture of the testbed in two versions. Figure 1(a) shows a mesh network that is implemented by configuring the Ad-Hoc mode on devices and the routing protocol for mesh networks - OLSR [19]. The testbed includes 4 devices based on a single-board computer Raspberry Pi 3 model B, which included an RPiPowerPack 2 power expansion board for a PL 104270 4000 mAh rechargeable battery. In Fig. 1(b), a network with a star topology was organized between the devices via a wireless communication channel based on the 802.11n standard. To provide the control interface for the Raspberry Pi, two types of connections were used: ssh - port 22, vnc - port 5900.

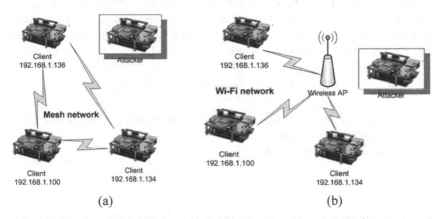

Fig. 1. An architecture of the experimental stand (a) for a mesh network (b) for a star topology.

We also emulated the transmission of a video stream using the HTTPS protocol - port 443. A group management system was implemented for this testbed. The group control algorithms were implemented: exchange of terrain map sections, joint terrain mapping, target allocation, collision avoidance, obstacle avoidance [20]. At the same time, as part of these tasks, the nodes exchanged data using the UDP protocol.

Further, active attacks were carried out on this experimental stand and the results were collected in the form of text files for analysis. Each test was run 10 times to improve the accuracy of the data collected. At the same time, the data was collected because of fixing: normal network operation when no attacks were carried out, a SYN-flood attack, a TCP-RST attack and a deauthentication attack. It was carried out two types of SYN-flood attacks. During "SYN -FLOOD attack on ports 22 and 5900 simultaneously" two ports were attacked intime, but for "SYN -FLOOD attack on ports 22 and 5900" ports were attacked one by one. Normal operation was recorded in several variants: when the network nodes exchanged with packets only according to the group control algorithms, and the video stream was added and periodically transmitted to the server.

4   Implementation of the Process of Forming Data Sets for Training a Neural Network

4.1   Development of a Data Normalization Method

Data is collected according to the following cyber-physical parameters: device temperature, CPU utilization level, number of received/sent TCP/UDP/ICMP/OLSR/802.11 packets. In previous works, it was found that attacks changed these parameters [21]. To receive changes in parameters from cyber-physical subsystems, a data collection module is used. The data collection module must contain several observer objects. Each of these objects monitors a certain cyber-physical parameter and collects data. The collected data is then transferred to the processing module. The data processing module solves the problem of analyzing the information and calculates the coefficients. In this paper, 2 types of probability distributions are used to normalize raw data: Gaussian distribution and Poisson distribution. Depending on the type of data, one or another distribution is applied. The Gaussian distribution is shown by formula 1:

$$p(x) = \frac{1}{\sigma * \sqrt{2\pi}} * e^{-\frac{1}{2}*(\frac{x-\mu}{\sigma})^2} \tag{1}$$

where p is the probability, x is the initial value, σ is the standard deviation, μ is the mathematical expectation, e is the Euler number.

The Poisson distribution is shown by (2):

$$p(x) = \frac{\lambda^x}{x!} * e^{-\lambda} \tag{2}$$

where p is the probability, x is the desired value, λ is the mathematical expectation, e is the Euler number.

Then, the obtained probability distributions are compared with each other using the Kullback-Leibler divergence [22]. In this work, a particular definition of this coefficient is used, which has the following form:

$$D_{KL}(P||Q) = \sum_{i=1}^{n} p_i \log \frac{p_i}{q_i} \tag{3}$$

where p_i is the i-th distributed value of the n-th measurement cycle, q_i - the i-th distributed value of the $(n-1)$-th measurement cycle.

4.2   Describing of Received Dataset

To receive a set of data for training, a research methodology was developed. This methodology included changing the timing of attacks, alternating between the attack and a normal connection, changing the ports to be attacked, etc. Thus, for each of the topologies, the following datasets were obtained, presented in Tables 1 and 2:

The data contains 48245 records, where each segment contains 10 samples. For training, each segment is converted from a two-dimensional array to a one-dimensional array so that it is convenient to read the data.

Table 1. Training datasets for mesh network

Type of attack	Segments	Samples
No anomalies	9394	93940
SYN -FLOOD attack on ports 22 and 5900	3874	38740
SYN -FLOOD attack on ports 22 and 5900 simultaneously	3570	35700
SYN-FLOOD attack on port 22	3388	33880

Table 2. Training datasets for star topology

Attack type for "wi-fi" network	Segments	Samples
No anomalies	6324	63240
SYN -FLOOD attack on ports 22 and 5900	4594	45940
SYN -FLOOD attack on ports 22 and 5900 simultaneously	3029	30290
SYN-FLOOD attack on port 22	4490	44900
SYN -FLOOD attack on port 22 using ssh	5263	52630
Deantiunification attack	4228	42280
Neighbor attack	68	408
Packet spoofing attack	23	138

5   Development of a Module for Detecting Attacks on a UAV Group

5.1   Analysis of DL Elements for Building an Attack Detection System

The MLP architecture was chosen to detect attacks on UAVs. The input data of the UAV attack detector is a one-dimensional array of 54 size, which is suitable for MLP [23]. The output layer defines the output of the network. The size of this layer depends on the problem that the model is solving. To solve the problem of a multi-class classification into 4 classes, there should be 4 perceptions in the output layer, each of which is responsible for the probability of the corresponding class.

Perceptron works according to the formula:

$$y = func\left(\sum\nolimits_{i=1}^{n} w_{ix_i} + b\right), \tag{4}$$

where *func* is the activation function, w_i – weight for the i-th input, x_i – ith input, b – bass. From formula (4) the perceptron performs a linear transformation, which limits the work with data that cannot be divided linearly.

5.2   Analysis of the Model Architecture for Detecting Attacks Aimed at UAVs

MLP was chosen as the architecture, which is shown in Fig. 2.

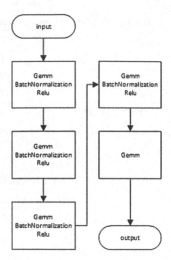

Fig. 2. Block diagram of the UAV attack detector architecture

The figure shows that the architecture consists of several blocks: gemm – multilayer perceptron layer, batchnormalization – batch normalization layer, relu – the layer using the ReLU activation function, the input and output blocks are the input and output layers, respectively. The input data for neural network is the number of input/output packets, as well as the temperature and percentage of the CPU. The output is a class label: presence/absence of an attack on the drone.

6   Assessment of Learning Outcomes

As the results, graphs of metrics on training/validation datasets will be presented: accuracy, precision, recall, f1score, learning error. To understand the metrics, consider TP, FP, TN, FN. When training a model, there are 2 sets of values: the prediction of the model, the "ideal" values that the model tends to. Let us assume that the presence of a class in a given position will have the value 1, the rest of the positions - the value 0, hence the name – positives (1) and negatives (0). Let us compare the sets: if the detector response and mask are equal to 1, then this point is TP, if the detector response is 1 and the wound mask is 0, then this point is FP, if the detector response and mask are 0, then this point is TN, if the detector response is 0 and the wound mask is 1, then this point is FN. Accuracy – allows you to find out the ratio of correct results relative to all results (Eq. 5):

$$Accuracy = \frac{TP + TN}{TP + FP + TN + FN} \qquad (5)$$

Precision – allows you to find out the ratio of correct positives relative to all positives of the detector (Eq. 6):

$$Precision = \frac{TP}{TP + FP} \qquad (6)$$

Recall – allows to find out the ratio of correct positives relative to all positives of the mask (Eq. 7):

$$Precision = \frac{TP}{TP + FP} \tag{7}$$

F1score – is a combination of completeness and sensitivity, calculating the average between them (Eq. 8):

$$F1\ score = \frac{Precision * Recall}{Precision + Recall} \tag{8}$$

Metrics plots during the learning phase presented on Figs. 4, 5, 6, 7 and 8. It can be seen from Figs. 4, 5, 6 and 7 that with an increase in the number of epochs, the accuracy metrics increase, and the learning error decreases. From the results described above, we can conclude that the system unambiguously detects attacks directed at UAVs (Fig. 3).

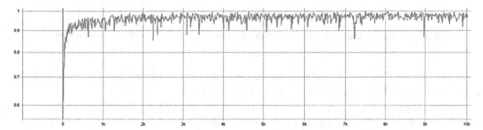

Fig. 3. Changing the Accuracy metric depending on the number of epochs (training phase).

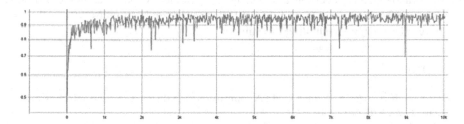

Fig. 4. Changing the F1score metric depending on the number of epochs (training phase).

7   Results

Thus, this paper deals with the issues of attack detection by a UAV group. The idea is that the UAVs analyze network traffic, CPU utilization and temperature. At the same time, he can detect if an attack is carried out on a neighbor. If a denial-of-service attack is carried out on a neighboring UAV, then the level of useful traffic for the UAV decreases, and the traffic picture changes significantly. Such a change is reflected in the value of

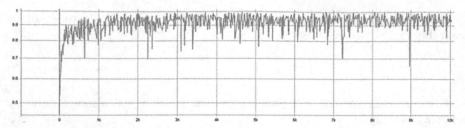

Fig. 5. Changing the Precision metric depending on the number of epochs (training phase)

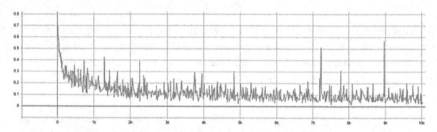

Fig. 6. Change in learning error depending on the number of epochs (learning phase)

entropy. Due to this, it is possible to train the neural network and record the fact of an attack not on the UAV itself, but on the neighboring one.

When implementing a SYN flood of an attack on the UAV itself, the traffic level increases and, accordingly, the entropy value increases significantly. It also increases the CPU utilization. In a deauthentication attack, on the contrary, the traffic level decreases and the CPU activity also decreases. In the first case, the CPU rises as the UAV must hold many half-open connections at the same time. In the second case, on the contrary, there is no network connection and the CPU does not spend resources on it. All these changes can be fixed by evaluating the value of entropy.

References

1. Biermann, E., Cloete, E., Venter, L.M.: A comparison of intrusion detection systems. Comput. Secur. **20**(8), 676–683 (2001)
2. Debar, H., Dacier, M., Wespi, A.: Towards a taxonomy of intrusion-detection systems. Comput. Netw. **31**(8), 805–822 (1999)
3. Lauf, A.P., Peters, R.A., Robinson, W.H.: A distributed intrusion detection system for resource-constrained devices in ad-hoc networks. Ad Hoc Netw. **8**(3), 253–266 (2010)
4. Tseng, C.-Y., Balasubramanyam, P., Ko, C., Limprasittiporn, R., Rowe, J., Levitt, K.: A specification-based intrusion detection system for AODV. In: Proceedings of the 1st ACM Workshop on Security of Ad Hoc and Sensor Networks, pp. 125–134 (2003)
5. Vaidya, V.: Dynamic signature inspection-based network intrusion detection, August 2001
6. Patcha, A., Park, J.-M.: An overview of anomaly detection techniques: existing solutions and latest technological trends. Comput. Netw. **51**(12), 3448–3470 (2007)
7. Wankhede, S., Kshirsagar, D.: DoS attack detection using machine learning and neural network. In: 2018 4th International Conference on Computing Communication Control and Automation (ICCUBEA). IEEE (2018)

8. Sharafaldin, I., Lashkari, A.H., Ghorbani, A.A.: Toward generating a new intrusion detection dataset and intrusion traffic characterization. In: Proceedings of the 4th International Conference on Information Systems Security and Privacy, SCITEPRESS. Science and Technology Publications (2018)
9. Hastie, T., Tibshirani, R., Friedman, J.: The Elements of Statistical Learning. Springer, New York (2009)
10. Diro, A.A., Chilamkurti, N.: Distributed attack detection scheme using deep learning approach for Internet of Things. Fut. Gener. Comput. Syst. **82**, 761–768 (2018). https://doi.org/10.1016/j.future.2017.08.043
11. Tavallaee, M., Bagheri, E., Lu, W., Ghorbani, A.A.A.: detailed analysis of the KDD CUP 99 data set. In: 2009 IEEE Symposium on Computational Intelligence for Security and Defense Applications. IEEE (2009)
12. Hadžiosmanović, D., Sommer, R., Zambon, E., Hartel, P.H.: Through the eye of the PLC: semantic security monitoring for industrial processes. In: Proceedings of the 30th annual Computer Security Applications Conference, pp. 126–135 (2014)
13. Abur, A., Exposito, A.G.: Expósito 2004 Power System State Estimation. CRC Press.https://doi.org/10.1201/9780203913673
14. Urbina, D.I., et al.: Limiting the impact of stealthy attacks on industrial control systems. In: Proceedings of the 2016 ACM SIGSAC Conference on Computer and Communications Security, pp. 1092–1105 (2016)
15. Goh, J., Adepu, S., Tan, M., Lee, Z.: Anomaly detection in cyber physical systems using recurrent neural networks. In: IEEE18th International Symposium on High Assurance Systems Engineering, HASE 2017, pp. 140–145 (2017)
16. Dán, G., Sandberg, H.: Stealth attacks and protection schemes for state estimators in power systems. In: 1st IEEE International Conference on Smart Grid Communications, Smart Grid Comm. 2010, pp. 214–219 (2010)
17. Monshizadeh, M., Khatri, V., Kantola, R., Yan, Z.: An orchestrated security platform for internet of robots. In: Man Ho Allen, Au., Castiglione, A., Choo, K.-K., Palmieri, F., Li, K.-C. (eds.) Green, Pervasive, and Cloud Computing, pp. 298–312. Springer, Cham (2017). https://doi.org/10.1007/978-3-319-57186-7_23
18. Mowla, N.I., Tran, N.H., Doh, I., Chae, K.: Federated learning-based cognitive detection of jamming attack in flying ad-hoc network. IEEE Access **8**, 4338–4350 (2020). https://doi.org/10.1109/ACCESS.2019.2962873
19. Mowla, N.I., Tran, N.H., Doh, I., Chae, K.: AFRL: adaptive federated reinforcement learning for intelligent jamming defense in FANET. J. Commun. Netw. **22**(3), 244–258 (2020). https://doi.org/10.1109/JCN.2020.000015
20. Basan, E., Makarevich, O., Basan, A.: Evaluating and detecting internal attacks in a mobile robotic network. In: The 10th International Conference on Cyber-Enabled Distributed Computing and Knowledge Discovery, Proceedings CyberC 2018, pp 1–9 (2018). 978-0-7695-6584-2/18/
21. Basan, E., Basan, A., Nekrasov, A.: Method for detecting abnormal activity in a group of mobile robots. Sensors **19**(18) 4007 (2019). https://doi.org/10.3390/s19184007. https://www.mdpi.com/1424-8220/19/18/4007
22. Basan, E., Medvedev, M., Teterevyatnikov, S.: Analysis of the impact of denial of service attacks on the group of robots. In: 10th International Conference on Cyber-Enabled Distributed Computing and Knowledge Discovery, pp. 63–71 (2018). 978-0-7695-6584-2/18/. https://doi.org/10.1109/CyberC.2018.00023

23. Basan, E., Basan, A., Makarevich, O.: Detection of anomalies in the robotic system based on the calculation of Kullback-Leibler divergence. In: International Conference on Cyber-Enabled Distributed Computing and Knowledge Discovery CyberC, pp. 337–341 (2019). https://doi.org/10.1109/CyberC.2019.00064

24. Nikolenko, S., Kadurin, A., Arkhangelskaya, E.: Deep Learning. Peter, Moscow (2018).(in Russian)

Machine Learning

NiaAML2: An Improved AutoML Using Nature-Inspired Algorithms

Luka Pečnik[✉], Iztok Fister[✉], and Iztok Fister Jr.[✉]

Faculty of Electrical Engineering and Computer Science,
Koroška ul. 46, 2000 Maribor, Slovenia
luka.pecnik@student.um.si, {iztok.fister,iztok.fister1}@um.si

Abstract. Using machine learning methods in the real-world is far from being easy, especially because of the number of methods on the one hand, and setting the optimal values of their parameters on the other. Therefore, a lot of so-called AutoML methods have emerged nowadays that also enable automatic construction of classification pipelines to users, who are not experts in this domain. In this study, the NiaAML2 method is proposed that is capable of constructing the classification pipelines using nature-inspired algorithms in two phases: pipeline construction, and hyper-parameter optimization. This method improves the original NiaAML capable of this construction in one phase. The algorithm was applied to four UCI ML datasets, while the obtained results encouraged us to continue with the research.

Keywords: Nature-inspired algorithms · Machine learning · AutoML · Classification pipeline

1 Introduction

Finding a Machine Learning (ML) method for solving a certain problem usually becomes a very difficult task due to several factors, like proper preparation and data preprocessing, the suitability of the selected ML method for a given problem and setting the optimal values of hyper-parameters for the selected ML methods [7,9]. The domain of Automated Machine Learning (AutoML) has emerged for the sake of simplification and automation of some of the mentioned steps. This domain deals with the problem of finding the optimal classification pipeline automatically. The classification pipeline is a sequence of methods or algorithms, necessary for the successful implementation of the entire ML process, from data preprocessing to the point, where the results are calculated [8]. By automating the individual pipeline discovery steps, ML has become more accessible to the wider community. Researches show that, using the AutoML methods, the classification pipelines can also be constructed by non-expert users [5,12].

For this purpose, many AutoML methods have been developed such as Auto-WEKA [11] and Hyperopt-sklearn [1] that both base on Bayesian optimization.

© Springer Nature Switzerland AG 2021
Y. Tan and Y. Shi (Eds.): ICSI 2021, LNCS 12690, pp. 243–252, 2021.
https://doi.org/10.1007/978-3-030-78811-7_23

Auto-sklearn [5] upgraded the idea of Bayesian optimization by adding a meta-learning step. In the field of AutoML, several methods have been developed that search for classification pipelines using stochastic population-based nature-inspired algorithms. The frameworks that use such methods are TPOT [12] and RECIPE [3], which use genetic algorithms to construct the classification pipeline. One of the more interesting methods is also the NiaAML method [7], which is used to find optimal classification pipelines using the stochastic population-based nature-inspired algorithms.

In this study, an extension of the NiaAML method is presented, i.e., the improved NiaAMLv2. Indeed, the proposed NiaAMLv2 eliminates the main weakness of the original method, where the hyper-parameters' optimization is performed simultaneously with construction of the classification pipelines in a single phase. This means that only one instance of nature-inspired algorithm is needed, where the ML methods and their hyper-parameters are searched for simultaneously in each generation. The improved version divides the construction of the pipeline and hyper-parameter optimization into separate phases, where two nature-inspired instances of algorithms are applied sequentially one after another, i.e., the former by covering the construction of pipeline and the later by optimizing the hyper-parameters of the proposed ML-methods. Results of experiments for both methods are also discussed and compared using the four datasets from the UCI machine learning repository [4].

The paper is organized as follows. In Sect. 2 stochastic population-based nature-inspired algorithms are described along with using the NiaAML. The improved NiaAMLv2 is proposed in Sect. 3, while Sect. 4 describes the experiments and the obtained results. The conclusion of the paper is given in Sect. 5, where we also outline directions for the future work.

2   Population-Based Nature-Inspired Algorithms for AutoML

Nowadays, the nature-inspired algorithms present one of the more popular groups for solving the optimization problems [14]. According to the metaphor taken from nature, they are divided into four groups [6], where the SI-based algorithms are a subset of algorithms in the field of biology. This subset also includes Evolutionary Algorithms (EA) working on the principle of Darwin's evolution [2], which proves that a generation of organisms can survive in the environment only if they are well adapted to the environmental conditions [2]. Together with SI-based algorithms, they form a class of nature-inspired algorithms.

A classification pipeline construction can be modeled as a continuous optimization problem that can be solved with a number of optimization algorithms. Similar as the NiaAML [7], this paper focuses on the interesting approach for constructing classification pipelines using stochastic population-based nature-inspired algorithms for solving this problem.

2.1   NiaAML

NiaAML is distinguished from the other AutoML methods by mapping of real-valued vectors to classification pipelines. The vectors are evolved using the stochastic population-based nature-inspired algorithms. Indeed, each individual of the population represents one of the potential classification pipelines [7]. According to [9], NiaAML belongs to the process of collecting metadata or meta-learning, which includes: (1) the construction of the classification pipeline, (2) optimization of the hyper-parameters and (3) evaluation of the model. The construction of the classification pipeline with the NiaAML method demands following three mandatory steps, i.e., selection of: (1) the feature selection, (2) the feature transformation, and (3) the classifier algorithms. The aforementioned ML method can be selected from any framework of tools [7].

Individuals representing possible classification pipelines in the NiaAML method are presented in the optimization algorithm as a real-valued vector [7]:

$$\mathbf{x}_i^{(t)} = \{\underbrace{x_{i,1}^{(t)}}_{\text{FSA}}, \underbrace{x_{i,2}^{(t)}}_{\text{FTA}}, \underbrace{x_{i,3}^{(t)}}_{\text{CLA}}, \underbrace{x_{i,4}^{(t)}, \ldots, x_{i,k}^{(t)}, x_{i,k+1}^{(t)}, \ldots, x_{i,D+3}^{(t)}}_{\text{hyperparameters}}\}, \tag{1}$$

where the first three elements are intended for the selection of pipeline components, while the variable length of vector D depends on the number of hyper-parameters used by the components included in the pipeline. Let us notice that the length of each solution is $D+3$. Because both observed methods, i.e., NiaAML and NiaAMLv2, operate on the same set of aforementioned selection components, the detailed description of these is illustrated in the next section.

Pseudo-code of the proposed algorithm for constructing the classification pipelines is presented in Algorithm 1, from which it can be see that the algorithm suits the general form of the SI-based algorithms. After initialization

---

Algorithm 1. A pseudo-code of the NiaAML method

1: *population* ← Initialize real valued vectors $\mathbf{x}_i$ regarding Eq. (1)
2: *best_pipeline* ← Evaluate and select the best (*population*)
3: **while** Termination condition not met **do**
4:    **for each** $\mathbf{x}_i \in$ *population* **do**
5:       $\mathbf{x}_{trial}$ ← Modify population using variation operators ($\mathbf{x}_i$)
6:       *pipeline* ← Construct pipeline ($\mathbf{x}_{trial}$)
7:       Evaluate (*pipeline*)
8:       **if** *pipeline* is better than 'Construct pipeline ($\mathbf{x}_i$)' **then**
9:          $\mathbf{x}_i$ ← $\mathbf{x}_{trial}$ ▷ Replace the worse individual
10:      **end if**
11:      **if** *pipeline* is better than *best_pipeline* **then**
12:         *best_pipeline* ← *pipeline*
13:      **end if**
14:   **end for**
15: **end while**
16: **return** *best_pipeline*

---

and evaluation of initial population (line 1 and 2), each individual undergoes acting the variation operators (line 5) by generation of trial solution. The pipeline is constructed by appropriate genotype-phenotype mapping based on collection of selection components (line 6) and evaluated according its quality (line 7). If the quality of constructed pipeline from the trial is better than the quality of the target individual, the target is replaced with the trial (lines 8–10).

3   NiaAMLv2

The improved version NiaAMLv2 divides the AutoML process into two phases: (1) classification pipeline construction, and (2) hyper-parameter optimization. In line with this, two algorithms are developed for each particular phase. However, both phases are implemented by a different PSO algorithm.

The first algorithm represents the solutions of the construction process as real-valued vectors, i.e.:

$$\mathbf{x}_i^{(t)} = \{ \underbrace{x_{i,1}^{(t)}}_{\text{FSA}}, \underbrace{x_{i,2}^{(t)}}_{\text{FTA}}, \underbrace{x_{i,3}^{(t)}}_{\text{CLA}} \}, \tag{2}$$

which consists of three elements denoting: the Feature Selection Algorithm (FSA), Feature Transformation Algorithm (FTA), and selection of the CLassification Algorithm (CLA). Indeed, the meaning of the three parameters is described in Table 1, from which it can be seen that the four stochastic nature-inspired population-based algorithms can be selected for feature selection, there are two feature transformation options beside the selection of the no transformation, and six classification algorithms. The implementations of all the aforementioned components were taken from the Scikit-learn Python library [13].

Table 1. Components of the original NiaAML method.

Component	Abbreviation	Framework of tools
Feature selection	FSA	$\{DE, PSO, GWO, BA\}$
Feature transformation	FTA	$\{NONE, SCALING, NORMAL\}$
Classificator	CLA	$\{MLP, LS\_SVM, ADA, RF, ERT, BAG\}$

The second algorithm is devoted to the hyper-parameter optimization and, therefore, represents the solution as a variable vector of real-values drawn from domains of feasible values randomly, in other words:

$$\mathbf{y}_i^{(t)} = \{ \underbrace{y_{i,1}^{(t)}, \ldots, y_{i,k}^{(t)}, y_{i,k+1}^{(t)}, \ldots, y_{i,D}^{(t)}}_{\text{Hyper-Parameters}} \}. \tag{3}$$

Due to the limitation of the paper, the hyper-parameters for the selected components that enter into the optimization are not illustrated explicitly.

Algorithm 2. A pseudo-code of the NiaAMLv2 method

1: $population_1 \leftarrow$ Initialize real valued vectors $\mathbf{x}_i$ from the Eq. (2)
2: $population_2 \leftarrow$ Initialize real valued vectors $\mathbf{y}_j$ from the Eq. (3)
3: $best\_pipeline \leftarrow$ Evaluate and select the best($population_1$,$population_2$)
4: **while** Termination condition$_1$ not met **do**
5:     **for each** $\mathbf{x}_i \in population_1$ **do**
6:         $\mathbf{x}_{trial} \leftarrow$ Modify using variation operators ($\mathbf{x}_i$)
7:         **while** Termination condition$_2$ not met **do**
8:             **for each** $\mathbf{y}_j \in population_2$ **do**
9:                 $\mathbf{y}_{trial} \leftarrow$ Modify using variation operators ($\mathbf{y}_j$)
10:                 $pipeline \leftarrow$ Construct pipeline ($\mathbf{x}_{trial}, \mathbf{y}_{trial}$)
11:                 Evaluate ($pipeline$)
12:                 $target \leftarrow$ Construct pipeline ($\mathbf{x}_i, \mathbf{y}_j$)
13:                 **if** $pipeline$ is better than $target$ **then**
14:                     $\mathbf{x}_i \leftarrow \mathbf{x}_{trial}$
15:                     $\mathbf{y}_j \leftarrow \mathbf{y}_{trial}$
16:                 **end if**
17:                 **if** $pipeline$ is better than $best\_pipeline$ **then**
18:                     $best\_pipeline \leftarrow pipeline$
19:                 **end if**
20:             **end for**
21:         **end while**
22:     **end for**
23: **end while**
24: **return** $best\_pipeline$

The concept of the NiaAMLv2 is presented in the Algorithm 2. from which it can be seen that both phases, i.e., algorithms' selection and hyper-parameter optimization are launched sequentially one after another by two different nature-inspired algorithms. Interestingly, both algorithms use the same evaluation function (function Evaluate), i.e., pipeline accuracy expressed as follows:

$$f\left(\mathbf{x}_i^{(t)}\right) = \text{Accuracy}\left(M\left(\mathbf{x}_i^{(t)}\right)\right), \tag{4}$$

where Accuracy(.) denotes the accuracy of model $M\left(\mathbf{x}_i^{(t)}\right)$ based on classification pipeline $\mathbf{x}_i$. Thus, the accuracy is a statistical measure reflecting the bias between the true and all number of cases, in other words:

$$\text{Accuracy}\left(M\left(\mathbf{x}_i^{(t)}\right)\right) = \frac{TP + TN}{TP + TN + FP + FN}, \tag{5}$$

where $TP =$ True Positive, $TN =$ True Negative, $FP =$ False Positive and $FN =$ False Negative [7]. The task of the optimization is to find models with the maximum value of accuracy [7].

After we obtained the best pipelines found in both phases, we tested them using the stratified 10-fold cross validation and compared their results using accuracy, precision, F1-score and Cohen's κ metrics.

4   Experiments and Results

The goal of our experimental work was to show that the proposed NiaAMLv2 is suitable for AutoML, on the one hand, and that it constructs classification pipelines of quality equal to, if not better than, the original NiaAML. In line with this, extensive experimental work was conducted, where all experiments were performed on an HP ProDesk 400 G6 MT computer running Microsoft Windows 10, an Intel (R) Core (TM) i7-9700 CPU @ 3.00 GHz processor, and 8 GB of installed physical memory.

When running the NiaAMLv2, we used a population of 15 individuals with 300 fitness function evaluations in both optimization steps, while we used a population of 20 individuals with 400 evaluations with the original NiaAML. Particle Swarm Optimization (PSO) [10] was used in all cases with the following parameters: cognitive component and social component set to 2.0, inertia weight had a value of 0.7 and both minimal and maximal velocities set to 1.5.

During the experimental work, we used four datasets from the UCI Machine Learning Repository [4]. The list and their characteristics are shown in Table 2, from which it can be seen that the first two consist of real-valued attributes, while the other two of mixed ones (i.e., real-valued, integer, and discrete). The Abalone dataset exposed the highest number of instances. Interestingly, the last Cylinder bands dataset also included the missing data. To impute missing data in this dataset, we used an average value for numeric features and a mode for categorical features. To encode categorical features into numeric, we used one-hot encoding.

Table 2. Datasets used in the experiment.

Dataset	Type of attributes	# Instances	# Features	Missing data
Ecoli	Real	336	8	No
Yeast	Real	1484	8	No
Abalone	Real, integer, categorical	4177	8	No
Cylinder bands	Real, integer, categorical	512	39	Yes

The results of the algorithms were compared according to four classification evaluation metrics: Accuracy, Precision, F_1, and Cohen's Kappa. The accuracy is already applied as part of the fitness function evaluation and is defined by using Eq. (5). The other metrics are defined as follows: The precision of the model $M(\mathbf{x}_i)$ is defined as ratio between true positive and all number of cases:

$$\text{Precision}\left(M\left(\mathbf{x}_i\right)\right) = \frac{TP}{TP + FP}. \tag{6}$$

The F_1 metric is calculated using Precision and Recall:

$$F_1\left(M\left(\mathbf{x}_i\right)\right) = 2 \cdot \frac{\text{Precision}\left(M\left(\mathbf{x}_i\right)\right) \cdot \text{Recall}\left(M\left(\mathbf{x}_i\right)\right)}{\text{Precision}\left(M\left(\mathbf{x}_i\right)\right) + \text{Recall}\left(M\left(\mathbf{x}_i\right)\right)}, \tag{7}$$

where Recall $= \frac{TP}{TP+FN}$. Actually, metrics Precision, Recall, and F_1 need to be calculated, when we have to do with the classification to two classe.

The Cohen's Kappa is defined as follows:

$$\kappa\left(M\left(\mathbf{x}_i\right)\right) = \frac{n\sum_{i=1}^{k} CM_{i.i} - \sum_{i=1}^{k} CM_{i,*} CM_{*,i}}{n^2 - \sum_{i=1}^{k} CM_{i,*} CM_{*,i}}, \qquad (8)$$

where $CM_{i,*}$ is sum of elements in the i-th row of confusion matrix CM and $CM_{*,i}$ the sum of the i-t columns of the same matrix.

In the remainder of the paper, the results obtained by constructing the optimal classification pipeline are presented in detail.

4.1 The Results of Constructing the Pipeline on UCI-ML Datasets

The presentation of the results obtained by optimization of both AutoML methods is divided into two parts: In the first part, the optimal classification pipelines are exposed according to each dataset per specific version of the AutoML method. The second part is devoted for comparing the results between the original NiaAML and the improved NiaAMLv2 according to four classification evaluation metrics (i.e., Accuracy, Precision, F1, and Cohen's Kappa), and depicted in the corresponding boxplots. The boxplots show the results of the aforementioned values of metrics obtained by stratified 10-fold cross validation over pipelines optimized by both methods. Each of them presents the average value calculated in each of ten folds.

The boxplots, comparing the results between NiaAml and NiaAMLv2, are presented in Figs. 1 and 2, where each figure is divided into diagrams illustrating one of the observed datasets. In the diagrams, the results of the NiaAML method are presented in the red boxes, while the results of the NiaAMLv2 method in the blue boxes.

In the case of the Ecoli dataset (Fig. 1(a)), both classification pipelines selected all features. Despite the fact that a few outliers are visible in both cases, the classification pipeline obtained by the NiaAMLv2 shows the better results on average. All features were also selected in the case of the Yeast dataset by both classification pipelines (Fig. 1(b)). On average, the results were very similar, but the pipeline obtained by the NiaAMLv2 method had a smaller interquartile range.

The results of constructing the optimal classification pipelines for the Abalone dataset by the NiaAMLv2 method removed one feature, while the NiaAML used only four features. The classification pipeline optimized with the NiaAMLv2 method showed better results overall (Fig. 2(a)). The pipeline optimized with the NiaAMLv2 method showed better results on average by also optimizing the Cylinder bands dataset, where there are also some visible outliers (Fig. 2(b)).

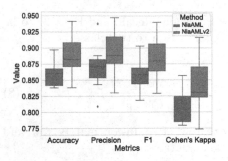
a: Ecoli dataset.

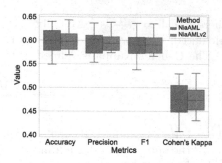
b: Yeast dataset.

Fig. 1. Results of the NiaAML and NiaAMLv2 pipelines for the Ecoli and Yeast datasets.

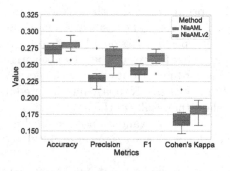
a: Abalone dataset.

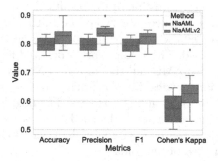
b: Cylinder bands dataset.

Fig. 2. Results of the NiaAML and NiaAMLv2 pipelines for the Abalone and Cylinder bands datasets.

4.2 Time Complexity

The time complexity of the NiaAMLv2 was much higher than the time complexity of the original NiaAML as can be seen in Table 3, where the time complexities for each observed database in seconds are presented for both AutoML methods. Moreover, the ratios between time complexities of NiaAML and NiaAMLv2 in percents have been added to the table. As can be seen from the table, the time complexity of the proposed AutoML method was more than 100 times higher than the original one. However, the reason for this must be searched for in separating of the hyper-parameter optimization into an independent phase.

4.3 Discussion

Based on the results of the experiment, we found out that the classification pipelines generated by the NiaAMLv2 were slightly better in treating the Ecoli,

Table 3. Time complexity of optimization.

Nr.	Dataset	Time [sec]		Ratio
		NiaAML	NiaAMLv2	[%]
1	Ecoli	27.48	29,031.30	0.0945
2	Yeast	93.49	33,127.02	0.2822
3	Cylinder bands	189.23	122,262.46	0.1548
4	Abalone	2,868.90	295,909.65	0.9695
$\sum$		3,538.37	55,4863.14	0.6377

Abalone and Cylinder bands datasets than the pipelines optimized by the original NiaAML. On the other hand, we were surprised due to observation that there were no noticeable differences in comparing the results obtained by the Yeast dataset. Indeed, we expected that the NiaAMLv2 would give much better results than NiaAML in all datasets, but this assumption was not justified in all cases. The reason for such results was most likely in the size of the search spaces introduced by continuous domains of hyper-parameters, and in the sensitivity of both methods to occasional discovery of a very good fitness value, presumably due to a random favorable distribution of data for training and testing pipelines.

5   Conclusion

In this article, we described the NiaAMLv2 method for automatic construction of classification pipelines that presents an extension of the original NiaAML. The main weakness of the NiaAML was that this method constructs a classification pipeline and optimizes the corresponding hyper-parameters simultaneously in one step. The proposed method separates both mentioned steps in two phases conducted sequentially one after another. The results of our experiments showed that the NiaAMLv2 outperformed the results of the original NiaAML by constructing the classification pipelines for three of the four UCI ML datasets in tests, while the achieved results on the other two datasets were similar.

In the further research, a stratified 10-fold cross validation could be used to provide less sensitivity to occasional discovery of good fitness values due to a random favorable distribution of data for training and testing pipelines. The better results would also likely be brought by discretizing the huge continuous domains of hyper-parameters into more discrete classes, and, thus, getting the search algorithm more chance to explore this search space more effectively.

Acknowledgements. The authors acknowledge the financial support from the Slovenian Research Agency (research core funding No. P2-0041 - Digital twin).

References

1. Bergstra, J., Komer, B., Eliasmith, C., Yamins, D., Cox, D.D.: HyperOpt: a python library for model selection and hyperparameter optimization. Comput. Sci. Discov. **8**(1), 014008 (2015)
2. Dasgupta, D., Michalewicz, Z.: Evolutionary algorithms in engineering applications. Springer Science and Business Media (2013). https://doi.org/10.1007/978-3-662-03423-1
3. de Sá, A.G.C., Pinto, W.J.G.S., Oliveira, L.O.V.B., Pappa, G.L.: RECIPE: a grammar-based framework for automatically evolving classification pipelines. In: McDermott, J., Castelli, M., Sekanina, L., Haasdijk, E., García-Sánchez, P. (eds.) EuroGP 2017. LNCS, vol. 10196, pp. 246–261. Springer, Cham (2017). https://doi.org/10.1007/978-3-319-55696-3_16
4. Dua, D., Graff, C.: UCI machine learning repository. http://archive.ics.uci.edu/ml (2017)
5. Feurer, M., Klein, A., Eggensperger, K., Springenberg, J., Blum, M., Hutter, F.: Efficient and robust automated machine learning. In: Cortes, C., Lawrence, N., Lee, D., Sugiyama, M., Garnett, R. (eds.) Advances in Neural Information Processing Systems, vol. 28, pp. 2962–2970. Curran Associates Inc, (2015)
6. Fister Jr, I., Yang, X.-S., Fister, I., Brest, J., Fister, D.: A brief review of nature-inspired algorithms for optimization. Elektrotehniški vestnik **80**(3), 116–122 (2013)
7. Fister Jr, I., Zorman, M., Fister, D., Fister, I.: Continuous optimizers for automatic design and evaluation of classification pipelines. In: Frontier Applications of Nature Inspired Computation, pp. 281–301 (2020)
8. Guyon, I., et al.: Design of the 2015 chalearn automl challenge. In: 2015 International Joint Conference on Neural Networks (IJCNN), pp. 1–8 (2015)
9. He, X., Zhao, K., Chu, X.: AutoML: a survey of the state-of-the-art. Knowl.-Based Syst. **212** 106622 (2020)
10. Kennedy, J., Eberhart, R.: Particle swarm optimization. In: Proceedings of ICNN 1995 - International Conference on Neural Networks, vol. 4, pp. 1942–1948 (1995)
11. Kotthoff, L., Thornton, C., Hoos, H.H., Hutter, F., Leyton-Brown, K.: Auto-WEKA 2.0: Automatic model selection and hyperparameter optimization in WEKA. J. Mach. Learn. Res. **18**(1), 826–830 (2017)
12. Olson, R.S., Bartley, N., Urbanowicz, R.J., Moore, J.H.: Evaluation of a tree-based pipeline optimization tool for automating data science. In: Proceedings of the Genetic and Evolutionary Computation Conference 2016, GECCO 2016. ACM, pp. 485–492, New York, NY, USA (2016)
13. Pedregosa, F., et al.: Scikit-learn: machine learning in Python. J. Mach. Learn. Res. **12**, 2825–2830 (2011)
14. Vrbančič, G., Brezočnik, L., Mlakar, U., Fister, D., Fister Jr, I.: NiaPy: Python microframework for building nature-inspired algorithms. J. Open Source Softw. **3**, **613** (2018)

Proof Searching in PVS Theorem Prover Using Simulated Annealing

M. Saqib Nawaz[1] , Meng Sun[2] , and Philippe Fournier-Viger[1](✉)

[1] School of Humanities and Social Sciences,
Harbin Institute of Technology (Shenzhen), Shenzhen, China
{msaqibnawaz, philfv}@hit.edu.cn
[2] School of Mathematical Sciences, Peking University, Beijing, China
sunmeng@math.pku.edu.cn

Abstract. The proof development process in PVS theorem prover is interactive in nature, that is not only laborious but consumes lots of time. For proof searching and optimization in PVS, a heuristic proof searching approach is provided where simulated annealing (SA) is used to search and optimize the proofs for formalized theorems/lemmas in PVS theories. In the proposed approach, random proof sequence is first generated from a population of frequently occurring PVS proof steps that are discovered with sequential pattern mining. Generated proof sequence then goes through the annealing process till its fitness matches with the fitness of the target proof sequence. Moreover, the performance of SA with a genetic algorithm (GA) is compared. Obtained results suggest that evolutionary/heuristic techniques can be combined with proof assistants to efficiently support proofs finding and optimization.

Keywords: PVS · Simulated annealing · Genetic algorithm · Proof searching · Proof sequences

1 Introduction

Interactive theorem provers (ITPs), also known as proof assistants, such as PVS [1], Coq [2] and HOL4 [3], follow a user driven proof development process. In ITPs, the user guides the proof process by providing the proof goal (theorem or lemma) and by applying proof commands and decision procedures to prove the goal. Generally, one needs to do lots of repetitive work to prove a nontrivial goal, which is laborious and consumes a large amount of time. Thus, proof guidance, proof searching and automation are some extremely desirable features in ITPs.

For proof guidance, a sequential pattern mining (SPM)-based proof process learning approach was proposed [4] for PVS. The approach can be used to find frequent proof steps/patterns that are used in the proofs, their relationships with each other, dependency of new proofs on already proved facts and to predict the next proof step(s). A high-utility itemset mining (HUIM)-based proof process learning approach was also proposed [5] for PVS. Moreover, an evolutionary approach was proposed [6], where a genetic algorithm (GA) with different crossover

© Springer Nature Switzerland AG 2021
Y. Tan and Y. Shi (Eds.): ICSI 2021, LNCS 12690, pp. 253–262, 2021.
https://doi.org/10.1007/978-3-030-78811-7_24

and mutation operators was used for proof searching and optimization in theories available in HOL4 library. This proof searching approach was quite efficient in evolving random proofs. However, there is still room to develop alternative proof searching approaches.

Based on previous studies [4,6], a heuristic-based approach is designed in this paper for proof searching and optimization in PVS theories. The basic idea is to use Simulated Annealing (SA) [7] for proof searching, where an initial population (a set of potential solutions) is first created from frequent PVS proof steps that are discovered with the approach in [4]. A random proof sequence is then generated from the population and is evolved with the annealing process. The annealing process continues till the fitness of the random proof sequence matches that of the original (target) proof for formalized theorems and lemmas. Moreover, the performance of the proposed SA is compared with GA [6] for various parameters. It is found that SA outperforms GA for proof finding and optimization.

In literature, evolutionary algorithms have been used in ITPs [8–10]. However, the approach in [8,9] for Coq can only be used to successfully find the proofs of easy theorems (that contain less number of proof steps). Similarly, the approach in [10] represented Isabelle's proofs with a tree structure, which are linearized. However, linearization leads to the loss of important connections (information) between different branches of the proofs due to which interesting patterns and tactics may be lost in the evolution process. A recent work [11] briefly discussed the applicability of evolutionary computation in improving the heuristics of automatic proof search in Isabelle/HOL theorem prover. Moreover, the framework based on GA in [12] can be used to find good search heuristics in the automated theorem prover (ATP) E. The dataset for the proof sequences in this work contains all the essential information needed to discover frequent PVS proof steps, which are used for the generation of initial population. Moreover, the SA-based proof searching approach can handle formalized proof goals of various length.

The remainder of this paper is organized as follows. Section 2 provides the proof searching approach based on SA for PVS theorem prover. Obtained results are discussed in Sect. 3, followed by the conclusion in Sect. 4.

2   Proof Searching in PVS with SA

The proof development process in PVS is interactive in nature and it follows the *sequent-style proof representation* [13]. A user first provides the property (in the form of a lemma or theorem) that is called a proof goal. Each proof goal is a *sequent* consisting of a sequence of formulas called *antecedents* (hypotheses) and a sequence of formulas called *consequents* (conclusions).

User then applies proof commands, inference rules and decision procedures to solve the proof goal. The action resulting from a proof command, inference rule or decision procedure is referred to as a PVS proof step (*PPS*) here. A *PPS* may either prove the goal or generates another sequent or divide the main goal into sub-goals. The proof development process for a theorem or lemma is completed when the sequent or all the sub-goals are proved.

After proof development, PVS saves the proof scripts of a theory in a separate proof file. These files contain *PPS* with some other information related to PVS. Inside a theory, a particular proof goal for a theorem or lemma depends on the specifications and it can be completed by applying the *PPS* in different orders. This makes it difficult to automatically find the proof for a goal or to carry out a brute force or pure random search approach for proof searching. However, evolutionary and heuristic algorithms have the potential to search for the proofs of theorems/lemmas due to their ability to handle black-box search and optimization problems. The schematic of using SA for proof searching in PVS is shown in Fig. 1.

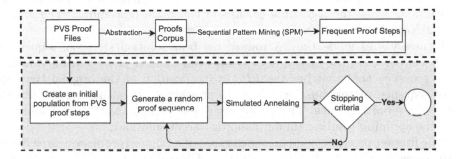

Fig. 1. Proof searching in PVS using SA

The data available in PVS proof files is first converted into a proper compu tational format so SA algorithms can be used. Moreover, the redundant infor- mation (related to PVS) that plays no part in proof searching and evolution is removed from the proof files. The complete proof for a goal can now be considered as a sequence of *PPS*. Let $PS = \{PPS_1, PPS_2, \ldots PPS_m\}$ represent the set of *PPS* proof steps. Let the notation $|PSS|$ denotes the set cardinality. For example, consider that $PS = \{skeep, expand, flatten, typepred, split, beta, iff, assert\}$. The set $\{skeep, expand, flatten, assert\}$ is a proof step set that contains four proof steps. A proof sequence is a list of proof step sets $S = \langle PSS_1, PSS_2, ..., PSS_n \rangle$, such that $PSS_i \subseteq PSS$ $(1 \leq i \leq n)$. For example, $\langle \{skeep, flatten\}, \{typepred\}, \{split, beta, iff, assert\} \rangle$ is a proof sequence which has three *PSS* and seven *PPS* that are used to prove a proof goal. A *proof dataset PD* is a list of proof sequences $PD = \langle S_1, S_2, ..., S_p \rangle$, where each sequence has an identifier (ID). For example, Table 1 shows a *PD* that has five proof sequences.

Algorithm 1 presents the pseudocode of the SA that is used to find the proofs in PVS theories. An initial population (*Pop*) is first created from frequent *PPS* (*FPPS*) that are discovered with various SPM techniques. In data mining, SPM can be used to find useful and interesting patterns (information) that are generally hidden in large corpora of sequential data [14]. From population, one

Table 1. A sample proof dataset

ID	Proof sequence
1	$\langle \{inst \ , \ grind\} \rangle$
2	$\langle \{skosimp, \ expand, \ flatten, \ assert\} \rangle$
3	$\langle \{skosimp, \ expand, \ propax\} \rangle$
4	$\langle \{skeep, \ expand, \ typepred, \ expand, \ flatten, \ expand, \ inst \ , \ assert\} \rangle$
5	$\langle \{induct\},\{expand, \ propax\},\{skosimp, \ expand, \ assert\} \rangle$

random proof sequence (PS) is generated. Random proof sequences goes through the annealing process (Steps 9–25 in Algorithm 1), where it is evolved till its fitness is equal to the fitness of the target proof sequence from PD. Besides annealing process, the algorithm consists of two main procedures: (1) *Fitness*, and (2) *Get_ Neighbor* (GN).

Fitness values guide the SA toward the best solution(s) (proof sequences here). For our case, the fitness value is the total number of PPS in the random proof sequence that matches the PPS in the position of the original (target) proof sequence. Algorithm 2 presents the procedure for calculating the fitness value of a proof sequence. This procedure evaluates how close a given solution is to the optimum solution (in our case, the target solution).

The fitness procedure compares each gene i of a random proof sequence ($Pseq$) with the genes of the target (P). The fitness of $PSeq$ is set to 0, and increased by 1 for each matching gene and if genes in both sequences are equal then fitness of 1 is assigned. For example, consider the random proof sequence (RP) and the target sequence (TP) are:

$$RP = skosimp, \ expand, \ lemma, \ grind, \ flatten, \ assert, \ skolem, \ assert$$

$$TP = skeep, \ expand, \ typepred, \ expand, \ flatten, \ expand, \ inst, \ assert$$

The *Fitness* procedure will return 3 as three PPS are same in both sequences (at positions 2, 5 and 8 respectively).

In the annealing process, a neighbor random sequence is first generated. Algorithm 3 presents the procedure for getting the neighbor solution. The selected location value is changed from its original value in *Get_ Neighbor*. For a proof sequence, a randomly chosen genes value i is replaced by a random PPS from the current population Pop. For example, a neighbor of the proof sequence RP is:

$$RP' = skosimp, \ expand, \ lemma, \ grind, \ flatten, \textbf{\textit{inst}}, \ skolem, \ assert$$

It is important to point out here that the standard mutation operator of GA and the get neighbor procedure in SA are same. The fitness of the neighbor solution is then calculated. The random generated proof sequence and the neighbor sequence is compared. If the fitness of the neighbor is better, then it is selected. Otherwise, an acceptance rate (Step 19 in Algorithm 1) is used to select one

---

Algorithm 1. SA proof searching

Input: *FPPS*: Frequent PVS proof steps, *PD*: proof sequences database, Temp, Temp_min, α

Output: Generated proof sequences

1: $Pop \leftarrow FPPS$
2: **for each** $P \in PD$ **do**
3:     $OF \leftarrow Fitness(P, P)$
4:     $PS \leftarrow$ randomseq(Pop, length(P))
5:     $BF \leftarrow Fitness(PS, P)$
6:     **if** $BF \geq OF$ **then**
7:         return PS
8:     **end if**
9:     **while** $(Temp > Temp\_min)$ **do**
10:         $NS \leftarrow get\_neighbor(PS)$
11:         $NF \leftarrow Fitness(NS, P)$
12:         **if** $NF == OF$ **then**
13:             return NS
14:         **end if**
15:         **if** $NF > BF$ **then**
16:             $PS \leftarrow NS$
17:             $BF \leftarrow NF$
18:         **end if**
19:         $ar \leftarrow exp(\frac{T}{1+T})$
20:         **if** $ar >$ randomuniform$(0, 10)$ **then**
21:             $PS \leftarrow NS$
22:             $BF \leftarrow NF$
23:         **end if**
24:         $Temp \leftarrow Temp \times \alpha$
25:     **end while**
26:     return PS
27: **end for**

---

sequence from the two sequences. The acceptance rate depends on temperature. In last, the temperature *Temp* is decreased with the following formula:

$$Temp = Temp \times \alpha \tag{1}$$

where the value of α is in the range of $0.8 < \alpha < 0.99$. The process of annealing is repeated (Steps 9–24 in Algorithm 1) until the random proof sequence fitness matches with the target proof sequence or *Temp* reaches to the minimum (*Temp_min*). In our case, we set the value of *Temp* such that the SA always terminates when the random proof sequence matches with the target proof sequence. The process that distinguishes SA from GA is the annealing process.

Besides the annealing process, another main concept in SA is the acceptance probability. SA checks whether the new neighbor solution is better than the present best solution. If the new neighbor solution is worse than the present best solution, SA may still select the news solution with probability known as

---

Algorithm 2. Fitness

Input: *Pseq*: A proof sequence, *P*: The current target proof sequence
Output: Integer that represents the fitness of a proof sequence (*Pseq*)

1: **procedure** FITNESS(*Pseq, P*)
2:     $i, f \leftarrow 0$
3:     **while** ($i \leq$ length(*Pseq*) - 1) **do**
4:       **if** (*Pseq*[*i*] = *P*[*i*]) **then**
5:         $f \leftarrow f + 1$
6:       **end if**
7:       $i \leftarrow i + 1$
8:     **end while**
9:     return f
10: **end procedure**

Algorithm 3. Get Neighbor

Input: P_1: A proof sequence
Output: A neighbor of P_1

1: **procedure** GN(P_1)
2:     *ind* $\leftarrow$ randomint(1, length(P_1))
3:     *alter* $\leftarrow$ randomsample(*Pop*, 1)   ▷ (*1-length* proof sequence form *Pop*)
4:     P_1[*ind*] $\leftarrow$ *alter*   ▷ (P_1[*ind*] $\neq$ alter)
5:     return P_1
6: **end procedure**

---

acceptance probability. *Acceptance probability* recommends whether to switch to the worst solution or not. This enable SA to avoid the local optimum (maybe) by exploring other solutions. This idea may seem unacceptable or worse at the start, but it could lead SA towards the global optimum. In the proposed SA for proof searching, the *acceptance probability* is calculated by using acceptance rate (AR) formula, that is:

$$AR = exp(\frac{Temp}{1 + Temp}) \tag{2}$$

3   Results and Discussion

The proposed SA for proof searching is implemented in Python. To evaluate the proposed approach, experiments were carried on a computer equipped with a fifth generation Core *i5* processor and 4 GB of RAM. For all experiments with SA, the values for *temp*, *min_temp* and α were set to 50,000, 0.0002, and 0.9954 respectively. Moreover, AR value (using Eq. 2) is compared with a random number generated within the range (2.71825400004040, 2.71825464604849). This range is selected after experimenting with the values of *temp*, *min_temp* and α.

    We investigate the performance of the proposed SA for finding the proofs of theorems, lemmas and proof obligations in two PVS theories [15, 16]. In total, we have 41 proof sequences and 23 distinct *PPS* in the *PD*. The SA was run ten times on considered theorems and obtained results are listed in Table 2. As discussed in previous section, SA can select the new solution (proof sequence in this work) obtained with the *GN* procedure that is not better than the present solution with the *acceptance probability*. The reason for this is that there is always a possibility that the new solution could lead the SA to the global optimum. In SA, we chose the *acceptance probability* (AR) using Eq. (2), which is then compared with a random number generated within the range (2.71825400004040, 2.71825464604849). If the value of AR is greater than the random number generated within the range, then the new solution (that is not better than present) is selected. We used a counter called acceptance rate counter (ARC) that keeps track of how many times the new neighbor solution that is not better than the present solution is picked. By simulation, it was found that this factor does not play huge role in the overall generation count or time. This is due to the fact that in our case, we do not have any local optimum. SA finds only one global solution for each random proof sequence based on the fitness value. The average generation count of SA for all proof sequences in the *PD* with and without the acceptance rate is listed in Table 2. The obtained result for ARC indicates that SA selected neighbor solution that is not better than previous solution 565 times.

Table 2. Performance of SA for proof searching in PVS theories

Results without AR		Results with AR ($ARC = 565$)	
Avg. Gen. Count	Time (S)	Avg. Gen. Count	Time (S)
41,569	0.71 s	42,134	0.72 s

Recently, a GA with different crossover and mutation operators was used for proof searching and optimization in HOL4 proof assistant [6]. Same like GA, an initial population for GA was first created using the SPM-based learning approach [4]. Random proof sequences from the population were then evolved by applying two GA operators (crossover and mutation). Both operators randomly evolve the random proof sequences by shuffling and changing the proof steps at particular points. This process of crossover and mutation continues till the fitness of random proof sequences matches the fitness of original (target) proofs for the different theorems and lemmas. Three crossover operators (single point crossover (SPC), multi point crossover (MPC) and uniform Crossover (UCO)) and two mutation operators (standard mutation (SM) and modified pairwsie interchange mutation (MPIM)) were used. The reason to use more than one crossover and mutation operators was to compare their effect on the overall performance of GAs in proof searching. We modified the GA [6] and run it on *PD* to compare its performance with SA. The average generations for the SA and GA with different crossover and mutation operators to reach the target proof sequences in the whole dataset (*PD*) are shown in Table 3. The time taken by algorithms and memory used is also listed in Table 3.

GA with different crossover and *MPIM* operator is approximately five times faster than the GA with different crossover operator and *SM*. Whereas, SA is four times faster than GA with *MPIM* and different crossover operators. One of the possible reason for this is that in SA, one procedure (*GN*) is called. On the other hand, in GA, two procedures (crossover and mutation) are called.

Table 3. Average total generation count for SA and GA

	Avg. generation count	Time	Memory
SA	41,569	0.71 s	3,614 Mb
GA (SPC/SM)	902,772	21.06 s	3,555 Mb
GA (MPC/SM)	851,609	20.67 s	4,085 Mb
GA (UCO/SM)	916,746	21.78 s	3,660 Mb
GA (SPC/MPIM)	182,679	9.85 s	3,617 Mb
GA (MPC/MPIM)	177,697	9.21 s	3,579 Mb
GA (UCO/MPIM)	173,055	9.05 s	3,618 Mb

In last, non-parametric statistical tests are used to compare the performance of evolutionary/heuristic-based proof searching approaches. First, the Friedman test [17] is used to statistically compare SA with different versions of GA for proof searching in PVS. The following hypotheses are tested:

H0: *The means for the generations taken by the proof searching approaches for each proof sequence in* PD *are same.*
H1: *The mean number of generations for at least one algorithm is different from the others.*

The results of the Friedman test indicate that the null hypothesis ($H0$) should be rejected and $H1$ should be accepted ($p < 0.05$) with result of 187.41. The Wilcoxon test [18] is further used to investigate SA and GAs more. The Wilcoxon test results for the number of generations for the two algorithms are listed in Table 4. The results in table show that SA is significantly better than different versions of GA. It can bee seen that GA with SM and different crossover operators (GA (SPC/SM, MPC/SM and UCO/SM)) are not significantly better than each other. The same is the case with GA with MPIM and different crossover operators (GA (SPC/MPIM, MPC/MPIM and UCO/MPIM)). Combining these statistical results with previous results suggests that SA is indeed significantly better than GAs. Moreover, the Wilcoxon test results on the means for the times taken by the SA and GAs to find the correct proofs for target proof sequences in the *PD* also indicated that SA is significiantly better than different versions of GA.

Overall, it was observed through various experiments that the SA is able to quickly optimize and automatically find the correct proofs for formalized theorems/lemmas in PVS theories. Obtained results in this paper are also consisted with the results for proof searching in HOL4 with SA [19, 20]. This clearly indicates that the developed approaches is not limited to only one proof assistant and can be used easily and effectively in different proof assistants, such as PVS, Coq, HOL4, Isabelle/HOL etc.

Table 4. Wilcoxon p-value matrix for generations

	SPC/SM	MPC/SM	UC/SM	SPC/MPIM	MPC/MPIM	UC/MPIM	SA
SPC/SM	—	4.88E−1	7.90E−1	2.61E−8	2.42E−8	2.42E−8	2.42E−8
MPC/SM	4.88E−1	—	4.88E−1	2.42E−08	2.42E−8	2.81E−8	2.42E−8
UC/SM	7.90E−1	4.88E−1	—	2.61E−8	2.42E−8	2.42E−8	2.42E−8
SPC/MPIM	2.61E−8	2.42E−08	2.61E−8	—	8.91E−1	4.64E−1	1.02E−5
MPC/MPIM	2.42E−8	2.42E−8	2.42E−8	8.91E−1	—	6.30E−1	6.05E−7
UC/MPIM	2.42E−8	2.81E−8	2.42E−8	4.64E−1	6.30E−1	—	2.46E−7
SA	2.42E−8	2.42E−8	2.42E−8	1.02E−5	6.05E−7	2.46E−7	—

4 Conclusion

Developing proofs for unproved theorems and lemmas (also known as conjectures) in PVS requires our interaction with the proof assistant. For long and complex proofs, the proof development process becomes a cumbersome and time consuming activity. This paper provides a proof searching approach, where a SA is used to optimize and find the correct proofs in PVS theories. Moreover, the performance of SA was compared with GA for proof searching and optimization. Obtained results suggests that SA was faster than GA and also indicate that the research direction of linking and integrating evolutionary and heuristic algorithms with proof assistants is worth pursuing.

There are several directions for future work. For example, making the proof searching process more general in nature to evolve PVS frequent proof steps to a compound proof strategies for guiding the proofs of new conjectures. Moreover, creating a large dataset by including more formalized theorems/lemmas for other PVS theories. We believe that such dataset can be used for the development of learning environments to search for proofs and to automate the interaction in different ITPs. In this regard, some work has been done in [21–24]. Last but not the least, by exploiting Curry-Howard isomorphism for sequent calculus [25], a SA or GA can be used for program (proof) writing and proof assistant can be used for simplification and verification by computationally evaluating the program, that represents the corresponding proof.

References

1. Owre, S., et al.: PVS system guide, PVS prover guide. PVS language reference. Technical report, SRI International, USA (2001)
2. Bertot, Y., Casteran, P.: Interactive Theorem Proving and Program Development Coq'Art: The Calculus of Inductive Construction. Springer (2003). https://doi.org/10.1007/978-3-662-07964-5
3. Slind, K., Norrish, M.: A brief overview of HOL4. In: Mohamed, O.A., Muñoz, C., Tahar, S. (eds.) TPHOLs 2008. LNCS, vol. 5170, pp. 28–32. Springer, Heidelberg (2008). https://doi.org/10.1007/978-3-540-71067-7_6
4. Nawaz, M.S., Sun, M., Fournier-Viger, P.: Proof guidance in PVS with sequential pattern mining. In: Hojjat, H., Massink, M. (eds.) FSEN 2019. LNCS, vol. 11761, pp. 45–60. Springer, Cham (2019). https://doi.org/10.1007/978-3-030-31517-7_4

5. Nawaz, M.S., et al.: Proof learning in PVS with utility pattern mining. IEEE Access **8**, 119806–119818 (2020)
6. Nawaz, M.Z., et al.: Proof searching in HOL4 with genetic algorithm. In: Proceedings of SAC, pp. 513–520. ACM (2020)
7. Delahaye, D., Chaimatanan, S., Mongeau, M.: Simulated annealing: from basics to applications. In: Gendreau, M., Potvin, J.-Y. (eds.) Handbook of Metaheuristics. ISORMS, vol. 272, pp. 1–35. Springer, Cham (2019). https://doi.org/10.1007/978-3-319-91086-4_1
8. Huang, S.I., Chen, Y.P.: Proving theorems by using evolutionary search with human involvement. In: Proceedings of CEC, pp. 1495–1502. IEEE (2017)
9. Yang, L.A., et al.: Automatically proving mathematical theorems with evolutionary algorithms and proof assistants. In: Proceedings of CEC, pp. 4421–4428. IEEE (2016)
10. Duncan, H.: The use of data-mining for the automatic formation of tactics. PhD thesis, University of Edinburgh, UK (2007)
11. Nagashima, Y.: Towards evolutionary theorem proving for Isabelle/HOL. In: Proceedings of GECCO (Poster), pp. 419–420. ACM (2019)
12. Schafer, S., Schulz, S.: Breeding theorem proving heuristics with genetic algorithms. In: Proceedings of GCAI, pp. 263–274. EasyChair (2015)
13. Nawaz, M.S., et al.: A survey on theorem provers in formal methods. arXiv:cs.SE/1912.03028 (2019)
14. Fournier-Viger, P., et al.: A survey of sequential pattern mining. Data Sci. Patt. Recogn. **1**(1), 54–77 (2017)
15. Nawaz, M.S., Sun, M.: Reo2PVS: formal specification and verification of component connectors. In: Proceedings of SEKE, pp. 391–396 (2018)
16. Nawaz, M.S., Sun, M.: Using PVS for modeling and verification of probabilistic connectors. In: Hojjat, H., Massink, M. (eds.) FSEN 2019. LNCS, vol. 11761, pp. 61–76. Springer, Cham (2019). https://doi.org/10.1007/978-3-030-31517-7_5
17. Friedman, M.: A comparison of alternative tests of significance for the problem of m rankings. Ann. Math. Stat. **11**(1), 86–92 (1940)
18. Wilcoxon, F.: Individual comparisons by ranking methods. Biometrics Bull. **1**(6), 80–83 (1945)
19. Nawaz, M.S., et al.: Proof searching and prediction in HOL4 with evolutionary/heuristic and deep learning techniques. Appl. Intell. **51**(3), 1580–1601 (2021)
20. Nawaz, M.S., et al.: An evolutionary/heuristic-based proof searching framework for interactive theorem prover. Appl. Soft Comput. (2021). https://doi.org/10.1016/j.asoc.2021.107200
21. Huang, D., et al.: GamePad: a learning environment for theorem proving. In: Proceedings of ICLR (Poster) (2019)
22. Yang, K., Deng, J.: Learning to prove theorems via interacting with proof assistants. In: Proceedings of ICML, pp. 6984–6994. PMLR (2019)
23. Bansal, K., et al.: HOList: an environment for machine learning of higher order logic theorem proving. In: Proceedings of ICML, pp. 454–463, PMLR (2019)
24. Sanchez-Stern, A., et al.: Generating correctness proofs with neural networks. In: Proceedings of MAPL@PLDI, pp. 1 10. ACM (2020)
25. Santo, J.E.: Curry-howard for sequent calculus at last! In: Proceedings of TLCA, pp. 165–179 (2015)

Deep Reinforcement Learning for Dynamic Scheduling of Two-Stage Assembly Flowshop

Xin Lin and Jian Chen[✉]

College of Economics and Management, Nanjing University of Aeronautics and Astronautics,
Nanjing 211106, China
jchen@nuaa.edu.cn

Abstract. Dynamic scheduling of jobs is increasingly needed in modern make-to-order manufacturing companies considering various inevitable uncertainties, such as dynamic arrivals or frequent changes of customer orders. To this end, we model a dynamic scheduling of two-stage assembly flowshop (AF) to minimize the total tardiness as a Markov decision process (MDP). Then we propose a deep reinforcement learning (DRL) approach to build a scheduler agent to make real-time decisions of dispatching rules according to current production environment. A proximal policy optimization (PPO) algorithm is developed to efficiently train agent using production data. Numerical experiments show that the trained agent can fully capture the knowledge in historical data so as to make a real-time scheduling quickly and effectively, and the proposed approach is both superior and general to each well-known dispatching rule in different production scenarios. The proposed approach is highly desirable for many practical scheduling situations that requires dynamic and quick decisions.

Keywords: Assembly flowshop scheduling · Deep reinforcement learning · Proximal policy optimization

1 Introduction

In manufacturing industries, many products are produced by assembling several components together where each of the components is processed on specific facilities, such as machines or production lines. We consider a two-stage AF scheduling problem where components are manufactured in the machining stage and then joined together in the assembly stage. The first stage consists of m machining machines $\{M_{11}, M_{12}, \ldots, M_{1m}\}$ the second stage has a assembly machines of $\{M_{21}, M_{22}, \ldots, M_{2a}\}$. The machines are dedicated machines in the first stage, each of which is specially used to process a specific type of component in the component set $CP = \{CP_1, CP_2, \ldots, CP_m\}$. The m components are assembled to a product through the second stage. We study the make-to-order production that each customer requires a customized product. The deadline of customer k is donted as d_k. All customer orders in a given planning horizon are accumulated to a production order described as the set $O = \{P_1 * n_1, P_2 * n_2, \ldots, P_j * n_j, \ldots\}$, where P_j means the product type j, n_j means the quantity of P_j. The set O is ordered by n

Y. Tan and Y. Shi (Eds.): ICSI 2021, LNCS 12690, pp. 263–271, 2021.
https://doi.org/10.1007/978-3-030-78811-7_25

customers, where each customer requires a customized product among the set O, thus $n = \sum_j n_j$.

The two-stage AF scheduling problem needs the following constraints: 1) different components of each product can only be processed once on its dedicated machine; 2) Each product can only be assembled once on one assembly machine; 3) Each product enters the second stage of assembly after all its m components are finished; 4) Each machining/assembly machine can only process one component/product at any time without interruption. The scheduling decision is to seek the optimal processing sequence for products in each O, in order to minimize the total tardiness time under the condition of considering the sequence-independent set-up time. The studied two-stage AF scheduling problem is abbreviated as $AF(M_{1m}, M_{2a})|ST_{si}|\sum_{k=1}^{n} T_k$, where $T_k = (C_k - d_k)^+$. Figure 1 shows an example of BOM lists for customized products. Triangles, rectangles, ovals, diamonds represent components of a product and cubes represent the final products.

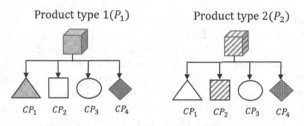

Fig. 1. Customized product BOM list

Exact and heuristic algorithms are mostly used in the two-stage AF scheduling problem. Hariri et al. (1997) propose a branch-and-bound (B&B) to solve the two-stage AF scheduling problem by combining four kinds of priority rules with the goal of completion time. The effectiveness of the algorithm is proved by the comparison of data simulation experiments [1]. Aiming at minimizing the total completion time, Sung et al. (2008) give a lower bound based on the SPT rule, and solved it through the B&B algorithm. At the same time, it was proved that the optimal solution contained one permutation optimal solution at least [2]. Xiong et al. (2014) study the two-stage AF scheduling problem with sequence-independent set-up time, aiming to minimize the total completion time, and proposed three meta-heuristic algorithms based on variable neighborhood search (VNS), genetic algorithm (GA) and differential evolution (DE) [3]. Mozdgir et al. (2013) aim to minimize the weighted sum of completion cycles and average completion times. In this paper, a hybrid VNS algorithm is proposed, which combines VNS with a new heuristic algorithm, and the results are compared with GAMS software [4]. Although the previous algorithms, whether exact or heuristic algorithms, can obtain good-quality solutions, it is difficult to deal with the real-time scheduling problems with dynamic environment. In this paper, deep reinforcement learning (DRL) is proposed to solve the real-time AF scheduling problem. As we know, deep learning (DL) and reinforcement learning (RL) have strong perceptual ability and decision-making ability, respectively. Therefore, DRL combines the advantages of the two and provides a solution to the perceptual decision-making problem of complex systems.

As an important technology for artificial intelligence (AI), DRL has attracted more and more attentions from many fields. In recent years, it is found with surprising applications in combinatorial optimization problems. Nazari et al. (2018) present an end-to-end framework of deep reinforcement learning to address VRP problems, the Pointer network is improved and the policy gradient algorithm is used to optimize the parameters. The results show that the Pointer network is better than OR-Tools in the medium scale [5]. Zhang et al. (2020) propose a novel multi-agent attention model (MAAM) and a RL algorithm to solve the multi-vehicle routing problem with soft time windows (MVRPSTW) problem. The results show that the proposed method is superior to OR-Tools and traditional methods with little computation time by tested on four networks of different scales [6]. Hu et al. (2020) build an architecture consisting of shared graph neural network and distributed policy network to learn the common policy for multiple traveling salesman problem (MTSP) [7].

Value-based DQN is proposed to solve the job shop scheduling problem (JSSP) with new job insertions by Luo (2020). The results show that the proposed method is superior to the proposed scheduling rules [8]. Shiue et al. (2018) combines two main mechanisms of off-line learning module and Q-learning based on RL to solve the real-time scheduling problem for flexible manufacturing system (FMS) [9]. Using the disjunctive graph representation of JSSP, Zhang et al. (2020) propose a graph neural network based scheme to embed the states encountered in the solution process and the generalization of large-scale instances is realized effectively [10]. To sum up, few scholars study real-time scheduling problem regarding AF using DRL. However, AF has a wide range of applications, such as in electronic or automobile industries. This paper attempts to study DRL to address a two-stage AF real-time scheduling problem derived from our collaborating automobile manufacturing company.

The reminders are organized as follows. Section 2 introduces the MDP of the two-stage AF scheduling problem. Section 3 proposes a PPO algorithm. Section 4 presents the details of experimental experiments and analyses the results. Section 5 concludes the research.

2 MDP Formulation

Generally speaking, the two-stage AF real-time scheduling problem can be modeled as an MDP, which is defined as a 5-tuple $\mathbb{C} = (S, A, P(s'|s, a), R(s, a), \gamma)$. S represents the state feature, A is limited action set, $P(s'|s, a)$ is the probability of transferred to the state s' after performing action under state s. $R(s, a)$ represents the reward for performing action a in the state s. γ represents the discount factor of values in the interval $[0,1]$. The MDP model is described as follows.

2.1 State Feature

In most RL-based scheduling problems, state features are defined either too redundant or simplified. Redundant features are easy to cause the loss of neural network and increase the training cost, and oversimplification may not be able to describe the structure of the problem accurately. To address this issue, we propose the following five skillfully-designed state features, which are extracted to serve as the input of PPO. Features 1–2 are obtained by the average machining/assembly time of the first/second stages, and they make the neural network to learn the bottleneck stage of the assembly shop at decision point. Features 3–4 are used to identify the bottleneck of m machining machines in the first stage. Besides, the products studied in this paper are customized products, and different customized products may have the same components, in view of this, feature 5 is proposed to express the serial number of the product being processed at decision point which can help the neural network learn the correlation between different customized product types. All features are shown in Table 1.

Table 1. State feature list.

No.	Formula	Description
1	$\frac{\sum_{i=1}^{m} PT_{1i}}{m}$	The average processing time of products assigned on m machines in the first stage at decision point
2	$\frac{\sum_{i'=1}^{a} AT_{2i'}}{a}$	The average assembly time of a machines in the second stage at decision point
3	$argmax\{i|PT_{1i}\}$	The machine with the longest assigned processing time of the first stage at decision point
4	$argmin\{i|PT_{1i}\}$	The machine with the shortest assigned processing time of the first stage at decision point
5	j	The product type being processed at decision point

2.2 Action

Traditional research on real-time scheduling widely adopt dispatching rules. As we all know, no dispatching rule can perform well for all environments of AF. Therefore, this paper takes advantage of the characteristics of DRL agent interacting with the scheduling environment and giving real-time feedback, selects the proper dispatching rules to minimize the total tardiness. The dispatching rules are defined as the action set of the RL. In addition to the classic four dispatching rules namely, SPT, TSPT, EDD and CR, we also elaborately design two composite dispatching rules, i.e., SST, SSD as action sets. In detail, for SST, the total current set-up times of product k is computed as $\sum_{i}^{m} st_{ik}$, where st_{ik} is current set-up time of the component CP_i of product k regarding the current component processing on machine i. For SSD, the combination of SST and EDD of product k is computed as $d_k \sum_{i}^{m} st_{ik}$. The considered dispatching rules and their descriptions are shown in Table 2.

Table 2. Action set.

No.	Action	Description
1	SPT	The product with the shortest total processing times for all components in the first stage is preferred
2	TSPT	The product with the shortest total of processing and assembly times is preferred
3	EDD	The product with the earliest due date is preferred
4	CR	The product with the lowest ratio of due date time to total processing and assembly time is preferred
5	SST	The product with the shortest total set-up times with regard to the current processing product is preferred
6	SSD	The product with the shortest combination of SST and EDD is preferred
7	Random	Randomly select a product for processing

2.3 Reward

The goal is to minimize the total tardiness time for the production order of a batch of customer orders. To this end, the reward R_t for the state–action (s_t, a_t) is designed to be related to the delay of the current scheduled product, the reward is −1 if the currently scheduled product is delayed, and 1 if it is not.

$$\begin{cases} R = 1, & C_k - d_k \leq 0 \\ R = -1, & C_k - d_k > 0 \end{cases}$$

3 DRL Algorithm

We train the DRL agent using PPO algorithm. RL mainly usually makes decisions, while DL is mainly used for perception. The deep learning network used in DRL adopts an actor-critic framework, where actor is responsible for interacting with the environment to collect samples, and critic is responsible for judging the action of actor. Actor network $\pi_{\theta old}$ consists of three fully connected layers with one input layer, one output layer and hidden layer, the number of nodes in hidden layer is 128. Critic network π_θ has the same structure as actor, except that the number of hidden layer node is 64. PPO adopts the off-policy training strategy, which greatly improves the sampling performance. The procedure of PPO is given in Algorithm 1.

Algorithm 1 Procedure of PPO

Input: Current state s_i, action a_i, reward R, next state s_{i+1}.

Output: Next action a_{i+1}.

1:     Initialize the AF environment and PPO network parameters, $\pi_{\theta old} = \pi_{\theta}$;

2: **for** *Episode* $= 1 : L$ **do**

3:      **for** $n = 1 : N$ **do**

4:          **for** $t = 1 : T$ (T is the terminal product) **do**

5:              Estimate advantages $\hat{A}_t = \sum_0^t \gamma^t r_{n,t} - V_\varphi(s_{n,t})$, $r_{n,t}(\theta) = \frac{\pi_\theta(a_{n,t}|s_{n,t})}{\pi_{\theta old}(a_{n,t}|s_{n,t})}$

6:                  if all the products are scheduled

7:                      break

8:          **end for**

9:              Calculate cumulative loss $Loss_n(\theta, \varphi)$

10:     **end for**

11:     **for** $k = 1 : K$ **do**

12:         update parameter θ, φ

13:     **end for**

14:     $\theta_{old} \leftarrow \theta$

15: **end for**

---

4   Numerical Experiments

The production order is released at time 0. The processing time of components at the first stage follows the uniform distribution between 200 and 500, and the assembly time at the second stage follows the uniform distribution between 300 and 600. The set-up time is set as a fixed value of 150, due date of each customer order is set from the uniform distribution [50%, 100%] of maximum average workload of each machine. Consider several machine combinations that the number of machines at the first stage, $m = 4, 6, 8$, the number of assembly machines at the second stage, $a = 1, 2, 3$.

The training data is randomly generated using small-scale production order. Both small-scale and large-scale scale production orders are used for testing. The instances are randomly generated using the parameters in Table 3. For each production order instance $O = \{P_1 * n_1, P_2 * n_2, \ldots, P_j * n_j, \ldots\}$, *Num_P* is the number of production types, n_j is the number of product type j, We randomly generate *Num_P* by the uniform distribution [3, 10] and [10, 25], for small-scale and large-scale instances, respectively. n_j follows the uniform distribution [1, 15] and [15, 25], for small-scale and large-scale instances, respectively. Thus, the number of small-scale and large-scale customer orders is subjected to $3 \leq n \leq 150$ and $150 \leq n \leq 625$, respectively. For example, if *Num_P* = 3, $n_j = [4, 6]$, one of the production order instance could be generated as $O = \{P_1 * 4, P_7 * 4, P_{23} * 6\}$.

10000 small-scale instances are randomly generated for training. 100 small-scale and large-scale instances are generated for testing. The proposed PPO is implemented in Python3.8 on a PC with Intel Core i5–5250 @ 1.60 GHz CPU and 4 GB RAM, and the DL framework uses Pytorch 1.6.0. We use fixed hyperparameters for training. For training, we train the policy network for 10000 iterations (each instance for one iteration), each of which contains 4 independent trajectories.

Table 3. Parameter settings of numerical experiments instances

Scale	Num_P	n_j
Small ($3 \leq n = \sum_j n_j \leq 150$)	[3, 10]	[1, 15]
Large ($150 \leq n = \sum_j n_j \leq 625$)	[10, 25]	[15, 25]

In this paper, the single dispatching rule and the neural network trained by PPO are compared. The average total tardiness of the testing instances is used as the performance indicator for comparison. Table 4 shows the experimental results of the average total tardiness 100 small-scale testing instances. PPO outperforms all other dispatching rules for small-scale production orders in all machine combinations.

Table 4. Results of the 100 small-scale testing instances

	PPO	SPT	TSPT	EDD	CR	SST	SSD
$m4\_a1$	**123527**	130948	127560	165070	166468	131355	128230
$m4\_a2$	**64478**	87357	87520	87112	88712	88666	81067
$m4\_a3$	**65522**	87613	87762	87163	74243	88928	128230
$m6\_a1$	**118872**	129757	130640	154646	156713	132900	124803
$m6\_a2$	**94402**	112514	112347	113003	114688	114692	102189
$m6\_a3$	**94599**	112774	112606	113063	114692	109846	124803
$m8\_a1$	**118751**	128588	127724	152454	155027	131718	123980
$m8\_a2$	**93957**	111792	110382	112856	114408	108736	100986
$m8\_a3$	**94308**	112064	110650	112891	114390	109012	101296

We use the PPO neural network trained by small-scale instances for large-scale testing instances. Table 5 shows that using the trained small-scale neural network can perform well in large-scale instances. It demonstrates that the PPO neural network has a strong generalization ability.

Although the PPO neural network training is time-consuming, using the network trained to make real-time scheduling is speedy. In addition to its generalization ability, PPO has strong practicality in many practical scheduling situations that requires dynamic and quick decisions.

Table 5. Generalization results of the 100 large-scale testing instances

	PPO	SPT	TSPT	EDD	CR	SST	SSD
$m4\_a1$	**4973726**	5140547	4849434	7557691	7719437	5228243	5156134
$m4\_a2$	**2841488**	3231287	3227634	3633275	3747908	3411909	2884288
$m4\_a3$	**2848809**	3231862	3228194	3633276	3748045	3412451	2885084
$m6\_a1$	**4177936**	4927260	4955798	6996077	7162356	5172136	4234799
$m6\_a2$	**3408957**	4479353	4509668	4877664	5024785	4466482	3508625
$m6\_a3$	**3418743**	4479992	4510316	4877618	5024828	4467192	3509869
$m8\_a1$	**4226555**	4832353	4912900	6833692	6983645	5074071	4448174
$m8\_a2$	**3653255**	4400596	4420942	4756692	4843826	4361402	3704384
$m8\_a3$	**3653111**	4401235	4421582	4756610	4844035	4361988	3705345

5   Conclusion

This paper proposes a DRL approach to address a real-time scheduling problem of two-stage AF. A policy-based PPO is used to train the deep neural network via the interaction between the agent and the production environment. The deep neural network enables a scheduler agent to make real-time decisions of dispatching rules. Numerical experiments show that the proposed DRL approach is stable and superior to several famous dispatching rules in both small-scale and large-scale instances in AF scheduling problem. In addition, the DRL approach is speedy and has great generalization ability, which is suitable for many practical scheduling situations that requires dynamic and quick decisions.

Acknowledgements. This research was supported by the National Natural Science Foundation of China (Nos. 52075259, 51705250), and China Postdoctoral Science Foundation (2019M661839).

References

1. Hariri, A.M.A., Potts, C.N.: A branch and bound algorithm for the two-stage assembly scheduling problem. Eur. J. Oper. Res. **103**, 547–556 (1997)
2. Sung, C.S., Kim, H.A.: A two-stage multiple-machine assembly scheduling problem for minimizing sum of completion times. Int. J. Prod. Econ. **113**(2), 1038–1048 (2018)
3. Xiong, X.F.K., Wang, F., Lei, H., Han, L.: Minimizing the total completion time in a distributed two stage assembly system with setup times. Comput. Oper. Res. **47**, 92–105 (2014)
4. Mozdgir, A., Fatemi, G.S.M.T., Jolai, F.: Two-stage assembly flow-shop scheduling problem with non-identical assembly machines considering setup times. Int. J. Prod. Res. **51**(12), 1–18 (2013)
5. Nazari, M., Oroojlooy, A., Takč, M., Snyder, L.V.: Reinforcement learning for solving the vehicle routing problem. In: 32nd Conference on Neural Information Processing Systems. Montreal, pp. 9839–9849 (2018)

6. Zhang, K., Fang, H., Zheng, C.Z., Xi, L., Meng, L.: Multi-vehicle routing problems with soft time windows: a multi-agent reinforcement learning approach. Trans. Res. Part C **121**, 102861 (2020)

7. Hu, Y.J., Yao, Y., Lee, W.S.: A reinforcement learning approach for optimizing multiple traveling salesman problems over graphs. Knowl. Based Syst. **204**, 106244 (2020)

8. Luo, S.: Dynamic scheduling for flexible job shop with new job insertions by deep reinforcement learning. Appl. Soft Comput. J. **91**, 106208 (2020)

9. Shiue, Y.R., Lee, K.C., Su, C.T.: Real-time scheduling for a smart factory using a reinforcement learning approach. Comput. Ind. Eng. **125**, 604–614 (2018)

10. Zhang, C., Song, W., Cao, J.G., Zhang, J, Tan, P.S., Xu, C.: Learning to dispatch for job shop scheduling via deep reinforcement learning. In: 34th Conference on Neural Information Processing Systems, pp. 1–18, Vancouver (2020)

A Hybrid Wind Speed Prediction Model Based on Signal Decomposition and Deep 1DCNN

Yuhui Wang, Qingjian Ni$^{(\boxtimes)}$, Shuai Zhao, Meng Zhang, and Chenxin Shen

School of Computer Science and Engineering, Southeast University,
Nanjing 211189, China
nqj@seu.edu.cn

Abstract. Wind speed prediction is a typical time series prediction and is of great importance in power generation. In order to deal with those problems of heavy resource consumption and complex hyperparameter selection in traditional methods, we propose a multidimensional prediction method based on decomposition methods. However, using a model to fit all subseries may lead to the model's performance degradation and error increasing, which is called "preference" error. To solve this problem, a one-dimensional CNN (1DCNN) is used to capture the relationships between subseries. As to better explore this problem and enhance the stability of the CNN model, the generative adversarial network (GAN) method is tried to generate and generalize this "preference" error and expand training samples for 1DCNN. This paper combines multiple methods including the decomposition method, RNN model, CNN model, and GAN method in order, and chooses the best combination in different datasets. The experiments on two real-world wind datasets demonstrate that this method can achieve excellent performance in wind speed prediction with the help of combining the above methods.

Keywords: Wind speed prediction · Signal decomposition · LSTM · CNN · WGAN

1 Introduction

Wind speed is typical non-stationary, nonlinear time series data. The development and utilization of wind energy have a great influence on human energy construction and meteorological science. As long and medium-term prediction needs stable conditions and short-term forecasting methods, this paper is dedicated to short-term wind speed prediction.

Short-term wind speed prediction models can be roughly divided into two types, one is the physical model, and the other is the statistical model. The physical model is based on numerical weather prediction (NWP) modeling [1], which is used to analyze wind speed based on dynamics and other meteorological factors.

© Springer Nature Switzerland AG 2021
Y. Tan and Y. Shi (Eds.): ICSI 2021, LNCS 12690, pp. 272–281, 2021.
https://doi.org/10.1007/978-3-030-78811-7_26

In the statistical model, current researchers generally try to use machine learning methods to predict wind speed. There are two main methods in meteorological forecasting, including wind speed. One is the application of different prediction models. Support Vector Regression (SVR) [2], and Long Short-Term Memory network (LSTM) [3] model are the typical model widely applied to wind speed prediction. Now, researchers are innovating prediction models with different structure, such as LSTNet [4] and Temporal Convolutional Network (TCN) [5]. The other is signal decomposition. It is a general method in wind speed prediction. The performance of the prediction model can be improved by decomposing the unstable nonlinear series into regular subseries. Intrinsic Time-scale Decomposition (ITD) [2] and Singular Spectrum Analysis (SSA) [6] are found in wind speed field. When analyzing those methods, a multidimensional prediction method based on signal decomposition is proposed to combine the advantages of both and reduce resource consumption.

Due to the outputs are multidimensional subseries in our prediction model, we consider that the inter-dimensional feature can be captured by CNN. The use of CNN structure to extract information within the structure has been mentioned in many articles, such as TextCNN [7], DnCNN [8,9].

Finally, to further enhance our effect, we set our sights on GAN. Researchers have experimented with GAN for series data recently. Two methods [10,11] were proposed to solve sequential problems using WGAN [12], but they were not suitable for wind speed time series data. To secure the better ability of GAN, the target function involved by SEGAN [13] and WGAN is used in our GAN method to learn the parameters.

In summary, this paper makes the following contributions in our framework to the short-term wind speed prediction problem:

- The multidimensional prediction method based on the signal decomposition method is tested, discussed, and used to reduce the error and resource consumption.
- A CNN model is proposed to process the multiple subseries results obtained by prediction models.
- An improved GAN method is adopted to generate more and effective training data.
- Our method is compared with the state of the arts on two real wind datasets and performs well.

2 Preliminary

2.1 Decomposition Method

SSA is a signal decomposition method combining time and frequency domain [14]. It can capture the periodicity of signals and is suitable for weather analysis. Based on the original series data, SSA constructs the trajectory matrix and the singular values obtained from the matrix calculation, then decomposes the series.

CEEMD [15] is different from SSA. It is a decomposition method to decompose the original series into different subseries in frequency domain.

2.2   Prediction Model

The prediction models used in these experiments are LSTM and TCN.

LSTM is to solve the gradient disappearance of the RNN model in the longer parameter passing with gate control and long memory unit. The LSTM used in this paper is a simple LSTM unit and a fully connected model structure.

TCN [5] has been proposed to solve the sequence problem using CNN structure in recent years. TCN solves the dependencies between sequences by using dilated convolutions and casual convolution. And residual connection is used to reduce the error of layers transmission.

3   Proposed Method

3.1   The Framework of Proposed Method

The framework of our method is divided into three parts, followed by the decomposition method, prediction model, and reconstruction model. And the three parts are sequentially used to process the series generated in the previous step. The first step is decomposition. Since some inner information is extracted by the decomposition methods, we are bound to get a better result. So this paper focuses on how to ensemble those methods.

In order to avoid being trapped in a large number of search space and heavy resource consumption in the mixed model, we intended to pretreat the original series to k subseries by decomposition method. At the second step, some basic prediction methods are utilized to forecast multidimensional subseries simultaneously and get k dimensional outputs. In order to further reduce "preference" error, CNN and GAN structure which can capture the internal trait of information is adopted to improve the accuracy of results from the prediction model. Finally, multiple different methods or models are tried in decomposition method and prediction models separately. The combinations getting the best result are suggested to select and compare with others in this framework. The order of our framework will be introduced in turn below:

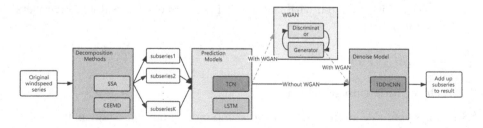

Fig. 1. The process of proposed prediction method. And a group components of route obtaining best result in validation set will be used in predicting ultimately.

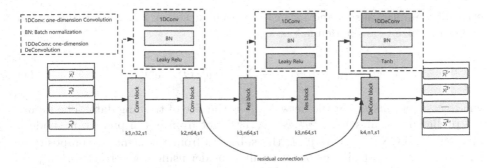

Fig. 2. The network architecture of 1DCNN. The input is preliminary result $\widehat{y}_i^k$ from initial prediction models, and the output is the final results $f(\widehat{y}_i^k; \Theta)$. The k, n and s are the number of kernels, filters and stride respectively. And the padding is 1 in the res block, is 0 in other block.

1. SSA or CEEMD was used to decompose the original wind speed series X_T to K dimension subseries set S, $S = \left\{ X_T^1, X_T^2, ..., X_T^k A \right\}$, K is the number of subseries.
2. The prediction model, LSTM or TCN is used to forecast those multidimensional subseries. The predictive error produced in this term is called "preference" error.
3. WGAN is used to generate and generalize the "preference" error caused by the prediction model using Gaussian noise. And the generated data is processed by 1DCNN with output from the prediction model.
4. Each output obtained from the prediction model is further processed to reduce the "preference" error and restructure the multidimensional results by 1DCNN.

The whole process of our method is shown in Fig. 1. Excluding the WGAN method, we can get four different models of a combination of two decomposition methods and two prediction methods with a 1DCNN model. After verifying these four models on the verification set, we select the best model for testing.

3.2   1DCNN Model for Reconstruction

Since the original series was decomposed into multiple subseries joint prediction, those subseries were relevant at the same time. However, the network may not be able to converge well to all subseries, which leads to a "preference" error in multidimensional joint prediction with some noise in the subseries. It is a shortcoming compared to predicting individually. However, it is difficult for us to understand the deviation of the results of the network training due to noise. Thus, a feature extraction network structure is necessary.

To capture this "preference" error, a 1DCNN was used to eliminate this error. Figure 2 shows the structure of 1DCNN. Since our goal is to further reduce the error of the initial result and reconstruct each subseries. The objective function to be minimized can be defined as:

$$L_{1DCNN}(\Theta) = \frac{1}{N} \sum_{i}^{N} \|f(\widehat{y}_i^k; \Theta) - y_i^k\|^2 \tag{1}$$

where Θ is parameters of the network, N is the size of training data. The inputs are preliminary results $\widehat{Y} = \{\widehat{y}_1^k, \widehat{y}_2^k, \ldots, \widehat{y}_t^k\}$, got from initial prediction models, like LSTM, TCN. And the y_i^k is the subseries from the signal decomposition. Since our number of dimensions is small, consider using a fewer network layers. The components of our network are one-dimension Convolution, Batch Normalization, LeakyRules as well as residual learning strategy, which are applied to improve convergence speed and performance. It is worth mentioning that 1DCNN is not to predict the noise inter subseries, but to reconstruct subseries.

We capture the feature inter subseries using one-dimension Convolution, and reconstruct subseries using one-dimension DeConvolution. The parameters of the layers are shown in the Fig. 2.

3.3 WGAN for Generalization

Since GAN was put forward in 2014, it has been widely applied in the field of image and got excellent performance. However, GAN had been inefficient in sequence analysis. The two methods [10,11] for solving the sequence problems using WGAN were introduced. In this paper, we used WGAN to generate more "preference" errors by the generator. It is able to provide more training data for the 1DCNN model and enhance its stability. In order to get the value of error conformed to the same magnitude, an objective function in SEGAN about generator was used as follows:

$$\min_{G} V_{SEGAN}(G) = \lambda \|G(z) - x\|^2$$
$$- E_{z \sim p_z(z)}[f_w(G(z))] \tag{2}$$

A hyper parameter λ was used to balance the WGAN loss and L_2 loss. In our experiments, set $\lambda = 1$. In WGAN, the structure of generator was used Fig. 2, discriminator used normal 1DCNN.

4   Experimental Results and Analysis

4.1 Experiment Setups

In order to show the performance with our method, we experimented in two different real wind speed datasets, shown in Table 1, including wind speed in a place in India in October 2018 and Turkey in January 2018. Wind speed data was recorded every 10 min.

The data in those datasets are incomplete, therefore we used cubic spline interpolation to supplement the original data value. All experiments use 90 min

Table 1. Related information about wind speed datasets

Region	Date	Max(m/s)	Min(m/s)	Ave(m/s)	Effective number
India	October	12.2149	1.04583	4.8852	4462
Turkey	January	22.4973	0	8.8567	3878

(H = 6) wind speed data to forecast the next step (P = 1) wind speed. And the original series was decomposed into 10 subseries, which are the input and output of the prediction models. In these experiments, the initial learning rates are set to 0.001, 0.01, and 0.005 in the prediction models, the 1DCNN model, and the WGAN model. The layer and hidden neurons are set 1 and 60 in LSTM, and the number of temporal blocks, the channel size, and the kernel size are set 3, 64, and 2 in TCN. The number of Conv Block and DeConv Block is set to 2, k in Conv Block and DeConv Block are [3, 2] and [3, 3]. Root mean squared error (RMSE), mean absolute error (MAE), and R-square (R2) was used to evaluate the prediction results.

4.2 Multidimension Prediction

We first researched whether the two decomposition methods have certain advantages in wind speed prediction. And secondly, the traditional method of predicting each subseries individually was compared with our multidimensional prediction method. The result was shown in Table 2. In Table 2, the name of the prediction model represents direct prediction without decomposition method, the name of "prediction model + decompositional method" indicates our multidimensional method and "Divide" is the means of predicting each subseries individually and adding it up eventually. Clearly seen in Table 2, prediction without any decomposition methods was the worst method in all the methods. And SSA method was more adaptable than CEEMD and performed well. Considering comparison in prediction individually and simultaneously, averaging the two datasets, multidimensional prediction using SSA reduced RMSE over 36.30%, decreased MAE 19.33%. And multidimensional prediction adopting CEEMD reduced RMSE and MAE obviously. In the other aspect, it takes several times

Table 2. Comparison of prediction without decomposition, multidimensional prediction with decomposition, and predicting each subseries individually

Method	RMSE	MAE	$R^2(\%)$	RMSE	MAE	$R^2(\%)$	Time(s)
	India			Turkey			
LSTM	0.4427	0.2756	89.41	1.4227	0.5597	82.09	422.84
SSA + LSTM	**0.1485**	**0.1245**	**98.81**	**0.3697**	**0.1895**	**98.81**	**310.54**
SSA + Divide + LSTM	0.2716	0.1854	95.81	0.4706	0.2012	98.07	3065.58
CEEMD + LSTM	0.3659	0.2861	92.76	1.7653	1.4275	72.80	380.83
CEEMD + Divide + LSTM	0.3910	0.3016	91.73	1.3572	0.8915	83.92	3035.44

longer to predict each subseries individually than our multidimensional method. Therefore, we got obtain advantages in all aspects. However, it is worth noting that CEEMD got the opposite result in the Turkey dataset.

In order to explore the reasons for the worse results of CEEMD in multi-dimensional prediction, the following experiments were designed. CEEMD is a frequency-based method, so different frequency subseries (can be called IMFs in CEEMD) may influence each other in the model. We extended experiments for discussing this characteristic in subseries generated by CEEMD. The multi-dimensional prediction method and single-dimensional prediction method are compared to predict these subsequences incrementally, and the results are shown in Fig. 3.

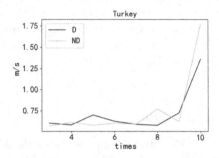

Fig. 3. The horizontal axis expresses IMFs number, "D" represents one-dimensional prediction subsequence in turn, and then accumulate the results, "ND" represents multi-dimensional prediction.

Experiments demonstrate that the prediction accuracy is affected from subseries close to the residual. On the other hand, the experiment verifies that the subsequence with a large frequency difference will affect the multi-dimensional prediction results. So, the multidimensional prediction should consider how to deal with the subseries with large frequency differences.

4.3   Subseries Reconstruction by 1DCNN

In the previous section, we tested and verified the superiority of the multidimensional prediction model inaccuracy and time consumption. However, when we simultaneously handle subseries with those mutual-dependence and similarity, we should focus on the mutual interference of each subseries during training. In order to further reduce the error, an elimination model of "preference" error is employed after the multidimensional prediction. Based on the dependencies between subseries, we proposed a 1DCNN in Sect. 3.1 to reduce error. The result is shown in Table 3. Table 3 shows that in the dataset of India, the addition of 1DCNN can reduce the RMSE average of the prediction model based on SSA by 20.79%, and MAE by an average of 28.15%. In the dataset of Turkey, 1DCNN reduced the RMSE and MAE average based on SSA by 0.34%, 7.37%. The result indicates that 1DCNN performs well on the correlation dependence of subseries, and further improves the prediction accuracy.

Table 3. Comparison of method with 1DCNN, with WGAN and without them, "Re" denotes the method applying 1DCNN to restructure prediction results, "WGAN" denotes the method applying 1DCNN and WGAN.

Method	RMSE	MAE	$R^2(\%)$	RMSE	MAE	$R^2(\%)$
	India			Turkey		
SSA + LSTM	0.1485	0.1245	98.81	0.3697	0.1895	98.81
SSA + Re + LSTM	**0.0954**	0.0714	**99.51**	**0.3584**	**0.1640**	**98.88**
SSA + WGAN + LSTM	0.1106	**0.0695**	99.34	0.4976	0.3457	97.84
SSA + TCN	0.1134	0.0843	99.30	0.3640	0.1797	98.84
SSA + Re + TCN	0.1068	0.0728	99.38	0.3700	0.1653	98.81

After the good performance from the reprocessing prediction model adopted 1DCNN, we tried to generate the "preference" error used WGAN for each model. The generator and discriminator are described in Sect. 3.2. According to the result of the above experiment, we only used SSA and LSTM to do some experiments. And the effect in experiments was shown in Table 3.

As is indicated in Table 3, this method can fit this error to a certain extent, but it is not stable. Moreover, the use of Gaussian noise to fit the prediction preference is easily affected by different waveforms at different moments, and the experiment cannot achieve the prediction effect. Our experiments demonstrate that the stable state of WGAN is obtained after 400 times (about 25epoch) in Turkey. However, we have not got an outstanding effect after the stable state in dataset of Turkey at that situation. In consideration, the pivotal problem is that the model separates sequence dependence and just fits a part of inner-subseries. Thus, it is hard for the noise generated by WGAN to be adapted in all times of series.

4.4 Comparison with State of the Art

After experimented and discussed some method in our framework, and in order to better demonstrate our performance, we compared with other prediction models including the Historical Average model (HA), Random Forest model (RF), and SVR for comparison (We used k-fold verification and grid search to find the best model among the different kernel functions and hyperparameters of SVR. And the number of a decision tree is 1000 in RF). In addition, some neural network methods also were used as contrasts including LSTM and TCN.

In Table 4, it is obviously observed that our method performs well in RMSE, MAE, and R^2. And comparing with SVR and RF, LSTM and TCN acquire better accuracy. So, our framework with the basic prediction model including LSTM and TCN is verified to be reasonable.

Table 4. Comparison of the proposed methodology with state of the art in wind speed prediction.

Method	RMSE	MAE	$R^2(\%)$	RMSE	MAE	$R^2(\%)$
	India			Turkey		
HA	0.5636	0.3500	82.70	1.3128	0.7956	84.96
RF	0.4657	0.3136	88.21	1.2724	0.5337	85.89
SVR	0.4483	0.3098	89.05	1.3988	0.6311	82.92
LSTM	0.4427	0.2756	89.41	1.4227	0.5597	82.09
TCN	0.4424	0.2924	89.28	1.3990	0.5367	82.92
Our Method(SSA + Re + LSTM)	**0.0954**	**0.0714**	**99.51**	**0.3697**	**0.1895**	**98.81**

5 Conclusion

In this paper, we proposed a hybrid prediction method. Two different decomposition methods are used to decompose the original signal into multiple subseries. Then, various prediction methods are applied to predict those multidimensional subseries. In order to make the results more precise, 1DCNN was added and used to process and restructure those subseries predicted after the prediction model. Finally, a combination performing well compared with others in the validation set is select to test. After experimental verification on two different data sets, time consumption and errors have been remarkably reduced. In addition, WGAN was tried to assist in generalizing predicted data. In summary, the method we proposed is effective and better than many states of the arts on two datasets.

Further research works will focus on the following aspects. It is worth studying that the suitable hyper-parameter optimization method. Furthermore, in order to avoid the instability of GAN, and also to make GAN suitable in prediction, CGAN may be a valuable method.

Acknowledgement. This paper is supported by National Key R&D Program of China (2018YFB1004300).

References

1. Lazić, L., Pejanović, G., Živković, M.: Wind forecasts for wind power generation using the eta model. Renew. Energy **35**(6), 1236–1243 (2010)
2. Zhang, L.L., Li, M.S., Ji, T.Y., Wu, Q.H.: Short-term wind power prediction based on intrinsic time-scale decomposition and ls-svm. In: The IEEE Innovative Smart Grid Technologies - Asia (ISGT-Asia), Melbourne, Asia, pp. 41–45 (2016)
3. Han, L., Zhang, R., Wang, X., Bao, A., Jing, H.: Multi-step wind power forecast based on VMD-LSTM. IET Renew. Power Gener. **13**, 1690–1700(10) (2019)
4. Lai, G., Chang, W.C., Yang, Y., Liu, H.: Modeling long-and short-term temporal patterns with deep neural networks. In: The 41st International ACM SIGIR Conference on Research and Development in Information Retrieval, New York, USA, pp. 95–104 (2018)

5. Bai, S., Zico Kolter, J., Koltun, V.: An Empirical Evaluation of Generic Convolutional and Recurrent Networks for Sequence Modeling. arXiv e-prints arXiv:1803.01271 (2018)
6. Yu, C., Li, Y., Zhang, M.: An improved wavelet transform using singular spectrum analysis for wind speed forecasting based on elman neural network. Energy Conv. Manag. **148**, 895–904 (2017)
7. Kim, Y.: Convolutional neural networks for sentence classification. In: Conference on Empirical Methods in Natural Language Processing, Doha, Qatar, pp. 1746–1751 (2014)
8. Zhang, K., Zuo, W., Chen, Y., Meng, D., Zhang, L.: Beyond a gaussian denoiser: residual learning of deep CNN for image denoising. IEEE Trans. Image Process. **26**(7), 3142–3155 (2017)
9. Chen, J., Chen, J., Chao, H., Yang, M.: Image blind denoising with generative adversarial network based noise modeling. In: The IEEE Conference on Computer Vision and Pattern Recognition (CVPR), Salt Lake City, US (2018)
10. Koochali, A., Schichtel, P., Dengel, A., Ahmed, S.: Probabilistic forecasting of sensory data with generative adversarial networks - forgan. IEEE Access **7**, 63868–63880 (2019)
11. Xu, Z., Du, J., Wang, J., Jiang, C., Ren, Y.: Satellite image prediction relying on GAN and LSTM neural networks. In: The IEEE International Conference on Communications (ICC), Shanghai, China, pp. 1–6 (2019)
12. Arjovsky, M., Chintala, S., Bottou, L.: Wasserstein gan. In: Proceedings of the 34th International Conference on Machine Learning, Sydney, Australia, vol. 70, pp. 214–223 (2017)
13. Pascual, S., Bonafonte, A., Serrà, J.: Segan: speech enhancement generative adversarial network. In: Conference of the International Speech Communication Association (INTERSPEECH), Stockholm, Sweden, pp. 3642–3646 (2017)
14. Mi, X., Liu, H., Li, Y.: Wind speed prediction model using singular spectrum analysis, empirical mode decomposition and convolutional support vector machine. Energy Conv. Manag. **180**, 196–205 (2019)
15. Torres, M.E., Colominas, M.A., Schlotthauer, G., Flandrin, P.: A complete ensemble empirical mode decomposition with adaptive noise. In: The IEEE International Conference on Acoustics, Speech and Signal Processing, Czech Republic, Prague, pp. 4144–4147 (2011)

A Cell Tracking Method with Deep Learning Mitosis Detection in Microscopy Images

Di Wu[1,3], Benlian Xu[2(✉)], Mingli Lu[3], Jian Shi[3], Zhen Li[2], Fei Guan[3], and Zhicheng Yang[3]

[1] School of Electrical and Power Engineering, China University of Mining and Technology, Xuzhou, People's Republic of China
[2] School of Mechanical Engineering, Changshu Institute of Technology, Changshu, People's Republic of China
xu_benlian@cslg.edu.cn
[3] School of Electrical and Automatic Engineering, Changshu Institute of Technology, Changshu, People's Republic of China
luml@cslg.edu.cn

Abstract. Cell motion analysis plays an important role in biomedical fields such as disease diagnosis and drug development. A crucial step in quantifying cell dynamics is to detect mitosis, which is the process that a mother cell divided into two daughter cells. To gain an accurate analysis of multiple cells, a deep framework combining U-net and convolution LSTM is proposed to simultaneously segment cells and detect mitosis, in which the spatiotemporal information of image sequences is fully utilized to predict the locations of cell and the occurrence of mitosis. With the obtained cell segmentation and mitotic event, a particle filter based tracking method is proposed to estimate individual state of cells, in which a two-step data association strategy is developed to handle the mitotic assignment. Simulation results are presented to show that the proposed method has favorable performance and can track cells effectively with mitosis.

Keywords: Cell tracking · Deep learning · Particle filter · Mitosis detection

1 Introduction

Cell migration is one of the key cellular processes, which is associated with a variety of biological phenomena such as inflammation and tumor development [1]. Accurate estimation of cell migration plays an important role in cellular and molecular biology research. Typically, cell migration analysis is performed manually in time-lapse microscopic image sequences, which can lead to significant errors in practical research. Therefore, it is necessary to use automated tracking methods to replace this time-consuming and tedious work. Automatic cellular

© Springer Nature Switzerland AG 2021
Y. Tan and Y. Shi (Eds.): ICSI 2021, LNCS 12690, pp. 282–289, 2021.
https://doi.org/10.1007/978-3-030-78811-7_27

tracking is challenging due to several factors such as low image contrast, variations in morphology, random migration, cell division, etc. As a result, numerous fully automatic cell tracking algorithms were proposed in recent years [2,3].

In this paper, we introduce a deep framework combining U-net [4] and convolutional LSTM [5] to simultaneously segment cells and to detect mitotic events. With the result of cell detections and mitotic detections, a particle filter based tracking method is then proposed for image sequences to achieve the dynamics of multi-cells. A two-step data association strategy is proposed to deal with the mitosis events. The rest of this paper is organized as follow: Sect. 2 details our cell tracking method with deep learning mitosis detection. We provide our experimental results in Sect. 3 and conclude this paper in Sect. 4.

2   Methods

Multi-object tracking generally deal with object detection and data association separately, i.e., tracking by detections approach. First we segment cells by a deep framework combining U-net and convolutional LSTM, and obtain mitosis detections simultaneously prepared for subsequent data association. After segmentation, the second step is cell tracking. Our tracking algorithm is based on estimating the distribution of each cell state by a particle filter. A two-step data association strategy is proposed to deal with the mitosis events.

2.1   A Deep Framework for Cell Segmentation and Mitosis Detection

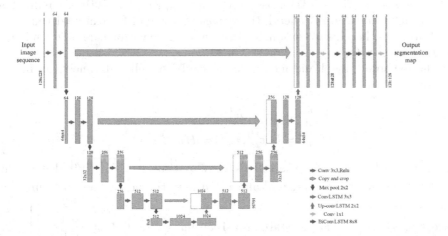

Fig. 1. Proposed network architecture. Each blue box corresponds to a multi-channel feature map. The number of channels is denoted on top of the box. The x-y-size is provided at the lower edge of the box. White boxes represent copied feature maps. The arrows denote the different operations as shown in the lower right corner of figure. (Color figure online)

U-Net operates with very few training images and yields more precise segmentations [4]. Cell motion is a process with temporal characteristics, such as a cell migrates in a certain area of vision, then decreases in size and becomes brighter, and finally divides into two daughter cells. Convolutional LSTM (ConvLSTM) is an extension of LSTM, which preserves both temporal and spatial information, thus it is suitable for image sequence processing.

The proposed network incorporates ConvLSTM blocks into the U-Net architecture. The introduction of ConvLSTM blocks into U-Net allows considering previous cell appearances at multiple scales by holding their compact representations in the ConvLSTM memory units. Figure 1 illustrates the proposed network architecture. We propose the incorporation of ConvLSTM layers in every scale of the decoder section of the U-Net. The output of U-net then fed into two bi-directional ConvLSTM (BiConvLSTM) lawyers. As mitotic events may happen in different stages of the sequence, the BiConvLSTM layer applies to get complementary cell division information from past (backwards) and future (forward) states simultaneously. In our method, the network expands from a binary classification to a multiple classification including background, common cells and mitotic cells, in which the mitotic cell denotes the cell that will divide into two daughters in the next frame. After obtaining the segmentation results, the watershed algorithm was used to segment the adhesive cells, and then the geometric center and edge of the connected regions are extracted as the center and contour of cell detections.

2.2 Multi-cell Tracking Based on Particle Filter

With the obtained cell segmentations and mitosis detections, a particle filter based tracking method is then used for image sequences to estimate individual state of multiple cells. In general, the implementation of PF can be decomposed into four steps, i.e., Initialization, Prediction, Update with resampling, and Output in sequence [6].

The dynamic model and measurement model of cell i at time t can be formulate as

$$
\begin{aligned}
x_{t+1}^{(i)} &= f(x_t^{(i)}, w_t^{(i)}) = A x_t^{(i)} + w_t^{(i)}, \\
z_t^{(i)} &= h(x_t^{(i)}, v_t^{(i)}) = H x_t^{(i)} + v_t^{(i)},
\end{aligned}
\tag{1}
$$

where $x_t^{(i)} \in \mathbb{R}^{n_x}$ denotes the cell i's state at time t, $z_t^{(i)} \in \mathbb{R}^{n_z}$ denotes the corresponding measurement, $w_t^{(i)}$ and $v_t^{(i)}$ denote the process and measurement noise respectively, A is the state transition matrix and $A = \begin{bmatrix} 1 & T & 0 & 0 \\ 0 & 1 & 0 & 0 \\ 0 & 0 & 1 & T \\ 0 & 0 & 0 & 1 \end{bmatrix}$ with sampling interval T and H is the observation matrix.

The cell i tracking process can be implemented as follows:

Step 1: Initialization: One draws N samples equally from the initial density function $\pi_{0|0}$, and yields $\{x_0^{(i,j)}, N^{-1}\}_{j=1}^N$. Let $\pi_{t-1|t-1}^N$ be the measure, so we define

$$\pi_{t-1|t-1}(dx_{t-1}^{(i)}|z_{t-1}^{(i)}) \sim \pi_{t-1|t-1}^N(dx_{t-1}^{(i)}|z_{t-1}^{(i)}) \triangleq \frac{1}{N} \sum_{j=1}^N \delta_{x_{t-1}^{(i,j)}}(dx_{t-1}^{(i)}), \quad (2)$$

where $\pi_{t-1|t-1}^N$ denotes the empirical distribution close to $\pi_{t-1|t-1}$, and $\delta_{x_{t-1}^{(i,j)}}(\cdot)$ denotes the delta-Dirac function centered at $x_{t-1}^{(i,j)}$. If we set $t = 1$, then Eq. (2) corresponds to the initial step.

Step 2: Prediction step: This step is carried out for each particle to obtain predicted particle $\overline{x}_t^{(i,j)}$ using the proposal density function

$$\overline{x}_t^{(i,j)} \sim q_t(\cdot|\overline{x}_{t-1}^{(i,j)}, z_t^{(i)}), j = 1, 2, ..., N. \quad (3)$$

Meanwhile, the individual weight for each particle is predicted and evaluated by

$$w_{t|t-1}^{(i,j)} = \frac{p_{t|t-1}(\overline{x}_t^{(i,j)}|x_{t-1}^{(i,j)})}{q_t(\overline{x}_t^{(i,j)}|\overline{x}_{t-1}^{(i,j)}, z_t^{(i)})} w_{t-1}^{(i,j)}. \quad (4)$$

Note that if we take $q_t(\overline{x}_t^{(i,j)}|\overline{x}_{t-1}^{(i,j)}, z_t^{(i)}) = p_{t|t-1}(\overline{x}_t^{(i,j)}|x_{t-1}^{(i,j)})$, then $w_{t|t-1}^{(i,j)} = w_{t-1}^{(i,j)}$. Let $\overline{\pi}_{t|t-1}^N$ be the measure, following Eq. (2), the empirical one-step predicted distribution is obtained by

$$\pi_{t|t-1}(dx_t^{(i)}|z_{t-1}^{(i)}) \sim \overline{\pi}_{t|t-1}^N(dx_t^{(i)}|z_{t-1}^{(i)}) \triangleq \sum_{j=1}^N w_{t|t-1}^{(i,j)} \delta_{\overline{x}_t^{(i,j)}}(dx_t^{(i)}). \quad (5)$$

Step 3: Update step: Once new measurement $z_t^{(i)}$ is available, and if Eq. (5) is substituted into Eq. (1), we have a simplified form of the Monte-Carlo approximation of $\pi_{t|t}(dx_t^{(i)}|z_t^{(i)})$:

$$\pi_{t|t}(dx_t^{(i)}|z_t^{(i)}) \sim \overline{\pi}_{t|t}^N(dx_t^{(i)}|z_t^{(i)}) \triangleq \sum_{j=1}^N w_t^{(i,j)} \delta_{\overline{x}_t^{(i,j)}}(dx_t^{(i)}), \quad (6)$$

$$w_t^{(i,j)} = \frac{p_t(z_t^{(i)}|\overline{x}_t^{(i,j)})}{\sum_{j=1}^N w_{t|t-1}^{(i,j)} p_t(z_t^{(i)}|\overline{x}_t^{(i,j)})} w_{t|t-1}^{(i,j)}. \quad (7)$$

Resampling step: Finally, the re-sampling step is done to obtain a set of unweighted particles $\{x_t^{(i,j)}, N^{-1}\}_{j=1}^N$, and these particles constitute the empirical distribution close to $\pi_{t|t}(dx_t^{(i)}|z_t^{(i)})$:

$$\pi_{t|t}(dx_t^{(i)}|z_t^{(i)}) \sim \pi_{t|t}^N(dx_t^{(i)}|z_t^{(i)}) \triangleq \frac{1}{N} \sum_{j=1}^N \delta_{x_t^{(i,j)}}(dx_t^{(i)}). \quad (8)$$

Step 4: Output: the final estimate for $\boldsymbol{x}_t^{(i)}$ of cell i at time t can be written as

$$\hat{\boldsymbol{x}}_t^{(i)} = E(\boldsymbol{x}_t^{(i)}|\boldsymbol{z}_t^{(i)}) = \sum_{j=1}^{N} N^{-1}\boldsymbol{x}_t^{(i,j)}. \tag{9}$$

The data association method here is a two-step strategy: First, we define the pairwise-distance matrix $\boldsymbol{D}_t \in \mathbb{R}_{N \times M}$, to calculate the distance of predicted state of all cells at time $t-1$ and all measurements at time t, where N denotes the number of predicted cells from time $t-1$ and M denotes the number of cell detections at time t. We use the Hungarian algorithm to estimate the optimal soft-assignment matrix $\boldsymbol{A}_t \in [0, 1]_{N \times M}$, where 1 means the measurement is assigned to the cell and 0 otherwise. Second, if there is any unassigned detection denoted by $d_t^{(i_1)}$ at time t and there is any cell that is a mitotic cell detected by Sect. 2.1 denoted by x_{t-1}^m at time $t-1$, $d_t^{(i_1)}$ will be assigned to x_{t-1}^m as one of its daughter cells when distance between $d_t^{(i_1)}$ and x_{t-1}^m is less than a threshold ε.

3   Experimental Results

In this section, we use two image sequences from publicly available website at http://www.celltrackingchallenge.net/ to evaluate the effectiveness of our method. The two cell image sequences include HeLa cells stably expressing H2b-GFP dataset including 28 frames and simulated nuclei of HL60 cells stained with Hoescht dataset including 30 frames.

The network in Sect. 2.1 is implemented using the open-source Keras and Tensorflow framework. Model training is performed on a NVIDIA 11 GB RTX 2080ti graphic card. The input of the network is set to be 128×128 because of hardware limitations. During the training process, a data augmentation method is used to avoiding overfitting. Rotation, translation, flipping and other operations are applied to each sequence in this experiments. We take three consecutive frames of images as a sequence and fed it to the network for training, so that the network could learn the spatio-temporal characteristics. The dataset contains 1043 sequences for the network training. The output of the network has three channels, including background, common cells and mitotic cells. The trained network has 98.2% mean IoU on the validation set and has 91.5% mean IoU on the test set. We show two channels including common cells and mitotic cells of the partial validation results in Fig. 2. In Fig. 3, three channels of the partial test results superimposed on one channel are presented in different colors.

Parameters in multi-cell tracking based on PF are listed below: Sampling interval is 30 min, system noise $Q = diag(0.01, 0.01, 0.01, 0.01)$, measurement noise $R = diag(0.01, 0.01, 0.01, 0.01)$, number of particles $N = 100$ (for each cell), and altogether 10 Monte-Carlo runs. The tracking results of two datasets at different times are shown in Fig. 4.

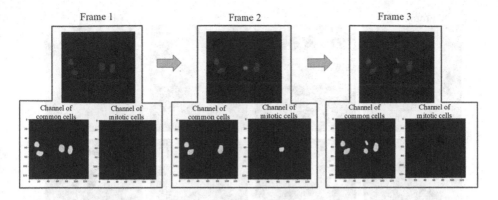

Fig. 2. Partial validation results.

To verify the effectiveness of our proposed method, we compare it with the Viterbi algorithm [2] and the JPDAF [9]. For performance evaluation, we use the complete tracks (CT) [7], which shows how good an approach is at reconstructing complete reference tracks and is defined as $CT = \dfrac{2T_{rc}}{T_{gt} + T_c}$, where T_{rc} denotes the number of completely reconstructed reference tracks, T_{gt} denotes the number of all reference tracks and T_c denotes the number of all computed tracks. To jointly evaluate the errors of label and cardinality of tracks, the metric of optimal sub-pattern assignment metric for track (OSPA-T) [8] is introduced as well. Our proposed method has better performance on two datasets, more specifically, getting the highest value on CT and the lowest value on OSPA-T, as shown in Table 1.

Table 1. Performance measure of different combinations

	Dataset	Metric	
		CT	OSPA-T
JPDAF	Hela	0.71	3.32
	HL60	0.69	3.77
Viterbi	Hela	0.86	2.81
	IIL60	0.81	3.24
Our proposed	Hela	0.94	2.26
	HL60	0.93	2.45

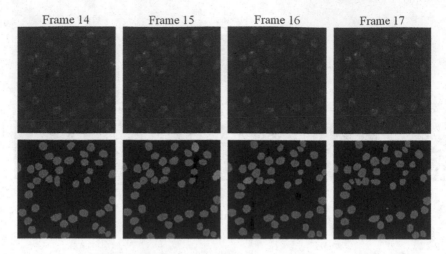

Fig. 3. Partial test results. Top row is the original images and bottom row is corresponding segmentation result. Black pixels denote the background, purple pixels denote the common cells and red pixels denote the mitotic cells in the bottom row. (Color figure online)

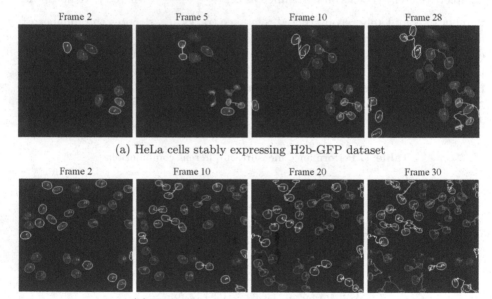

(a) HeLa cells stably expressing H2b-GFP dataset

(b) HL60 cells stained with Hoescht dataset

Fig. 4. Tracking results based on particle filter.

4   Conclusion

In this paper, a new particle-filter-based multi-cell tracking algorithm is proposed, which is based on a deep framework combining U-net and convolution LSTM to simultaneously segment cells and detect mitosis. It is demonstrated that our proposed approach shows better performance and has capability to automatically track multiple cells for real image sequences.

Acknowledgement. This work was supported by National Natural Science Foundation of China (No. 61876024) , 333 Project of Jiangsu Province (No. BRA2019284), and Project of talent peak of six industries (2017-DZXX-001).

References

1. Li, F., Zhou, X., Ma, J., Wong, S.T.C.: Multiple nuclei tracking using integer programming for quantitative cancer cell cycle analysis. IEEE Trans. Med. Imaging **29**(1), 96–105 (2009)
2. Magnusson, K.E.G., Jaldén, J., Gilbert, P.M., Blau, H.M.: Global linking of cell tracks using the viterbi algorithm. IEEE Trans. Med. Imaging **34**(4), 911–929 (2014)
3. Benlian, X., Mingli, L., Shi, J., Cong, J., Nener, B.: A joint tracking approach via ant colony evolution for quantitative cell cycle analysis. IEEE J. Biomed. Health Inf. (2020). https://doi.org/10.1109/JBHI.2020.3032592
4. Ronneberger, O., Fischer, P., Brox, T.: U-Net: convolutional networks for biomedical image segmentation. In: Navab, N., Hornegger, J., Wells, W.M., Frangi, A.F. (eds.) MICCAI 2015. LNCS, vol. 9351, pp. 234–241. Springer, Cham (2015). https://doi.org/10.1007/978-3-319-24574-4_28
5. Shi, X., Chen, Z., Wang, H., Yeung, D.Y., Wong, W.K., Woo, W.: Convolutional LSTM network: A machine learning approach for precipitation nowcasting (2015). arXiv preprint arXiv:1506.04214
6. Hu, X., Schon, T.B., Ljung, L.: A basic convergence result for particle filtering. IEEE Trans. Signal Proces. **56**(4), 1337–1348 (2008)
7. Ulman, V., et al.: An objective comparison of cell-tracking algorithms. Nature Methods **14**(12), 1141 (2017)
8. Ristic, B., Vo, B.-N., Clark, D., Vo, B.-T.: A metric for performance evaluation of multi-target tracking algorithms. IEEE Trans. Signal Process. **59**(7), 3452–3457 (2011)
9. Rezatofighi, S.H., Milan, A., Zhang, Z., Shi, Q., Dick, A., Reid, I.: Joint probabilistic data association revisited. In: Proceedings of the IEEE International Conference on Computer Vision, pp. 3047–3055 (2015)

A Knowledge Graph Enhanced Semantic Matching Method for Plan Recommendation

Rupeng Liang$^{(\boxtimes)}$, Shaoqiu Zheng, Kebo Deng, Zexiang Mao, Wei Ma, and Zhengwei Zhang

CETC28th Key Laboratory of Information System Requirement, Nanjing 210007, China

Abstract. In order to solve the problem of rapid matching and optimization of the plan, a semantic feature-based smart matching method of the plan is proposed, which can be used to solve the problem of the typical small sample date for recommendation of the best plan in the military field. In this method, the semantic feature of the battle plan is established to describe the combat scenario through a military knowledge graph. The semantic feature annotation of the plan is constructed based on the military knowledge map too. So the semantic features corresponding to each matching target plan object are described, which realize the explicit definition of the hidden knowledge of the combat plan. Based on knowledge enhancement technology, the similarity measurement of semantic features is calculated, realizing the intelligent semantic matching of combat plans, so as to solve the problem of low matching efficiency and accuracy based on pragmatic level features such as index or keywords, satisfy the rapid matching and precise recommendation of plans.

Keywords: Knowledge graph enhanced · Combat plan recommendations · Semantic feature · Semantic similarity · Feature vector · Semantic matching

1 Introduction

As the pace of war gradually accelerates, plan matching methods can improve the rapid response of combat and increase the speed of combat. Combat plan matching is essentially a search problem, that is, how to quickly match the corresponding scenarios with similar scenarios in the plan library according to the current situation, and then recommend the best plans.

At present, hard matching methods based on keywords or restriction rules cannot solve the problem of ambiguity in the description of battle scenes. It is difficult to achieve quantitative evaluation for the optimization of plans [1, 2], which require a large amount of manual participation. And it is difficult to meet the requirements of wartime quick match demand [3]. For the matching problem of the combat plan library, the method of combining keywords is usually used. This method is more suitable for fully formalized and deterministic target matching. It requires a highly formatted and quantitative description of the combat plan. At the same time, the combat plans mostly

© Springer Nature Switzerland AG 2021
Y. Tan and Y. Shi (Eds.): ICSI 2021, LNCS 12690, pp. 290–299, 2021.
https://doi.org/10.1007/978-3-030-78811-7_28

exist in the form of text. The traditional keyword matching method is difficult to meet the needs of rapid and accurate matching of the combat plan.

At the same time, the current mainstream artificial intelligence recommendation algorithm based on "Deep Neural Network" [4, 5] requires a large number of data samples for deep learning and training. But currently there are fewer data samples related to combat plan matching. The knowledge graph can realize the formal description of knowledge to a certain extent and meet the needs of intelligent automated reasoning. For typical small sample problems, by constructing a military knowledge graph, it provides a formal description of combat knowledge and matches queries for refined combat plans, which can be used to realize the knowledge enhanced intelligent matching recommendation of the plan.

In order to solve the problem of rapid matching and optimization of plan, this paper proposes an intelligent matching method of plan based on semantic features, which describes the semantic features of combat scenarios through military knowledge graphs, and defines the correspondence of each matching target plan. Based on the similarity measurement of semantic features, intelligent matching of semantic-level combat plans is realized, which simulate the process of commander's selection of plans.

There are many researches on feature-based matching and recommendation methods. The difficult problem lies in the definition and description of features [6, 7]. For military issues and other small sample feature extraction problems, a semantic feature description solution based on knowledge graphs is proposed. On the one hand, it solves the inherent shortcomings of low recall rate based on keywords and other hard matching methods. On the other hand, through the feature expression based on the military knowledge graph, the intelligent plan matching method based on the knowledge graph is constructed to improve the matching efficiency and accuracy.

2 Military Knowledge Graph Construction

The knowledge graph is a kind of structured formal description of concepts, entities and their relationships in the objective world, which can describe the information in a form close to the human cognition, and provide a better ability to organize, manage and understand massive amounts of information [8]. Knowledge graph brings vitality to the semantic search of massive information, and at the same time shows strong power in intelligent question and answer, and has become a knowledge-driven intelligent application infrastructure. Together with big data and deep learning, knowledge graphs have become one of the core driving forces that promote the development of Internet nuclear artificial intelligence.

With the explosive growth of battlefield information, traditional search methods have been unable to meet the needs of capturing and understanding battlefield information. And knowledge graph technology has shown great advantages in the accuracy and scalability of solving knowledge queries. It has become a hot issue of current research [9].

The existing knowledge graphs are mostly general knowledge graphs, and the domain knowledge with strong pertinence, especially the knowledge graphs in the military field, does not have a better construction and representation method. Therefore, the establishment of military knowledge graphs is of great significance. Military knowledge covers a wide range. The military knowledge map defined in this article is mainly for the field of combat planning.

In the process of constructing the military knowledge map, technologies such as data collection, knowledge extraction, knowledge disambiguation, and knowledge reasoning need to be involved [10]. The overall process is shown in Fig. 2.

Based on the knowledge graph, the formal definition of the semantics of the combat plan concept can be realized, which is used in obtaining the description of the semantic features of the plan, and realizing intelligent plan matching based on the semantic features (Fig. 1).

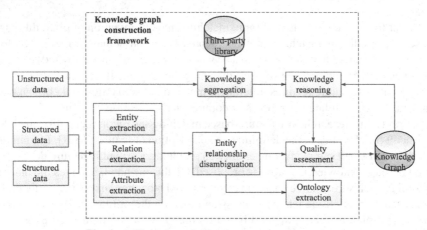

Fig. 1. Military knowledge map construction process.

The military knowledge map in this article is mainly used for the semantic information labeling of combat plans. The military knowledge map used in this article is based on "Chinese People's Liberation Army Military Terms" (2011 edition). Protégé tools are used in the construction of military knowledge map to achieve formal modeling of basic concepts in the military field. The term definition is used as the description information of the conceptual entity, and defines the parent-child inclusion relationship (SubClass Of) between the conceptual entities. For example, the battle plan contains sub-conceptual objects such as firepower plan, support plan, and coordinated plan, and the military map contains a list of conceptual objects.

3   Semantic Feature Annotation of Combat Plan

3.1   Semantic Feature Definition

Semantic features are the description of entity features through knowledge graphs to realize the semantic labeling of entities [11]. In order to obtain semantic information about the description of the plan, it is necessary to create a semantic description of the features of the plan. Through the mapping with the knowledge map, the structured metadata of the features of the plan is defined. Based on semantic reasoning tools, semantic annotation supports plan semantic matching and discovery, such as defining attributes for targets as shown in Fig. 2.

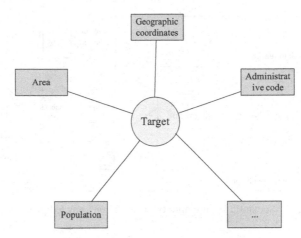

Fig. 2. Semantic feature entity samples.

3.2 Semantic Feature Annotation of Plan

Semantic accurate feature annotations. Semi-automatic feature semantic annotation algorithms can be constructed to realize the sorting of concept entities and concept relationships in the knowledge graph. Through manual intervention, a semi-automatic semantic annotation process for plan features is constructed to realize semantic feature mapping.

The semantic annotation algorithm provides the conceptual entities or relationships related to the semantic features of the plan, by analyzing the query keywords of the combat plan input by the user and mapping them to the corresponding nodes of the knowledge graph. After manual intervention, the graphical human-computer interaction interface is used to assist in the realization of the characteristic semantics [12].

In view of the large workload of manual labeling of combat plans, errors are easy to occur, and the current automatic plan semantic feature labeling algorithm is not mature. In order to improve the degree of automation of semantic feature labeling, it can be based on the basic algorithm of military knowledge map concept term matching. Considering the graphical features of the knowledge map, an algorithm optimization scheme which takes into account the conceptual map structure of the military knowledge map can be used. And the PageRank algorithm is integrated to achieve more efficient semantic feature concept matching and feature annotation [12].

In order to improve the efficiency of semantic feature annotation, this paper adopts the manual intervention method of semi-automatic semantic annotation of combat plan features, and realizes related concept entities and concepts in the knowledge graph based on the semi-automatic semantic annotation algorithm of concept matching. The matching recommendation of the relationship is to establish the relationship between the knowledge graph and the feature of the plan through manual selection to improve the reliability of semantic annotation. The specific process is shown in Fig. 3.

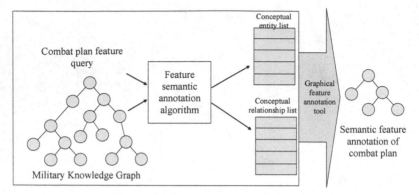

Fig. 3. Semantic feature annotation based on knowledge graph.

4   Combat Plan Matching Based on Semantic Features

4.1   Semantic Feature-Based Matching Mechanism

Compared with the traditional plan matching method based on database index and keyword matching, the feature semantic annotation associated with the knowledge graph supports logical reasoning. In the plan query process, the semantic reasoning engine can analyze the inclusion relationship between concepts, expand the search scope. And through calculation based on the similarity of semantic features of plan, a method for matching combat plans for specific combat scenarios is established to achieve semantic-level matching and discovery to ensure more accurate query results. The semantic feature matching mechanism of combat plans is shown in Fig. 4.

The core of realizing the semantic matching of the combat plan is to construct the matching rules and the semantic feature similarity measurement algorithm. First, based on combat requirements, construct restrictive matching rules and optimized matching rules to achieve preliminary screening of operations. Taking the pre-screened plans as the matching target set, semantic feature vectors is extracted based on the semantic annotation of each plan. And, a mixed-mode plan matching mechanism of sexual measurement and inclusive reasoning rules is established, to as to realize the semantic matching of combat plans and improve the efficiency and accuracy of plan matching.

First, the semantic feature matching of the combat plan needs to analyze the plan query request. Same to the plan query generalization method used in the plan semantic annotation algorithm, the query object is mapped to a set of TD-IDF vectors. the second step, using the semantic annotation generated in the plan annotation database, the semantic feature vector of each combat plan is extracted, which can be mapped to multiple TD-IDF vectors. And by using the feature vector similarity calculation method provided in the plan labeling algorithm, the similarity between the query object and the

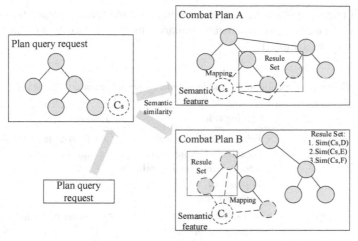

Fig. 4. Semantic feature-based matching mechanism

feature vector of each plan is calculated, through which the association relationship with the target query is determined to realize the sorting of the plan.

4.2 Combat Plan Matching Rules

Compared with the traditional plan matching method based on database index and keyword matching, the feature semantic annotation associated with the knowledge graph supports logical reasoning. In the plan query process, the semantic reasoning engine can analyze the inclusion relationship between concepts, expand the search scope. And through calculation based on the similarity of semantic features of plan, a method for matching combat plans for specific combat scenarios is established to achieve semantic-level matching and discovery to ensure more accurate query results. The semantic feature matching mechanism of combat plans is shown in Fig. 4.

(1) Restriction rules

Match the combat plans in the combat plan library according to restrictive rules, it is used to exclude the combat plans that do not meet the conditions, so as to narrow the set of matching combat plans. Restrictive rules mainly include several aspects: target type, our strength state, etc., through the 0–1 matching algorithm, the set of plans that meet the conditions is matching.

(2) Optimized matching rules

Considering the importance of the target, the degree of damage to the target, the probability of penetration, etc., the similar matching algorithm based on feature vectors is used to recommend the plans according to priority, mainly including target similarity, damage matching degree, etc. At the same time, according to the penetration probability, the combat forces can be automatically sorted, and the manual selection of combat weapons is supported [13–15].

Through expert evaluation, sample training, etc., the feature parameters are determined and normalized, and the cosine similarity methodis used to sort the similarity [16–19], as shown in Table 1.

Table 1. Matching rules and input and output tables of combat plans

Input	Matching rules	Output
Matching target vector	• Feature vector (casualty population, economic loss, political importance, military importance, penetration success rate)	• Combat plan matching similarity ranking
Combat plan feature vector	• Weight matrix (the importance of each value), which may be adjusted according to the situation	• Sorting of component factors (such as success rate, combat efficiency, etc.)

4.3 Combat Plan Matching Process

A three-level matching mechanism is adopted, as shown in Fig. 5. According to the matching rules, the similarity calculation method of semantic feature vector is used to realize the calculation of feature vector sorting. And based on the matching factor to realize the plan matching, the secondary screening of the combat plan is realized according to the current combat power state, and finally form the combat plan matching set.

(1) Semantic feature definition, that is, to create a semantic description of the features of the combat plan, provide semantic feature information for the matching of the plan, and establish the association mapping between the combat plan and the military knowledge map through graphical semantic annotation tools, thereby realizing the structuring of the semantic features of the plan Metadata definition.

(2) Semantic feature annotation of the plan. Before the plan is matched, the semantic feature definition of the plan needs to be established in advance, that is, semantic labeling. Here, the semantic annotation algorithm based on concept matching is used to establish the mapping relationship between the plan feature and the knowledge map node. First, analyze the user input plan feature query information, and map the feature input to the query feature vector; at the same time, map the concept node of the military knowledge map to the target feature vector; realize the concept recommendation queue by calculating the distance between the query feature vector and the target feature vector The sorting of, using the graphical human-computer interaction interface, through manual intervention, the features of the plan are mapped to the corresponding nodes of the knowledge graph, which assists in the realization of the definition of semantic features.

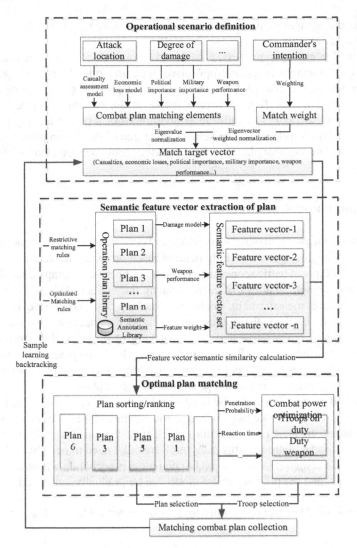

Fig. 5. Plan matching process

(3) Combat plan matching rules include two types of rules. One is restricted restriction rules, which mainly include: target type, our power status, etc. Restricted rules effectively reduce the set of matching targets; the second is optimized matching rules. Including the importance of the target, the degree of damage to the target, the probability of penetration, etc., the optimized matching rules realize the sorting of plans.

(4) Combat plan sorting based on semantic features. In the process of combat plan query, the screening of recommended combat plans is realized according to the matching rules of the construction of combat plans, and based on the calculation of semantic feature similarity of combat plans, a sorting method of combat plans

for specific combat scenarios is established to achieve recommendation. The plan queue is generated.

(5) Combat plan matching process. First, based on operational requirements, construct restrictive matching rules and optimized matching rules to achieve preliminary combat screening; take the pre-screened combat plans as the matching target set, and extract semantic feature vectors based on the semantic feature annotations of each combat plan. Through the establishment of a plan matching mechanism to measure the similarity of the semantic features of the fusion plan, the semantic matching of the combat plan is realized, and the recommended plan list after sorting is output.

5   Summary and Conclusions

In order to solve the problem of rapid matching and optimization of emergency plans, this paper proposes an intelligent matching method based on semantic features. It describes the semantic features of battle scenes through military knowledge graphs, and defines the corresponding semantic features of each matching target plan. Based on the similarity of semantic features measurement, intelligent matching of semantic-level combat plans is realized, which simulates the process of commander's selection of plans, and meets the requirements of quantitative and rapid matching of plans. And the method solves the problem of low matching efficiency and accuracy based on pragmatic features such as indexes or keywords. Intelligent semantic matching of combat plans with semantic features meets the needs of rapid and precise matching of plans.

The method in this paper can be used to solve the inherent shortcomings of hard matching methods based on keywords, and provide feature expressions based on military knowledge bases. In view of the training problem of the small sample of the combat plan matching model, there are still many problems that need to be solved. The feature annotation data and training samples can be continuously accumulated during the combat training process, and the plan matching model can be gradually optimized.

Acknowledgments.   The paper is supported by the National Natural Science Foundation of China (No. 41401463).

References

1. Liu, X., Zhao, H., Yang, H.: Overview of operational plan evaluation method. J. Ordnance Equipment Eng. **39**(8), 79–84 (2018)
2. Teng, Z., Jiang, N.: Study on structure of ship formation of combat plan's elements. Ship Electron. Eng. **38**(2), 1–4 (2018)
3. Cheng, H., Wang, K., Li, B.: Efficient friend recommendation scheme for social networks. Comput. Sci. **45**(6), 433–436 (2018)
4. Huang, L., Jiang, B., Lu, S.: Study on deep learning based recommender systems. Chinese J. Comput. **41**(7), 1619–1644 (2018)
5. Wang, Y., Tang, J.: Deep learning-based personalized paper recommender system. J. Chin. Inf. Process. **32**(4), 114–119 (2018)

6. Chen, Z., Lv, M., Wu, L., Xu, Y.: Space scene similarity metrics based on feature matrix and associated graph. Geomat. Inf. Sci. Wuhan Univ. **42**(7), 956–962 (2017)
7. Tian, H., Gu, J., Chen, Q.: Shape correspondence analysis based on feature matrix similarity measure. J. Comput. Appl. **37**(6), 1763–1767 (2017)
8. Ding, J., Zhao, Q., Xia, B: The method study of armament knowledge graph's establishment based on open source data. Command Control Simul. **40**(2), 22–26 (2018)
9. Ge, B., Tan, Z., Zhang, C.: Military knowledge graph construction technology. J. Command Control **2**(4), 308–320 (2016)
10. Yang, T., Liu, Z., Zhu, X.: Combat system operation mechanism description method. Command Control Simul. **40**(2), 15–20 (2018)
11. Zhou, Z.: Machine Learning. Tsinghua University Press, Beijing (2016)
12. Liang, R., Li, H.: Utilizing PageRank in geographical conceptual graph-based semantic annotation Algorithm Optimization. Geogr. Geo-Information Sci. **30**(2), 1–4 (2014)
13. Wu, L., Wang, G., Yang, S., Yu, X.: CBR-based anti-missile combat plan generation technology. J. Air Force Eng. Univ. **12**(5), 45–49 (2011)
14. Zeng, Q., Zhang, H., Pan, S.: Study on the technology of intelligent emergency plan engine. Softw. Eng. Appl. **5**(4), 237–242 (2016)
15. Zhao, Y., Qi, Z., Wang, H.: Preplan matching method for maritime assaultive operations. Command Inf. Syst. Technol. **9**(1), 56–61 (2018)
16. Lu, Q., Liu, X.: Semantic matching model of knowledge graph in question answering system based on transfer learning. J. Comput. Appl. **38**(7), 1846–1852 (2018)
17. Lin, J., Zhou, Y., Yang, A.: Emotion feature vector extraction method based on semantic similarity. Comput. Sci. **44**(10), 296–301 (2017)
18. Cardoso-Cachopo, A., Oliveira, L.A.: Empirical evaluation of centroid-based models for single-label text categorization. INSEC-ID Technical Report (2006)
19. Qin, C., Zhu, H.S., Zhuang, F.Z., et al.: A survey on knowledge graph-based recommender systems (in Chinese). Sci. Sin. Inform. **50**(7), 937–956 (2020)

Classification of Imbalanced Fetal Health Data by PSO Based Ensemble Recursive Feature Elimination ANN

Jun Gao[1], Canpeng Huang[1], Xijie Huang[1], Kaishan Huang[2(✉)], and Hong Wang[1]

[1] College of Management, Shenzhen University, Shenzhen 518060, China
[2] Greater Bay Area International Institute for Innovation, Shenzhen University, Shenzhen 518060, China
hks@szu.edu.cn

Abstract. Electrocardiogram (CTG) is a simple and low-cost option to assess the health of the fetus. However, the number of normal fetuses is larger than the number of abnormal fetuses, leading to imbalances in CTG data. Existing studies have attempted to optimize the data processing or model training process by integrating machine learning methods with optimization algorithms. However, the effectiveness of features and appropriate selection of machine learning method creates new challenges. This study proposed an comprehensive method that considers the feature effectiveness and data imbalance issue. The proposed method uses the Particle Swarm Optimization (PSO) algorithm to optimize the parameters of the Edited Nearest Neighbours (ENN), Recursive Feature Elimination (RFE), and Artificial Neural Network (ANN) algorithms to find the optimal combination of the parameters of the three algorithms to further improve the accuracy of the fetal health prediction and reduce the cost of tuning. Experimental results show that the algorithm proposed in this paper can effectively solve the imbalance of CTG data, with a classification accuracy of 0.9942 and a kappa measure of 0.9783, which can effectively assist doctors in diagnosing fetal health and improve the quality of hospital visits.

Keywords: Fetal health · Edited nearest neighbours · Recursive feature elimination · Artificial neural network · Particle swarm optimization

1 Introduction

With the continuous improvement of the quality of life, the public's awareness of personal health importance is increasing. Fetus is a special group whose health is vital for the family and society. Currently, electrocardiogram (CTG) [20] is a simple and low-cost option for assessing fetal health, and assisting healthcare professionals in taking action to prevent fetus and maternal deaths. Among them,

Y. Tan and Y. Shi (Eds.): ICSI 2021, LNCS 12690, pp. 300–312, 2021.
https://doi.org/10.1007/978-3-030-78811-7_29

fetal electrocardiogram and fetal movement [13, 25] are the most important and intuitive physiological indicators, which are critical for the diagnosis of fetal health. The statistics and analysis of daily fetal heart rate, fetal ECG waveform changes, and the number of fetal movements can also indicate the health of the fetus, and can be used as a reference for doctors' diagnosis during pregnancy examination. **Machine learning methods can be used in processing the CTG data for health prediction** [16]. However, since the cases of fetal health far exceed cases of fetal abnormalities, the prediction accuracy remains low for practical application.

There is a certain research foundation about fetal health prediction. In view of the weak predictive ability of CTG data, Signorini et al. [19] proposed to add the FHR parameters obtained from the experiment to a classifier such as neural network. Huang et al. [9] used eleven continuous attributes in fetal CTG data to establish models by Discriminant Analysis (DA), Decision Tree (DT) and ANN. Research shows that ANN has the highest classification accuracy, but the decision tree shows a concise and effective decision rule that can help doctors effectively judge the health of the fetus. The author also proposes that the efficiency of data analysis can be improved by selecting features. Sahin et al. [16] evaluated the performance of eight different machine algorithms on CTG datasets. They eliminated the suspicious class to classify the health and pathology of the data and used 10-fold cross validation to measure the classification performance. It is found that RF has the best effect on this problem, with F-Measure of 99.2%. Ocak [12] proposed to use genetic algorithm (GA) to select the optimal feature subset combined with SVM to distinguish fetal health and pathology. Das et al. [6] processed the fetal CTG dataset through six automated feature reduction techniques and two manual selection methods to obtain eight feature subsets and a complete feature set, and used four machine learning algorithms to classify them separately. The study finds that when using Minimum Redundancy Maximum Relevance (MRMR) to select features, Random forest gives 99.91% accuracy and kappa value of 0.997. Lisen et al. [7] proposed the LS-SVM-PSO-BDT algorithm, based on the binary decision tree model using LS-SVM to classify the three types of labels and the model parameters are optimized by using PSO. The overall classification accuracy rate reaches 91.62%. Kadhim et al. [10] combined the Firefly algorithm with NBC to classify fetal health. They used the Firefly algorithm to find a subset of features to improve the performance of NBC. The accuracy of the method reaches 86.547%. Subasi et al. [23] used Bagging ensemble algorithm to classify fetal CTG data. It is found that the algorithm gets good performances, and the Bagging with RF reaches a classification accuracy of 99.02%. Chen et al. [5] proposed a weighted random forest algorithm for the imbalance of fetal data. Pivot is used to assess the relevance between features and labels, and selecting ten important features for modeling. The F1-score of the model is 97.85%, which can effectively solve the imbalance problem.

In existing studies, most researchers have not paid attention to the imbalance of fetal health data, and only add optimization algorithms in the process of training the classifier. In addition, all the features are considered in their method, which may not achieve the optimal prediction accuracy of the model.

To sum up, this paper proposes an PER-ANN (Particle Swarm Optimization (PSO) Based Ensemble Recursive Feature Elimination ANN) model. The model considers the imbalance characteristics and diverse feature effectiveness in fetal data. By improving the prediction accuracy, this paper has significant practical contribution for preliminary diagnosis of fetal health condition.

The content framework of this paper is as follows. In Sect. 2, the algorithms used are introduced in the process of innovation. Section 3 briefly describes strategies and procedures of the innovative algorithm in this paper. In Sect. 4, the data used is described, along with the analysis and discussion of the experimental results obtained. Section 5 presents the conclusion of this study.

2   Related Works

2.1   Edited Nearest Neighbours (ENN)

ENN is one of the data cleaning methods. This kind of method mainly cleans overlapping data through certain rules, so as to achieve the purpose of under-sampling. The basic idea of the ENN algorithm [2] is to classify the existing sample set into S classes. If K/2 of the K adjacent samples of a sample of the majority class does not belong to the majority class, this sample will be rejected.

ENN algorithm steps are as follows:

a) The sample set X is divided into S subsets $X_1, X_2, \cdots, X_s$
b) Using the nearest neighbor method and (X_{i+1} mod s) as the reference set, the samples in X_i are classified, where i = 1, 2, $\cdots$, s
c) Remove the wrong samples from each sample
d) Combine the remaining samples and replace the original sample X

2.2   Recursive Feature Elimination (RFE)

Generally, Feature selection has two goals [3]: one is to find minimum feature subset to maximize model accuracy, and the other is to Eliminating redundant features. GUYON et al. [8] proposed the recursive feature elimination algorithm to show good performance through moderate calculation work. The main idea of this algorithm is to select the best (or worst) features based on the weight coefficients through repeated construction of the model, and then repeat the above process until the iteration condition is satisfied. Commonly used classifiers include Support Vector Machines (SVM), Random Forests (RF), and Naive Bayes (NB). Relevant studies [15,24,27] show that using random forests on some realistic tasks performs better and has stronger generalization capabilities. Therefore, this experiment chooses RF as the Selector of RFE.

The steps of the RFE algorithm are as follows:

a) All features are given a certain weight, and the input selector is used to train the features.
b) Take the absolute value of the feature weight and eliminate the minimum value.
c) Repeat a) and b), and recurse continuously until the iteration condition is satisfied.

2.3   Artificial Neural Network (ANN)

Neural network [28] is a complex nonlinear model, which is abstracted from human brain neural network. The basis of neural networks is neurons. The characteristics and functions of a neuron have two important characteristics. First, its typical characteristic is multiple input and single output. Second, its functional characteristic is that it can process nonlinear information. Therefore, it has functions similar to our brain, capable of processing a large amount of non-linear dynamic information. We also often use its features and functions to build models.

The mathematical expression of neuron is as follows:

$$U_i = \sum_{j=1}^{n} \omega_{ij} x_i - \theta_i \qquad (1)$$

$$Y_i = f(U_i) \qquad (2)$$

Among them ω is the weight coefficient, θ is a bias, $f(U_i)$ of neuron activation function. The activation function mainly controls the sensitivity of the neuron, so that small changes in the weight coefficient ω and bias θ only produce small output changes.

Common activation functions: Sigmoid, tanh, etc.

1) sigmoid:

$$\varphi(v) = \frac{1}{1 + exp(-\alpha v)} \qquad (3)$$

2) tanh:

$$\varphi(v) - \frac{1 - exp(-v)}{1 + exp(v)} \qquad (4)$$

In addition, the structure of the neural network also includes an input layer, a hidden layer, and an output layer. Input data to the input layer, the hidden layer analyzes the data to update the weight and transfer information, and the output layer returns the results.

2.4   Particle Swarm Optimization (PSO)

Particle Swarm Optimization (PSO) [11] is based on a flock of birds looking for a random food characteristics derived optimization algorithm. They search for food in a collaborative way, adjusting their search methods based on their own and other members' information. The algorithm initializes the particle swarm, characterizes the particles with three indicators of position, speed and fitness value, and uses the fitness function to indicate the quality of the particles. In the solution space, an iterative method is used to find the optimal solution. In each iteration, the particle will update its position and velocity based on the individual best value and the global best value. The individual best value and the global best value respectively refer to the best position of the fitness calculated in the individual experience position and the best position of the fitness searched by each particle.

Update speed and position formula:

$$v_{id} = \omega * v_{id} + c_1 * r_1 * (pbest_{id} - x_{id}) + c_2 * r_2 * (gbest - x_{id}) \tag{5}$$

$$x_{id} = x_{id} + v_{id} \tag{6}$$

Among them, x_{id} is the particle position, v_{id} is the particle velocity, ω is the inertia weight, c_1, c_2 are the learning factors, r_1, r_2 are random numbers, $pbest$ is the individual optimal value, and $gbest$ is the global optimal value (Table 1).

2.5   Evaluation Index

Table 1. Introduction to symbols

Symbol	Description
TP	Positive samples with positive prediction
FP	Negative samples with positive prediction
TN	Negative samples with negative prediction
FN	Positive samples with negative prediction

The experiment uses accuracy, weight_F1_score, and kappa as evaluation indicators. Due to the imbalance of the data set, kappa is used to evaluate the experimental results.

Accuracy

$$accuracy = \frac{TP + TN}{TP + TN + FP + FN} \tag{7}$$

Weight_F1

$$weight\_F1 = \frac{2 * Precision_{weight} * Recall_{weight}}{Precision_{weight} + Recall_{weight}} \tag{8}$$

$$Precision_{wight} = \frac{\sum_{i=1}^{L} Precision_i * \omega_i}{|L|} \tag{9}$$

$$Recall_{wight} = \frac{\sum_{i=1}^{L} Recall_i * \omega_i}{|L|} \tag{10}$$

$$Prccision = \frac{TP}{TP + FP} \tag{11}$$

$$Recall = \frac{TP}{TP + FN} \tag{12}$$

Taking into account the problem of unbalanced categories, weight_F1 based on macro_F1 is used to evaluate the results.

Kappa

$$kappa = \frac{p_o - p_e}{1 - p_e} \tag{13}$$

$$p_e = \frac{x_1 * y_1 + x_2 * y_2 + \cdots + x_c * y_c}{n * n} \tag{14}$$

p_e is the total number of correctly classified samples in each category divided by the total number of samples. n is the total number of samples. $x_1, x_2, \cdots, x_c$ is the actual number of samples in each category. $y_1, y_2, \cdots, y_c$ is the number of samples predicted for each category.

3   PSO Based Ensemble Recursive Feature Elimination ANN (PER-ANN)

This paper proposes a Particle Swarm Optimization (PSO) based ensemble recursive feature elimination ANN. The process of PER-ANN is shown in Fig. 1. In general, the process of building a model requires preprocessing the data, selecting features, and training the classifier. Researchers need to adjust the parameters of each process or a certain process. However, because the algorithms are not optimized for each process at the same time, a global optimal result may not be achieved. The method proposed in this paper optimize the parameters of the three processes simultaneously to maximize the performance of the model.

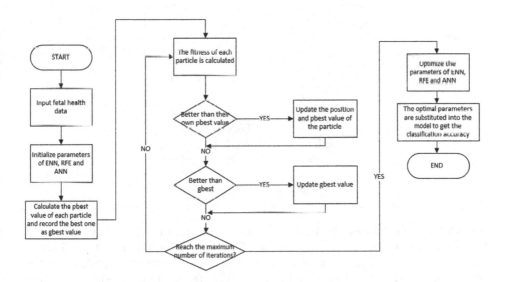

Fig. 1. Ensemble optimized model

In this model, ENN (edited nearest neighbors) is used to deal with the imbalance of fetal health data. Random forest is used as the model of RFE (recurrent feature elimination) for feature selection, and ANN (artistic neural network) is used as the classifier.

PSO is a commonly used optimization algorithm. The advantage of PSO lies in the simplicity of the algorithm, the speed of searching for the best particles, and the few parameters that need to be set. The objective function of PSO is the accuracy of the model. The parameter 'number of nearest neighbors' in ENN algorithm, the parameter 'select feature number' in RFE algorithm and the number of four hidden layer nodes in ANN are used as variables of the objective function of PSO. For PSO implementation, the optimization goal is to maximize the prediction accuracy according to Eq. 8. In each iteration of PSO, the *gbest* and *pbest* value is updated when a better set of parameters is found. This process repeats until the maximum number of iterations is reached. Finally, the parameter combination found in the above process that can maximize the accuracy of the model is used by the three processes of the model. The composed model shall produce the best prediction result.

4   Experimental Results and Discussion

4.1   Dataset Description

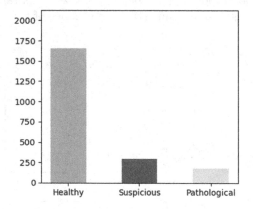

Fig. 2. Data description

We obtain CTG data on fetal health classification from the Kaggle platform [1]. There are 21 features in CTG data, 11 of which are continuous and 10 are discrete. Use three labels (healthy, suspicious, pathological) to classify the condition of the fetus. There are a total of 2126 samples in CTG data. As shown in Fig. 2, there are 1655 healthy fetus samples, 295 suspicious samples, and 176 pathological samples. There is a problem of data imbalance. In addition, we also adopted a correlation analysis, and the results show that some features are found to be highly correlated. Therefore, we will perform imbalance processing and feature selection on the data to solve the above two problems.

4.2   Experiments and Results

The experiment used six machine learning algorithms (logistic regression (LGR) [22], random forest (RF)) [21], support vector machine (SVM) [17], artificial neural network (ANN), naive bayes(NBC) [14], decision tree (DT)) [4] as classifiers. After preprocessing the data, use PSO, genetic algorithm (GA) [18], and differential evolution (DE) [26] as optimization algorithms to optimize parameters of each classifier and compare the results of the method proposed in this paper.

The parameters setting of PSO are followed: The ω is 0.9, the c_1, c_2 are both 2.05, the r_1, r_1 are 0.8 and 0.5 respectively, the number of particles is 50, and the number of iterations is 30.

As shown in Fig. 3, when no processing is done, among the classification algorithms, the effect of random forest is best. Its accuracy and Weight_F1 are 0.9201 and 0.9188 respectively, but kappa measure is only 0.7767, because healthy sample is the majority.

After that, we add feature selection and imbalance processing in the experiment, and use algorithm to optimize the classifier. Obviously, under these processing conditions, it gets a effective promotion in the classification effect. As shown in Fig. 4, Fig. 5 and Fig. 6, three kinds of optimal algorithms we use can assist the classifier to attain the optimal parameters. In these models, the classification accuracy of GA-ANN and DE-ANN, whose accuracy, weight_f1 and kappa are 0.989, 0.989, 0.960, is the best.

When combine the method proposed in this article with each classification algorithm, it can effectively improve the classification effect of each model. As shown in Table 2, compared with the model which is no processing, the accuracy and weight_f1 of Each model increase by 6%–11%. Kappa measure increases by 16%–24%. And in contrast to the model that is only optimized the classifier, it also obtain a bit of increase in the accuracy. The most evident change is the rise of kappa. Each model grows by 3%–11%. The improvement of kappa measure shows that the method proposed in this article can effectively deal with the imbalance problem. We also apply other optimization algorithms except PSO to prove the effectiveness of the method in this passage. Among them, it gets the best effect in the PER-ANN. Its accuracy, weight_f1 and kappa measure are 0.9942, 0.9942 and 0.9783, and the kappa measure is close to 0.98, which shows that the classification

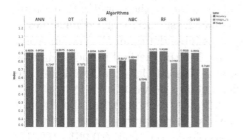

Fig. 3. Single classifier

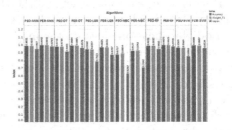

Fig. 4. Comparison of PSO optimization model indexes

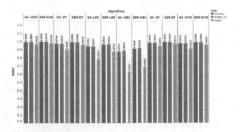

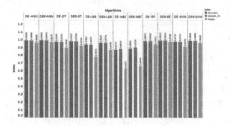

Fig. 5. Comparison of GA optimization model indexes

Fig. 6. Comparison of DE optimization model indexes

method can better deal with the imbalanced data and accurately classify the three categories of the fetal health.

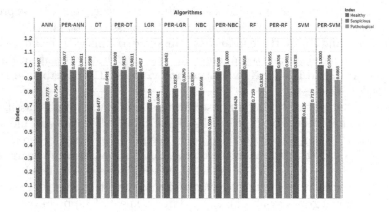

Fig. 7. Classification of the three categories by algorithms

Although the overall classification accuracy of most classifiers is better, the data of majority class are far more than the minority class label, which causes that the overall accuracy does not reflect classifier performance. In order to show the performance of each classifier, use the Recall to indicate the classification status of each classifier. Figure 7 shows that the original classification algorithm achieves a good classification effect on the majority class label data, but the classification of the minority class label data is poor. And using the method proposed in this paper, the classifiers except NBC can classify the minority label data well, and the classification effect of random forest, artificial neural network, and decision tree on the three types of label data is relatively balanced. Among them, The overall performance of artificial neural network is the best, and the classification accuracy of the three categories (healthy, suspicious, pathological) is 0.9977, 0.9615, 0.9811.

Table 2. Comparison of classification accuracy

	Algorithms	Accuracy	F1_score	Kappa
Single	LGR	0.8934	0.8947	0.7080
	RF	0.9201	0.9188	0.7767
	SVM	0.9028	0.8991	0.7185
	ANN	0.9028	0.9038	0.7347
	NBC	0.8072	0.8244	0.5545
	DT	0.9075	0.9051	0.7371
PSO	PSO-LGR	0.9382	0.9377	0.7823
	PSO-RF	0.9831	0.9832	0.9415
	PSO-SVM	0.9569	0.9569	0.8469
	PSO-ANN	0.9831	0.9831	0.9407
	PSO-NBC	0.8727	0.8854	0.6267
	PSO-DT	0.9738	0.9739	0.9091
	PER-LGR	**0.9625**	**0.9631**	0.8707
	PER-RF	**0.9925**	**0.9925**	**0.9738**
	PER-SVM	0.9869	0.9867	0.9534
	PER-ANN	**0.9942**	**0.9942**	**0.9783**
	PER-NBC	**0.9177**	**0.9243**	**0.7042**
	PER-DT	**0.9884**	**0.9885**	**0.9576**
GA	GA-LGR	0.9382	0.9377	0.7823
	GA-RF	0.9831	0.9832	0.9415
	GA-SVM	0.9757	0.9752	0.9125
	GA-ANN	0.9888	0.9887	0.9608
	GA-NBC	0.8727	0.8854	0.6267
	GA-DT	0.9700	0.9697	0.8939
	GER-LGR	**0.9625**	**0.9631**	**0.8713**
	GER-RF	0.9922	0.9923	0.9713
	GER-SVM	**0.9888**	**0.9886**	**0.9601**
	GER-ANN	0.9940	0.9939	0.9758
	GER-NBC	0.9096	0.9181	0.6887
	GER-DT	**0.9884**	**0.9885**	**0.9576**
DE	DE-LGR	0.9382	0.9377	0.7823
	DE-RF	0.9831	0.9832	0.9415
	DE-SVM	0.9757	0.9752	0.9125
	DE-ANN	0.9888	0.9886	0.9600
	DE-NBC	0.8727	0.8854	0.6267
	DE-DT	0.9700	0.9703	0.8966
	DER-LGR	**0.9625**	**0.9631**	0.8707
	DER-RF	**0.9925**	**0.9925**	**0.9738**
	DER-SVM	**0.9888**	**0.9886**	0.9600
	DER-ANN	0.9906	0.9906	0.9670
	DER-NBC	0.8895	0.9026	0.6668
	DER-DT	0.9787	0.9790	0.9234

"Single" means no data preprocessing and optimization. "PSO-LGR", "GA-LGR", "DE-LGR", etc. means PSO, GA, DE only optimizes the classifier, as are other algorithms. "PER-LGR", "GER-LGR", "DER-LGR", etc. means PSO, GA, DE combined with the methods proposed in this article, as are other algorithms.

4.3 Discussion

This article proposes a Particle Swarm Optimization (PSO) based ensemble recursive feature elimination ANN. The model uses the pso optimization algorithm to optimize imbalanced data processing, feature selection and model parameters at the same time to improve the accuracy of model classification. Compared with the model classification result without any processing, it has a positive effect on the improvement of model classification accuracy. Besides, other algorithm is used to replace the PSO, but the result is that the effect of PER-ANN is better than other models. Its classification accuracy of the test set reaches 0.9942, which can meet the needs of hospital fetal health diagnosis. According to the experiment, we conclude that PSO is suitable for multi-parameter model and gain a better effect. During the experiment, the accuracy of PER-ANN reached 0.997. But, it is a pity that we cannot extract the parameter.

This experiment also has certain limitations. Although the classification accuracy of PER-ANN ensemble optimized model can reach 0.9942, its classification accuracy of suspect and pathological is only 0.9615 and 0.9811, which is slightly lower than the overall accuracy. The ultimate cause is the imbalanced data sets, where the total quantity of "suspicious" and "pathological" is twice less than that of "Healthy". We have dealt with the problem, but it still influences the classification consequence. Thus, subsequent research in this area can be carried out. Consider categorizing suspicious and pathological into one category. Firstly classify 'healthy' and this category, and then classify 'suspicious', 'pathological' in detail to improve the classification accuracy of 'suspicious' and 'pathological'.

Certainly, it is practicable to seek other relevant variables to enhance the classification effect. The examination index of the pregnant women is thinkable.

5 Conclusions

In this experiment, the classification accuracy of LGR, RF, ANN, SVM, NBC, DT models after optimized is calculated, and it concludes that PER-ANN gets the best classification result, which is used in the classification of fetal health and obtain an accuracy of 0.9942 and a kappa value of 0.9783 Therefore, Particle Swarm Optimization (PSO) based ensemble recursive feature elimination ANN can accurately predict the health of the fetus, assist doctors in the hospital in making decisions, provide scientific data analysis for them, and improve the quality of hospital visits.

Acknowledgements. This study is supported by National Natural Science Foundation of China (71901150, 71702111, 71971143, 71901152), the Natural Science Foundation of Guangdong Province (2020A151501749), Shenzhen University Teaching Reform Project (Grants No. JG2020119) as well as Guangdong Basic and Applied Basic Research Foundation (Project No. 2019A1515011392).

References

1. de Campos, D.A., et al.: Sisporto 2.0: a program for automated analysis of cardiotocograms. J. Maternaletal Med. **9**(5), 311–18 (2000)
2. Bach, M., Werner, A., Ywiec, J., Pluskiewicz, W.: The study of under- and oversampling methods' utility in analysis of highly imbalanced data on osteoporosis. Inf. Sci. **384**, 174–190 (2016)
3. Brezočnik, L., Fister, I., Podgorelec, V.: Swarm intelligence algorithms for feature selection: a review. Appl. Sci. **8**(9), 1521 (2018)
4. Chang, C.L., Chen, C.H.: Applying decision tree and neural network to increase quality of dermatologic diagnosis. Exp. Syst. Appl. **36**(2 Part 2), 4035–4041 (2009)
5. Wei, J., et al.: Imbalanced cardiotocography multi-classification for antenatal fetal monitoring using weighted random forest. In: Chen, H., Zeng, D., Yan, X., Xing, C. (eds.) ICSH 2019. LNCS, vol. 11924, pp. 75–85. Springer, Cham (2019). https://doi.org/10.1007/978-3-030-34482-5_7
6. Das, S., Mukherjee, H., Obaidullah, S.M., Roy, K., Saha, C.K.: Ensemble based technique for the assessment of fetal health using cardiotocograph – a case study with standard feature reduction techniques. Multimedia Tools Appl. **79**(47), 35147–35168 (2020). https://doi.org/10.1007/s11042-020-08853-2
7. Ersen, Y., Kilikçier, K.: Determination of fetal state from cardiotocogram using LS-SVM with particle swarm optimization and binary decision tree. Comput. Math. Meth. Med. **2013**, 487179 (2013)
8. Guyon, I., Weston, J., Barnhill, S., Vapnik, V.: Gene selection for cancer classification using support vector machines. Mach. Learn. **46**(1–3), 389–422 (2002)
9. Huang, M.L., Hsu, Y.Y.: Fetal distress prediction using discriminant analysis, decision tree, and artificial neural network. J. Biomed. Sci. Eng. **05**(9), 526–533 (2012)
10. Kadhim, N., Abed, J.K.: Enhancing the prediction accuracy for cardiotocography (CTG) using firefly algorithm and Naive Bayesian classifier. IOP Conf. Ser. Mater. Sci. Eng. **715**(1), 012101 (2020)
11. Nguyen, B.H., Xue, B., Zhang, M.: A survey on swarm intelligence approaches to feature selection in data mining. Swarm Evol. Comput. **54**, 100663 (2020)
12. Ocak, H.: A medical decision support system based on support vector machines and the genetic algorithm for the evaluation of fetal well-being. J. Med. Syst. **37**(2), 9913 (2013)
13. Ohno, Y., et al.: Assessment of fetal heart rate variability with abdominal fetal electrocardiogram: changes during fetal breathing movement. Asia-Oceania J. Obstet. Gynaecol. **12**(2), 301–304 (2010)
14. Rana, R., Pruthi, J.: Naive Bayes classification (2014)
15. Richhariya, B., Tanveer, M., Rashid, A.H.: Diagnosis of Alzheimer's disease using universum support vector machine based recursive feature elimination (USVM-RFE). Biomed. Sig. Process. Control **59**, 101903 (2020)
16. Sahin, H., Subasi, A.: Classification of the cardiotocogram data for anticipation of fetal risks using machine learning techniques. Appl. Soft Comput. **33**(C), 231–238 (2015)
17. Saunders, C., et al.: Support vector machine. Comput. Sci. **1**(4), 1–28 (2002)
18. Shah, S., Kusiak, A.: Cancer gene search with data-mining and genetic algorithms. Comput. Biol. Med. **37**(2), 251–261 (2007)
19. Signorini, M.G., Magenes, G., Cerutti, S., Arduini, D.: Linear and nonlinear parameters for the analysis of fetal heart rate signal from cardiotocographic recordings. IEEE Trans. Biomed. Eng. **50**(3), 365–374 (2003)

20. Smith Jr., J.F.: Fetal health assessment using prenatal diagnostic techniques. Curr. Opin. Obstet. Gynecol. **20**(2), 152–156 (2008)
21. Statistics, L.B., Breiman, L.: Random forests. In: Machine Learning, pp. 5–32 (2001)
22. Steyerberg, E.W., Eijkemans, M., Harrell, F.E., Habbema, J.: Prognostic modeling with logistic regression analysis: in search of a sensible strategy in small data sets. Med. Decis. Making **21**(1), 45–56 (2001)
23. Subasi, A., Kadasa, B., Kremic, E.: Classification of the cardiotocogram data for anticipation of fetal risks using bagging ensemble classifier. Procedia Comput. Sci. **168**, 34–39 (2020)
24. Sylvester, E.V., et al.: Applications of random forest feature selection for fine-scale genetic population assignment. Evol. Appl. **11**, 153–165 (2018)
25. Wheeler, T., Gennser, G., Lindvall, R., Murrills, A.J.: Changes in the fetal heart rate associated with fetal breathing and fetal movement. BJOG: Int. J. Obstet. Gynaecol. **87**(12), 1068–1079 (2010)
26. Zhao, F.Q., Zou, J.H., Yang, Y.H.: A hybrid approach based on artificial neural network (ANN) and differential evolution (DE) for job-shop scheduling problem. Appl. Mech. Mater. **26–28**, 754–757 (2010)
27. Zhou, Q., Hao, Z., Zhou, Q., Fan, Y., Luo, L.: Structure damage detection based on random forest recursive feature elimination. Mech. Syst. Sig. Process. **46**(1), 82–90 (2014)
28. Zou, J., Han, Y., So, S.S.: Overview of artificial neural networks. Meth. Mol. Biol. **458**(458), 15 (2009)

Evolutionary Ontology Matching Technique with User Involvement

Xingsi Xue[1,2]([✉]) [iD], Chaofan Yang[1,2] [iD], Wenyu Liu[1,2] [iD], and Hai Zhu[3] [iD]

[1] Intelligent Information Processing Research Center, Fujian University of
Technology, Fuzhou 350118, Fujian, China
[2] School of Computer Science and Mathematics, Fujian University of Technology,
Fuzhou 350118, Fujian, China
[3] School of Network Engineering, Zhoukou Normal University, Zhoukou 466001,
Henan, China

Abstract. Ontology matching is able to identify the entity correspondences between two heterogeneous ontologies, which is an effective method to solve the data heterogeneous problem on the Semantic Web. Traditional fully-automatic ontology matching techniques suffers from the limitation of similarity measure, whose alignment's quality can not be ensured. To overcome this drawback, in this work, an Evolutionary Ontology Matching technique with User Involvement (EOM-UI) is proposed, which utilizes both the Compact Evolutionary Algorithm and user knowledge to improve the algorithm's performance and the alignment's quality. In addition, an optimization model is established to formally define the ontology entity matching problem, and an efficient interacting strategy is proposed to reduce the user's workload and maximize his working value. The experiment uses Ontology Alignment Evaluation Initiative (OAEI)'s benchmark to test our proposal's performance. The experimental results show that our approach is able to make use of the user knowledge to improve the alignment's quality, and it also outperforms OAEI's participants.

Keywords: Ontology matching · Interaction · Compact Evolutionary algorithm

1 Introduction

The sharing and integration of resources in Semantic Web (SW) is hindered by the data heterogeneity problem largely. Ontology is recognized as an effective method to solve data heterogeneity problem through the standardized concept specification. However, due to the heterogeneity of ontology itself, it is not always satisfactory to use ontology to solve the data heterogeneity problem. Ontology

This work is supported by the Natural Science Foundation of Fujian Province (No. 2020J01875) and the National Natural Science Foundation of China (Nos. 61801527 and 61103143).

Y. Tan and Y. Shi (Eds.): ICSI 2021, LNCS 12690, pp. 313–320, 2021.
https://doi.org/10.1007/978-3-030-78811-7_30

matching, a method of finding the semantic relationship between heterogeneous entities in different ontologies, is the great solution of ontology heterogeneity [16]. At the same time, many studies have been done in various field and a number of intelligent algorithms emerged to address the complex problems [2–5,7]. Then, some of these algorithms have been applied to address the ontology matching issue, i.e. many Swarm Intelligence Algorithms (SIAs) [9–12,17–20] have been developed to solve complex and large-scale ontology matching problems. In recent years, researchers has put their interests in the interactive ontology matching technique, which aims at improving the alignment quality through get the user involved in the automatically matching process [15], i.e. Interactive Evolutionary Tabu Search Algorithm (IETSA) [14], Interactive Compact Hybrid Evolutionary Algorithm (ICHEA) [21]. However, Due to the limitation of the similarity measure and the scalability of the algorithm, it is difficult to improve the quality of the alignments obtained by EAs based ontology matching techniques, especially when the heterogeneity is high and the scale of the entities is very large. This work proposes an Evolutionary Ontology Matching technique with User Involvement (EOM-UI), which allows the user to take part in the evolving process and guide EA's search direction. The contributions made in this paper are listed as follows: (1) An optimization model is established to formally define the ontology entity matching problem; (2) An IEA is proposed to interactively match the ontologies, which is able to improve the alignment's quality; (3) An efficient interacting strategy is proposed, which is able to reduce the user's workload and maximize his working value. The rest of the article is arranged according to the following structure: the second section is preparatory knowledge; the third section is the method proposed in this paper; the fourth section is experiment and discussion; and the last section is conclusion.

2   Preliminaries

An ontology O is considered to be a five-tuple [1] whose elements are class, properties, instance, axioms, and annotations, which are represented by C, P, I, Ax and An respectively. Ontology matching is a function, which has five inputs, i.e. the source ontology O_1, the target ontology O_2, a partial matching results A', the parameters P, and the external resource R, and one ouput, i.e. a matching result A to be obtained. The alignment A is a set, and each element in A is called a correspondence C, which is a four-tuple whose elements includes an entity e_1 in the source ontology, an entity e_2 in the target ontology, the similarity value n and the relationship r between e_1 and e_2 [13].

3   Methodology

The ontology entity matching issue is addressed by ICEA in this work. Compared with the original EA, the proposed CEA has higher efficiency, and the user's knowledge is introduced into the evolving process to guide the algorithm's searching direction. The flow chart of this algorithm is shown in Fig. 1.

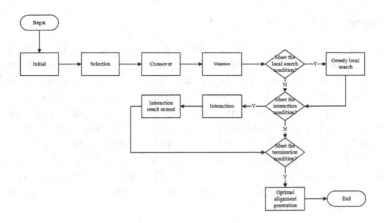

Fig. 1. The flow chart of interactive rvolutionary algorithm

3.1 Ontology Entity Matching Problem

Given an alignment A, its quality is measured by the following equation:

$$f(A) = 2 \cdot (\beta \cdot \phi(|A|) + (1 - \beta) \cdot g(A)) \tag{1}$$

where $\beta \in [0, 1]$ is the a aggregating weight, which represents the relative weight of $\phi(|A|)$ and $g(A)$. A is the final alignment. $\phi(|A|)$ and $g(A)$ are two indexes to evaluate the final alignment. Their expressions are defined as follows:

$$\phi(|A|) = \frac{|A|}{min(O_1, O_2)} \tag{2}$$

$$f(A) = \frac{\sum_{i=1}^{|A|} \eta_i}{|A|} \tag{3}$$

where $|O_1|$ and $|O_2|$ represent the number of source ontology and target ontology respectively, which is consistent with the above. $|A|$ indicates the number of matching results. η_i represents the similarity of $i - th$ correspondence in the alignment A. On this baisi, the optimization model of ontology entity matching problem is defined as follows:

$$\begin{cases} max \ f(A) \\ s.t. \quad X = (x_1, x_2, ..., x_{|O_1|})^T \\ \quad\quad x_i \in \{1, 2, ..., |O_2|, |O_2| + 1\}, \ i = 1, 2, ..., |O_1| \end{cases} \tag{4}$$

where X is the decision variable whose dimension is $|O_1|$, i.e. the cardinality of source ontology; x_i is the $i-th$ correspondence, which represent the $i-th$ source entity is mapped with the x_{i-th} target entity. In particular, $x_i = -1$ means that no entity in the target ontology is mapped with the $i - th$ entity in the source ontology. Supposing A is the alignment corresponding to X, $f'(X)$ is equal to $f(A)$, and the objective is to maximize its value.

3.2 Interaction Process

When the elite solution keeps unchanged for 20 generations, the interaction is executed and a decision pool is shown to the user. The decision pool is extracted from the alignment generated by the optimal solution in the current population, which is categorized into three parts: the conflicted correspndences P_{cof}, the correspndences that might be wrong P_{wro}, and the correspndences that might be correct but not in the alignment P_{cor}. P_{cof} mean that more than one entity in the source ontology matches the same entity in the target ontology. The determination of conflict correspondence is shown in Algorithm 1, where C_b is the best chromosome in current generation. P_{wro} and P_{cor} are all obtained by comparing the similarity in the similarity matrix, whose determinations are respectively shown in Algorithm 2 and Algorithm 3.

Algorithm 1. Conflicted correspondences pool

1: **for** $(int\ i = 0; i < (|C_b|); i + +)$ **do**
2:    **for** $(int\ j = 0; j < (|C_b|); j + +)$ **do**
3:       **if** $(i \neq j)$ *and* $(C_b(i) = C_b(j))$ **then**
4:          $add[O_1(i), O_1(j), O_2(C_b(i))]$ *into* P_{cof}
5:       **end if**
6:    **end for**
7: **end for**

Algorithm 2. Wrong correspondences pool

1: **for** $(int\ i = 0; i < (|A|); i + +)$ **do**
2:    *entity of* O_1 *in* A_i *is* e_x, *entity of* O_2 *in* A_i *is* e_y
3:    **if** $sim(x, y) < threshold + \delta$ **then**
4:       $add[e_x, e_y]$ *into* P_{wro}
5:    **end if**
6: **end for**

Algorithm 3. Correct correspondences pool

1: **for** $(int\ i = 0; i < (|O_1|); i + +)$ **do**
2:    **for** $(int\ j = 0; j < (|O_2|); j + +)$ **do**
3:       **if** $(sim(x, y) > threshold - \delta)$ *and* $([e_x, e_y]\ not\ in\ A)$ **then**
4:          $add[e_x, e_y]$ *into* P_{cor}
5:       **end if**
6:    **end for**
7: **end for**

---

where *threshold* and δ are the parameters of IEA which initialized in the begining of algorithm. the formula of δ is presented as follow:

$$\delta = min(\delta_t, \delta_0 + \frac{generation}{100})$$ (5)

When all the correspondences in the decision pool are validated by the user, population's elements and the similarity matrix's corresponding elements are modified as 1 for correct correspondence or 0 for incorrect correspondence.

3.3 Validating Result's Propagation

The propagation of validating results can effectively enhance the value of user's knowledge. The motivation here is that, the similarity value between two entities will increase if their neighbor entities are validated as correct correspondence[6]. In the following, we show the formula that updates the correspondence's similarity value during the propagating process:

$$sim_{(i,j)} = \begin{cases} sim_{(i,j)} + \frac{1}{|sub_i| + |sub_j|}, & \text{if their } sub_{entities} \text{ are approved} \\ sim_{(i,j)} + \frac{1}{|sib_i| + |sib_j|}, & \text{if their } sup_{entities} \text{ or } sib_{entities} \text{ are approved} \\ sim_{(i,j)}, & \text{else} \end{cases}$$ (6)

where $sim_{(i,j)}$ represents the similarity between the $i-th$ entity in the source ontology and the j-th entity in the target ontology, $|sub_i|$ and $|sub_j|$ represent the number of sub-entities and $|sib_i|$, $|sib_j|$ represent the number of siblings owned by entity e_i and entity e_j, respectively.

4 Experiment

4.1 Experiment Configuration

The testing case used in this paper is the benchmark provided by the famous OAEI (Ontology Alignment Evaluation Initiative), which is a competition dedicated to evaluating the performance of ontology matching systems. These testing case can be classified into four categories according to the heterogeneity with the default source ontology case, i.e. 101–104, 201–210, 221–247 and 248 266.

The evaluation on ontology alignment usually starts from two aspects, *recall* and *precision*, which represent the completeness and accuracy of the solution respectively. To be specific, they are defined as follows:

$$recall(A) = \frac{|R \bigcap A|}{|R|}$$ (7)

$$precision(A) = \frac{|R \bigcap A|}{|A|}$$ (8)

where A represents the final alignment, and R is the reference alignment (usually provided in the test case set). $|A|$ and $|R|$ represent the cardinality of the final alignment and the reference alignment, respectively. *recall* and *precision* are two important metrics to evaluate the alignment quality, but conflict with each other in most conditions. The balance of them is needed, and the common method is to use $f-measure$, which is the weighted harmonic mean of *recall* and *precision*:

$$f-measure(A) = \frac{recall(A) \cdot precison(A)}{\alpha \cdot recall(A) + (1-\alpha) \cdot precison(A)} \tag{9}$$

where α is a decimal that represents the relative weights of *recall* and *precision*, with values ranging from 0 to 1. In this work, the value of α is set as 0.5 to prefer none of them.

4.2   Results and Analysis

We compare IEA based matching technique among EA based matching technique, i.e. its non-interactive version, and OAEI's participants to testing the performance of IEA based matching technique. The results of the control experiment are shown in Fig. 2, which compares IEA with EA. As can be seen from the figure, the matching system with user interaction has a significant improvement in each type of heterogeneous test case compared with the original system, thus verifying the effectiveness of the interaction.

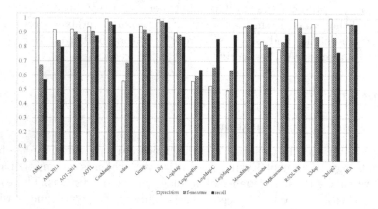

Fig. 2. Comparison between EA and IEA based ontology matching techniques in terms of f-measure, precision and recall

Further, we compare IEA with the participants of OAEI in 2014, 2015 and 2016 in terms of the overall mean value, as shown in Fig. 3. It can be found that IEA is better than most of the participants in the overall mean, and is equal to a few participants (CroMatch and Lily[8]), which shows the effectiveness of IEA.

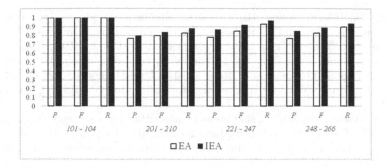

Fig. 3. Comparison among IEA based ontology matching technique and OAEI's participants

5   Conclusion and Future Work

EA based ontology matching technique is a ground methodology for solving the problem of ontology heterogeneity, however, due to the limitation of similarity measure, it is difficult to determine high-quality alignments by using the fully-automated ontology matching technique. To overcome this drawback, an IEA is proposed in this paper, which gets the user to participate the EA's evolving process and uses his knowledge to guide the algorithm's searching direction. The experimental results show that IEA based ontology matching technique can significantly improve the alignment's quality, which outperforms other state-of-the-art ontology matching techniques. In the future, we are interested in applying IEA to match the ontologies in different domains to further test its performance. When dealing with large-scale matching tasks, the scalability of our approach should be taken into consideration. Since the user is not able to deal with large quantity of problematic correspondences, it is necessary to further reduce the workload of user and improve the efficiency of interacting process.

References

1. Acampora, G., Loia, V., Vitiello, A.: Enhancing ontology alignment through a memetic aggregation of similarity measures. Inf. Sci. **250**, 1–20 (2013)
2. Chen, C.H.: An arrival time prediction method for bus system. IEEE Internet Things J. **5**(5), 4231–4232 (2018)
3. Chen, C.H.: A cell probe-based method for vehicle speed estimation. IEICE Trans. Fund. Electron. Commun. Comput. Sci. **103**(1), 265–267 (2020)
4. Lin, J.C.W., Shao, Y., Djenouri, Y., Yun, U.: Asrnn: a recurrent neural network with an attention model for sequence labeling. Knowl.-Based Syst. **212**, 106548 (2021)
5. Lin, J.C.W., Srivastava, G., Zhang, Y., Djenouri, Y., Aloqaily, M.: Privacy preserving multi-objective sanitization model in 6G IoT environments. IEEE Internet Things J. (2020)

6. Melnik, S., Garcia-Molina, H., Rahm, E.: Similarity flooding: a versatile graph matching algorithm and its application to schema matching. In: Proceedings 18th International Conference on Data Engineering, pp. 117–128. IEEE (2002)

7. Pan, J.S., Song, P.C., Chu, S.C., Peng, Y.J.: Improved compact cuckoo search algorithm applied to location of drone logistics hub. Mathematics 8(3), 333 (2020)

8. Wang, P., Wang, W.: Lily results for OAEI 2016. In: OM@ISWC, pp. 178–184 (2016)

9. Xue, X., Wang, Y., et al.: Using memetic algorithm for instance coreference resolution. IEEE Trans. Knowl. Data Eng. **28**(2), 580–591 (2016)

10. Xue, X., Chen, J., Pan, J.: Evolutionary algorithm based ontology matching technique (2018)

11. Xue, X.: A compact firefly algorithm for matching biomedical ontologies. Knowl. Inf. Syst. **62**(7), 2855–2871 (2020). https://doi.org/10.1007/s10115-020-01443-6

12. Xue, X., Chen, J.: Optimizing ontology alignment through hybrid population-based incremental learning algorithm. Memetic Comput. **11**(2), 209–217 (2018). https://doi.org/10.1007/s12293-018-0255-8

13. Xue, X., Chen, J.: Using compact evolutionary tabu search algorithm for matching sensor ontologies. Swarm Evol. Comput. **48**, 25–30 (2019)

14. Xue, X., Hang, Z., Tang, Z.: Interactive biomedical ontology matching. PloS one **14**(4), e2015147 (2019)

15. Xue, X., Liu, J.: Collaborative ontology matching based on compact interactive evolutionary algorithm. Knowl.-Based Syst. **137**, 94–103 (2017)

16. Xue, X., Pan, J.S.: An overview on evolutionary algorithm based ontology matching. J. Inf. Hiding Multimed. Signal Process **9**, 75–88 (2018)

17. Xue, X., Wang, Y.: Optimizing ontology alignments through a memetic algorithm using both matchfmeasure and unanimous improvement ratio. Artif. Intell. **223**, 65–81 (2015)

18. Xue, X., Wang, Y.: Using memetic algorithm for instance coreference resolution. IEEE Trans. Knowl. Data Eng. **28**(2), 580–591 (2015)

19. Xue, X., Wu, X., Jiang, C., Mao, G., Zhu, H.: Integrating sensor ontologies with global and local alignment extractions. Wirel. Commun. Mob. Comput. **2021** (2021)

20. Xue, X., Yang, C., Jiang, C., Tsai, P.W., Mao, G., Zhu, H.: Optimizing ontology alignment through linkage learning on entity correspondences. Complexity **2021** (2021)

21. Xue, X., Yao, X.: Interactive ontology matching based on partial reference alignment. Appl. Soft Comput. **72**, 355–370 (2018)

Sequential Stacked AutoEncoder-Based Artificial Neural Network and Improved Sheep Optimization for Tool Wear Prediction

Fei Ding, Mingyan Jiang[✉], Dongfeng Yuan, Falei Ji, and Haiyan Yu

School of Information Science and Engineering, Shandong University,
Qingdao 266237, China
jiangmingyan@sdu.edu.cn

Abstract. Real-time tool wear prediction is one of the key problems to be solved in the field of intelligent manufacturing. In recent years, with the help of powerful feature extraction capability of AutoEncoder(AE), the Stacked AutoEncoder-based Deep Neural Network(SAE-DNN) model has achieved good results on tool wear prediction. However, the SAE-DNN model ignores the time series characteristics between original samples. Besides, training the neural network with a gradient-based algorithm is prone to fall into local optimum. To solve the problems in the SAE-DNN, we propose a new improved model: Sequential Stacked AutoEncoder-based Artificial Neural Network(SSAE-ANN). First, the SSAE-ANN model uses the Sequential Stacked AutoEncoder(SSAE), which can not only fuse the features between different channels of a single sample, but also extract the time series features between adjacent samples. Second, the SSAE-ANN model uses the Improved Sheep Optimization(ISO) algorithm to train the neural network. Compared with the gradient algorithm, the ISO algorithm has stronger robustness and global optimization ability. At the end of the paper, we perform experiments on PHM2010 dataset and verify the superiority of SSAE-ANN model by comparing with other algorithms.

Keywords: Autoencoder · Neurel network · Feature selection fusion · Tool wear · Sheep optimization algorithm

1 Introduction

With the transformation of modern mechanical processing industry to digitalization, Computer Numerical Contral(CNC) cutting tools as an important part of modern digital machining technology, its state directly affects the quality of machining products and the reliability and stability of machining system operation. Therefore, the application of tool wear status monitoring technology in modern digital machining systems is of great significance [1, 2].

© Springer Nature Switzerland AG 2021
Y. Tan and Y. Shi (Eds.): ICSI 2021, LNCS 12690, pp. 321–330, 2021.
https://doi.org/10.1007/978-3-030-78811-7_31

There are two main methods to detect tool wear: direct methods and indirect methods. The direct methods are to use sensors, such as proximity sensors, radioactive sensors and visual sensors, which can directly obtain accurate dimensional changes. These works are usually performed when the processing system is offline [3]. The indirect methods obtain tool wear information through various sensor signals, such as cutting force, torque, vibration, acoustic emission, spindle power, current and surface roughness. Compared with direct methods, they can be measured while the system is running, and they are easier to install and implement, but the predicted tool wear value is not as accurate as direct methods [4].

Indirect tool wear measurement refers to the method of monitoring tool wear value indirectly by monitoring tool status signal in machining process. Early researchers directly established physical models for the degradation mechanism of specific equipment [5], but with the increasing complexity of machining equipment, data-driven models based on machine learning have gradually become the mainstream of academic research [6]. Machine learning methods generally include the following steps: signal collection, signal preprocessing, feature extraction and selection, model prediction [7]. After preprocessing the signal collected by the sensor, the time-domain, frequency-domain and time-frequency-domain features are extracted by Fourier transform and wavelet transform to form high-dimensional feature vector. Due to the problems of dimensionality curse[8] and redundant features, it is necessary to reduce the dimension of original feature vectors and select features.

In recent years, more and more researchers began to use the Stacked AutoEncoder based Deep Neural Networks(SAE-DNN) model to deal with the predictive maintenance of machine tools [9, 10], SAE-DNN model is a two-stage algorithm: in the first stage, Stacked AutoEncoder is used to extract low-dimensional fusion features from original high-dimensional feature vector, and in the second stage, deep neural network is used for fitting.

However, there are still deficiencies in these methods.

(1) In the first stage of SAE-DNN, SAE is used to extract low-dimensional features from high-dimensional features of a single sample, but the temporal dependence between samples is ignored. The features extracted by SAE-DNN are only abstracted at the single sample level, so the inherent time series characteristics of the data can not be extracted.

(2) In the second stage of SAE-DNN model, DNN is used as the fitter and BP algorithm is used for training. DNN based on gradient optimization is prone to problems such as gradient disappearance, gradient explosion and Interval Covariate Shift(ICS) [11]. At the same time, since SAE has extracted high-level abstract features in the first stage, it is easy to produce over fitting problems when using DNN(a large VC dimension model), in the second stage.

In order to solve the problems of SAE-DNN, we propose a new model: Sequential Stacked AutoEncoder-based Artificial Neural Network(SSAE-ANN). In the first stage of SSAE-ANN, Sequential Stacked AutoEncoder is used to extract the temporal features between samples; in the second stage, because a simple single hidden layer neural network can approximate any continuous function on a compact set [12,13], to reduce the VC dimension of the model, we use a single hidden layer network as a fitter. At the same time, for the BP algorithm in shallow network training that is sensitive to initial values and easy to fall into local extreme values, we use a swarm intelligence optimization algorithm with stronger robustness and global optimization capabilities to train single hidden layer networks. In this paper, we choose the Sheep Optimization(SO) algorithm which is excellent in recent years as the base algorithm, and propose the Improved Sheep Optimization(ISO) algorithm according to the shortcomings of the herding algorithm. The structure of this paper is as follows: Sect. 1 introduces the background and significance of tool wear prediction; Sect. 2 introduces the improved sheep optimization algorithm, and verifies the performance of ISO algorithm through test functions; Sect. 3 introduces the specific structure of SSAE-ANN; Sect. 4 introduces the specific details of the experiment and comparative analysis of the results; Sect. 5 is the conclusion of this paper.

2   Improved Sheep Optimization Algorithm

2.1   Basic Sheep Algorithm

The basic sheep algorithm [14] consists of three parts. The first part is leading by the leader. The individual with the best fitness value is selected as the leader, and all individuals approach the leader. The second part is the flock interaction. Each sheep randomly selects another sheep from the entire flock for comparison. If the latter is in a better position, move to it, otherwise, move away. The third part is the supervision of shepherd dog. When the difference between the fitness value of the leader of this generation and the previous generation is less than a prescribed threshold, it is judged that the group may fall into a local optimum. By resetting some individuals, it is possible to jump out of the local optimum.

2.2   Bicentric Individuals

In the leading part of the sheep algorithm, all sheep approach the leader, and individual sheep in the group will be attracted to the area near the global extremum, so the global extremum plays a vital role in the entire search process. Compared with the global extremum, the overall center of the sheep population and the individual extremum centers of all sheep may be closer to the optimal solution. To improve the global extremum, we introduce the General Central Sheep (GCS) and the Special Central Sheep (SCS). x^{GCS} and x^{SCS} respectively represent the individual extreme center of all sheep individuals and the overall center of the sheep population. These two individuals are essentially the same as other individuals, such as participating in the cooperation and competition of the flock, and comparing the advantages and disadvantages between individuals. The most important one is to participate in the global extreme value competition. The update method of the two central individuals is:

$$x^{GCS} = \frac{1}{n-2} \left(\sum_{i=1}^{n-2} pbest_i \right), \tag{1}$$

$$x^{SCS} = \frac{1}{n-2} \left(\sum_{i=1}^{n-2} x_i \right), \tag{2}$$

$$f(gbest) = \min\left(f(pbest_1), f(pbest_2), \cdots, \right.$$
$$\left. f(pbest_{n-2}), f(x^{GCS}), f(x^{SCS})\right). \tag{3}$$

Although these two individual sheep may be insignificant compared to the overall number of the entire flock, these two central individuals may have a very important impact on the global extremum of the population. By improving the global extremum, the quality of each individual sheep in the flock is greatly improved.

2.3 Combined with Dfferential Evolution Algorithm

For the flock interaction part of the sheep algorithm, three evolutionary operators of mutation, crossover and selection in the differential evolution algorithm are introduced.

Mutation operation: the basic idea is to start with a randomly generated initial population, use the difference vector of two individuals randomly selected from the population as the source of random change for the third individual, and then add the difference vector to the third individual according to certain rules to generate mutation individuals.

$$v_i = x_{r1} + F(x_{r2} - x_{r3}), \tag{4}$$

where r_1, r_2 and r_3 are integers between $[1, N]$ that is different from i and different from each other, and $F \in (0, 1)$ is the scaling factor.

Crossover operation: According to a certain probability, crossover operation is performed between the random vector individual v_i obtained by the mutation and the parent individual x_i to generate a new individual u_i.

$$u_{ij} = \begin{cases} v_{ij}, & rand \leq C_R \text{ or } j = rn_j \\ x_{ij}, & else \end{cases}, \tag{5}$$

where x_{ij}, v_{ij} and u_{ij} respectively represent the jth dimension component of the solution vector, $C_R \in [0, 1]$ is the crossover probability, $rn_j \in [0, NP]$, $rand$ is the random number between $[0, 1]$.

Selection operation: The selection operation adopts a "greedy" strategy to ensure that individuals with better fitness enter the next generation. Comparing u_i with x_i, the better of the two will be retained and evolve to the next generation.

2.4   Performance Test of Improved Algorithm

Select eight standard test functions for testing, and the parameter settings are as follows:

The number of individuals in the SO algorithm, the ISO algorithm and the Particle Swarm Optimization (PSO) algorithm are all 50. The reset probability of shepherd dog supervision is 0.2, the initial value of mutation operator F is 0.4, and the crossover probability is 0.1. The dimensions of the test functions are all 100, and the number of iterations is 10,000. Each algorithm runs 30 times independently, and the results are shown in Fig. 1.

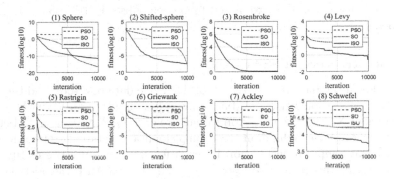

Fig. 1. Convergence curve of PSO, SO, ISO on benchmark problems over 30 runs.

It can be seen from Fig. 1 that compared with the original SO algorithm and PSO algorithm, the ISO algorithm has greatly improved its convergence speed and accuracy in the process of optimizing multiple commonly used test functions.

The leading part of the sheep algorithm represents the rapid approach of the individual herd to the position of the global extremum, which makes the sheep algorithm have better search efficiency. However, the global extremum at this time is likely to be a local extremum, and the herd as a whole falls into the wrong direction. The introduction of bicentric individuals effectively improves the quality of the global extremum, thereby affecting the quality of the entire flock. For high-dimensional optimization problems, the limited number of herd individuals is not enough to represent the diversity of the entire search space. At this time, it is particularly important to maintain and be able to increase the diversity of the population. Combining the sheep algorithm with the differential evolution algorithm can fully increase the diversity of the herd and further enhance the algorithm's global search capabilities.

3   Sequential Stacked Autoencoder-Based Artificial Neural Network

3.1   Database

The experimental data comes from the open data of cutting tool health prediction competition of high-speed CNC machine tools of New York Prediction and Health Management Association(PHM2010) [15]. C1 and C4 knife data sets with labels are selected as experimental data, in which C1 and half of C4 samples are used as training sets and the other half of C4 is used as test sets. The data set of each knife contains 315 samples, and each sample contains 7 sensor signals, among which 3 vibration signals are selected for feature extraction, and each signal extracts 10 features in time domain, frequency domain and time-frequency domain for wear value prediction. We extract features of the Root-Mean-Square(RMS), Variance, Maximum, Skewness, Kurtosis and Peak-to-Peak in the time domain, the Spectral Skewness, Spectral Kurtosis and Spectral Power in the frequency domain, the Wavelet Energy in the time-frequency domain. The dimension of feature matrix extracted from each subset is $315 \times 3 \times 10$ because of the 3 signals.

3.2   The Structure of SSAE-ANN

The samples in the data set are the whole life cycle data of a tool, and there is a time series correlation between adjacent samples. In addition, for each sample, the features of X, Y and Z vibration signals are extracted, and the data between different sensors have redundant and complementary data. Therefore, in order to extract the temporal correlation between features and fuse the feature data of multi-sensors, this paper proposes Sequential Stacked Auto Encoder-based to extract and fuse the original features, and at the same time complete the task of feature dimension reduction. We stack the extracted features. The process is shown in Fig. 2. Firstly, the three sensor features of each sample are vertically superimposed, and a feature matrix with a size of 315×30 is obtained to extract the correlation between multiple sensors. Secondly the features of the adjacent samples are stacked to extract the temporal correlation. Finally we obtain the stacked feature matrix, and its dimension is 315×60.

We stack the extracted features and input them into SSAE to further extract the correlation features of multi-sensor signals and the temporal correlation features between adjacent samples, and at the same time realize feature dimension reduction. The structure of SSAE is shown in Fig. 3. The whole framework is in the form of AutoEncoder, and AutoEncoder is constructed by superposition of linear layers. The data dimension size of each layer is shown in detail in the Fig. 3.

The SSAE is trained by the data of the test set, and the data of the training set and the test set are input into the trained model, and the results of the intermediate hidden layer are taken as the output, so that the original features can be fused with multi-sensor data, the time series correlation between samples can be extracted, and the feature dimension can be reduced. Finally, each sample can get a 5-dimensional feature vector for subsequent wear value prediction.

3.3   Single Hidden Layer Network Trained by ISO

Single hidden layer neural network can approximate any continuous function on a compact set and prevent the over-fitting on small data sets because of the

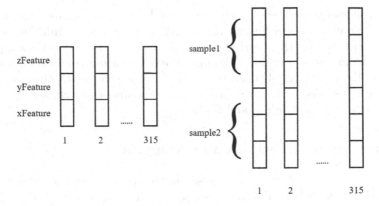

Fig. 2. Combine the features of adjacent samples into a feature vector containing sequential information.

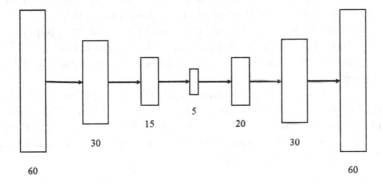

Fig. 3. The specific structure of SSAE model.

small structural complexity. Back propagation is sensitive to the initial values of network parameters and easily falls into local optimum, so we use swarm intelligence optimization algorithm with stronger robustness and global optimization ability to train single hidden layer network. The training process of the ANN-ISO model is as follows:

(1) Determine the structure of neural network. This process includes the number of nodes in the input layer, hidden layer and output layer of neural network. According to the structure of neural network, the dimension of individuals in Improved Sheep Optimization algorithm is determined.
(2) Determine the fitness function. The fitness function is used as the evaluation standard for every position update of flock, and the training error of neural network is used as the fitness value of function in this model.
(3) Initialize the population. According to the neural network structure, the flock is initialized randomly, and each sheep individual represents a set of the weights and biases of neural networks.

(4) The randomly generated sheep individuals are brought into the network for calculation. According to the fitness value, a new population is generated, and the new population is brought into the neural network again for calculation and iterated in turn until the termination condition is met.

(5) The optimal individual generated by the Improved Sheep Optimization algorithm is decoded as the weight and bias of neural network, and finally the trained neural network model is obtained.

4   Experimental Results and Analysis

The SSAE-ANN-ISO algorithm proposed in this paper is compared with SAE-DNN, SAE-ANN, SSAE-ANN-BP, Support Vector Regression(SVR), ElasticNET algorithm.

In the first half of the model, the autoencoder network uses the original features and the time series stacked features to train 5000 times, and takes the output of hidden layer as the feature after dimension reduction.

The dimension of input layer of ANN model used in this paper is 5, the dimension of hidden layer is 10 and an output layer dimension of 1. The input layer dimension of DNN model is 5, the hidden layer dimension is 10 and 20, and the output layer dimension is 1.

The super parameter learning rate is 0.0005, the number of iterations is 10000 and the regularization parameter is 0.1. All algorithms run for 30 times, and calculate the Root Mean Squared Error(RMSE) of the predicted value and the actual label value of the training set and test set as the evaluation standard. The results are shown in the Table 1.

Table 1. Performance of SAE-ANN, SAE-DNN, RBF-SVR, ElasticNET, SSAE-ANN-BP, SSAE-ANN-ISO on PHM data set.

Model	Training result	Test result
RBF-SVR	16.0402	23.3708
ElasticNET	15.3882	22.1999
SAE-ANN	17.0355 ± 2.1439	21.2288 ± 4.2732
SAE-DNN	16.3345 ± 1.9907	23.5418 ± 6.7709
SSAE-ANN-BP	15.2608 ± 1.6399	19.5918 ± 3.0882
SSAE-ANN-ISO	12.8021 ± 0.2474	14.4492 ± 0.7596

The following conclusions can be drawn from Table 1: (1) the SSAE-ANN-ISO model proposed in this paper performs best among all algorithms, which proves the effectiveness of SSAE-ANN-ISO model. (2) According to the results of SAE-ANN and SAE-DNN, since the autoencoder has extracted low-dimensional features, the over complex deep network model can achieve good results in the training set, but it does not perform well in the test set, resulting in the problem of over fitting. (3) The SSAE-ANN-BP model proposed in this paper has better results than the SAE-ANN-BP model. The prediction models used by the two methods are the same, and the optimization methods are consistent. The difference lies in the input features after dimensionality reduction of the prediction model. Compared with the features extracted by the simple Stacked AutoEncoder, the features extracted by the Sequential Stacked AutoEncoder proposed by this paper include the time-series correlation features of the vibration signal and the correlation features between multiple sensors, so it can get better prediction results. (4) Compared with the network SSAE-ANN-BP optimized by backpropagation, the network SSAE-ANN-ISO optimized by ISO has better prediction results. The inputs of the prediction models of the two methods are all features extracted by SSAE. The prediction models are both single hidden layer neural networks. The BP neural network is sensitive to the initial value of the network parameters and easily falls into the local optimum. The swarm intelligence optimization algorithm has a stronger global optimization ability and can optimize the network convergence faster and more accurately, so it can get better results.

5   Conclusion

In this paper, we propose a two-phase model SSAE-ANN for tool wear detection. In the first phase of the model, SSAE is used to extract features. Compared with traditional AutoEncoder, SSAE incorporates a time-series structure to extract time-series features from the original samples. In the second phase of the model, we use the ISO algorithm to optimize the single-hidden layer network. Compared with the traditional back-propagation optimization algorithm, the swarm intelligence optimization algorithm can prevent the neural network from falling into the local optimum and has a stronger robustness. Experiments on PHM2010 dataset verify the superiority of SSAE-ANN-ISO algorithm by comparing with RBF-SVR, ElasticNET, SAE-DNN, SAE-ANN and SSAE-ANN-BP algorithms. In future work, we plan to study more effective feature extraction methods and prediction models to get better results.

Acknowledgement. This study is supported by the Shandong Province Science Foundation of China (Grant No.ZR2020MF153) and Key Innovation Project of Shandong Province (Grant No.2019JZZY010111).

References

1. Fatemeh, A., Antoine, T., Marc, T.: Tool condition monitoring using spectral sub traction and convolutional neural networks in milling process. Int. J. Adv. Manuf. Technol. **98**, 3217–3227 (2018)
2. Zhang, C., Yao, X., Zhang, J., Jin, H.: Tool condition monitoring and remaining useful life prognostic based on a wireless sensor in dry milling operations. Sensors (Basel, Switzerland) **16**(6), 795 (2016)

3. Du, D., Zhang, J., Si, X., Hu, C.: Remaining useful life estimation: A review on stochastic process-based approaches. Recent Pat. Eng. **15**(1), 69–76 (2021)
4. Serin, G., Sener, B., Ozbayoglu, A.M., Unver, H.O.: Review of tool condition monitoring in machining and opportunities for deep learning. Int. J. Adv. Manuf. Technol. **109**(3), 953–974 (2020). https://doi.org/10.1007/s00170-020-05449-w
5. Shamsaei, N., Fatemi, A.: Small fatigue crack growth under multiaxial stresses. Int. J. Fatigue **58**(1), 126–135 (2014)
6. Gouarir, A., Martınez-Arellano, G., Terrazas, G., Benardos, P., Ratchev, S.: Inprocess tool wear prediction system based on machine learning techniques and force analysis. Procedia CIRP **77**, 501–504 (2018)
7. Wang, G., Zhang, Y., Liu, C., Xie, Q., Xu, Y.: A new tool wear monitoring method based on multi-scale PCA. J. Intell. Manuf. **30**(1), 113–122 (2016). https://doi.org/10.1007/s10845-016-1235-9
8. Novak, E., Ritter, K.: The Curse of Dimension and a Universal Method For Numerical Integration. In: Nürnberger, G., Schmidt, J.W., Walz, G. (eds) Multivariate Approximation and Splines. ISNM International Series of Numerical Mathematics, vol. 125, pp 177-187. Birkhäuser, Basel (1997). https://doi.org/10.1007/978-3-0348-8871-4_15
9. Chen, L., Wang, Z.Y., Qin, W.L., Ma, J.: Fault diagnosis of rotary machinery components using a stacked denoising autoencoder-based health state identification. Sig. Processing **130**(1), 377–388 (2017)
10. Sun, J., Yan, C., Wen, J.: Intelligent bearing fault diagnosis method combining compressed data acquisition and deep learning. IEEE Trans. Inst. Measur. **67**(99), 185–195 (2017)
11. Arpit, D., Zhou, Y., Kota, B., Govindaraju, V.: Normalization propagation: A parametric technique for removing internal covariate shift in deep networks. In: International Conference on Machine Learning, pp. 1168–1176. PMLR (2016)
12. Cybenko, G.: Approximation by superpositions of a sigmoidal function. Mathematics of Control, Signals and Systems **2**(4), 303–314 (1989). https://doi.org/10.1007/BF02551274
13. Hornik, K.: Approximation capabilities of multilayer feedforward networks. Neural Netw. **4**(2), 251–257 (1991)
14. Qu, D., Xu, L., Lu, Y., Yuan, X., Huang, M., Wang, X.: A new swarm intelligence algorithm for simulating herd behavior. Acta Electron. Sinica **46**(6), 1300–1305 (2018)
15. Wang, J., Xie, J., Zhao, R., Zhang, L., Duan, L.: Multisensory fusion based virtual tool wear sensing for ubiquitous manufacturing. Robot. Comput. Integr. Manuf. **45**, 47–58 (2017)

Application of Internet Plus: TCM Clinical Intelligent Decision Making

Jun Xie[1], Sijie Dang[1], Xiuyuan Xu[2], Jixiang Guo[2], Xiaozhi Zhang[2], and Zhang Yi[2(✉)]

[1] Sichuan Institute of Traditional Chinese Medicine (Sichuan Second Hospital of T.C.M), Chengdu, China
[2] Machine Intelligence Laboratory, College of Computer Science, Sichuan University, Chengdu, China
zhangyi@scu.edu.cn

Abstract. To improve the electronic medical record database for traditional Chinese medicine (TCM) and pass on the experience of distinguished TCM practitioners using artificial intelligence technology for TCM clinical decision making. The clinical intelligent assisted decision-making system independently developed by hospital joint enterprises was used to standardize clinical decision making. To create a decision-making support tool, it adopted the approach of disease and syndrome differentiation of TCM; it also analyzed and modeled the empirical approach of distinguished TCM practitioners. Internet Plus information technology supported TCM clinical decision making and allowed more people to benefit from TCM services. Non-expert doctors can understand the system functions quickly; they can grasp a patient's actual situation more efficiently and issue high-level prescriptions for TCM. Through the development and application of the intelligent decision-making system for TCM, the diagnostic efficiency and ability of doctors has greatly improved. The system has attained its expected goal.

Keywords: Informatization of TCM · Characteristics of TCM · Disease · Syndrome differentiation · Aid decision making

1 Background

With the rapid development of cloud computing, big data, artificial intelligence, and other new technologies, there are increasing demands for high-quality diagnosis and treatment in traditional Chinese medicine (TCM). How to use new technologies to improve diagnosis and treatment as well as the efficiency of medical institutions have become key areas in medical informatization [1].

The Outline of Healthy China 2030 Plan states that it is necessary to emphasize the unique advantages of TCM and improve its service capacity. The plan indicates that in township health centers and community health service centers, it is necessary to promote appropriate technologies in this regard; all primary medical and health institutions should be able to provide TCM services. The plan stipulates that China should promote the

© Springer Nature Switzerland AG 2021
Y. Tan and Y. Shi (Eds.): ICSI 2021, LNCS 12690, pp. 331–338, 2021.
https://doi.org/10.1007/978-3-030-78811-7_32

development of ethnic medicine. According to the plan, TCM should by 2030 be playing a leading role in disease treatment: that amounts to TCM being a key element in the synergistic treatment of major diseases and TCM occupying a central role in disease rehabilitation.

China's 13th Five-Year Plan states that the information-based development of TCM requires the following: (1) enhance the information platform for TCM; (2) coordinate the development of business application systems for TCM; (3) reinforce service guarantees for TCM medical informationization; (4) promote the application and development of big data for TCM; (5) promote services related to Internet Plus with TCM; (6) reinforce the development of specialties at county-level TCM hospitals, improve the capability of TCM with respect to special diagnoses, treatment, and comprehensive service as well as consolidating the basis for graded diagnosis and treatment. The plan stipulates that among all diagnoses and treatments, the target volume accounted for by TCM should be 30%. The plan stated that by 2020, all community health service institutions and township hospitals and 70% of village clinics should offer TCM services.

With regard to promoting the development of Internet Plus and medical health with respect to TCM, the General Office of the State Council made the following statements. (1) It is necessary to support TCM with respect to syndrome differentiation and treatment. (2) It is essential to apply "smart Chinese medicine pharmacy" and improve the quality of decoction pieces. (3) It is necessary to apply artificial intelligence technology to promote telemedicine services.

It is clearly required in the "Code of Basic Functions of Hospital Information System of Traditional Chinese Medicine (Revised) (Draft for Soliciting Opinions)" that an auxiliary diagnosis and treatment system for TCM should be established (Sect. 4). That code also states that a system should be created to promote TCM (Sect. 5).

The Opinions of the CPC Central Committee and the State Council on Promoting the Innovative Development of TCM Inheritance and Inheritance clearly state that it is necessary to create a database using TCM electronic medical records. It is also imperative to develop Internet hospitals and create TCM intelligent auxiliary diagnosis and treatment systems.

2   Current Problems with Informatization for TCM

2.1   TCM Informatization Standards Urgently Needed

Medical information standard is a unified standard that has been widely applied and observed in medicine; it is the basis for developing hospital information. Unfortunately, the standardization of hospital information for TCM began late and progressed slowly; that has hampered the development of TCM information in China. Thus, it is difficult to achieve standardization. With respect to long-term development, the standardization of medical information for TCM is a prerequisite for sustainable, rapid development of informatization for TCM; the standardization of TCM information is extremely urgent.

2.2   Development Gap with Hospital Informatization

The gap in hospital informationization for TCM is mainly evident in two areas. The information system in TCM hospitals is the same as that with general hospitals. However, with TCM hospitals, there is comparatively low investment and a higher professional degree regarding TCM informatization.

2.3 Westernization of TCM and Insufficient Informatization

Currently, clinicians—especially non-expert doctors (e.g., young doctors, grassroots doctors, those with Western learning)—have a low level of skills for clinical diagnosis and treatment with TCM; they are not generally accustomed to treating patients using TCM syndrome differentiation [2]. Syndrome differentiation and treatment is a basic principle in TCM for understanding and treating diseases; in diagnosis and treatment, it is the most significant feature of TCM and constitutes its greatest difference from Western medicine [3]. Applying the clinical decision-making system of TCM in other areas of medicine could help physicians achieve better diagnosis and treatment through information technology. It would also assist in promoting informatization for TCM in other hospitals.

2.4 Lack of TCM Diagnosis and Treatment Data

First, TCM is characterized by inheritance. In addition to oral transmission, it is necessary that TCM should be properly recorded. That recording should incorporate informatization related to doctors' visits; data should be stored such that they can later be effectively retrieved [4]. Second, at present, there are technical restrictions to the accumulation and development of data related to diagnosis and treatment with TCM. Currently, TCM clinical decision making is mainly limited to a single workstation in a medical institution; the knowledge base applied is often limited to that particular institution and is not shared on the Internet.

3   System Overview and Functions

By applying its syndrome differentiation and treatment methods and using the experience and knowledge of veteran practitioners, TCM employs an intelligent decision making system; the system aims to clarify diagnosis, disease identification, and prescription formulation for young doctors in hospitals [5].

The functions of the system are described in the following sections.

3.1   TCM Intelligent Decision Making

The system makes clinical decisions depending on a patient's symptoms or signs and other information input. Doctors enter the patient's chief complaint, history of current diseases and other information. Those data are structured by the system, which automatically accesses the standard symptom and sign terminology database using TCM natural language-processing technology [6]. According to the knowledge map of symptoms and signs, the corresponding refined symptoms, associated symptoms, and supplementary symptoms appear. After comprehensively collecting patient information under the doctor's guidance, the system distinguishes among diseases (for TCM or Western medical treatment) and syndromes; it recommends diagnosis points for suspected diseases and a diagnosis and treatment plan with TCM.

The auxiliary decision-making process is as follows.

1. Enter basic information about the patient.
2. Enter the patient's chief complaint and history of present illness.
3. The system displays the symptoms (signs) entered by the user and prompts the user to supplement and refine the symptoms (signs).
4. Users supplement and refine symptoms and confirm symptoms (signs).
5. System output. The results of "disease differentiation + syndrome differentiation" (disease differentiation is divided into western medicine disease and traditional Chinese medicine disease syndrome, and give the diagnosis key points) and the corresponding treatment and recommended prescription. The result of simple syndrome differentiation contains the corresponding treatment and recommended prescription.
6. Users can modify and perfect the prescription according to their own experience (adding, reducing and combining prescription).
7. Confirm the prescription.
8. Complete auxiliary diagnosis and treatment.

3.2 Medical Records

Physicians can directly input basic information for the patients' medical records (such as chief complaint and history of current illness) using the system's interface; that content is then accessible on the doctor's workstation [7]. All patients' medical records can be directly accessed from a physician's workstation: that avoids secondary input by other doctors, reduces the workload, and improves efficiency.

3.3 Medical Record Analysis and Symptom Association

Medical Record Analysis. The system can automatically interpret the chief complaint and history of current illness from the medical records. It presents standardized descriptions of TCM symptoms and displays them in a structured manner using natural language recognition technology.

Symptom Association. Through an analysis of medical records and using its TCM knowledge map and artificial intelligence algorithm, the system automatically indicates possible patient symptoms. The system indicates the symptoms using different colors for further consultation and confirmation between doctors and patients.

3.4 Symptom Entry

The physician can enter standard symptoms manually or select an input according to system prompts. To facilitate effective consultation between doctors and patients, the system provides comprehensive, targeted symptom guidance for consultation. The system includes diagnostic references for tongue, facial, and pulse images. It distinguishes between primary and concurrent symptoms, and allows the introduction of supplementary content related to the primary symptoms at any time.

3.5 Symptom Refinement and Supplementation

If the inputted symptoms require further refinement, the system automatically makes suggestions among which the physician can choose. The system supports the grouping of supplementary symptoms; it makes real-time calculations and displays of recommended symptoms based on selected symptoms.

3.6 Syndrome Differentiation and Prescription

Syndrome Differentiation. According to the symptoms inputted by the doctor and using its knowledge map and TCM diagnosis and treatment knowledge base, the system indicates the TCM syndrome diagnosis and appropriate treatment through an artificial intelligence algorithm.

Square Root. The system assesses TCM syndrome type based on the inputted symptoms; it makes TCM diagnosis and recommends treatment using syndrome differentiation. The system displays the multidimensional diagnosis by means of inputted information about treatment knowledge and experience from textbooks, guides, and famous TCM practitioners. It also provides visualization of the relationship between previous and the current syndrome; some of the data are displayed as numerical values. Recommendations are made based on the highest degree of matching with inputted information and clearly indicated (Figs. 1 and 2).

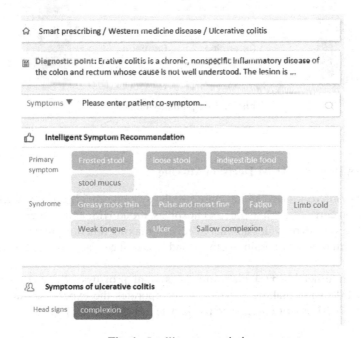

Fig. 1. Intelligent prescription

Fig. 2. Drug indication

3.7 Rational Drug Use

Rational drug use for TCM is monitored with respect to the four aspects of performance, compatibility (18 to 19), contraindications, and dosage. Drug contraindications are suggested during prescription; they mainly include such factors as compatibility contraindications, toxicity, dose compliance, and pregnancy contraindications. In that way, prescriptions are made in accordance with compatibility rules. The system matches drug contraindications according to such matters as the patient's gender and age, and it makes corresponding suggestions. Thus, for males, it does not suggest pregnancy contraindications (Fig. 3).

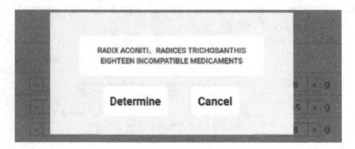

Fig. 3. Compatibility taboos

3.8 Internet Plus and TCM Decision Making

Using Internet micro-service architecture [8], the system provides function scheduling for information carriers, such as Internet hospitals, cloud HIS, and medical confederate platforms. The system provides auxiliary decision-making functions for medical institutions in the form of a cloud brain platform, and it introduces famous TCM practitioners online.

With the Internet hospital platform, the system provides physicians with clinically assisted decision-making services; it incorporates the experience data of eight distinguished TCM practitioners, who participate in the platform [9].

4 System Evaluation

4.1 Standardization of TCM Clinical Decision Making

The results with the TCM clinical decision making may vary as a result of doctors' experience and knowledge, schools of TCM, and other factors. However, the standardization and processing with regard to disease and syndrome differentiation achieve standardization of the TCM clinical decision-making process.

4.2 Improving TCM Clinical Diagnosis and Treatment by Nonexperts

The knowledge base of the system combines TCM classic texts, textbooks, and experience of famous TCM practitioners: that forms a firm foundation for TCM decision making. The system greatly reduces the difficulty with TCM diagnosis and syndrome differentiation, allowing young doctors to prescribe at the level of masters.

4.3 Improving TCM Electronic Medical Record Database

The system builds a knowledge map related to high-quality diagnosis and treatment conducted by medical institutions; it supports clinical auxiliary diagnosis and treatment. In that way, a two-way interactive mechanism develops, whereby the experience of distinguished TCM practitioners is passed on to young doctors. That constitutes a valuable asset to medical institutions [10]. Big data are stored in a structured form; that helps improve TCM electronic medical records: it compensates for the situation whereby electronic medical records are dominantly those of Western medicine and TCM having almost none [11]. By recording and accumulating detailed data related to diagnosis and treatment, the medication habits and prescription rules of doctors can be accessed to better support clinical decisions, research [12], teaching, and clinical practice [13].

4.4 Famous Practitioners' Services Provided

With the proposed system, doctors can provide professional TCM prescriptions for patients in different regions and even countries through Internet follow-up, online outpatient, and other services. The Internet referral service also enables doctors to provide immediate or non-immediate consultation and prescribing services to patients at different times. Internet Plus allows the TCM decision-making system to provide cross-regional and cross-temporal services for patients.

4.5 Discussion

During the Novel Coronavirus pneumonia epidemic, doctors can use the platform for remote diagnosis and treatment, avoiding the risk of infection caused by direct contact; The manual input/writing of clinical manifestations (symptoms) has been transformed into the selection of check boxes and drop-down boxes, which saves the input time. Moreover, a large number of doctors are using this platform together to form a database, which can help users to make more accurate clinical diagnosis, which is equivalent to many experienced experts helping to make decisions at the same time.

5 Conclusions

Through the development and application of the TCM decision-making system, a hospital's TCM information structure greatly improves; the TCM decision-making level of clinicians advances; clinical diagnostic thinking becomes standardized; a hospital accumulates considerable structured, standardized data related to TCM diagnosis and treatment. Further, using Internet Plus allows patients to access convenient, intelligent medical services. We believe that with the combination of Internet Plus and artificial intelligence, informatization for TCM clinical decision making will enter a new era.

References

1. Wang, Z.: Computational medicine – responding to the challenge of big data and transforming it to clinical practice. J. Bengbu Med. Coll. **39**(1), 1–2 (2014)
2. Gu, Z., Xin, F.: Chinese patent medicine application status and countermeasures of non-traditional Chinese medicine background physicians. J. Community Med. **9**(14), 82–83 (2011)
3. Liu, B.: Reflections on the clinical evaluation of syndrome differentiation. J. Chin. Med. **48**(1), 12–14 (2007)
4. Lu, P., et al.: Development and application of software of TCM inheritance assistant system. Chin. J. Exp. Formulae **18**(09), 1–4 (2012)
5. Sun, X., et al.: Technology and application of clinical decision-making system in traditional Chinese medicine. New Era Sci. Technol. (04), 56–66 (2017)
6. Wang, H., Wang, P., Wang, F., Huang, R., Li, L., Huang, Y.: Research and application of multivariate data fusion in clinical decision making system. Digit. Med. China **14**(11), 18–20 (2019)
7. Liu, B., Liu, B., He, L., Bai, W., Zhao, Y., Luo, W.: Demand analysis and preliminary assumption of constructing decision-making model of Chinese patent medicine in syndrome differentiation. Proprietary Chin. Med. **35**(05), 1085–1087 (2013)
8. Yang, Y., Yan, X., Deng, X.: Construction of hospital information system based on cloud computing and service-oriented middle platform architecture. Chin. J. Health Inf. Manag. **16**(06), 749–754 (2019)
9. Ren, X., Ji, Q., Wang, X., Zhang, Y.: Discussion on the informationization construction of the heritage laboratory of famous elderly Chinese medicine in China. China Materia Medica and Clinics **18**(03), 467–468 (2018)
10. Xiao, Y.: Research status of experience inheritance and application of intelligent analysis system of TCM prescription. China J. Inf. Tradit. Chin. Med. **17**(05), 1–3 (2010)
11. Zhou, X., Cao, S.: Ontology-based medical knowledge acquisition. Computer Sci. **30**(10), 35–39 (2003)
12. Lin, L., Zhu, S.: Analysis of the rule of prescription for the treatment of bone metastasis based on traditional Chinese medicine inheritance assistant system. Chin. J. Exp. Formulae **20**(17), 219–222 (2014)
13. Zhu, B., Bao, L., Chen, X., Zhou, Q., Lou, Y.: The realization of the construction concept of modern TCM prescription research service platform. Shanghai J. Tradit. Chinese Med. **43**(03), 49–52 (2009)

Parallel Random Embedding with Negatively Correlated Search

Qi Yang[ID], Peng Yang[ID], and Ke Tang[(✉)][ID]

Southern University of Science and Technology, Shenzhen, China
11930392@mail.sustech.edu.cn, {yangp,tangk3}@sustech.edu.cn

Abstract. Evolutionary algorithm (EA) is proved to be a promising way for parameter optimization in deep reinforcement learning (RL) in recent years. However, it still suffers from the curse of dimensionality when dealing with high-dimensional inputs. Based on experiments, we observe that only a few variables contribute significantly to the performance of large-scale RL policy. Intuitively, we propose a parallel random embedding framework to optimize strategies on multiple parameter subspaces to incorporate classical Evolution algorithms and techniques for the million-scale RL policy optimization. Experiments show that our approach has outperforming performance with Negatively Correlation Search (NCS) in the framework.

Keywords: Evolution algorithm · Reinforcement learning

1 Introduction

In recent years, there has been a surge of deep reinforcement learning (DRL) studies that successfully solve sequential decision problems in various fields, such as video games, physical simulation, computer vision, natural language processing, autonomous driving. (see [15] for a comprehensive review). With high-dimensional parametric policy representations and massive training data, deep neural network shows its excellent capability to approximate the target policy or the value function in end-to-end RL, also posts a significant challenge on large-scale non-convex decision variables optimization [12].

Depending on whether gradient information is used in parametric optimization, existing methods are grouped into two categories : gradient-based methods [13,14,21] and derivative-free methods [4,5,19,22]. Derivation-free algorithms such as Evolution Strategy (ES) is an alternative to existing mainstream

This work is supported by the Natural Science Foundation of China (Grant No. 61806090 and Grant No. 61672478), Guangdong Provincial Key Laboratory (Grant No. 2020B121201001), the Program for Guangdong Introducing Innovative and Entrepreneurial Teams (Grant No. 2017ZT07X386), Shenzhen Science and Technology Program (Grant No. KQTD2016112514355531), the Program for University Key Laboratory of Guangdong Province (Grant No. 2017KSYS008).

Y. Tan and Y. Shi (Eds.): ICSI 2021, LNCS 12690, pp. 339–351, 2021.
https://doi.org/10.1007/978-3-030-78811-7_33

RL methods for the competitive performance in non-convex RL problems and several additional merits in practical including stability, exploration ability and scalability on modern distributed systems [19]. With the input (e.g. images) at an increasing scale, RL policy's network size continuously grows, which leads to the deteriorated performance of derivative-free methods, i.e., the curse of dimension. For example, a simple 5-layers network with 84 × 84 pixels input contains 1.7 million weights and biases [13].

To apply existing methods on this million-scale problem, the current idea is mainly to transform the problem into an equivalent small-scale problem or set of problems by decomposition or embedding [8,9]. Decomposition methods exploit a divide-and-conquer strategy, which follows three steps : divide the original problem into multiple sub-problems, solve separately and combine solutions of sub-problems [10,25,26]. Embedding methods aim at optimizing in the most efficient subspace of the original problem, which is obtained by dimensionality reduction techniques [1,2,6,17,20,24]. The key to both types of methods is that problem transformation and solution recovery could guarantee completeness and convergence. In this article, we propose an embedding method to optimize high-dimensional policy parameters of the RL problem in multiple embedding searching spaces with an incorporated EAs. In comparative experiments, the parallel random embedding framework with negatively correlated search (NCSRE) outperforms four state-of-arts baselines in gradient-based and derivative-free methods. There are three main merits of our methods. Firstly, it embeds a large-scale problem into a small-scale problem set through a simple implementation of embedding techniques, allowing established derivative-free algorithms to be drawn upon in the RL domain. Secondly, unlike previous work which focuses on finding solutions, the proposed algorithm also finds a more efficient subspace to allocate computational resources. Thirdly, the framework reduces the inter-dependencies between subproblems, facilitating their application in modern distributed systems and parallelization.

The remainder of this article is organized as follows. Section 2 formulates our problem with a perspective of RL and reviews the related works. Section 3 introduces the random embedding background and provides a general view on the parallel random embedding framework, also incorporates a specific algorithm into our proposed framework. Section 4 gives the experiment details, results and discussion of the instance. At last, the conclusion and future work will be drawn in Sect. 5.

2   Backgrounds

2.1   Problem Formulation

We formulate this problem as a black-box optimization problem. As Fig. 1 shows, a predefined policy network with millions of trainable parameters interacts continuously with the environment to collect data in the general DRL framework. In an interaction, policy receive a visual state information from environment and pick up an action, then receive an instant reward as a feedback. Given a large number of interactions, policy parameters are optimized towards its objective that maximizes the cumulative score when it makes decisions automatically [7]. From a perspective of optimization, we consider all the policy parameters as a D-dimensions optimization variable x, and the objective function is Eq. 1.

$$f(x) = \sum_{i=0}^{T_{max}} c(\pi(x)) \tag{1}$$

where c is the one-step reward of policy π with a set of parameters x at t step, T_{max} is the maximum episode length, that is, the number of interactions in an episode. It is noted that f is an implicit objective function behind the dynamics of environment (i.e., Markov Decision Process).

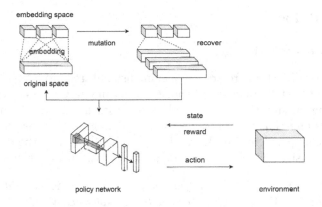

Fig. 1. Overview

2.2 Related Works

Then we briefly introduce related works focusing on large-scale RL policy optimization especially decomposition and embedding. Back in 2012, compression sensing was introduced in the linear bandit problem, which randomly projects the high dimension sparse linear bandit problem into random isotropic directions for efficient exploration [3]. Similarly, the random embedding (RE) technique was proposed under the Bayesian framework [24]. With random matrix theory and probability theory, some works provides a solid theoretical basis for the validity of random embedding [6,24,27] and inspires to cope with large-scale black-box optimization with RE. For example, some recent studies combine RE with a variety of derivative-free optimization approaches, including estimated distribution algorithms [6,20], a series of state-of-art single-objective algorithms [1,2,16,17], and multi-objective optimization [18]. The multiple-run sequential, random embedding framework proposed in [16] merges the idea of decomposition and dimension reduction by performing random embedding at each iteration and run a basic

EA to search in embedding space for several iterations. They also provide a comprehensive sensitivity analysis on embedding space's size and the convergence speeds on synthetic functions and classification tasks. Nonetheless, the next stage of optimization strongly depends on the previous iteration results and is not suitable for massive parallelization or distribution training. Besides, this work is a good attempt to optimize 10,000 decision variables, but a larger search space of parameters is also tackled to obtain satisfactory solutions in a reasonable number of iterations. As for the decomposition directions of EA, Ma's reviews provided a broader horizon of a series of EA in a cooperative co-evolution perspective [10]. We further extend the random embedding technique and the co-evolution algorithm [16] to a more independent and parallel framework and make it possible for distributed large-scale training for practical needs.

3   Methodology

3.1   Random Embedding

Random embedding(RE) refers to a dimensional reduction technique that project data into lower dimensional space with a randomly generated matrix [24]. The underlying assumption is that only a few dimensions are dominant in the search process and thus we can solve the embedding problem as an equivalence to the large-scale problem. Similar with [16] and [24], we give the definition 1 of ϵ-effective dimension, which formally define the contribution of dimensionality.

Definition 1. *For any $\epsilon > 0$, a function $f : \mathbb{R}^D \rightarrow \mathbb{R}$ is said to have an ϵ-effective subspace V_ϵ, if there exists a linear subspace $V_\epsilon \subseteq \mathbb{R}^D$ s.t. for all $x \subseteq \mathbb{R}^D$, we have $|f(x) - f(x_\epsilon)| \leq \epsilon$, where $x_\epsilon \in V_\epsilon$ is the orthogonal projection of x onto V_ϵ.*

Further, we define the optimal ϵ-effective dimension as dimension of the least linear subspace V_ϵ, among the collection of effective dimensions $\mathbb{V}_\epsilon$, denoted as d_ϵ. Random embedding defines a map vector $x \in V_\epsilon$ from the original space to a corresponding vector $y \in V_\epsilon$ that satisfies $x = Ay$ by matrix multiplication, where A is a $\mathbb{R}^{D \times d}$ Gaussian random matrix. Lemma 1 shows provable completeness guarantee to optimize on any one of effective subspace in following [16].

Lemma 1. *Given a function $f : \mathbb{R}^D \rightarrow \mathbb{R}$ with optimal ϵ-effective dimension d_ϵ, and any random matrix $\mathbf{A} \in \mathbb{R}^{D \times d}(d > d_\epsilon)$ with independent entries sampled from $\mathcal{N}$, then, with probability 1, for any $x \in \mathbb{R}^D$, there exists $y \in \mathbb{R}^d$, s.t. $|f(x) - f(\mathbf{A}y)| \leq 2\epsilon$.*

3.2   Parallel Embedding Framework

Intuitively, we intend to search a satisfactory solution on several sub-spaces rather than the million-scale search space. Furthermore, we prefer the sub-spaces contain the decision variables that contribute most to the final performance (i.e., the effective dimensions). For that purpose, we develop a parallel multiple-run random embedding framework to optimize both of the embedded sub-space and solution. We initialize with λ processes and each one runs on a randomly embedding subspace defined by a generated matrix. On the embedding space, we use a traditional small-scale derivative-free algorithm as an internal search process to obtain the best candidate. The final candidate is then recovered as a complete solution to the original problem after aggregation over all search processes. The framework can be divided into three steps at each stage: embedding, optimizing separately and updating together. The pseudo-code for the general RE framework is given in Algorithm 1.

1. At embedding steps, it generates μ random embedding matrix $A \in \mathbb{R}^{D \times d}$, each element of which is drawn from $\mathcal{N}(0, I)$ at the beginning of every phase.
2. At optimization steps, λ sub-processes of a specific evolution algorithm (i.e., populations) search separately. For each individual, it optimizes on a defined sub-space for L_{phase} iterations, then return an optimal solution y^*. The fitness function of individuals in the embedding space is as $g(y_{i,j}) = f(Ay_{i,j})$. The individuals are evaluated by corresponding solutions $x^* \in \mathbb{R}^D$.
3. At update steps, λ optimal solutions y in each subspace are recovered to x and competed with each other to decide the best solutions for this subspace (i.e., Localbest solution at line 11). The optimal output (i.e., Globalbest at line 12) is selected among Localbest candidates and the best individual of the previous generations (i.e., elitism).

Algorithm 1. Parallel Random Embedding Framework

Input:
    Number of subprocess λ; Population size of subprocesses μ;
    ϵ-Effective embedding dimension d; Origin optimization dimension D;
    The number of iterations in a phase L_{phase}; Fitness evaluation limitation T_{max};

1: Initialize λ individuals $x_{i,j}^0 \in \mathbb{R}^D$ uniformly where $i = 1 \ldots \lambda, j = 1, \ldots, \mu$;
2: Set current phase $m = 0$;
3: **while** $t < T_{max}$ **do**
4:     **for** $j = 1$ to μ **do**
5:         Generate a random matrices $A_j \in \mathbb{R}^{D \cdot d}$ with $\mathcal{N}(0, I)$;
6:         Initialize an evolution algorithm with x_j^{m-1} where $j = 1 \ldots, \mu$;
7:         Embedding with $y_j^{m-1} = A_j^{-1} \cdot x_j^{m-1}$ where $j = 1 \ldots, \mu$;
8:         Optimize $g(y_{i,j}) = f(A_j \cdot y_{i,j}^*)$, where $i = 1, \ldots \lambda$;
9:         Output $y_{i,j}^*$ for L_{phase} iterations;
10:     **end for**
11:     **for** $i = 1$ to λ **do:**
12:         Update the LocalBest solution $x_j^m = \arg \max_i f(A_j \cdot y*_{i,j})$
13:     **end for**
14:     Update the GlobalBest solution in $x^* = \arg \max_j f(x_j)$;
15: **end while**
16: **return** BestFound x^*

It then goes back to step 1 until the fitness evaluation limit is reached. After the number of evaluations has been exhausted, the algorithm returns the currently found Globalbest solution. In a general EAs, some details (such as whether to directly replace the BestFound, or how many solutions are output) are flexible.

3.3 Negatively Correlated Search with Random Embedding

While the search space is embedded independently in the above framework, cooperation between search processes can further improve efficiency. Negatively Correlation Search (NCS) is a novel framework for cooperative EAs proposed in 2016 that guides each search process in a promisingly different direction by introducing behavioral diversity. That is, search processes consider both fitness and correlation rather than considering fitness independently. The replace criterion, defined in our algorithm as the ratio of fitness to relevance, means that we prefer to choose a new solution with lower correlation and higher fitness. Instead of fine-tuning the hyper-parameters, we can then balance quality and diversity by an adaptive factor of the ratio. The Correlation term for individuals is calculated as Eq. 2, where the p_i is distributions of a search process, D_B is the minimum Bhattacharyya distance in Eq. 3 between each other in the population.

$$Corr(p_i) = \min_j \{D_B(p_i, p_j) | j \neq i|\} \qquad (2)$$

$$D_B(p_i, p_j) = -\ln(\int \sqrt{p_i(\boldsymbol{x})p_j(\boldsymbol{x})d\boldsymbol{x}}) \qquad (3)$$

Notably, there are a few additional details in the full algorithms. Firstly, withdraw variables α are borrowed from [16] to control greediness in D-dimension space. Second, the relatively narrow bound on the d-dimensional space should be carefully set to ensure the validity of the recovery solution. Since the optimization step size scales up dramatically on D-dimension and $D \gg d$, small bounded regions on the effective subspace may lead to the omission of the global optimum. Therefore, we use the two mechanisms described above to prevent excessive greed in the embedding space. Hence, we employ the above two mechanisms to prevent over-greedy in embedding space. Third, at the beginning of each stage (i.e., every L_{epoch} epoch), the coefficient matrix of the Gaussian operator for NCS individuals is reset in our version to encourage exploration. Alternatively, the Gaussian distribution learned in the previous stage can be inherit-ed directly to accelerate convergence. The details depend heavily on the characteristics of the problem.

4   Experiments

To investigate the redundancy of decision variables in RL, we took statistical analysis and sensitivity analysis on parameters of neural network (Sect. 4.1). Then, we employed NCS-C in the parallel random embedding framework denoted as NCSRE, and compared it against three state-of-art baselines (i.e., A3C [14],

CES [4], NCS-C [23]), representing gradient-based methods, evolution-based methods, and evolution-based methods under NCS, respectively (Sect. 4.2). We conducted all experiments on several most popular benchmarks on OpenAI gym and strictly followed the standard scheme in previous DRL research [11].

4.1   Redundancy in RL

Statistical Analysis. In a first experiment, we analyzed how the values of 1.7 million variables changed during the training process. The results show that most of the trainable parameters have little impact on the final performance of the strategy and are redundant in large-scale problems. Figure 2(b) visualizes the statistical distribution of parameter changes. For example, in Atari, a well-trained network (scoring 1000) had up to 61% of the parameter variations less than 1% of its maximum value, and 50% of the dimensional variations less than 0.1% during full training, compared to a randomly initialized network (scoring 234). These variables are the weights and biases of a well-trained network working (see Fig. 1(a)) with 84×84 raw pixels as input and 18 action options as output. The network parameters were initialized with a Gaussian distribution and trained to play an Atari game (i.e. Alien, Bowling, Freeway) via CES (Canonical Evolution Strategy).

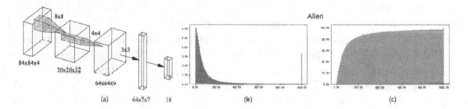

Fig. 2. (a)Network Architecture of DQN. (b) shows the frequency distribution of variations amplitude. (c) shows the cumulative frequency distribution. The y-axis measures how many parameters change (i.e., the percentage of parameters that change amplitude in that interval) while the x-axis measures how much they change (i.e., the increment percentage), with an interval of 5%.

Sensitivity Analysis. We also tested the sensitivity of the parameters by significantly varying their values and re-evaluating them. More specifically, we randomly selected 50 variables from the network's weights and biases, with the value of each variable being adjusted to a set of values in the range $[-10,10]$ at intervals $i = 0.1$. The manually modified policy network was then retested for 30 episodes on the corresponding games to measure the effect of the selected variables on the average performance. The results showed that the final performance curve did not vary with perturbations in certain dimensions.

4.2   Protocols

Environment. We selected five representative games on the Atari 2600 to evaluate the algorithms, that are, Bowling, Double Dunk, Freeway, Frostbite and Kangaroo. Firstly, the game portfolio contains 5 basic task types (i.e. navigation, obstacle avoidance, object recognition, 2-player gaming and global strategy), different random attributes and different action space sizes. In general, games with larger action spaces or more basic task types are more complex. Secondly, we mainly selected games that suf-fered from instability of gradient-based approach. All games were NoFrameskip-v4 versions.

Performance Metric. To estimate the performance of each candidate, we roll out the policy in the environment and used empirical results, i.e. the average test performance over a game launched 30 times. Test performance refers to the cumulative reward (i.e. score) that a policy network receives for a complete episode that starts after a random k frames ($0 < k < 30$) of no action and ends when a game over signal is received, or when the length of the episode exceeds the limit (i.e. 100,000 game frames).

Training Protocol. Other experimental protocols in training are described below. The network architecture shared by the baseline and the proposed algorithm is one of the most widely used networks proposed in [13] (details see Appendix). In the initial iteration, the weights and biases of this network architecture were randomly sampled from a uniform distribution of $[-10,10]$. For equivalence, we trained 0.1B derivative-free algorithms (i.e., NCSRE, CES, and NCS-C) and 0.04B gradient-based algorithms (i.e., A3C) because the ratio of execution time $\frac{t_{CES}}{t_{A3C}}$ on the same hardware is 2.5. On the same hardware was 2.5. During training, we processed the raw observations as follows [13]. First, we resized the observations to 84 × 84 pixels to fix the input size of the net-work. Second, we stacked consecutive k frames into a single input ($k = 4$) to approximate the human frequency (60 Hz). Third, we used the maximum pixel color value between two adjacent frames to eliminate flicker caused by the simulator. We roughly tuned the hyper-parameters in NCSRE and adopted the setting from the original papers and source codes in baselines (A3C, NSC-C and CES). Details of the recommended hyperparameter settings are given in the Appendix. The implement code of our algorithm was published on Github.

4.3   Results and Discussions

In each game, we trained an agent to perform 3 runs using the above algorithm to remove bias. As shown in Table 1, for each game, the first row Aver denotes the average score of the 3 runs, and the second row Percent denotes the percentage of the above score to the highest score of all algorithms. For each run, the final score is the average of 30 repeated evaluations of the best candidate output by the corresponding algorithm in the T_{max} frame. We have highlighted the best average scores from the 3 runs in bold. In summary, the average score of NCSRE outperforms NCS-C, CES and A3C across different games. In particular, in the dynamic obstacle avoidance games (Freeway and Frostbite), the best score in the existing baseline is only 64%, compared to 56% for our proposed method. In a 2-player game (Double Dunk), higher scores and positive scores are preferred, with negative scores implying agent failure. However, all algorithms fail to achieve a satisfactory solution, with NCSRE slightly outperforming the others (i.e. losing fewer scores).

Table 1. Detail of performance in terms of mean score across 3 runs over A3C CES NCS-C and NCSRE. Higher score is preferred in all 5 games.

Game	CES	A3C	NCS-C	NCSRE
Freeway	15.9	0.0	7.0	21.8
	12.7	0.0	9.4	23.0
	14.1	0.0	3.7	22.1
Aver	14.2	0.0	6.7	**22.3**
Percent	63.68%	0.00%	30.04%	100.0%
Bowling	28.0	0.0	65.4	63.8
	27.0	0.0	70.2	76.3
	20.0	0.0	58.0	91.0
Aver	25.0	0.0	64.5	**77.0**
Percent	32.46%	0.00%	83.76%	100.0%
Kangaroo	200.0	0.0	249.0	428.0
	200.0	0.0	504.0	737.0
	200.0	0.0	537.0	669.0
Aver	200.0	0.0	463.7	**557.5**
Percent	35.87%	0.00%	83.17%	100.0%
Frostbite	304.5	210.6	607.1	1069.2
	360.0	222.8	274.8	954.0
	324.0	312.3	790.7	976.2
Aver	329.5	248.6	557.5	**999.8**
Percent	32.96%	24.86%	55.76%	100.0%
Double dunk	−3.3	−2.0	−1.5	−1.7
	−2.8	−1.5	−0.8	−0.3
	−4.1	−2.8	−1.7	−1.9
Aver	−3.4	−2.1	−1.8	**−1.3**
Percent	0.00%	66.6%	47.05%	100.0%

From an optimization perspective, we further investigated the convergence rate of the Evolution-based algorithm on two deterministic games (i.e. Freeway and Bowling) to avoid stochastic effects. In the figure below, the training curves for the four algorithms are depicted in Fig. 3, which presents the variation of the

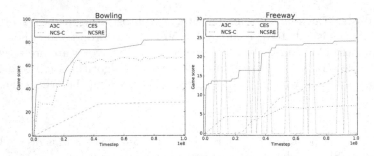

Fig. 3. Train curves of two games (Bowling and Freeway).

output model performance score with the number of frames consumed (positively correlated with the number of evaluations in RL). Overall, it can be clearly seen that the proposed algorithm converges faster to a good strategy in the early stages of training for both games. Compared to NCS-C, it searches on a hundred-scale optimization space. The superior convergence speed to NCS-C allows finding a relatively good enough solution with less interaction with the environment and fewer computer resources, which is very attractive in practice for improving the data efficiency of reinforcement learning. It also has more robust performance than gradient-based methods, due to the fact that gradient-based methods optimise the strategy by gradient descent, and the estimation of the strategy gradient is unstable and highly dependent on the quality of the offline samples. However, the derivative-free methods will not be guided to a local optimum by the gradient information.

5   Conclusion

In conclusion, this paper proposes a parallel random embedding algorithm incorporated with NCS works to address the high-dimension and hard-exploration problems and have a good performance on benchmarks. Compared with origin NCS or CES, the proposed approach is verified to be better on the final result and convergence speed, also alleviate the instability of state-of-art gradient-based algorithms A3C. There are several critic features to put NCSRE into practice, that are easy to implement, easy to parallelize, easy to accelerate, and easy to acquire a robust policy.

One of the main limitations of our method is the computational burden of embedding. It is cost-efficient to adopt NCSRE when the time cost of extra embedding computations is acceptable or reducible. For example, the overhead can be ignoble in expensive high-dimension optimization problems or can be largely accelerated with GPU or other professional hardware. In the future, a possible improvement of NCSRE to reduce that by acceleration techniques of large-scale matrix multiplications or engineering techniques, which allows NCSRE to be extended on a population of larger effective sub-spaces and further improve the optimization performance.

Appendix

Hyper-parameters. In NCSRE, the populations numbers is set to 7 and the population size of each sub-process is set to 2. For a network with 1.7 million parameters, we set embedding dimension to 100 considering the computational resource limitation and optimization effect. Generally speaking, algorithms 1000-dimensions effective space can achieve better performance if matrix production operation accelerates techniques are adopted. For NCS-C, the learning rate of sigma and adaptive negatively correlated coefficient are set the same as its origin paper, and the initial value of sigma is searched by grid method and set to 0.01 when the bound is [−0.1,0.1]. The bound of parameters defined our search space and should be carefully confirmed in different RL problems. We determined the bound of D-dimension space according to existing research and most parameters in well-performance network lie on [−10,10]. In our experiment, alpha is set to a constant for convenience. Highlightly, the length of the phase and epoch are corresponding to each other to match the sub-process and random embedding framework. In NCS-C, each sigma is adapted for every epoch iterations and epoch are usually set to 5 or 10. As for L_{phase}, there are two-edges effects that too large may lead to less embedding spaces, but too short phase leads to insufficient optimization in each subspace. Considering these factors, we set to 30 and 5. What's more, each network is re-evaluated with t times to reduce noise in games and take average scores as fitness. The initial value of t is set to 3 and is adaptively increased to 6 as $3 * \frac{t}{t_{max}}$ (Table 2).

Implemention. The implement code was published on Github https://github.com/Desein-Yang/NCS-RL.

Table 2. Recommended hyper-parameters of NCSRE in our experiments.

Item	Value	Description
λ	7	The number of populations (i.e. sub-processes)
μ	2	The population size of sub-processes
d	100	The size of embedding subspace
L_{phase}	30	The length of phase, i.e., update embedding sub-spaces every L_{phase} iterations
L_{epoch}	5	The length of epoch, i.e., update sigma every L_{epoch} iterations
α_0	1	Withdraw variable to control greediness
$[l, h]$	$[-0.1, 0.1]$	The bound of parameters in optimization bound
t	3 to 6	Repeated times in evaluation

Network Architecture. The parameters about the policy network architecture in our experiments are listed in Table 3.

Table 3. The network architecture of RL policy

	Input	Output	Kernel size	Stride	Filters	Activation
Conv1	$4 \times 84 \times 84$	$32 \times 20 \times 20$	8×8	4	32	ReLU
Conv2	$32 \times 20 \times 20$	$64 \times 9 \times 9$	4×4	2	64	ReLU
Conv3	$64 \times 9 \times 9$	$64 \times 7 \times 7$	3×3	1	64	ReLU
Fc1	$64 \times 7 \times 7$	512	–	–	–	ReLU
Fc2	512	Actions	–	–	–	–

References

1. Al-Dujaili, A., Suresh, S.: Embedded bandits for large-scale black-box optimization. In: Singh, S.P., Markovitch, S. (eds.) Proceedings of the Thirty-First AAAI Conference on Artificial Intelligence, February 4–9, 2017, San Francisco, California, USA. pp. 758–764. AAAI Press, New York (2017)

2. Binois, M., Ginsbourger, D., Roustant, O.: On the choice of the low-dimensional domain for global optimization via random embeddings. J. Global Optim. **76**(1), 69–90 (2019). https://doi.org/10.1007/s10898-019-00839-1

3. Carpentier, A., Munos, R.: Bandit theory meets compressed sensing for high dimensional stochastic linear bandit. Proc. Mach. Learn. Res. **22**, 190–198 (2012)

4. Chrabaszcz, P., Loshchilov, I., Hutter, F.: Back to basics: Benchmarking canonical evolution strategies for playing atari. In: Proceedings of the 27th International Joint Conference on Artificial Intelligence, pp. 1419–1426 (2018)

5. Conti, E., Madhavan, V., Such, F.P., Lehman, J., Stanley, K.O., Clune, J.: Improving exploration in evolution strategies for deep reinforcement learning via a population of novelty-seeking agents. In: Advances in Neural Information Processing Systems 31: NeurIPS 2018, December 3–8, 2018, Montreal, Canada. pp. 5032–5043 (2018)

6. Kaban, A., Bootkrajang, J., Durrant, R.J.: Towards large scale continuous EDA: a random matrix theory perspective. In: Proceeding of the Fifteenth Annual Conference on Genetic and Evolutionary Computation Conference, GECCO 2013, p. 383. ACM Press, New York (2013)

7. Kakade, S.M.: A natural policy gradient. In: Dietterich, T.G., Becker, S., Ghahramani, Z. (eds.) Advances in Neural Information Processing Systems 14 [Neural Information Processing Systems: Natural and Synthetic, NIPS 2001, December 3–8, 2001, Vancouver, British Columbia, Canada], pp. 1531–1538. MIT Press, Cambridge, MA (2001)

8. Knight, J.N., Lunacek, M.: Reducing the space-time complexity of the CMA-ES. In: Lipson, H. (ed.) Genetic and Evolutionary Computation Conference, GECCO 2007, Proceedings, London, England, UK, July 7–11, 2007, pp. 658–665. ACM Press, New York (2007)

9. Loshchilov, I.: A computationally efficient limited memory CMA-ES for large scale optimization. In: Arnold, D.V. (ed.) Genetic and Evolutionary Computation Conference, pp. 397–404. ACM Press, New York (2014)

10. Ma, X., et al.: A survey on cooperative co-evolutionary algorithms. IEEE Trans. Evol. Comput. **23**(3), 421–441 (2019)

11. Machado, M.C., Bellemare, M.G., et al.: Revisiting the arcade learning environment: evaluation protocols and open problems for general agents. J. Artif. Intell. Res. **61**(1), 523–562 (2018)
12. Müller, N., Glasmachers, T.: Challenges in high-dimensional reinforcement learning with evolution strategies. In: Auger, A., Fonseca, C.M., Lourenço, N., Machado, P., Paquete, L., Whitley, D. (eds.) PPSN 2018. LNCS, vol. 11102, pp. 411–423. Springer, Cham (2018). https://doi.org/10.1007/978-3-319-99259-4_33
13. Mnih, V., et al.: Human-level control through deep reinforcement learning. Nature **518**(7540), 529–533 (2015)
14. Mnih, V., Badia, A.P., et al.: Asynchronous methods for deep reinforcement learning. Proc. Mach. Learn. Res. **48**, 1928–1937 (2016)
15. Mousavi, S.S., Schukat, M., Howley, E.: Deep reinforcement learning: an overview. In: Bi, Y., Kapoor, S., Bhatia, R. (eds.) IntelliSys 2016. LNNS, vol. 16, pp. 426–440. Springer, Cham (2018). https://doi.org/10.1007/978-3-319-56991-8_32
16. Qian, H., Hu, Y.Q., Yu, Y.: Derivative-free optimization of high-dimensional non-convex functions by sequential random embeddings. In: Proceedings of the Twenty-Fifth International Joint Conference on Artificial Intelligence, IJCAI 2016, pp. 1946–1952. AAAI Press, New York (2016)
17. Qian, H., Yu, Y.: Scaling simultaneous optimistic optimization for high-dimensional non-convex functions with low effective dimensions. In: Proceedings of the Thirtieth AAAI Conference on Artificial Intelligence, AAAI 2016, pp. 2000–2006. AAAI Press, New York (2016)
18. Qian, H., Yu, Y.: Solving high-dimensional multi-objective optimization problems with low effective dimensions. In: Proceedings of the Thirty-First AAAI Conference on Artificial Intelligence, AAAI 2017, pp. 875–881. AAAI Press, New York (2017)
19. Salimans, T., Ho, J., et al.: Evolution strategies as a scalable alternative to reinforcement learning. arXiv:1703.03864 (2017)
20. Sanyang, M.L., Kabán, A.: REMEDA: random embedding EDA for optimising functions with intrinsic dimension. In: Handl, J., Hart, E., Lewis, P.R., López-Ibáñez, M., Ochoa, G., Paechter, B. (eds.) PPSN 2016. LNCS, vol. 9921, pp. 859–868. Springer, Cham (2016). https://doi.org/10.1007/978-3-319-45823-6_80
21. Schulman, J., Wolski, F., Dhariwal, P., Radford, A., Klimov, O.: Proximal policy optimization algorithms. arXiv:1707.06347 (2017)
22. Such, F.P., Madhavan, V., et al.: Deep neuroevolution: Genetic algorithms are a competitive alternative for training deep neural networks for reinforcement learning. arXiv:1712.06567 (2018)
23. Tang, K., Yang, P., Yao, X.: Negatively correlated search. IEEE J. Sel. Areas Commun. **34**(3), 542–550 (2016)
24. Wang, Z., Zoghi, M., et al.: Bayesian optimization in high dimensions via random embeddings. In: Proceedings of the Twenty-Third International Joint Conference on Artificial Intelligence, IJCAI 2013, pp. 1778–1784. AAAI Press (2013)
25. Yang, P., Tang, K., Yao, X.: A parallel divide-and-conquer-based evolutionary algorithm for large-scale optimization. IEEE Access **7**, 163105–163118 (2019)
26. Yang, Z., Tang, K., Yao, X.: Large scale evolutionary optimization using cooperative coevolution. Inf. Sci. **178**(15), 2985–2999 (2008)
27. Zhang, L., Mahdavi, M., Jin, R., Yang, T., Zhu, S.: Recovering the optimal solution by dual random projection. In: Proceedings of the 26th Annual Conference on Learning Theory, vol. 30, pp. 135–157 (2013)

Value-Based Continuous Control Without Concrete State-Action Value Function

Jin Zhu⬤, Haixian Zhang$^{(\boxtimes)}$⬤, and Zhen Pan⬤

Machine Intelligence Laboratory, College of Computer Science,
Sichuan University, Chengdu, China
zhanghaixian@scu.edu.cn

Abstract. In the value-based reinforcement learning continuous control, it is apparent that actions with higher expected return (state-action value, also as Q) will be selected as the action decision. But limited by the expression of deep Q function, researchers mostly introduce an independent policy function for approximating the preference of Q function. These methods, named actor-critic, implement value-based continuous control in an effective but compromise way.

However, the policy function and the Q function are highly correlated in Maximum Entropy Reinforcement Learning, so that these two have a close-form solution on each other. By this fact, we propose to implement a value-based continuous control algorithm without concrete Q function, which infers a temporary Q function from policy when needed. Compare to the current maximum entropy actor-critic method, our method saves a Q network needing training and a step of policy optimization, which results in an advance in time efficiency, while remains state of art data efficiency in experiments.

Keywords: Reinforcement learning · Continuous control · Value-based.

1 Introduction

Currently, reinforcement learning (RL) with deep learning is becoming a promising solution for variable applications, such as chess [25,28], video game [2], and robots [21].

Applying to different action spaces, deep reinforcement learning (DRL) is divided intoto discrete control and continuous control. When action space is finitely divided, a Q function (e.g. Q value table [34], or Q network [20]) is adequate to present the expected return of all decisions encountering a state, naturally the best policy at this state is to take the action with the highest Q value. But in facing infinitely divided action space, deep Q network cannot present infinite Q values, thus it is non-trivial for agent to acquire the preference of Q function.

To solve continuous instance, policy-based approaches, such as policy gradient (PG) [31], trust region policy optimization (TRPO) [26], and proximal policy

© Springer Nature Switzerland AG 2021
Y. Tan and Y. Shi (Eds.): ICSI 2021, LNCS 12690, pp. 352–364, 2021.
https://doi.org/10.1007/978-3-030-78811-7_34

optimization (PPO) [27], which directly learn a return-maximizing parameterized distribution policy conditioned on state, have proven practical, but suffer sample inefficiency [23].

Value-based solutions, more close to Q-learning, usually keep a Q function (critic) to record Q value, but different to original Q-learning, they introduce a policy (actor), to frame a circle—the actor interacts with environment to obtains experiences for critic improving, and the learned critic transforms its action preferences into the actor. These methods with the actor-critic architecture perform well in simulation [9] and real robots [14], currently become the baseline in continuous control.

However, compare to discrete instance, actor-critic algorithm needs an additional procedure for policy optimization, which results in **additional algorithm steps and computation resource consumption**. If value-based method could seamlessly acquire the optimal policy from value function, like discrete Q-learning, it may boost the whole agent learning progress.

In the study of maximum Entropy Reinforcement Learning (MERL) [35, 36], especially soft Q-learning [12], we noticed that the Q function is highly correlated to the policy function, so much as, a Q function uniquely ensures an optimal policy. But owing to the intractability of deep Q function on continuous control, the expressed policy is not available on interacting with environment, therefore, soft Q-learning has to keep a proxy policy (i.e. an actor).

But soft Q-learning reveals the straightforward correlation between the actor and the critic, now we are capable to inversely express an available Q function from a policy. By the light of soft Q-learning, we propose to implement a value-based continuous control algorithm which is of no concrete state-action value function, but to infer and express a Q function through a policy, moreover, the policy gets better as the inferred Q function close to the optimal, without the need of an procedure to optimize the policy.

Our method, Value-based Continuous control without Concrete state-action Value function (VCWCV), contributes on providing a concise way to implement the entropy regularized value-based continuous control algorithm, which removes an concrete state-action value function and saves a step of policy optimization. Then we demonstrate by experiments that, VCWCV is more time-efficient than the current maximum entropy actor-critic method (SAC), while remaining its state-of-art performance.

In the rest of this work, it will go as follows: Firstly, we introduce some basic idea about Reinforcement Learning and Maximum Entropy Reinforcement Learning, especially the soft Q-learning; Then, we explain the constraint correlation between actor and critic in soft Q-learning, then utilize this correlation to propose our inferred Q function and a practical algorithm; At last, we demonstrate the sample and time efficiency of our method on a popular benchmark.

2   Related Work

Besides actor-critic framework, researchers previously explored different methods to implement value-based continuous control.

The first approach predetermines the parameterization form of Q-function, which ensures that the optimization problem is trivial. For instance, we could discretize the continuous state and action spaces then introduce a tabular Q-function representation. However, discretization would be generally coarse with the expanding action dimensionality, finally resulting in unstable control. It could be avoided by averaging discrete actions weighted by their Q-values [19]; Wire-fitting [10] proposed to approximate Q functions piecewise-linearly over a discrete set of points, thus the optimal action is one of the extreme points; Normalised Advantage Functions (NAF [3,11]) unified critic and actor by assuming the state-action advantage is quadratic on action; And Input-convex neural network [1] ensures the Q function is concave, hence the optimal decision must be at the extreme point. However, these methods naturally limits the form of state-action value function, and may degrade performance while the domain does not conform to the expected structure.

Secondly, some approaches solve the inner maximization of the (optimal) Bellman residual loss with global nonlinear optimizers. QT-Opt [15] searches the best policy with cross-entropy method (CEM); Actor-expert [17] learns the best action by gradient ascent (GA); And discretizing action space to find the maximum action in finite selected options [16,29,32]. However, these approaches do not guarantee optimally. More currently, Continuous action Q learning (CAQL) [24] proposes to directly solve the optimal decision on deep Q network, and performs state-of-the-art sample efficiency, without an actor, but CAQL introduces a non-trivial Mixed-Integer Programming step to find the solution.

At last, an entropy regularizer in MERL, while encouraging efficient exploration, also ensures that both the optimal Q-function and policy have closed-form solutions. SAC and its predecessor SQL, propose to approximate the optimal policy recovered from the Q function by a proxy policy, which form an entropy regularized actor-critic algorithm. Quinoa [6] is most close to our work, which provides an inferred Q function with regard to its policy, however, Quinoa utilizes the Normalising Flow [7] to approximate the potential multi-modal policy, thus introduces inevitable complexity and results in disadvantage on data efficiency and final performance, now our work further explores this idea and proves that a MERL algorithm with the inferred Q function from a Beta-distribution formed policy function could achieve competitive performance.

3   Preliminaries

3.1   Reinforcement Learning

We consider a standard reinforcement learning (RL) setting [30], in a discrete-time *Markov decision process* [22] (MDP), defined by $(\mathcal{S}, \mathcal{A}, \mathcal{P}, \rho_0, r)$, where $\mathcal{S} \subseteq \mathbb{R}^n$ is a set of n-dimensional states, $\mathcal{A} \subseteq \mathbb{R}^m$ is a set of m-dimensional continuous actions, $\mathcal{P} : \mathcal{S} \times \mathcal{A} \times \mathcal{S} \rightarrow [0,1]$ is the state transition probability distribution, $\rho_0 : \mathcal{S} \rightarrow [0,1]$ is the distribution over initial states, and $r : \mathcal{S} \times \mathcal{A} \rightarrow \mathbb{R}$ is the reward function, we later abbreviate as: $r_t \triangleq r(s_t, a_t)$ to simplify notation.

Then a stochastic policy $\pi : \mathcal{S} \times \mathcal{A} \to [0,1]$ to maximize the expected return:

$$J(\pi) = \mathbb{E}_{(s_t, a_t) \sim \rho_\pi} \left[\sum_{t=0}^{\infty} \gamma^t r(s_t, a_t) \right] \tag{1}$$

Where $\rho_\pi(s_t)$ and $\rho_\pi(s_t, a_t)$ denote the state and state-action marginals of the trajectory distribution induced by the policy π. While the trajectory $(s_0, a_0, s_1 \ldots)$, follows $s_0 \sim \rho_0, a_t \sim \pi(\cdot|s_t), s_{t+1} \sim \mathcal{P}(s_{t+1}|s_t, a_t)$, and $\gamma \in (0, 1)$ is the discount factor.

For a state $s_t \subseteq \mathcal{S}$, Q-learning [34] and its successors [20,33] defines the state-action value function Q:

$$Q_\pi(s_t, a_t) = \mathbb{E}_{(s_{t+1}, a_{t+1}, \ldots) \sim \rho_\pi} \left[\sum_{l=0}^{\infty} \gamma^l r(s_{t+l}, a_{t+l}) \mid s_t, a_t \right] \tag{2}$$

In this work, we often use the abbreviate Q value for state-action value and abbreviate Q function for state-action value function.

3.2   Maximum Entropy Reinforcement Learning

Entropy regularization can introduce some advantages on environment exploration, which result in a better sample efficiency [14] and robustness [13]. Except the expected sum of rewards, Maximum Entropy Reinforcement Learning (MERL) also maximizes the policy entropy $H(\pi(\cdot|s_t))$ at the same time [35]:

$$J(\pi) = \mathbb{E}_{(s_t, a_t) \sim \rho_\pi} \left[\sum_{l=0}^{\infty} \gamma^l \left(r(s_{t+l}, a_{t+l}) + \alpha H(\pi(s)) \right) \mid s_t, a_t \right] \tag{3}$$

Where α is the temperature coefficient to determine the relative importance of entropy and reward.

To solve the entropy regularised objective (3), previous work [8,36] and soft Q-learning [12] proposed to directly find the optimal Q function, where the optimal policy can be recovered. The corresponding state-action value function Q is defined as:

$$Q_\pi(s_t, a_t) = \mathbb{E}_{(s_{t+1}, a_{t+1}, \ldots) \sim \rho_\pi} \left[\sum_{l=0}^{\infty} \gamma^l \left(r(s_{t+l}, a_{t+l}) + \alpha H(\pi(s)) \right) \mid s_t, a_t \right] \tag{4}$$

And the state value function:

$$V(s_t) \triangleq \alpha \log \int_A \exp \left(\frac{1}{\alpha} Q_{soft}(s_t, a) \right) da \tag{5}$$

For policy, MERL chooses to represent the corresponding policy by the form of energy-based policy:

$$\pi(a_t|s_t) \propto \exp\left(-\mathcal{E}\left(s_t, a_t\right)\right) \tag{6}$$

Where $\mathcal{E}$ is an energy function which satisfies : $\mathcal{E}(s_t, a_t) = -\frac{1}{\alpha}Q(s_t, a_t)$.

Based on the setting of MERL above, soft Q-learning and Soft Actor-Critic [13,14] formulates an actor-critic method, which builds an independent policy function for approaching the intractable energy-based policy. In this work, we follow the same framework, but to inversely recover Q function from the policy, that results in a concise implementation of soft Q-learning.

4   Infer Implicit Q Function from Policy Distribution

The relationship of the state value V, the Q value Q, and the policy π can be summarised as: When a Q value function is know, an optimal state value and a optimal policy can be uniquely ensured; Inversely, if the policy function and state value are given, the corresponding optimal Q value function can be recovered. The detail will be explained later in this section.

4.1   From Q Value to Policy

The definition of the state value (5) has revealed that the Q function can fully ensure a state value function. And the theorem (Lemma 1) in soft Q-learning shows that the Q function is capable of inferring the policy.

Lemma 1. *While Q function is defined by (4), and the value function is satisfied (5), then the optimal policy for the maximum entropy objective (3) is given by:*

$$\pi(a_t|s_t) \triangleq \frac{\exp\left(\frac{1}{\alpha}Q(s_t, a_t)\right)}{\exp\left(\frac{1}{\alpha}V(s_t)\right)} \tag{7}$$

proof. see soft Q-learning [12].

Derived from Eqs. (7) and (5), in each state, the optimal policy is directly recovered from the Q function. Ideally, soft Q-learning could consider only the optimization of the Q function, and interact with environments according to the policy expressed by the Q function.

However, the expressed policy (7) is not available for continuous control. The deep neural network Q function can be queried for only one state-action value $Q(s, a)$ for a state-action pair (s, a) each time, but it needs all the Q values in the infinite action space, to obtain a complete action distribution for sampling (i.e. policy distribution).

To interact with environments according to the optimal policy, SQL proposes to approximate the intractable policy by approximating inference [18], thus, adds some unscheduled complexity and instability. Subsequently, the approximating inference was replaced by direct KL-divergence reduction in Soft Actor-Critic (SAC) [13], but also introduce an independent policy function.

4.2   From Policy to Q Function

Since it is necessary to keep an available policy in continuous control, and the Q function is highly correlated to the policy, now we consider to inversely infer the Q function from a policy.

Theorem 1. *While Q function is defined by definition (4), and the value function is satisfied (5), it can be derive that, if the policy and value function is known, for each state-action pair, the corresponding Q value is given by:*

$$Q\left(s_t, a_t\right) = \alpha \cdot \ln \pi\left(a_t | s_t\right) + V(s_t) \tag{8}$$

proof. Directly transform on the optimal policy Eq. (7):

$$\pi(a_t|s_t) \triangleq \frac{\exp\left(\frac{1}{\alpha}Q(s_t, a_t)\right)}{\exp\left(\frac{1}{\alpha}V(s_t)\right)}$$

$$\rightarrow \exp\left(\frac{1}{\alpha}Q(s_t, a_t)\right) = \pi(a|s) \cdot \exp\left(\frac{1}{\alpha}V(s_t)\right)$$

$$\rightarrow \frac{1}{\alpha}Q(s_t, a_t) - \ln \pi(a_t|s_t) + \left(\frac{1}{\alpha}V(s_t)\right)$$

$$\rightarrow Q\left(s_t, a_t\right) = \alpha \cdot \ln \pi\left(a_t | s_t\right) + V(s_t)$$

Now we infer Q function from policy, but introduce the state value function $V(\cdot)$, it seemed that we did a superfluous thing. However fortunately, we are allowed to concentrate on the optimization of the inferred Q function (8), and the state value function and policy will be improved incidentally in the Q function optimization. We construct an implicit Q function, compare to frequently-used concrete Q function in continuous control, we construct a shared network to approximate both the policy function and the value function, then temporarily infer a Q function when updating Q function (Fig. 1).

4.3   Implicit Q Function in Soft Q-Learning

In this section, we implement a practical algorithm with the inferred Q function (8). Based on soft Q-learning algorithm, we remove the determination of state value function and the approximation of the energy-based policy, then, we replace the independent critic into our combined Q function. As a result, deriving a value-based method that fully focuses on the Q function.

Soft Bellman Equation and Soft Value Iteration. The main idea of soft Q-learning algorithm is to repeatedly determine the optimal Q value function. The procedure named soft value iteration is followed as:

Firstly, approaching the optimal Q function by the Bellman equation [4]:

$$Q\left(s_t, a_t\right) = r_t + \gamma \mathbb{E}_{s_{t+1} \sim p_s}\left[V\left(s_{t+1}\right)\right] \tag{9}$$

Secondly, ensuring the state value function by Q function with the definition of optimal state value (5). But the second step is not necessary now, since the state value function is a part of the inferred Q function (8), and it inversely determines the Q function.

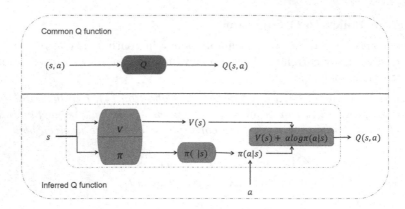

Fig. 1. *Common Q function (upper).* Traditional Q function in continuous control receives a state-action pair (s, a), and outputs a scalar as the Q value $Q(s, a)$; *Inferred Q function (below).* Our method has not a concrete Q function. To get a Q value for (s, a), the shared function simultaneously outputs the state value $V(s)$ and the policy $\pi(\cdot|s)$ according to the state s; Then, following the correlation in MERL, it combines the state value $V(s)$ and the action probability $\pi(a|S)$ to get a Q value $Q(s, a)$.

Soft Value Iteration with Inferred Q Function. Now, we concentrate on the Q value iteration. According to the definition of new Q function (8), the Bellman Equation (9) is equivalent to:

$$V(s) + \alpha \cdot \ln \pi (a_t|s_t) = r_t + \gamma \mathbb{E}_{s_{t+1} \sim p_s} [V (s_{t+1})] \tag{10}$$

Also as:

$$Q(s, a|V, \pi) = r_t + \gamma \mathbb{E}_{s_{t+1} \sim p_s} [V (s_{t+1})] \tag{11}$$

While we define: $Q(s, a|V, \pi) = V(s) + \alpha \cdot \ln \pi (a_t|s_t)$.

On the overt side, the new value iteration Eq. (11) aims to find the optimal Q function; On the other side, the optimization skips the process of reading the preference of state-action value function, directly acts on the policy and state value function. As a result, the policy, the value function, and the implicit Q function improve at the same time, in the procedure of determining the optimal Q function.

Implementation of Soft Value Iteration. We utilize a universal function approximator (i.e. deep neural network) to approximate the state value function v parameterized by ϕ, and a policy π parameterized by ω. From the Bellman Equation (10), we are going to find the optimal Q function by solving an objective:

$$J_Q(\phi, \omega) = \|V_\omega(s_t) + \alpha \cdot \ln \pi_\phi (a_t|s_t) - r_t - \gamma \mathbb{E}_{s_{t+1} \sim p_s} [V_{\bar{\omega}} (s_{t+1})] \|_1 \tag{12}$$

To better present its effect, we could regard policy π and state value function V as parts of the Q function and denote the parameter θ as: $\theta = \phi \cup \omega$, then the objective is equivalent to:

$$J_Q(\theta) = \|Q_\theta(s_t, a_t) - r_t - \gamma \mathbb{E}_{s_{t+1} \sim p_s}[V_{\bar{\theta}}(s_{t+1})]\|_1 \quad (13)$$

where $Q_\theta(s, a) = V_\theta(s) + \alpha \ln \pi_\theta(a \mid s)$, with regard to (8).

In practical implementation, we construct π and V as a state conditioned function with shared parameter θ, thus we optimize these two parts in a single step, with stochastic gradients:

$$\hat{\nabla}_\theta J_Q(\theta) = \nabla_\theta \|Q_\theta(s_t, a_t) - r(s_t, a_t) - \gamma V_{\bar{\theta}}(s_{t+1})\|_1 \quad (14)$$

4.4   Algorithm

We present our algorithms as (Algorithm 1). Like other value-based continuous control algorithm, the algorithm alternates between collecting experience by the current policy and updating the function approximators through the stochastic gradients according to batched replays sampled from a experience replay pool, but different from Actor-Critic algorithms which respectively update Q function and policy function, VCWCV improves a temporarily combined Q function and this improvement directly affects the shared approximator of policy function and value function, hence the parameter updating is in one step.

Algorithm 1. Value-based Continuous control without Concrete state-action Value function

---

Initialize a policy function π and a state value function V parameterized by θ
Initialize a replay buffer $\mathcal{D}$
for each iteration **do**
　　for each environment step **do**
　　　　select an action as current policy: $a_t \sim \pi_\theta(a_t \mid s_t)$
　　　　interact with the environment: $s_{t+1} \sim p(s_{t+1} \mid s_t, a_t)$
　　　　add new experience into the replay buffer: $\mathcal{D} \leftarrow \mathcal{D} \cap (s_t, a_t, r(s_t, a_t), s_{t+1})$
　　end for
　　for each gradient step **do**
　　　　sample a batch of experiences from replay buffer $\mathcal{D}$
　　　　update parameter by one step of gradient descent using:
　　　　　　$\nabla_\theta \|Q_\theta(s_t, a_t) - r(s_t, a_t) - \gamma V_{\bar{\theta}}(s_{t+1})\|_1$
　　end for
end for

Followed the same idea—MERL, our method is different from soft Q-learning and Soft Actor-Critic on the inverse inference from policy to Q function, which bypasses the problem that Q function expressed policy is not workable when practical interacting. We keep a controllable policy to interact with environments, and recover an available Q function when needed in soft value iteration, thus we do not execute an additional procedure to read the preference of state-action value function.

5   Experiments

We will demonstrate by experiments that: 1. Our method, without a concrete Q function, could remain the competitive sample efficiency compared to state of art value-based algorithms; 2. The method is more time efficient than current actor-critic method which also maximizes a same entropy regularised objective (3). Experiments go on popular challenging locomotion tasks from OpenAI gym benchmark suite [5].

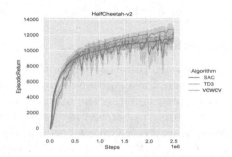

Fig. 2. *left.* HalfCheetah environment, where algorithms train a robot to run as fast as possible without losing balance. *right.* Training curve of sample efficiency compare experiments on continuous control benchmark. The experiments prove that TD3 (orange), SAC (blue) and VCWCV (green) all perform considerable and similar sample efficiency on the locomotion task. (Color figure online)

5.1   Sample Efficiency Experiments

We choose two state-of-the-art algorithms, that stably show ideal sample efficiency in common-used testing environment (Fig. 2 left), as the baseline. The first is Soft Actor-Critic (SAC) [14], an effective and stable value-based solution with remarkable robustness on hyperparameters and environments, with the original author-provided hyperparameter; And Twin Delayed Deep Deterministic policy gradient algorithm (TD3) [9], also with the author-provided hyperparameter. The experiment detail is in Appendix (A).

Figure 2 (right) shows the episodic return of evaluation rollouts during training for three algorithms. We train each algorithm for five different instances under different random seeds. The solid curves corresponds to the mean and the shaded region to the minimum and maximum returns over the five trials.

5.2   Time Efficiency Experiments

To demonstrate the time efficiency of our method, we compare it with SAC which is an actor-critic algorithm while also maximizes the same entropy regularized objective (3), on the wall clock time consumption. Experiments run on three simulated control tasks (InvertedPendulum, Reacher, HalfCheetah) with previous hyperparameter setting (Appendix A) and identical hardware condition, and we train two algorithm for the same environment steps in three simulated environments.

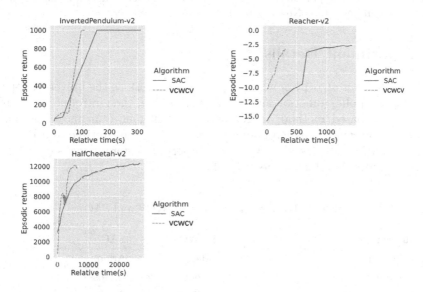

Fig. 3. *Training curve of time efficiency compare experiments on continuous control benchmark.* In the comparison of the dimension of time consumption, VCWCV (orange) runs out the predefined total environment steps earlier than SAC (blue) in 3 environments, but achieve similar episodic return.

Figure 3 shows the total episodic return of evaluation rollouts during the training time. We also run five different instances of each algorithm conditioned on different random seeds, and separately select the trial with the best performance for both two algorithms. Compare to SAC, our method earlier runs out the same total environment steps, and VCWCV mostly achieve higher episodic return in a less time, which prove that VCWCV is more efficient than SAC in time dimension.

6   Conclusion

In this paper, we discuss the constraint correlation between the action-state value function and the policy function in Maximum Entropy Reinforcement Learning. Then we utilize this correlation to present an inferred Q function from the policy function, and propose a value-based continuous control algorithm without a concrete state-action value function (VCWCV), but temporarily infer a Q function when value iteration. Moreover, the policy will immediately improve as the implicit Q function getting better, without additional attention to optimize the policy. Eventually, we demonstrate by experiments that, VCWCV keeps a competitive performance on sample efficiency compare to state-of-the-art value-based continuous control algorithms, and advances in time efficiency.

Acknowledgement. This work was supported by National Natural Science Foundation of China under Grant No.61836011.

A   Experiment Details

We implement our algorithm with deep neural network as the universal approximator for the policy function and the value function, then introduce a common trick in value-based method, target network. Different from most continuous control method with Gaussian policy, our policy is parameterized as a Beta distribution conditioned on state. VCWCV's hyperparameter setting is mostly from the SAC's, more detail is on the Table 1.

Table 1. Hyperparameters of VCWCV and SAC

Parameters	VCWCV	SAC
Optimizer	Adam	Adam
Loss function	Smooth L1 loss	L2 loss
Learning rate	$2 \cdot 10^{-4}$	$3 \cdot 10^{-4}$
Discount (γ)	0.99	0.99
Replay buffer size	10^6	10^6
Number of hidden layers	2	2
Number of hidden units per layer	256	256
Number of samples per minibatch	256	256
Temperature coefficient (α)	0.2	Auto-tuned
Entropy target	Null	$-dim(\mathcal{A})$ (e.g., -6 for HalfCheetah)
Nonlinearity	ReLU	ReLU
Update interval	2	1
Target smoothing coefficient (τ)	0.004	0.005
Target update interval	1	1
Gradient steps	1	1
Double network	False	True

References

1. Amos, B., Xu, L., Kolter, J.Z.: Input convex neural networks. In: International Conference on Machine Learning, PMLR, pp. 146–155 (2017)
2. Badia, A.P., et al.: Agent57: outperforming the atari human benchmark. In: International Conference on Machine Learning, PMLR, pp. 507–517 (2020)
3. Baird, L.C.: Reinforcement learning in continuous time: advantage updating. In: Proceedings of 1994 IEEE International Conference on Neural Networks (ICNN 1994), vol. 4, pp. 2448–2453. IEEE (1994)
4. Bellman, R.: The theory of dynamic programming. Technical report, Rand RAND Corporation Santa Monica, CA (1954)
5. Brockman, G., et al.: OpenAI gym. arXiv preprint arXiv:1606.01540 (2016)
6. Degrave, J., Abdolmaleki, A., Springenberg, J.T., Heess, N., Riedmiller, M.A.: Quinoa: a q-function you infer normalized over actions. CoRR abs/1911.01831 (2019)
7. Dinh, L., Sohl-Dickstein, J., Bengio, S.: Density estimation using real NVP. In: 5th International Conference on Learning Representations, ICLR 2017, Toulon, France, 24–26 April 2017, Conference Track Proceedings. OpenReview.net (2017)
8. Fox, R., Pakman, A., Tishby, N.: Taming the noise in reinforcement learning via soft updates. In: Proceedings of the Thirty-Second Conference on Uncertainty in Artificial Intelligence, pp. 202–211 (2016)
9. Fujimoto, S., Hoof, H., Meger, D.: Addressing function approximation error in actor-critic methods. In: International Conference on Machine Learning, PMLR, pp. 1587–1596 (2018)
10. Gaskett, C., Wettergreen, D., Zelinsky, A.: Q-learning in continuous state and action spaces. In: Australasian Joint Conference on Artificial Intelligence, pp. 417–428. Springer (1999)
11. Gu, S., Lillicrap, T., Sutskever, I., Levine, S.: Continuous deep q-learning with model-based acceleration. In: International Conference on Machine Learning, pp. 2829–2838 (2016)
12. Haarnoja, T., Tang, H., Abbeel, P., Levine, S.: Reinforcement learning with deep energy-based policies. In: International Conference on Machine Learning, PMLR, pp. 1352–1361 (2017)
13. Haarnoja, T., Zhou, A., Abbeel, P., Levine, S.: Soft actor-critic: off-policy maximum entropy deep reinforcement learning with a stochastic actor. In: International Conference on Machine Learning, PMLR, pp. 1861–1870 (2018)
14. Haarnoja, T., et al.: Soft actor-critic algorithms and applications. CoRR abs/1812.05905 (2018)
15. Kalashnikov, D., et al.: Scalable deep reinforcement learning for vision-based robotic manipulation. In: Conference on Robot Learning, PMLR, pp. 651–673 (2018)
16. Lazaric, A., Restelli, M., Bonarini, A.: Reinforcement learning in continuous action spaces through sequential Monte Carlo methods. Adv. Neural Inf. Process Syst. 20, 833–840 (2007)
17. Lim, S.: Actor-Expert: A Framework for using Q-learning in Continuous Action Spaces. Ph.D. thesis, University of Alberta (2019)
18. Liu, Q., Wang, D.: Stein variational gradient descent: a general purpose bayesian inference algorithm. In: Advances in Neural Information Processing Systems, pp. 2378–2386 (2016)

19. Millán, J.D.R., Posenato, D., Dedieu, E.: Continuous-action q-learning. Mach. Learn. **49**(2–3), 247–265 (2002)
20. Mnih, V., et al.: Human-level control through deep reinforcement learning. Nature **518**(7540), 529–533 (2015)
21. Nair, A.V., Pong, V., Dalal, M., Bahl, S., Lin, S., Levine, S.: Visual reinforcement learning with imagined goals. In: Advances in Neural Information Processing Systems, pp. 9191–9200 (2018)
22. Puterman, M.L.: Markov Decision Processes: Discrete Stochastic Dynamic Programming. Wiley, New York (2014)
23. Quillen, D., Jang, E., Nachum, O., Finn, C., Ibarz, J., Levine, S.: Deep reinforcement learning for vision-based robotic grasping: a simulated comparative evaluation of off-policy methods. In: 2018 IEEE International Conference on Robotics and Automation (ICRA), pp. 6284–6291. IEEE (2018)
24. Ryu, M., Chow, Y., Anderson, R., Tjandraatmadja, C., Boutilier, C.: CAQL: continuous action q-learning. In: 8th International Conference on Learning Representations, ICLR 2020, Addis Ababa, Ethiopia, 26–30 April 2020. OpenReview.net (2020)
25. Schrittwieser, J., et al.: Mastering Atari, go, chess and Shogi by planning with a learned model. Nature **588**(7839), 604–609 (2020)
26. Schulman, J., Levine, S., Abbeel, P., Jordan, M., Moritz, P.: Trust region policy optimization. In: International Conference On Machine Learning, pp. 1889–1897 (2015)
27. Schulman, J., Wolski, F., Dhariwal, P., Radford, A., Klimov, O.: Proximal policy optimization algorithms. arXiv preprint arXiv:1707.06347 (2017)
28. Silver, D., et al.: Mastering the game of go without human knowledge. Nature **550**(7676), 354–359 (2017)
29. Smart, W.D., Kaelbling, L.P.: Practical reinforcement learning in continuous spaces. In: ICML pp. 903–910. Citeseer (2000)
30. Sutton, R.S., Barto, A.G.: Reinforcement Learning: An Introduction. MIT Press, Cambridge (2018)
31. Sutton, R.S., McAllester, D.A., Singh, S.P., Mansour, Y.: Policy gradient methods for reinforcement learning with function approximation. In: Advances in Neural Information Processing Systems, pp. 1057–1063 (2000)
32. Uther, W.T., Veloso, M.M.: Tree based discretization for continuous state space reinforcement learning. In: Aaai/iaai, pp. 769–774 (1998)
33. Van Hasselt, H., Guez, A., Silver, D.: Deep reinforcement learning with double q-learning. In: Proceedings of the AAAI Conference on Artificial Intelligence, vol. 30 (2016)
34. Watkins, C.J., Dayan, P.: Q-learning. Mach. Learn. **8**(3–4), 279–292 (1992)
35. Ziebart, B.D.: Modeling purposeful adaptive behavior with the principle of maximum causal entropy (2010)
36. Ziebart, B.D., Maas, A.L., Bagnell, J.A., Dey, A.K.: Maximum entropy inverse reinforcement learning. In: AAAI, vol. 8, pp. 1433–1438, Chicago, IL, USA (2008)

Exploring the Landscapes and Emerging Trends of Reinforcement Learning from 1990 to 2020: A Bibliometric Analysis

Li Zeng[1]([⊠]) [iD], Xiaoqing Yin[1], Yang Li[2], and Zili Li[3]

[1] College of Advanced Interdisciplinary Studies, National University of Defense Technology, Changsha 410073, China
[2] Hunan Stringle Technology Co., Ltd., Changsha 410073, China
[3] High-Tech Research Institute, Hunan Institute of Traffic Engineering, Changsha 410073, China

Abstract. Reinforcement Learning (RL) becomes increasingly important in recent years as the huge success of AlphaGo and AlphaZero. However, this technique is a not a newly born research topic, which originates from the well-developed dynamic programming method. In this paper, we explore the history of RL from the bibliometric perspective for the last 30 years, to capture its landscapes and emerging trends. We conduct comprehensive assessments of the RL technology according to articles related to RL in SCI database from 1990 to 2020, and extensive results indicate that reinforcement learning research goes up significantly in the past three decades, including a total of 9344 articles covering 96 countries/territories. Top five most productive countries are USA, China, England, Japan, Germany and Canada. There are 4507 research institutes involved in the field of RL and among them the top five productive ones are Chinese Academy of Sciences, University College London, Beijing University of Posts and Telecommunications, Tsinghua University and Northeastern University and Princeton University. Besides, top frequently adopted keywords with strongest citation burst are Genetic Algorithm, Dynamic Programming, Q-Learning, Mobile Robot, Wireless Sensor Network, Smart Grid, Big Data, Inverse Reinforcement Learning and Cognitive Radio, which demonstrate the emerging trends of this field. We claim the results shown in this paper provide a dynamic view of the evolution of "Reinforcement Learning" research landscapes and trends from various perspectives that is able to serve as a potential future research guide, and the way we demonstrate could also be adopted to analyze other research topics.

Keywords: Reinforcement Learning · Bibliometrics · Citespace · Research trends

1 Introduction

As AlphaGo system developed by Google DeepMind prevailed Mankind's best expert Ke Jie in the Go game, the reinforcement learning technology behind it has attracted

L. Zeng and X. Yin—Contributed equally to this work.

© Springer Nature Switzerland AG 2021
Y. Tan and Y. Shi (Eds.): ICSI 2021, LNCS 12690, pp. 365–377, 2021.
https://doi.org/10.1007/978-3-030-78811-7_35

increasing attention from researchers with various backgrounds. Reinforcement Learning (RL) is one of three basic machine learning algorithms, the other two are supervised learning and unsupervised learning. Different from these two learning paradigms which take data features as input and output data labels, RL instead focuses on how agents should take actions in order to maximize the cumulative reward [1].

Reinforcement learning technology has developed rapidly in the last three decades and is widely used in fields such as computer science [2, 3], telecommunications [4], automation control [5], robotics [6], transportation [7], energy fuels [8], chemistry [9, 10], graph mining [11] and et al. There are many survey papers about reinforcement learning [12, 13], while there are few ones from the view of bibliometric, and we will show with evidences that this is an effective tool for evaluating research landscapes and emerging trends for many research topics [14]. In order to bridge the gap, we adopt a comprehensive bibliometric approach to analyze the RL literature with the intention to quantitatively assess current landscape and emerging trends of reinforcement learning, which consists of scientific outputs analysis (document number, subject categories, countries, institutions, and keywords), network analysis (co-occurrence frequency, betweenness centrality [15]) and evolution analysis with burst detection which may serve as a valuable reference for future research.

The remainder of the current paper is organized as follows: The data collection and analysis methods section summarizes the procedure of data collection and the methods used in this study. Subsequently, results are analyzed. Finally, we conclude the paper with a discussion.

2   Data Collection and Analysis Methods

2.1   Data Collection

Based on the Web of Science Database [16], we collected the bibliographic records related to reinforcement learning on January 30, 2021, and ultimate search query string about reinforcement learning as follows:

DATABASE: Web of Science Core Collectio
Indexes: SCI-EXPANDED
Theme: "Reinforcement Learning"
Timespan: 1990–2020
DOCUMENT TYPES: (ARTICLE OR PROCEEDINGS PAPER OR REVIEW

The above search branch captures the relevant literatures on reinforcement learning in the Web of Science core database. In order to focus on the core research literatures, the type of literatures are limited to ARTICLE, PROCEEDINGS or REVIEW.

2.2   Analysis Methods

Scientific outputs analysis of RL research are conducted by Microsoft Excel and python scripts, while network analysis are performed by bibliometric tools such as Citespace [17] and VOSviewer [18], which were usually employed to reveal the distinct patterns and emerging trends in many science fields.

3   Results

3.1   Growing Trend

To understand the growth trends of publications in reinforcement learning, a statistical analysis of the scientific literature over the 31 years from 1990 to 2020 was conducted, Fig. 1 shows the annual numbers of articles and cumulative articles between 1990 and 2020 in the field of reinforcement learning. The red one is the annual number of articles curve. As shown from the curve, a substantial research interest in reinforcement learning research did not emerge until 2002, although a few articles related to reinforcement learning were published previously. The highest number of articles arrived at 2020, with 1888 articles, accounting for 19.87% of the total number and the average number of articles was 306.4 per year.

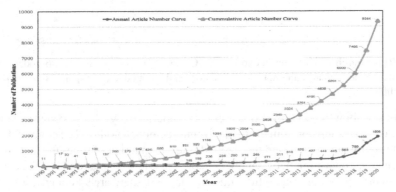

Fig. 1. Growing trend of RL research from 1990 to 2020.

3.2   Countries/Territories Distribution

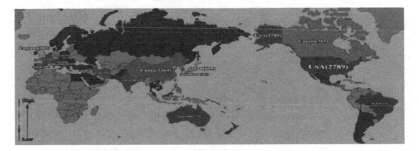

Fig. 2. Geographic distributions (countries or territories) of RL researchers.

Figure 2 shows the geographic distribution of countries/territories in the field of reinforcement learning. The number after the country name in the figure represents the number of publications correspondingly. Obviously, USA and China take the lead in the field of reinforcement learning research. Table 1 lists the top ten most productive countries/territories in the field of reinforcement learning. Over all, USA is the first most productive and also first most influential country in this field, with a total amount of 2798 papers, 1208 internationally collaborated papers, 110838 citations, 1266 institutes and its H-Index is 148.

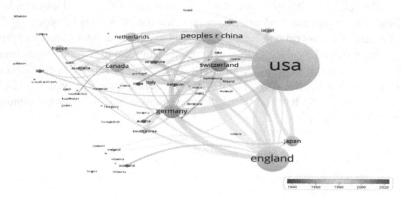

Fig. 3. RL research cooperation network w.r.t, Country/Territory, including 96 nodes and 714 edges. Each node indicates a country or a territory and each edge corresponds to a cooperation relationship.

Figure 3 shows the cooperation network generated by the VOSviewer software between countries/territories in the field of reinforcement learning. The color of the node represents publication time, and the size indicates the number of times the node is referenced, and the connection between nodes represents the cooperation between countries and their corresponding cooperation intensity. In total, there are 94 countries/territories in the field of reinforcement learning. An obvious characteristic is that there are close cooperation among China, USA, Germany, Canada and other countries.

Table 1. Top 10 most productive countries for RL research.

No.	NM	TP	CP	TC	HI	TI
1	USA	2789	1208	110838	148	1266
2	CHINA	2364	927	23332	66	1317
3	England	889	569	41780	87	765
4	Japan	609	182	11299	51	454
5	Germany	623	368	17319	64	334
6	Canada	585	349	13196	60	537
7	South Korea	382	134	3079	28	339
8	France	355	219	7152	44	198
9	Spain	280	120	3847	30	198
10	Iran	252	68	2724	25	277

No., Rank By TP; C/T, Country/Territory; TP, Total papers; IP, independent papers; CP, Internationally collaborated articles; TC, Total citations counts; HI, H Index; TI, Total Institutes numbers.

3.3 Research Institutes Distribution

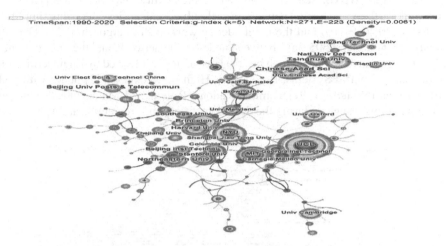

Fig. 4. RL research cooperation network, w.r.t, university/institute, including 271 nodes and 223 edges.

Overall, there are 4507 research institutes engaged in the field of RL during 1990 and 2020, and the top 5 most productive institutes are Chinese Academy of Sciences, University College London, Beijing University of Posts and Telecommunications, Tsinghua University, Northeastern University and Princeton University. Figure 4 demonstrates the core institutional cooperation network with 271 nodes and 223 links generated by Citespace. The size of the node is proportional to the number of publications, and different colors indicate the year of publication. The nodes in the network with red border are the institutes with strong citation burstiness [19]. Table 2 lists the top ten institutes with strong citation burstiness. Obviously, the University of Southern California with the highest citation burstiness 14.87 during the year of 2007 and 2013.

Table 2. Top 10 institutes with citation bursts.

Institutions	Strength	Begin	End
Univ So Calif	14.87	2007	2013
UCL	14.17	2011	2016
Univ Arizona	13.44	2007	2012
Univ Elect Sci & Technol China	12.55	2018	2020
Univ Washington	11.23	2009	2013
Tokyo Inst Technol	10.94	1998	2010
Univ Texas	10.09	2004	2007
Yale Univ	10.09	2007	2016
Chinese Acad Sci	8.64	2014	2018
Google DeepMind	6.47	2017	2020

3.4 Subject Category Distribution

Each article in the Web of Science database is labeled with at least one subject category. Figure 5 shows the co-occurring network of such subject categories after being simplified by g-index [20] (k = 10) and the pathfinder network scaling algorithm [21]. Overall, there are 258 nodes and 605 links between nodes in the network, and the most common category is computer science, which has the largest circle, followed by Engineering, and Automation & Control Systems, Psychology. High betweenness [22] nodes with pink border such as Psychology, Telecommunications, Robots and Mathematical represent the turning points category during the development of reinforcement learning.

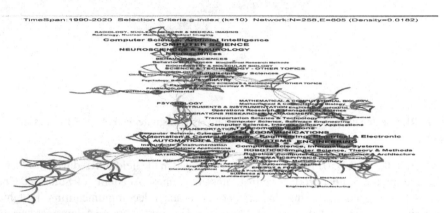

Fig. 5. RL research cooperation network w.r.t., subject category, including 258 nodes and 605 edges.

3.5 Research Hotspots

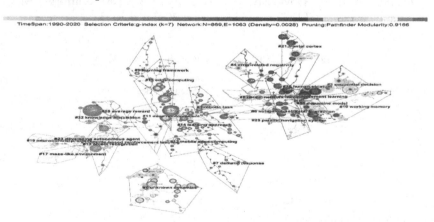

Fig. 6. Document co-cited network for reinforcement learning research, including 869 nodes and 1063 edges.

In order to probe the research hotspots of reinforcement learning, Citespace was utilized to analyze the co-citation network of reinforcement learning, and the special clustering algorithm [23] was used to divide the co-citation network into different clusters. Figure 6 is the document co-citation network after being simplified by g-index (k = 7) and the pathfinder network scaling algorithm, over all there are 869 nodes, 1063 links and 26 co-citation clusters in the network. Table 3 lists details of top five major clusters by their size, and the silhouette coefficient [24] was used to assess the quality of the cluster. Each cluster is labelled by keywords of the citing articles of cluster, and labels are chosen by the log-likelihood ratio test method (LLR) [25]. The average years of publication of a cluster are used to represents its novelty while the silhouette value is used to represent the quality of cluster. In order to identify the landscape of Reinforce Learning research, five representative clusters are selected and discussed in detail.

Table 3. Five main clusters ordered by size and mean year.

Cluster ID	Size	Silhouette	Mean (Year)	Label (LLR)
0	73	0.986	2012	Optimal control; adaptive dynamic programming; adaptive critic designs
1	39	0.98	1995	Subgoals; feature space; module-based rl
2	38	1	2016	Deep learning in robotics and automation; learning from demonstration; robot sensing systems
7	35	0.981	2016	Smart grid; adaptive traffic signal control; multi-agent system
12	30	1	2017	Mobile edge computing; servers; computation offloading

The silhouette coefficient was used to assess the quality of the cluster. Each cluster is labelled by keywords of the citing articles of cluster, and labels are chosen by the log-likelihood ratio test method.

Cluster#0 is the largest cluster, containing 73 references across a 7-year period from 2011 to 2017. The median year of all references in this cluster is 2012 and the silhouette value of the cluster is 0.986, indicating the cluster has a relatively high level of homogeneity. The labels of this cluster selected by LLR are optimal control, adaptive dynamic programming and adaptive critic designs. Among 73 members of this cluster, the top 5 references with high burstiness are surveyed as follows. Lewis, F. et al. [26] described the use of principles of reinforcement learning to design feedback controllers for discrete- and continuous-time dynamical systems. Liu, D. et al. [27] proposed a new discrete-time policy iteration adaptive dynamic programming (ADP) method for solving the infinite horizon optimal control problem of nonlinear systems. Bhasin, S. et al. [28] proposed a novel actor-critic-identifier architecture for approximate optimal control of uncertain nonlinear systems. Kiumarsi, B. et al. [29] proposed a novel approach based on the q-learning algorithm to solve the infinite-horizon linear quadratic tracker (LQT) for unknown discrete-time systems in a causal manner. Vamvoudakis, K. et al. [30]

bought together cooperative control, reinforcement learning, and game theory to present a multi-agent formulation for the online solution of team games.

Cluster#1 is the second largest cluster, containing 39 references across a 13-year period from 1990 to 2002. The median year of all references in this cluster is 1995 and the silhouette value of the cluster is 0.98. The labels of this cluster selected by LLR are optimal control, adaptive dynamic programming and adaptive critic designs. Among 39 members of this cluster, the top 5 references with high burstiness are surveyed as follows. Mahadevan, S. et al. [31] described a general approach for automatically programming a behavior-based robot. Tsitsiklis, J. N. et al. [32] discussed the temporal-difference learning algorithm, as applied to approximating the cost-to-go function of an infinite-horizon discounted Markov chain. Jaakkola, T. et al. [33] provided a rigorous proof of convergence of these DP-based learning algorithms by relating them to the powerful techniques of stochastic approximation theory via a new convergence theorem. Asada, M. et al. [34] presented a method of vision-based reinforcement learning by which a robot learns to shoot a ball into a goal. Tsitsiklis, J. et al. [35] developed a methodological framework and present a few different ways in which dynamic programming and compact representations can be combined to solve large scale stochastic control problems.

Cluster#2 is the third largest cluster, containing 38 references across a 9-year period from 2012 to 2020. The median year of all references in this cluster is 2016 and the silhouette value of the cluster is 1. The labels of this cluster selected by LLR are deep learning in robotics and automation, learning from demonstration and robot sensing systems. Among 38 members of this cluster, the top 5 references with high burstiness are surveyed as follows. Gu, S. et al. [36] demonstrated that a recent deep reinforcement learning algorithm based on off-policy training of deep Q-functions can scale to complex 3D manipulation tasks and can learn deep neural network policies efficiently enough to train on real physical robots. Todorov, E. et al. [37] described a new physics engine MuJoCo tailored to model-based control. Zhu, Y. et al. [38] speeded up softmax computations in DNN-based large vocabulary speech recognition by senone weight vector selection. Levine, S. et al. [39] described a learning-based approach to hand-eye coordination for robotic grasping from monocular images. Tobin, J. et al. [40] discussed domain randomization for transferring deep neural networks from simulation to the real world.

Cluster#7 contains 35 references across an 8-year period from 2012 to 2019. The median year of all references in this cluster is 2016 and the silhouette value of the cluster is 0.981. The labels of this cluster selected by LLR are smart grid, adaptive traffic signal control and multi-agent system. Among 35 members of this cluster, the top 5 references with high burstiness are surveyed as follows. Wu, J. et al. [41] researched continuous reinforcement learning of energy management with deep q network for a power split hybrid electric bus. Li, L. et al. [42] proposed a set of algorithms to design signal timing plans via deep reinforcement learning. Mocanu, E. et al. [43] explored for the first time in the smart grid context the benefits of using deep reinforcement learning, a hybrid type of methods that combines reinforcement learning with deep learning, to perform on-line optimization of schedules for building energy management systems. El-Tantawy, S. et al. [44] presented the development and evaluation of a novel system of multi-agent reinforcement learning for integrated network of adaptive traffic signal controllers. Ruelens, F. *et al.* [45] extended fitted Q-iteration, a standard batch RL technique, to the situation when a forecast of the exogenous data is provided.

Cluster#12 contains 30 references across a 7-year period from 2013 to 2019. The median year of all references in this cluster is 2017 and the silhouette value of the cluster is 1. The labels of this cluster selected by LLR are mobile edge computing, servers and computation offloading. Among 35 members of this cluster, the top 5 references with high burstiness are surveyed as follows. Lewis, F. L. *et al.* [46] described mathematical formulations for reinforcement learning and a practical implementation method known as adaptive dynamic programming. Al-Tamimi, A. *et al.* [47] gave the convergence proof of the discrete-time nonlinear HJB solution using approximate dynamic programming. Vamvoudakis, K. G. *et al.* [48] discussed an online algorithm based on policy iteration for learning the continuous-time (CT) optimal control solution with infinite horizon cost for nonlinear systems with known dynamics. Zhang, H. *et al.* [49] proposed a novel data-driven robust approximate optimal tracking control scheme for unknown general nonlinear systems by using the adaptive dynamic programming (ADP) method. Modares, H. *et al.* [50] developed an online learning algorithm to solve the linear quadratic tracking (LQT) problem for partially-unknown continuous-time systems.

3.6 Keywords Distribution

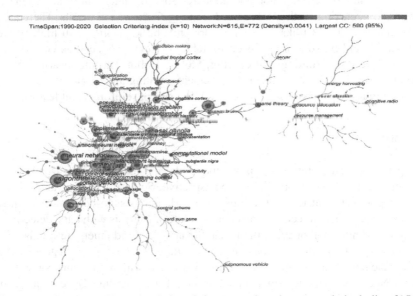

Fig. 7. Keyword co-occurring network for reinforcement learning research, including 615 nodes and 772 edges. Top 5 frequently used keywords are reinforcement learning, reward, neural network, q-learning, and representation.

Table 4. Top 20 keywords with high betweenness centralities ordered by starting year.

Keywords	Frequents	Betweenness	Strength	Begin	End	1990 - 2020
Neural Network	555	0.17	35.83	1992	2006	
Genetic Algorithm	102	0.02	15	1994	2008	
Dynamic Programming	55	0.15	11.33	1994	2002	
Q-Learning	384	0.14	6.77	1994	2006	
Mobile Robot	64	0.02	7.95	1995	2003	
Markov Decision Processes	64	0.01	13.47	1999	2014	
Temporal Difference	7	0.09	3.89	2002	2010	
Reward	624	0.14	25.38	2003	2012	
Representation	134	0.02	6.08	2005	2011	
Multi-Agent System	114	0.08	4.16	2010	2013	
Feedback Control	40	0.01	14.31	2013	2018	
Policy Iteration	61	0.07	8.82	2013	2015	
Evolution	68	0.00	4.36	2013	2015	
Zero Sum Game	28	0.04	8.11	2015	2020	
Wireless Sensor Network	56	0.01	6.64	2017	2020	
Particle Swarm Optimization	15	0.00	6.79	2018	2020	
Smart Grid	28	0.01	6.44	2018	2020	
Big Data	29	0.01	5.29	2018	2020	
Inverse Reinforcement Learning	26	0.01	4.74	2018	2020	
Cognitive Radio	77	0.05	4.35	2018	2020	

In order to find the research hotspots about reinforcement learning research in detail, a keyword co-occurring method were used, and Fig. 7 shows the results. There are 615 keyword nodes and 772 links in the network and the keywords with high betweenness centrality and burstiness are listed in Table 4. Keywords such as Genetic Algorithm, Dynamic Programming, Q-Learning, Mobile Robot, Particle Swarm Optimization, Smart Grid, Big Data, Inverse reinforcement learning and Cognitive Radio which can be used to represent evolving the research frontier of reinforcement learning.

4   Results

In this paper, we quantitatively explore the landscape and emerging trends of reinforcement learning from the bibliometric pespective. Scientific outputs analysis (year, subject categories, Countries, Institutions,), network analysis (co-occurrence frequency, betweenness centrality, and community) and evolution analysis with burst detection was conducted to vividly demonstrate the research landscape and emerging trends of reinforcement learning. Results indicate that reinforcement learning research goes up significantly for the past three decades. In addition, keywords with highly burst strength such as Neural Network, Genetic Algorithm, Dynamic Programming, Q-Learning, Mobile Robot, Policy Iteration, Wireless Sensor Network, Particle Swarm Optimization, Smart Grid, Big Data, Inverse Reinforcement Learning and Cognitive Radio predicts the emerging trends of reinforcement learning research, these valuable conclusions could serve as a potential guide for future RL research, and the tools we developed are also helpful to other research topics analysis.

References

1. Sutton, R.S., Barto, A.G.: Reinforcement Learning: An Introduction (1988)
2. Selvaraju, R.R., Das, A., Vedantam, R., Cogswell, M., Parikh, D., Batra, D.: Grad-CAM: visual explanations from deep networks via gradient-based localization. Int. J. Comput. Vis. **128**(2), 336–359 (2020)
3. Dai, H., Khalil, E.B., Zhang, Y., Dilkina, B., Song, L.: Learning combinatorial optimization algorithms over graphs. In: NIPS 2017 Proceedings of the 31st International Conference on Neural Information Processing Systems, vol. 30, pp. 6351–6361 (2017)
4. Luong, N.C., et al.: Applications of deep reinforcement learning in communications and networking: a survey. IEEE Commun. Surv. Tutor. **21**(4), 3133–3174 (2019)
5. Levine, S., Finn, C., Darrell, T., Abbeel, P.: End-to-end training of deep visuomotor policies. J. Mach. Learn. Res. **17**(1), 1334–1373 (2016)
6. Kober, J., Andrew Bagnell, J., Peters, J.: Reinforcement learning in robotics: a survey. Int. J. Robot. Res. **32**(11), 1238–1274 (2013)
7. Abdulhai, B., Pringle, R., Karakoulas, G.J.: Reinforcement learning for true adaptive traffic signal control. J. Transp. Eng.-ASCE **129**(3), 278–285 (2003)
8. Xiong, R., Cao, J., Yu, Q.: Reinforcement learning-based real-time power management for hybrid energy storage system in the plug-in hybrid electric vehicle. Appl. Energy **211**, 538–548 (2018)
9. Olivecrona, M., Blaschke, T., Engkvist, O., Chen, H.: Molecular de-Novo design through deep reinforcement learning. J. Cheminformatics **9**(1), 48 (2017)
10. Zhou, Z., Li, X., Zare, R.N.: Optimizing chemical reactions with deep reinforcement learning. ACS Cent. Sci. **3**(12), 1337–1344 (2017)
11. Fan, C., Zeng, L., Sun, Y., Liu, Y.-Y.: Finding key players in complex networks through deep reinforcement learning. Nat. Mach. Intell. **2**, 317–324 (2020)
12. Kaelbling, L.P., Littman, M.L., Moore, A.W.: Reinforcement learning: a survey. J. Artif. Intell. Res. **4**(1), 237–285 (1996)
13. Busoniu, L., Babuska, R., De Schutter, B.: A comprehensive survey of multiagent reinforcement learning. Syst. Man Cybern. **38**(2), 156–172 (2008)
14. Pritchard, A.: Statistical bibliography or bibliometrics. J. Documentation **25**, 348 (1969)
15. Fan, C., Zeng, L., Ding, Y., Chen, M., Sun, Y., Liu, Z.: Learning to identify high betweenness centrality nodes from scratch: a novel graph neural network approach. In: Proceedings of the 28th ACM International Conference on Information and Knowledge Management, pp. 559–568 (2019)
16. Garfield, E.: Citation indexes for science: a new dimension in documentation through association of ideas. Science **122**(3159), 108–111 (1955)
17. Chen, C.: CiteSpace II: detecting and visualizing emerging trends and transient patterns in scientific literature. J. Am. Soc. Inf. Sci. **57**(3), 359–377 (2006)
18. van Eck, N.J., Waltman, L.: Software survey: VOSviewer, a computer program for bibliometric mapping. Scientometrics **84**(2), 523–538 (2010)
19. Kleinberg, J.: Bursty and hierarchical structure in streams. Data Min. Knowl. Disc. **7**(4), 373–397 (2003)
20. Egghe, L.: Theory and practise of the g-index. Scientometrics **69**(1), 131–152 (2006)
21. Schvaneveldt, R.W.: Pathfinder Associative Networks: Studies in Knowledge Organization (1990)
22. Brandes, U.: A faster algorithm for betweenness centrality. J. Math. Sociol. **25**(2), 163–177 (2001)

23. Aryadoust, S.V., Tan, H.A.H., Ng, L.Y.: A scientometric review of Rasch measurement: the rise and progress of a specialty. Front. Psychol. **10**, 2197 (2019)
24. Ng, A.Y., Jordan, M.I., Weiss, Y.: On spectral clustering: analysis and an algorithm. Adv. Neural. Inf. Process. Syst. **14**, 849–856 (2001)
25. Rousseeuw, P.: Silhouettes: a graphical aid to the interpretation and validation of cluster analysis. J. Comput. Appl. Math. **20**(1), 53–65 (1987)
26. Lewis, F.L., Vrabie, D., Vamvoudakis, K.G.: Reinforcement learning and feedback control: using natural decision methods to design optimal adaptive controllers. IEEE Control Syst. Mag. **32**(6), 76–105 (2012)
27. Liu, D., Wei, Q.: Policy iteration adaptive dynamic programming algorithm for discrete-time nonlinear systems. IEEE Trans. Neural Netw. **25**(3), 621–634 (2014)
28. Bhasin, S., Kamalapurkar, R., Johnson, M., Vamvoudakis, K.G., Lewis, F.L., Dixon, W.E.: A novel actor-critic-identifier architecture for approximate optimal control of uncertain nonlinear systems. Automatica **49**(1), 82–92 (2013)
29. Karimpour, A., Naghibi-Sistani, M.B.: Reinforcement Q-learning for optimal tracking control of linear discrete-time systems with unknown dynamics. Automatica **50**(4), 1167–1175 (2014)
30. Vamvoudakis, K.G., Lewis, F.L., Hudas, G.R.: Multi-agent differential graphical games: online adaptive learning solution for synchronization with optimality. Automatica **48**(8), 1598–1611 (2012)
31. Mahadevan, S., Connell, J.: Automatic programming of behavior-based robots using reinforcement learning. Artif. Intell. **55**(2), 311–365 (1992)
32. Tsitsiklis, J.N., Van Roy, B.: An analysis of temporal-difference learning with function approximation. IEEE Trans. Autom. Control **42**(5), 674–690 (1997)
33. Jaakkola, T., Jordan, M.I., Singh, S.P.: On the convergence of stochastic iterative dynamic programming algorithms. Neural Comput. **6**, 1185–1201 (1994)
34. Asada, M., Noda, S., Tawaratsumida, S., Hosoda, K.: Purposive behavior acquisition for a real robot by vision-based reinforcement learning. Mach. Learn. **23**(2), 279–303 (1996)
35. Tsitsiklis, J.N., van Roy, B.: Feature-based methods for large scale dynamic programming. Mach. Learn. **22**(1), 59–94 (1996)
36. Gu, S., Holly, E., Lillicrap, T., Levine, S.: Deep reinforcement learning for robotic manipulation with asynchronous off-policy updates. In: 2017 IEEE International Conference on Robotics and Automation (ICRA), pp. 3389–3396 (2017)
37. Todorov, E., Erez, T., Tassa, Y.: MuJoCo: a physics engine for model-based control. In: 2012 IEEE/RSJ International Conference on Intelligent Robots and Systems, pp. 5026–5033 (2012)
38. Zhu, Y., Mak, B.: Speeding up softmax computations in DNN-based large vocabulary speech recognition by senone weight vector selection. In: 2017 IEEE International Conference on Acoustics, Speech and Signal Processing (ICASSP), pp. 5335–5339 (2017)
39. Levine, S., Pastor, P., Krizhevsky, A., Ibarz, J., Quillen, D.: Learning hand-eye coordination for robotic grasping with deep learning and large-scale data collection. Int. J. Robot. Res. **37**(4–5), 421–436 (2017)
40. Tobin, J., Fong, R., Ray, A., Schneider, J., Zaremba, W., Abbeel, P.: Domain randomization for transferring deep neural networks from simulation to the real world. In: 2017 IEEE/RSJ International Conference on Intelligent Robots and Systems (IROS), pp. 23–30 (2017)
41. Wu, J., He, H., Peng, J., Li, Y., Li, Z.: Continuous reinforcement learning of energy management with deep Q network for a power split hybrid electric bus. Appl. Energy **222**, 799–811 (2018)
42. Li, L., Lv, Y., Wang, F.-Y.: Traffic signal timing via deep reinforcement learning. IEEE/CAA J. Automatica Sinica **3**(3), 247–254 (2016)
43. Mocanu, E., et al.: On-line building energy optimization using deep reinforcement learning. IEEE Trans. Smart Grid **10**(4), 3698–3708 (2019)

44. El-Tantawy, S., Abdulhai, B., Abdelgawad, H.: Multiagent reinforcement learning for integrated network of adaptive traffic signal controllers (MARLIN-ATSC): methodology and large-scale application on downtown Toronto. IEEE Trans. Intell. Transp. Syst. **14**(3), 1140–1150 (2013)

45. Ruelens, F., Claessens, B.J., Vandael, S., De Schutter, B., Babuska, R., Belmans, R.: Residential demand response of thermostatically controlled loads using batch reinforcement learning. IEEE Trans. Smart Grid **8**(5), 2149–2159 (2017)

46. Lewis, F.L., Vrabie, D.: Reinforcement learning and adaptive dynamic programming for feedback control. IEEE Circuits Syst. Mag. **9**(3), 32–50 (2009)

47. Al-Tamimi, A., Lewis, F.L., Abu-Khalaf, M.: Discrete-time nonlinear HJB solution using approximate dynamic programming: convergence proof. Syst. Man Cybern. **38**(4), 943–949 (2008)

48. Vamvoudakis, K.G., Lewis, F.L.: Online actor-critic algorithm to solve the continuous-time infinite horizon optimal control problem. Automatica **46**(5), 878–888 (2010)

49. Zhang, H., Cui, L., Zhang, X., Luo, Y.: Data-driven robust approximate optimal tracking control for unknown general nonlinear systems using adaptive dynamic programming method. IEEE Trans. Neural Netw. **22**(12), 2226–2236 (2011)

50. Modares, H., Lewis, F.L.: Linear quadratic tracking control of partially-unknown continuous-time systems using reinforcement learning. IEEE Trans. Autom. Control **59**(11), 3051–3056 (2014)

Data Mining

NiaClass: Building Rule-Based Classification Models Using Nature-Inspired Algorithms

Luka Pečnik, Iztok Fister, and Iztok Fister Jr.[✉]

Faculty of Electrical Engineering and Computer Science, Koroška ul. 46,
2000 Maribor, Slovenia
luka.pecnik@student.um.si, {iztok.fister,iztok.fister1}@um.si

Abstract. Searching for a set of rules, with which the knowledge hidden in data is extracted, can also be applied for multi-class classification. In line with this, a collection of nature-inspired algorithms are selected for determining the set of rules capable of classifying the samples into three or more classes. This set is encoded into representation of individuals and undergoes acting the variation operators. The results of various nature-inspired algorithms, obtained after their application on more UCI classification databases, are compared with each other, and revealed that some of them can be potential candidates for real-world applications.

Keywords: Nature-inspired algorithms · Machine learning ·
AutoML · Classification pipeline

1 Introduction

A multi-class classification is a problem of classifying the instances (samples) into one of three or more classes. This is an extension of more known binary classification, where samples are classified into two classes only. The rule-based classification models identify and utilize a set of rules in the form of IF-THEN rules that, together, represent the knowledge captured by the algorithm. In the multi-class classification, the specific class to which a definite sample belongs, is determined by the consequent of the rule. Although there are a lot of traditional methods for solving this problem (e.g., Bayesian networks) [1], this study proposes the application of nature-inspired algorithms to this hard nut to crack. The main advantages of these algorithms over the traditional methods are that they are typically less computationally expensive [13]. Most of the statistical machine learning methods are described as complex mathematical functions, and they are rather incomprehensible and opaque to humans [15]. On the other hand, nature-inspired algorithms can ensure accuracy and comprehensibility – a fact that is important in Explainable Artificial Intelligence [2].

Studying the approaches based on the population-based metaheuristics for discovering the classification rules is still very popular in the research community [14]. Although the first approaches appeared more than 15 years ago with

© Springer Nature Switzerland AG 2021
Y. Tan and Y. Shi (Eds.): ICSI 2021, LNCS 12690, pp. 381–390, 2021.
https://doi.org/10.1007/978-3-030-78811-7_36

the rise of big data, exploring and developing new approaches is still very fruitful. One of the first algorithms for classification rule discovery was proposed by Parpinelli et al. and was called Ant-Miner [16]. Ant-miner was based on Ant Colony Optimization (ACO). On the contrary, the development of algorithms that were based on other inspirations of nature-inspired algorithms was also at full pace. The first algorithm based on Particle Swarm Optimization was proposed by Sousa back in 2004 [17]. A plethora of newer nature-inspired algorithms, in combination with some conventional machine learning methods, are also used for solving classification tasks [3,12]. It is worth mentioning that there is a very thin line between algorithms for Numerical Association Rule Mining (NARM) [9] and algorithms for discovering the classification rules. In other words, NARM algorithms with smaller modifications can also be applied for discovering classification rules.

This paper presents an extension of an approach that was published last year at the ICCS conference [8], where a new method for discovering classification rules based was proposed on the Firefly Algorithm (FA). This method was tailored only for solving binary classification problems. In this paper we step further, and extend this method for also coping with multi-class classification problems.

The main contributions of this paper are the following:

- extension of binary classification to multi-class classification,
- inspired by the recent NARM algorithm [10], new measures are modeled in the fitness function,
- extensive experimental comparison among different nature-inspired algorithms are conducted.

The structure in the remainder of the paper is as follows. Section 2 discusses the basic information needed for understanding the subjects that follow. In Sect. 3, the design of the proposed algorithm for three-class classification is explained in detail. The experiments and results are the subjects of Sect. 4, while the paper concludes with Sect. 5, that also outlines the directions for the future development.

2 Basic Information

This section is devoted to discussing information potential readers need to understand the subjects treated in the remainder of the paper. In line with this, the basics are discussed about the concept of nature-inspired algorithms. The section is concluded with a description of the NiaClass concept, within which the algorithm began to be developed.

2.1 Nature-Inspired Algorithms

Primarily, the nature-inspired population-based algorithms comprise two families: (1) Evolutionary Algorithms (EA) [7], and (2) Swarm Intelligence (SI) based algorithms [4]. The former are inspired by the Darwinian struggle for existence, where, similarly as in nature also in simulated evolution, only the fittest individuals (i.e., solutions) can survive in the hard environmental conditions (simulated by the fitness function). One of the more prominent members of this class is undoubtedly Differential Evolution (DE) [18]. The latter mimics the behavior of social living insects and animals revealing some kinds of optimization process. Obviously, the social living swarm of birds present good examples of these behaviors (inspiration for the Particle Swarm Optimization (PSO) [11]) and swarm of insects, precisely fireflies (inspiration for the Firefly Algorithm (FA) [19]).

Indeed, this paper focuses on three nature-inspired algorithms for solving rule-based classification models, i.e., DE, PSO, and FA, operating using real-valued representation. The DE is well known evolutionary algorithm especially appropriate for continuous optimization [5]. The PSO and FA are SI-based algorithms suitable for solving problems arising in almost all application domains. Among the mentioned algorithms, the FA was successfully applied also for classification problems [8].

2.2   NiaClass - A Classification Platform in Python

The basic concept of the NiaClass classifier is to use stochastic nature-inspired population-based algorithms implemented in Python programming language in order to find the optimal set of classification rules for a given datasets [8]. The full source code of NiaClass software is available on Github[1]. Actually, the authors plan to extend a collection of those algorithms in the future as well.

3   Proposed Method

Although the proposed method enables multi-class classification of an arbitrary number of classes, here, we are focused on three-class classification problem due to its simplicity. The problem can be defined formally as follows: Let us assume a 3-class classification problem on database Db consisting of M features is given, where each feature attribute is a tuple:

$$Attr_k = \langle type, off, cont, [\mathbf{D}|\mathbf{c}] \rangle, \quad \text{for } k = 1, \ldots, M, \tag{1}$$

and

$$type \; - \; \text{set of attribute types, i.e. } type \in \{num, cat\},$$
$$off \; - \; \text{attribute offset within solution vector } \mathbf{x}_i,$$
$$cont \; - \; \text{control random variable,}$$
$$\mathbf{D} \; - \; \text{vector of numeric 3-class attribute domains,}$$
$$\mathbf{c} \; - \; \text{vector of discrete 3-class attributes.}$$

Thus, the vector of the three-class attribute domain is defined as $\mathbf{D} = (D_1, D_2, D_3)$, where $D_l = [lb, ub]$ for $l = 1, \ldots, 3$, and the vector of discrete

---

[1] https://github.com/lukapecnik/NiaClass.

3-class attributes as $\mathbf{c} = (c_1, c_2, c_3)$ with elements $c_l \in AttrSet_k$ for $l = 1, \ldots, 3$. Indeed, attributes consist of rules for classifying the sample into the corresponding class. Then, the solution vector of the nature-inspired algorithm is represented as follows:

$$\mathbf{x}_i = (x_{i,1}, \ldots, x_{i,L}), \quad \text{for } i = 1, \ldots, N, \tag{2}$$

where N is the population size, and L denotes the number of elements, calculated as:

$$L = (1 + 2 \cdot class \cdot \#numeric\_attr) + (1 + class \cdot \#category\_attr) + 1, \tag{3}$$

where $class$ denotes the number of classes (in our case, $class = 3$), $\#numeric\_attr$ the number of numerical features, and $\#category\_attr$ is the number of categorical features. Interestingly, the numerical attributes are represented as a tuple $\langle x_{Attr_k.off}, \ldots, x_{Attr_k.off+6} \rangle$ consisting of seven, and the categorical attributes as a tuple $\langle x_{Attr_k.off}, \ldots, x_{Attr_k.off+3} \rangle$ consisting of four elements.

The variable $cont \in \{0, 1\}$ of the k-th attribute is mapped from the search space to the problem space as follows:

$$Attr_k.cont = \begin{cases} 0, & \text{if } x_{Attr_k.off \leq threshold}, \\ 1, & \text{otherwise.} \end{cases} \tag{4}$$

However, when the variable $cont = 0$, the feature of this class is omitted from the classification.

The k-th element representing the numeric 3-class attribute domains is mapped from the solution vector as follows:

$$Attr_k^{(num)}.D_l.lb = \frac{ub_l - lb_l}{UB_l - LB_l} \cdot lb_l + lb_l.$$
$$\quad \text{for } l = 1, \ldots, 3, \tag{5}$$
$$Attr_k^{(num)}.D_l.ub = \frac{ub_l - lb_l}{UB_l - LB_l} \cdot ub_l + lb_l,$$

where $ub_l = Attr_k.off.x_{2(l-1)+2}$ and $lb_l = Attr_k.off.x_{2(l-1)+1}$. Obviously, the relation $ub_k > lb_k$ must be ensured by the equation.

Similarly, the k-th element representing the categorical attribute is mapped from the solution vector as follows:

$$Attr_k^{(cat)}.c_l = \lfloor Attr_k.off.x_{2(l-1)+1} * |Attr_k^{(cat)}.c_l| \rfloor, \tag{6}$$

where the term $|Attr_k^{(cat)}.c_l|$ denotes the number of the attributes describing the feature.

Finally, the attributes are composed into IF-THEN rules as follows:

$$P(Attr_1, S_1) \wedge \ldots \wedge P(Attr_k, S_k), \wedge \ldots \wedge (Attr_M, S_M) \Rightarrow class, \tag{7}$$

where $P(.)$ denotes a predicate returning either $true$ or $false$, $Attr_k$ the attribute of the k-th feature and S_k the corresponding value of the feature in the sample, and $class = \{class_1, class_2, class_3\}$ is the three-class classification set. If the k-th attribute is categorical, the predicate verifies if the discrete value of the attribute is equal to one of the three predicted values of the $Attr_k^{(cat)}$, while, in the case of a numerical attribute, it verifies if the sample value is higher than the lower bound and lower than the higher bound of the numeric attribute $Attr_k^{(num)}$ in each of the observed three-classes. This means that there are three classification rules decoded per one solution vector.

The mapping from the search to the problem space is illustrated in Fig. 1, where the values of variables *cont* are represented in black, *threshold* in red, while all the other values are painted in the corresponding class colors, i.e., a set of rules for $class_1$ is represented in green, rules for $class_2$ are represented in blue and rules for $class_3$ in brown.

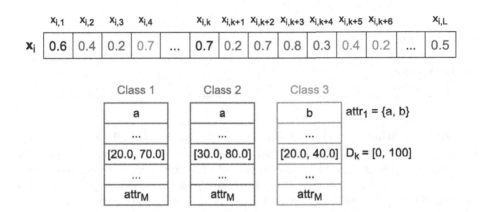

Fig. 1. Search space to problem space mapping. (Color figure online)

The fitness function in the optimization process of the proposed method is calculated after applying the mapped solution vector *Attr* to the classification problem on problem database *Db*, and it is composed of three terms as follows:

$$f(Attr, Db) = -score() + 0.5 \cdot ldc - 0.5 \cdot oc, \tag{8}$$

where the first term is *score* function, that can be any classification metric such as accuracy, precision, F1-score or Cohen's κ, in other words:

$$score = \{accuracy, precision, \text{F1-}score, \kappa\}.$$

The value *ldc* denotes the length differences' coefficient, and *oc* denotes the coefficient of intervals' overlapping. Both coefficients were multiplied by the weight of 0.5, as we expect that the metric should still be the most important factor of the final fitness.

The pseudo-code of the proposed method is presented in the Algorithm 1, from which it can be see that the algorithm suits the general form of the nature-

---

Algorithm 1. NiaClass for building rule-based classification models

1: $\langle \mathbf{D}, \mathbf{c} \rangle \leftarrow$ preprocess characteristics of numeric/discrete features (Db)
2: $population \leftarrow$ initialize real valued vectors $\mathbf{x}_i$ randomly
3: $\langle best\_fitness, Attr\_best \rangle \leftarrow$ find the best ($population$)
4: **while** not termination condition met **do**
5:      **for each** $\mathbf{x}_i \in population$ **do**
6:         $\mathbf{x}_{trial} \leftarrow$ modify individual using variation operators ($\mathbf{x}_i$)
7:         $Attr \leftarrow$ decode ($\mathbf{x}_{trial}$)
8:         $fitness = f(Attr, Db))$
9:         **if** $fitness$ is better than 'decode ($\mathbf{x}_i$)' **then**
10:            $\mathbf{x}_i \leftarrow \mathbf{x}_{trial}$ ▷ Replace the worse individual
11:         **end if**
12:         **if** $fitness$ is better than $best\_fitness$ **then**
13:            $best\_fitness \leftarrow fitness$
14:            $Attr\_best \leftarrow Attr$
15:         **end if**
16:      **end for**
17: **end while**
18: Return best set of rules $Attr\_best$

---

inspired algorithms. After initialization, evaluation of initial population, and finding the best solution (line 1 and 3), each individual undergoes acting the variation operators (line 6) by generation of trial solution. The rule-based classification model is constructed by appropriate genotype-phenotype mapping as described in Eq. 7 (line 7) and evaluated according to Eq. 8 (line 8). If the quality of constructed pipeline from the trial is better than the quality of individual, the individual is replaced with the trial (lines 9–11). In similar way, the best solution with corresponding fitness is determined in lines 12–15.

4   Experiments and Results

The goal of our experimental work was to show that the proposed NiaClass classifier is suitable for both, binary and multi-class classification problems, and that by adding certain information obtained through preprocessing of the fitness function, we can further improve the final results. In line with this, extensive experimental work was conducted, where all experiments were performed on an HP ProDesk 400 G6 MT computer running Microsoft Windows 10, an Intel (R) Core (TM) i7-9700 CPU @ 3.00 GHz processor, and 8 GB of installed physical memory.

A list of datasets from the UCI Machine Learning Repository [6] used in the experiments and their characteristics are shown in Table 1. The missing values of the numerical attributes were replaced by the mean of the feature, while the missing categorical values were replaced by the mode of the feature in the case of the Cylinder Bands dataset.

Table 1. Datasets used in the experiments.

Dataset	Type of attributes	Instances	Features	Missing data
Haberman	Integer	306	3	No
Ecoli	Real	336	8	No
Yeast	Real	1484	8	No
Cylinder Bands	Real, integer, categorical	512	39	Yes

When fitting the NiaClass classifier, we used a population size of 90 individuals using 5000 evaluations of the fitness function. In the optimization process, we used the Differential Evolution (DE), Firefly Algorithm (FA) and Particle Swarm Optimization (PSO). Settings of their parameters are shown in Table 2.

Table 2. Parameter settings of the optimization algorithms in the experiment.

Algorithm	Parameters
DE	$F = 1,\ CR = 0.8$
FA	$alpha = 0.5,\ betamin = 0.2,\ gamma = 1.0$
PSO	$C1 = 2.0,\ C2 = 2.0,\ w = 0.7,\ vMin = -1.5,\ vMax = 1.5$

The final results were obtained after 25 independent runs of each algorithm for each dataset using: (1) The basic fitness function that does not take into account the information about the numerical intervals, and (2) The extended fitness function capable of using the information from the dataset preprocessing step. Before each run, the dataset was divided into training and test sets in a ratio of 0.2, and after fitting, the classification was performed using the test set. The results are shown in Table 3.

Interestingly, nature-inspired algorithms using the extended fitness function failed to dominate their counterparts using the basic fitness function on the observed datasets. Although we expected that using the extended fitness function would improve the results significantly, this claim holds only for the results obtained on the Haberman dataset. Indeed, the best results were achieved by the PSO algorithm when comparing in terms of the optimization algorithms.

4.1 Discussion

As can be seen from the results of our experiments, we were unable to improve the results obtained by the nature-inspired algorithms using the basic fitness function compared with those achieved by using the the extended fitness function. The cause for this was most likely due to the fact that the fitness function for determining the class of an individual from the training set was very simple,

Table 3. Detailed results of classification according to accuracy.

Dataset	Fitness	Algorithm	Min	Max	Mean	Median	Std
Haberman	Basic	DE	0.6452	0.8226	0.7323	0.7258	0.0434
		FA	0.6290	0.8065	0.7381	0.7419	0.0462
		PSO	0.6290	0.7903	0.7335	0.7419	0.0422
	Upgraded	DE	0.6613	0.8387	0.7310	0.7419	0.0451
		FA	0.6613	0.8065	**0.7400**	0.7419	0.0346
		PSO	0.6290	**0.8710**	0.7245	0.7258	0.0605
Ecoli	Basic	DE	0.5441	0.7794	0.6476	0.6618	0.0803
		FA	0.3235	0.6471	0.5347	0.5441	0.0765
		PSO	0.5735	**0.8088**	**0.7006**	0.7059	0.0750
	Upgraded	DE	0.4412	0.7941	0.6335	0.6471	0.0717
		FA	0.3971	0.6618	0.5376	0.5441	0.0800
		PSO	0.3088	**0.8088**	0.6429	0.6471	0.1168
Cylinder Bands	Basic	DE	0.6019	0.7222	0.6778	0.6852	0.0334
		FA	0.5833	0.7685	0.6607	0.6574	0.0416
		PSO	0.6111	**0.7870**	**0.6985**	0.7037	0.0437
	Upgraded	DE	0.5370	0.7593	0.6419	0.6389	0.0483
		FA	0.5278	0.7037	0.6274	0.6204	0.0502
		PSO	0.5833	0.7130	0.6563	0.6574	0.0366
Yeast	Basic	DE	0.3367	0.4646	0.3954	0.3939	0.0335
		FA	0.2593	0.4276	0.3390	0.3401	0.0402
		PSO	0.3333	**0.4983**	**0.4168**	0.4141	0.0389
	Upgraded	DE	0.2626	0.3670	0.3244	0.3266	0.0260
		FA	0.2290	0.4074	0.3088	0.3131	0.0404
		PSO	0.3131	0.4478	0.3636	0.3603	0.0363

and, therefore, additional coefficients considered in the extended fitness function did not contain enough information about the quality of the rule based classifiers, or even the selected weights were inappropriate. It was also difficult to determine the parameter setting, and the suitability of the selected optimization algorithms for the experimental datasets is also questionable.

Indeed, the purpose of the study was to show that the problem of rule-based classification models could be solved successfully using nature-inspired algorithms beside the traditional methods. This preliminary work has proven this hypothesis. In line with this, the FA has exposed the best results among the algorithms in test, while the DE algorithm did not turn out too well. Therefore, new experiments must be conducted in order to show the best characteristics of this kind of algorithms necessary for achieving better results in solving the problem.

5   Conclusion

The rule-based multi-class classification is very interesting area of machine learning that was typically solved using traditional methods, like Bayesian networks. In this paper, the classification problem was solved using the nature-inspired algorithms, where each set of rules was encoded into a representation of individuals and indicated the property of a specific class. Three different nature-inspired algorithms were employed in this study, i.e., DE, PSO, and FA. The first algorithm belongs to a class of EAs, while the other two to the SI-based algorithms' family. Normally, all three algorithms operate with real-valued representation, and are obviously included into the NiaClass repository.

The proposed algorithms were applied to four UCI Machine Learning Repository datasets. The results were observed according to two fitness functions, i.e., the basic and extended, where the latter also explores the preprocessing information. Although the results of experiments showed that using the extended measure by the nature-inspired algorithms did not improve the results of the classification significantly, the study revealed the potential of the proposed algorithms especially in the sense of their complexity.

Although the authors are aware that the preliminary results are slightly worse than the results achieved by the traditional algorithms, they found many potential directions for improving the results in the future. For instance, better results could be achieved by finding the optimal algorithms, their parameter settings and fitness function weights using a separate optimization process. We could also try to improve the fitness function of the NiaClass classifier's fitting method by mining other potentially useful information from datasets that we could consider in the fitness value, or even by implementing a separate procedure that would literally, in some way, build a fitness function for each data set separately.

Acknowledgement. The authors acknowledge the financial support from the Slovenian Research Agency (research core funding No. P2-0041 - Digital twin).

References

1. Aly, M.: Survey on multiclass classification methods. Technical report, Caltech (2005)
2. Arrieta, A.B., et al.: Explainable Artificial Intelligence (XAI): concepts, taxonomies, opportunities and challenges toward responsible AI. Inf. Fusion **58**, 82–115 (2020)
3. Banchhor, C., Srinivasu, N.: Integrating cuckoo search-grey wolf optimization and correlative naive Bayes classifier with map reduce model for big data classification. Data Knowl. Eng. **127**, 101788 (2020)
4. Blum, C., Merkle, D.: Swarm Intelligence: Introduction and Applications, 1st edn. Springer Publishing Company Incorporate, Heidelberg (2008). https://doi.org/10.1007/978-3-540-74089-6
5. Das, S., Suganthan, P.N.: Differential evolution: a survey of the state-of-the-art. IEEE Trans. Evol. Comput. **15**(1), 4–31 (2011)

6. Dua, D., Graff, C.: UCI machine learning repository (2017). http://archive.ics.uci.edu/ml

7. Eiben, A.E., Smith, J.E.: Introduction to Evolutionary Computing, 2nd edn. Springer Publishing Company Incorporated, Heidelberg (2015). https://doi.org/10.1007/978-3-662-44874-8

8. Fister, I., Fister, I., Fister, D., Vrbančič, G., Podgorelec, V.: On the potential of the nature-inspired algorithms for pure binary classification. In: Krzhizhanovskaya, V.V., et al. (eds.) ICCS 2020, Part V. LNCS, vol. 12141, pp. 18–28. Springer, Cham (2020). https://doi.org/10.1007/978-3-030-50426-7_2

9. Fister, I. Jr., Fister, I.: A brief overview of swarm intelligence-based algorithms for numerical association rule mining. arXiv preprint arXiv:2010.15524 (2020)

10. Fister Jr., I., Podgorelec, V., Fister, I.: Improved nature-inspired algorithms for numeric association rule mining. In: Vasant, P., Zelinka, I., Weber, G.-W. (eds.) ICO 2020. AISC, vol. 1324, pp. 187–195. Springer, Cham (2021). https://doi.org/10.1007/978-3-030-68154-8_19

11. Kennedy, J., Eberhart, R.: Particle swarm optimization. In: Proceedings of ICNN 1995 - International Conference on Neural Networks, vol. 4, pp. 1942–1948 (1995)

12. Kulhari, A., Pandey, A., Pal, R., Mittal, H.: Unsupervised data classification using modified cuckoo search method. In 2016 Ninth International Conference on Contemporary Computing (IC3), pp. 1–5. IEEE (2016)

13. Liu, B., Abbass, H.A., McKay, B.: Density-based heuristic for rule discovery with ant-miner. In: The 6th Australia-Japan Joint Workshop on Intelligent and Evolutionary System, vol. 184. Citeseer (2002)

14. Martens, D., Baesens, B., Fawcett, T.: Editorial survey: swarm intelligence for data mining. Mach. Learn. **82**(1), 1–42 (2011). https://doi.org/10.1007/s10994-010-5216-5

15. Martens, D., De Backer, M., Haesen, R., Vanthienen, J., Snoeck, M., Baesens, B.: Classification with ant colony optimization. IEEE Trans. Evol. Comput. **11**(5), 651–665 (2007)

16. Parpinelli, R.S., Lopes, H.S., Freitas, A.A.: Data mining with an ant colony optimization algorithm. IEEE Trans. Evol. Comput. **6**(4), 321–332 (2002)

17. Sousa, T., Silva, A., Neves, A.: Particle swarm based data mining algorithms for classification tasks. Parallel Comput. **30**(5–6), 767–783 (2004)

18. Storn, R., Price, K.: Differential evolution - a simple and efficient heuristic for global optimization over continuous spaces. J. Global Optim. **11**, 341–359 (1997). https://doi.org/10.1023/A:1008202821328

19. Yang, X.-S.: Nature-Inspired Optimization Algorithms. Academic Press, Cambridge, MA, United States (2020)

Mining Neighbor Frames for Person Re-identification by Global Optimal Tracking

Kai Han[1,2], Jinho Lee[3], Lang Huang[1], Fangcheng Liu[1], Seiichi Uchida[4], and Chao Zhang[1(✉)]

[1] Peking University, Beijing, China
{hankai,laynehuang,equation,c.zhang}@pku.edu.cn
[2] Noah's Ark Lab, Huawei Technologies, Beijing, China
[3] PLK Technologies, Seoul, South Korea
[4] Kyushu University, Fukuoka, Japan
uchida@ait.kyushu-u.ac.jp

Abstract. Person re-identification is a challenging task aiming to identify the same person across different cameras. However, most of existing image-based person re-identification methods neglect the spatial and
temporal constraint, the information in neighbor frames of each person
image is rarely exploited by previous studies. In this paper, we propose a novel neighbor frames mining framework (NFM) to exploit the
spatial-temporal information. For each gallery image, we use a dynamic
programming-based global optimal tracking method to search images of
the same person in its neighbor frames. From those images, the image
features extracted by the shared convolutional neural network (CNN) in
the constructed neighbor sequence are merged via an attention weighted
averaging technology. To this end, a novel supervised attention mechanism is designed for dealing with tracking errors. The final feature with
multi-view and robust information is used for matching. Experimental
results show the superiority and efficiency of the proposed method on
two benchmark datasets including DukeMTMC-reID and PRW.

Keywords: Person Re-ID · Optimal tracking · Neighbor frames
mining

1 Introduction

Person re-identification is a realistic task in surveillance video, which aims to
find occurrences of the query person across different cameras. According to the
type of retrieval item, it can be divided into two categories: image-based person
re-identification and video-based person re-identification. In image-based person
re-identification, given a query image of one specific person, the task is to return
all the images of the same person in gallery database. The image is replaced by
video (or say image sequence) in video-based person re-identification.

© Springer Nature Switzerland AG 2021
Y. Tan and Y. Shi (Eds.): ICSI 2021, LNCS 12690, pp. 391–406, 2021.
https://doi.org/10.1007/978-3-030-78811-7_37

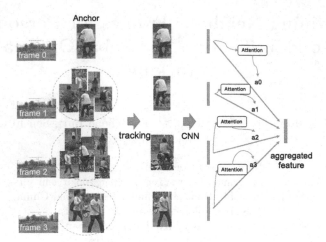

Fig. 1. Feature extraction pipeline for anchor image in our method (frame $-3, -2, -1$ are omitted for simplicity).

Research on image-based person re-identification has achieved impressive progress in the last several years, especially boosted by deep learning methods [3,7,15,32,35,45,49,52,53]. However, single image contains only one view of the person, which means the visual information available is limited and biased for this person. Instead, video sequence which consists of many frames can provide more information. Thus video-based person re-identification has appealed much investigation recently [12,17,18,43,48,50]. However, video-based methods usually need carefully annotated video, that is, person image tracklets. In addition, operations on video need much more computational and storage resources. Moreover, the person image sequences are produced by object detector, thus video-based methods may fail when the detector fails. And people consisting of few images are usually discarded in video-based methods.

In this paper, we propose to utilize person images in neighbor frames to help image-based person re-identification. The key idea is to get multiple images of the target person by a visual object tracking method and aggregate those images to extract a feature vector that represents the person. If the tracking result gives sufficient appearance variations of the person, we can expect more reliable re-identification results. Visual object tracking itself is still active research area; this means that object tracking is also a difficult task. Fortunately, for person re-identification, we do not need to track the target person from the starting frame to the end. As shown by a later experimental result, it is enough to track it just for a short period. This fact is also beneficial for computational cost because a longer-period tracking requires more computations. After the neighbor sequence construction stage, we get the neighbor target images for each person image.

We will also introduce a novel attention mechanism to select reliable target person images from the tracking result. Although we use a dynamic programming (DP)-based globally optimal tracking method, its result sometimes contains an image of a wrong person due to essential difficulty of visual object tracking (e.g., occlusion). The proposed attention mechanism is trained with person identity labels and automatically selects (weights) appropriate images, as shown in Fig. 1. The image features in neighbor sequence are merged by linear combination using attention weights, forming the aggregated representation for each person image.

We term the proposed framework neighbor frames mining (NFM) method, which can utilize information of neighbor frames efficiently to learn more robust feature of persons, thus improve the re-identification performance. Experiments on two large-scale benchmarks demonstrate the robustness and effectiveness of our method with largely improved state-of-the-art performance. The main contributions of this paper include: 1) we present a novel and efficient NFM framework for person image representation learning; 2) our architecture adopts dynamic programming algorithm to track images of the same person as anchor image, and the idea of utilizing neighbor frame images has seldom been studied before in the context of person re-identification; 3) in order to avoid the negative impact of false images in tracking, a novel supervised attention mechanism is proposed to suppress noisy images and preserve positive images.

The rest of the paper is organized as follows. In the next section, relevant prior works are briefly reviewed. And then, the proposed NFM method is described, detailed experiments and discussion of the results are presented. Finally, we conclude the paper in the last section.

2 Related Work

During the last several years, CNN [9,11,29,33] based methods have achieved much progress on person re-identification. These works use CNN to learn a discriminative feature for each person image. Classification loss [37,52] or metric learning loss [3,35,45,52] is used to supervise the training process. Classification based methods treat each person as one category and train the model in classification manner. In metric learning methods, image pairs [35,45] or triplets [3,47] are fed into CNN learn an optimized metric. Some works use generative adversarial network (GAN) to generate new data for training, which can be viewed as a type of data augmentation [53,54]. Aside from them, another line of research proposed to enhance the robustness by local feature learning. These works divide the input image or the feature map into small patches [30] or stripes [32] to extract region features from the local patches or stripes. Some other works incorporated pose [39] or human mask [7,15] to extract local feature. These works are orthogonal to ours.

Recently, video-based person re-identification has received much attention [12,48,50] since video contains more information than single image. The operation item is video sequence and some methods are proposed to incorporate time flows. Max/average pooling is widely used manner to merge features of images in video [6,50]. Some papers [23,42] used model based on recurrent neural network (RNN) to learn the feature of person from video sequence while another line of works proposed to use attention [17,18,41] or Graph Convolutional Networks

(GCN) [43] to aggregate spatial-temporal features. The disadvantages of video-based methods include needing careful annotation and much more computational and storage costs.

Idea in person tracking is adopted in this work, and the tracking-by-detection approach is the most related [1,14,16,25,34]. Tracking is performed directly on person bounding boxes in tracking-by-detection methods. One common formulation for person tracking is DP based methods [1,14]. Tang et al. [34] used person re-identification to help improving person tracking. Compared to tracking methods with network flow, recurrent neural network, and so on, DP is faster and more efficient. Thus, tracking by DP is used for finding useful person images in neighbor frames in our method.

Attention mechanism has been widely used in computer vision and natural language processing, such as machine translation [36], image recognition [4,10], visual question answering [38,44], image captioning [40], semantic segmentation [13,46] and object detection [2]. Many works [7,8,15,19,49] have also proposed to use attention schemes to align human part representation or non-human part representation [7] in Person Re-identification. Attention modules in those works are learned by latent supervision of the final loss function. Here we focus on soft attention mechanism which computes a weighted combination of the factors to focus on the important parts when performing a particular task. Within the image sequence in our task, the importance weights of true images should be large and false images should be zero. In this paper, we propose a novel supervised attention mechanism for feature aggregation of image sequence.

3   Proposed Method

Suppose we have the cropped person images from the raw frames extracted by hand or existing person detection method, such as DPM [5], Faster R-CNN [26], SSD [22], etc.. For each person image, its neighbor sequence consisting of the images of the same person is constructed. We use the sequence rather than a single image to represent the person, then extract a multi-view and robust feature. This feature is used for re-identification. We describe the details of the proposed method in the following.

3.1   Neighbor Sequence Construction

For each person image A, we can construct its neighbor sequence containing images of the same person from its previous T frames and next T frames. Here we describe how to construct the sequence of the next T frames, and the previous one can be similarly constructed. Suppose there are K^τ person bboxes $\{A_j^\tau, j = 1, \cdots, K^\tau\}$ in the τ-th next frame ($\tau = 1, \cdots, T$), we have to select one bbox in each frame for each anchor. First, the cropped image A is fed into a shared CNN $g(\cdot)$ for feature extraction, which is formulated as $g(A)$. Then, we use dynamic programming algorithm to construct neighbor sequence.

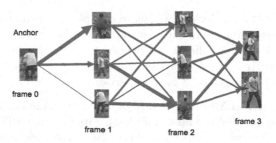

Fig. 2. Neighbor sequence construction by DP tracking (frame $-T, \cdots, -1$ are omitted for simplicity), where thinner line means lower cost. The optimal solution with the lowest cumulative cost is the red path. (Color figure online)

Dynamic Programming. As shown in Fig. 2, we can create a directed acyclic graph (DAG) with each image as a node. We connect nodes in successive frames with fully connected manner. Define the probability that every person image which is the same person as the anchor as

$$p_j^\tau = \frac{g(A)g(A_j^\tau)^T}{\|g(A)\|\|g(A_j^\tau)\|}, \tag{1}$$

and the probability that the two images connected by every edge belong to the same person as

$$p_{i,j}^\tau = \frac{g(A_i^{\tau-1})g(A_j^\tau)^T}{\|g(A_i^{\tau-1})\|\|g(A_j^\tau)\|}, \tag{2}$$

where $\|\cdot\|$ is the ℓ_2 norm. On the one hand, we want to select the person images which are the same person as the anchor; on the other hand, the appearance of the same person may change largely from the anchor over time, so we need to take the similarity between images in subsequent frames into consideration. The cost on each edge is defined as the negative log likelihood:

$$c(A_i^{\tau-1}, A_j^\tau) = -\log\left(p_{i,j}^\tau \cdot p_j^\tau\right). \tag{3}$$

In Fig. 2, thinner line means lower cost. To get a tracklet of the anchor image A, we can use dynamic programming algorithm to find an optimal path with the lowest cumulative cost in this graph [14]. For the previous T frames, we construct the sequence by reverse way, and the final neighbor sequence is denoted as $S_i = \{I_{-T}^i, \cdots, I_0^i, \cdots, I_T^i\}$.

3.2 Representation Learning with Attention

Now we have neighbor sequence $S_i = \{I_{-T}^i, \cdots, I_0^i, \cdots, I_T^i\}$ for each anchor[1] I_0^i and the corresponding label vectors $\{y_{-T}^i, \cdots, y_0^i, \cdots, y_T^i\}$. The images in the

---

[1] Where A and I_0^i are the same.

tracking path provide multi-view and more complete information. We design a model to learn representation of the sequence, replacing the single image feature as shown in Fig. 1. For simplicity, we denote $g(I_\tau^i)$ as $\boldsymbol{f}_\tau^i$. One simple way to merge the features in the sequence is averaging: $\bar{\boldsymbol{f}}_i = \frac{1}{2T+1} \sum_\tau \boldsymbol{f}_\tau^i$, which is served as average baseline model for comparison.

Supervised Attention Mechanism. In practice, tracking is not so perfect that some images in the tracklet are false samples. Simple average will include the features of false samples, which will lower down the recognition performance. Thus we propose a novel supervised attention mechanism to suppress noisy images. As shown in Fig. 3, the attention module is a simple neural network with two fully connected layers $g_{att}(\cdot)$, taking the concatenation of $\boldsymbol{f}_\tau^i$ and $\boldsymbol{f}_0^i$ as input. For each image in the sequence, the attention weight is calculated as follow:

$$\hat{a}_\tau^i = g_{att}([\boldsymbol{f}_\tau^i, \boldsymbol{f}_0^i]). \tag{4}$$

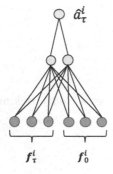

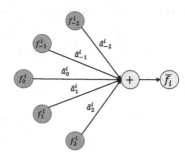

Fig. 3. The attention network $g_{att}(\cdot)$, including two fully connected layers.

Fig. 4. The aggregated feature is a linear combination of raw features.

As shown in Fig. 4, the final aggregated feature is obtained by the weighted average of features:

$$\bar{\boldsymbol{f}}_i = \frac{\sum_\tau \hat{a}_\tau^i \boldsymbol{f}_\tau^i}{\sum_\tau \hat{a}_\tau^i}. \tag{5}$$

Different from the attention mechanism in prior literature which is only supervised latently by the final loss, the proposed supervised attention mechanism can be supervised directly. We can get the ground-truth weights by the label of images, that is, $a_\tau^i = 1$ if I_τ^i and I_0^i belong to the same person, $a_\tau^i = 0$ otherwise. Mean average error loss is used to supervise the learning of attention weights:

$$\mathcal{L}_{mse} = \sum_{\tau=-T}^{T} \left(a_\tau^i - \hat{a}_\tau^i \right)^2. \tag{6}$$

3.3 Training and Testing

Total Loss. For each image and each sequence, we use a shared fully connected layer for classification. The loss for classification is softmax cross-entropy loss

$$\hat{y}^i_\tau = \text{softmax}(f^i_\tau W + b), \tag{7}$$

$$\mathcal{L}_{img} = -\sum_{\tau=-T}^{T} \sum_{j=1}^{C} y^i_{\tau,j} \log \hat{y}^i_{\tau,j}, \tag{8}$$

where $\hat{y}^i_\tau$ is the predicted category probability vector, $W \in \mathbb{R}^{d \times C}$, $b \in \mathbb{R}^C$ are the weight matrix and bias vector. Similarly, the loss for sequence classification is softmax cross-entropy loss

$$\hat{y}_i = \text{softmax}(\bar{f}_i W + b), \tag{9}$$

$$\mathcal{L}_{seq} = -\sum_{j=1}^{C} y_{i,j} \log \hat{y}_{i,j}, \tag{10}$$

where $\hat{y}_i$ is the predicted category probability.

The total loss in training is

$$\mathcal{L} = \mathcal{L}_{seq} + \alpha \mathcal{L}_{img} + \beta \mathcal{L}_{mse}, \tag{11}$$

where α and β are the trade-off hyper-parameters.

End-to-End Training. The entire model in our method is differentiable and can be trained in end-to-end manner using stochastic gradient descent (SGD) algorithm via back-propagation. Particularly, the gradient computation of our attention modules is quite simple via the chain rule. Let $\partial \mathcal{L}_{seq}/\partial \bar{f}_i$ be the gradient of the loss function $\mathcal{L}_{seq}$ w.r.t. $\bar{f}_i$, then we have

$$\frac{\partial \mathcal{L}}{\partial \hat{a}^i_\tau} = \frac{\partial \mathcal{L}_{seq}}{\partial \hat{a}^i_\tau} + \beta \frac{\partial \mathcal{L}_{mse}}{\partial \hat{a}^i_\tau} \tag{12}$$

$$= \frac{\partial \mathcal{L}_{seq}}{\partial \bar{f}_i} \left(\frac{\partial \bar{f}_i}{\partial \hat{a}^i_\tau} \right)^T + \beta \frac{\partial \mathcal{L}_{mse}}{\partial \hat{a}^i_\tau} \tag{13}$$

$$= \frac{\partial \mathcal{L}_{seq}}{\partial \bar{f}_i} \left(\frac{f^i_\tau}{\sum_\tau \hat{a}^i_\tau} - \frac{\sum_\tau \hat{a}^i_\tau f^i_\tau}{(\sum_\tau \hat{a}^i_\tau)^2} \right)^T + 2\beta \left(\hat{a}^i_\tau - a^i_\tau \right). \tag{14}$$

For the gradient w.r.t. f^i_τ, we have

$$\frac{\partial \mathcal{L}}{\partial f^i_\tau} = \frac{\partial \mathcal{L}_{seq}}{\partial f^i_\tau} + \alpha \frac{\partial \mathcal{L}_{img}}{\partial f^i_\tau} + \beta \frac{\partial \mathcal{L}_{img}}{\partial f^i_\tau} \tag{15}$$

$$= \sum_j \frac{\partial \mathcal{L}_{seq}}{\partial \bar{f}_{i,j}} \frac{\partial \bar{f}_{i,j}}{\partial f^i_\tau} + \alpha \frac{\partial \mathcal{L}_{img}}{\partial f^i_\tau} + \beta \frac{\partial \mathcal{L}_{mse}}{\partial \hat{a}^i_\tau} \frac{\partial \hat{a}^i_\tau}{\partial f^i_\tau} \tag{16}$$

$$= \frac{\hat{a}^i_\tau}{\sum_\tau \hat{a}^i_\tau} \frac{\partial \mathcal{L}_{seq}}{\partial \bar{f}_i} + \alpha \frac{\partial \mathcal{L}_{img}}{\partial f^i_\tau} + 2\beta \left(\hat{a}^i_\tau - a^i_\tau \right) \frac{\hat{a}^i_\tau}{\partial f^i_\tau}, \tag{17}$$

where $\partial \mathcal{L}_{img}/\partial \boldsymbol{f}_\tau^i$ is standard gradient of softmax loss, $\partial \hat{a}_\tau^i/\partial \boldsymbol{f}_\tau^i$ is the gradient of the attention value $\hat{a}_\tau^i$ (that is $g_{att}([\boldsymbol{f}_\tau^i, \boldsymbol{f}_0^i]))$ $w.r.t.$ $\boldsymbol{f}_\tau^i$. We can compute the gradients in other parts in a similar way, and if we get more than one gradients $w.r.t.$ the same variable, summing them forms the final gradient. After all the gradients computed, the model is updated by stochastic gradient descent method.

Testing Stage. For the query image I_q, we construct its sequence by repeating itself, that is, $S_q = \underbrace{\{I_q, \cdots, I_q\}}_{2T+1}$. For the gallery image, its sequence is constructed using the above method. After feature extraction, we use the concatenation of single image feature and sequence merged feature as the final representation. We compute and sort the cosine similarity between the query and all the gallery descriptors to get the ranking result.

4   Experiments

4.1   Datasets and Evaluation Protocol

PRW. PRW dataset [51] is a large-scale person search dataset extracted from a 10-hour video captured in campus, including 11,816 video frames (picked every 25 frames from raw video), 34,304 bounding boxes belonging to 932 identities. It provides 5,134 frames (482 persons) for training, and the testing set contains 2,057 query images and a gallery of 6,112 frames.

DukeMTMC-reID. DukeMTMC-reID [53] is a subset of the DukeMTMC [27] for person re-identification, which is captured by 8 cameras. Pedestrian images are cropped from the videos every 120 frames, yielding in total 16,522 training images of 702 identities, 2,228 query images of the other 702 identities and 17,661 gallery images.

Evaluation Protocol. Following [51] and [53], we adopt the mean averaged precision (mAP) and the top-k matching accuracy ($k = 1, 5, 20$) as evaluation metrics. For PRW dataset, Faster R-CNN is used to conduct person detection. We evaluate our method under different number of detected bounding boxes per image. True positives are defined by IoU > 0.5, that is, a candidate window is positive if the overlap with a ground truth is larger than 0.5, following the setup in previous works [21,37].

4.2   Implementation Details

Our models are implemented using PyTorch [24]. We use ResNet-50 pre-trained on ImageNet as base network and fine-tune it on target dataset for 50 epochs. Stochastic gradient descent (SGD) with batch size of 32 is used to update the parameters of the network. The learning rate is initialized as 0.001 and then set to 0.0001 in the final 10 epochs. We use a momentum of $\mu = 0.9$ and weight decay $\lambda = 5 \times 10^{-4}$. The size of input image is 224×224 for all the experiments. Random crop and random flip are used for data augmentation.

We set $\alpha = 1.0$ and $\beta = 0.5$ empirically. If person detection is needed, we use Faster R-CNN based on VGG16 [26], where the confidence threshold is set as 0.01. All the experiments are conducted on an NVIDIA Pascal TITAN X GPU and Intel Xeon CPU E5-2620.

4.3   PRW Dataset

Comparison with Baselines. We use simple ID classification CNN model as baseline. Another baseline model is simple average instead of attention mechanism. We use 9 frames ($T = 4$) in the experiment. From the results in Fig. 5, we can see that simple average of 9 neighbor selected images outperforms the baseline with a large margin, which indicates that the features from neighbor frames provide additional multi-view and robust information. Our model with attention mechanism can further improve simple average model. It benefits from the attention mechanism, which activates the same person images and suppress the noisy images (see examples in Sect. 4.5). We also observe that #bbox has important impact on re-identification performance. When #bbox is few, images of the target person may be not included with high probability. And when too many bboxes are detected, more junk gallery images will have a negative influence impact. The observations generally conform with results in prior works [51].

Comparison with SOTA. We compared our model with recent state-of-the-art methods considering 5 bboxes per frame and the results are listed in Table 1. Our method outperforms other methods on both mAP and top-1 accuracy.

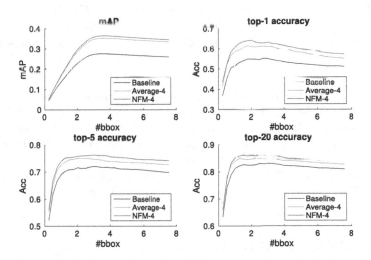

Fig. 5. Results of baseline, average model and our model on PRW dataset. The number in legend represents the value of T.

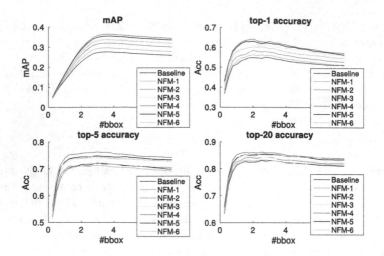

Fig. 6. Performance *w.r.t.* T on PRW dataset. The number in legend represents value of T, and #bbox means the number of bboxes extracted per frame.

Table 1. Comparison with state-of-the-art methods on PRW dataset.

Methods	Feature-CNN	#bbox	mAP (%)	Top-1 (%)
DPM-Alex+LOMO+XQDA [20]	-	5	13.0	34.1
DPM-Alex+IDE$_{det}$ [51]	Alex	5	20.3	47.4
DPM-Alex+IDE$_{det}$+CWS [51]	Alex	5	20.5	48.3
OIM [37]	ResNet-50	5	21.3	49.9
NPSM [21]	ResNet-50	-	24.2	53.1
Baseline	ResNet-50	5	27.01	52.46
Average ($T = 4$)	ResNet-50	5	34.66	58.14
NFM ($T = 4$)	ResNet-50	5	**35.79**	**59.80**

Impact of Parameter T. The hyper-parameter T controls the number of neighbor frames to be mined. It is an important parameter to our model. We conduct experiments to investigate the sensitiveness of our model with respect to T. The results are shown in Fig. 6. From the results, we can see that the best performance occurs when $T = 4$. When hyper-parameter T increase from 1 to 4, mAP and accuracy increase. When T continues to increase, the performance decrease gradually. This process is the balance of information and noise from neighbor frames.

4.4   DukeMTMC-ReID Dataset

Comparison with State-of-the-Art and Baselines. We compare our model with other state-of-the-art methods, and the results are listed in Table 2. All the methods based on deep learning defeat traditional hand-crafted method in

[20]. The compared methods utilize strategies such as advanced loss [37,52], data augmentation [53,54], revised network [31], *etc.* Our method with neighbor frames mining can get the best mAP and competitive top-1 accuracy among all the methods. We also test the baseline model and simple average model. The simple average model outperforms baseline ResNet-50 by a large margin due to multi-view feature from more selected images. Our NFM model with attention mechanism improves simple average model significantly, since attention mechanism can preserve the correct images and suppress the noisy images, which can be seen intuitively in Sect. 4.5.

Table 2. Comparison with state-of-the-art and baselines on DukeMTMC-reID dataset.

Methods	Feature-CNN	mAP (%)	Top-1 (%)
LOMO+XQDA [20]	-	17.04	30.75
LSRO [53]	ResNet-50	47.13	67.68
OIM [37]	ResNet-50	-	68.9
Verif + Identif [52]	ResNet-50	49.3	68.1
ACRN [28]	GoogleNet	51.96	72.58
SVDNet [31]	ResNet-50	56.8	**76.7**
Baseline	ResNet-50	41.08	60.41
Average ($T = 4$)	ResNet-50	55.67	70.92
NFM ($T = 4$)	ResNet-50	**58.32**	72.68

Table 3. Performance *w.r.t.* T on DukeMTMC-reID dataset.

Method	Baseline	Average	NFM					
		$T = 4$	$T = 1$	$T = 2$	$T = 3$	$T = 4$	$T = 5$	$T = 6$
mAP (%)	41.08	55.67	45.08	52.30	55.11	**58.32**	56.25	49.52
Top-1 (%)	60.41	70.92	62.57	68.13	70.20	**72.68**	70.15	66.74

Impact of Parameter T. On DukeMTMC-reID dataset, we also analyze the impact of hyper-parameter T. As shown in Table 3, the phenomenon here is similar to that in PRW dataset, that is, the best performance occur when $T = 4$ and smaller or larger T decrease the accuracy gradually.

4.5 Visualization

We visualize one example of the tracking result using DP and its attention values in Fig. 7. From the tracking result, DP tracking success if there exist the target person in current time step. However, it often happens that no target

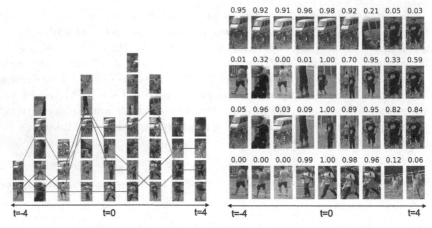

(a) tracking result, where one line represents one tracking path.

(b) attention result, where one row represents one tracking path and the values on
top of images are attention weights.

Fig. 7. Visualization of tracking and attention in PRW dataset.

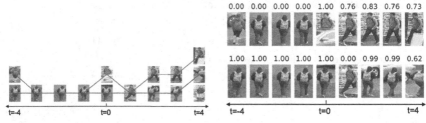

(a) tracking result, where one line represents one tracking path.

(b) attention result, where one row represents one tracking path and the values on
top of images are attention weights.

Fig. 8. Visualization of tracking and attention in DukeMTMC-reID dataset.

person appear in some frames. The attention mechanism is used to suppress
the track-failing images. There are 4 anchor images when $t = 0$, so we get 4
tracking paths. The anchor image in the first path is a car which means a junk
gallery image, though, the attention values are pretty large for correct images
and small for incorrect images. The anchor images in the second and the third
paths are the same person, and the attention results are similar. The images
from $t = 1$ to $t = 4$ are target person and their attention weights are large
enough, while attention weights at $t = -4, -2, -1$ (not target person) are small.
A bad case occur at $t = -3$ where the person look extremely similar to the
target person. The attention result in the forth path is almost perfect as we can
see. Visualization result of DukeMTMC-reID dataset is shown in Fig. 8. Since
the number of persons in each frame is fewer than that of PRW dataset, the

situation is simpler. The tracking result and attention result are pretty good with almost perfect tracking and reasonable attention weights.

4.6 Inference Time

In large-scale person re-identification task, inference time is an important property aspect. Here, we compare the inference time of our method and single image baseline of ResNet-50 on PRW dataset. Compared to single image based methods, the additional time of our method is DP tracking and attention weighted average. Since DP is an fast and well designed algorithm and attention module is a small neural network, the extra time is negligible compared to that of deep CNN forward pass. The results in Table 4 show that the total time to get the final feature per gallery image in our method is ~1.4× that in baseline, which is acceptable.

Table 4. Feature extraction time per image in gallery on PRW dataset

Methods	Time (ms)
CNN	2.430
CNN + tracking	2.845
CNN + tracking + attention	3.298

5   Conclusions

In this paper, we propose a novel neighbor frames mining framework for person representation learning. Person tracking technology based on dynamic programming is used to find out the images of the same person as anchor image. Simple averaging the features of the tracking images can form a more discriminative representation than single image feature. Through the novel supervised attention mechanism, merging the features via attention weighted averaging can preserve positive neighbor images and suppress negative ones, thus further improve the performance of the final feature. Experimental results show our method achieves the state-of-the-art performance and the speed of our model is on the same level as single image based CNN.

Acknowledgements. This work is supported in part by National Key R&D Program of China under Grant 2018AAA0100300 and National Nature Science Foundation of China under Grant 62071013 and 61671027.

References

1. Berclaz, J., Fleuret, F., Fua, P.: Robust people tracking with global trajectory optimization. In: CVPR, vol. 1, pp. 744–750. IEEE (2006)

2. Carion, N., Massa, F., Synnaeve, G., Usunier, N., Kirillov, A., Zagoruyko, S.: End-to-end object detection with transformers. In: Vedaldi, A., Bischof, H., Brox, T., Frahm, J.-M. (eds.) ECCV 2020. LNCS, vol. 12346, pp. 213–229. Springer, Cham (2020). https://doi.org/10.1007/978-3-030-58452-8_13

3. Ding, S., Lin, L., Wang, G., Chao, H.: Deep feature learning with relative distance comparison for person re-identification. Pattern Recogn. **48**(10), 2993–3003 (2015)

4. Dosovitskiy, A., et al.: An image is worth 16x16 words: transformers for image recognition at scale. In: ICLR (2021)

5. Felzenszwalb, P.F., Girshick, R.B., McAllester, D., Ramanan, D.: Object detection with discriminatively trained part-based models. TPAMI **32**(9), 1627–1645 (2010)

6. Gao, J., Nevatia, R.: Revisiting temporal modeling for video-based person reid. In: BMVC (2018)

7. Guo, J., Yuan, Y., Huang, L., Zhang, C., Yao, J.G., Han, K.: Beyond human parts: dual part-aligned representations for person re-identification. In: ICCV, pp. 3642–3651 (2019)

8. Han, K., Guo, J., Zhang, C., Zhu, M.: Attribute-aware attention model for fine-grained representation learning. In: ACMMM, pp. 2040–2048 (2018)

9. Han, K., Wang, Y., Tian, Q., Guo, J., Xu, C., Xu, C.: GhostNet: more features from cheap operations. In: CVPR, pp. 1580–1589 (2020)

10. Han, K., Xiao, A., Wu, E., Guo, J., Xu, C., Wang, Y.: Transformer in transformer. arXiv:2103.00112 (2021)

11. He, K., Zhang, X., Ren, S., Sun, J.: Deep residual learning for image recognition. In: CVPR, pp. 770–778 (2016)

12. Hirzer, M., Beleznai, C., Roth, P.M., Bischof, H.: Person re-identification by descriptive and discriminative classification. In: Heyden, A., Kahl, F. (eds.) SCIA 2011. LNCS, vol. 6688, pp. 91–102. Springer, Heidelberg (2011). https://doi.org/10.1007/978-3-642-21227-7_9

13. Huang, L., Yuan, Y., Guo, J., Zhang, C., Chen, X., Wang, J.: Interlaced sparse self-attention for semantic segmentation. arXiv:1907.12273 (2019)

14. Jiang, H., Fels, S., Little, J.J.: A linear programming approach for multiple object tracking. In: CVPR, pp. 1–8. IEEE (2007)

15. Kalayeh, M.M., Basaran, E., Gökmen, M., Kamasak, M.E., Shah, M.: Human semantic parsing for person re-identification. In: CVPR, pp. 1062–1071 (2018)

16. Lee, J., Iwana, B.K., Ide, S., Hayashi, H., Uchida, S.: Globally optimal object tracking with complementary use of single shot multibox detector and fully convolutional network. In: Paul, M., Hitoshi, C., Huang, Q. (eds.) PSIVT 2017. LNCS, vol. 10749, pp. 110–122. Springer, Cham (2018). https://doi.org/10.1007/978-3-319-75786-5_10

17. Li, J., Wang, J., Tian, Q., Gao, W., Zhang, S.: Global-local temporal representations for video person re-identification. In: ICCV, pp. 3958–3967 (2019)

18. Li, S., Bak, S., Carr, P., Wang, X.: Diversity regularized spatiotemporal attention for video-based person re-identification. In: CVPR, pp. 369–378 (2018)

19. Li, W., Zhu, X., Gong, S.: Harmonious attention network for person re-identification. In: CVPR, pp. 2285–2294 (2018)

20. Liao, S., Hu, Y., Zhu, X., Li, S.Z.: Person re-identification by local maximal occurrence representation and metric learning. In: CVPR, pp. 2197–2206 (2015)

21. Liu, H., et al.: Neural person search machines. In: ICCV (2017)

22. Liu, W., et al.: SSD: single shot multibox detector. In: Leibe, B., Matas, J., Sebe, N., Welling, M. (eds.) ECCV 2016. LNCS, vol. 9905, pp. 21–37. Springer, Cham (2016). https://doi.org/10.1007/978-3-319-46448-0_2

23. McLaughlin, N., del Rincon, J.M., Miller, P.: Recurrent convolutional network for video-based person re-identification. In: CVPR, pp. 1325–1334. IEEE (2016)
24. Paszke, A., et al.: Pytorch: an imperative style, high-performance deep learning library. In: NeurIPS, vol. 32, pp. 8026–8037 (2019)
25. Pirsiavash, H., Ramanan, D., Fowlkes, C.C.: Globally-optimal greedy algorithms for tracking a variable number of objects. In: CVPR, pp. 1201–1208. IEEE (2011)
26. Ren, S., He, K., Girshick, R., Sun, J.: Faster R-CNN: towards real-time object detection with region proposal networks. In: NeurIPS, pp. 91–99 (2015)
27. Ristani, E., Solera, F., Zou, R., Cucchiara, R., Tomasi, C.: Performance measures and a data set for multi-target, multi-camera tracking. In: Hua, G., Jégou, H. (eds.) ECCV 2016. LNCS, vol. 9914, pp. 17–35. Springer, Cham (2016). https://doi.org/10.1007/978-3-319-48881-3_2
28. Schumann, A., Stiefelhagen, R.: Person re-identification by deep learning attribute-complementary information. In: CVPR, pp. 1435–1443. IEEE (2017)
29. Simonyan, K., Zisserman, A.: Very deep convolutional networks for large-scale image recognition. In: ICLR (2015)
30. Subramaniam, A., Chatterjee, M., Mittal, A.: Deep neural networks with inexact matching for person re-identification. In: NeurIPS, pp. 2675–2683 (2016)
31. Sun, Y., Zheng, L., Deng, W., Wang, S.: SVDNet for pedestrian retrieval. In: CVPR, pp. 3800–3808 (2017)
32. Sun, Y., Zheng, L., Yang, Y., Tian, Q., Wang, S.: Beyond part models: person retrieval with refined part pooling (and a strong convolutional baseline). In: Ferrari, V., Hebert, M., Sminchisescu, C., Weiss, Y. (eds.) ECCV 2018. LNCS, vol. 11208, pp. 501–518. Springer, Cham (2018). https://doi.org/10.1007/978-3-030-01225-0_30
33. Tan, M., Le, Q.: EfficientNet: rethinking model scaling for convolutional neural networks. In: ICML, pp. 6105–6114. PMLR (2019)
34. Tang, S., Andriluka, M., Andres, B., Schiele, B.: Multiple people tracking by lifted multicut and person re-identification. In: CVPR, pp. 3539–3548 (2017)
35. Varior, R.R., Shuai, B., Lu, J., Xu, D., Wang, G.: A siamese long short-term memory architecture for human re-identification. In: Loibe, B., Matas, J., Sebe, N., Welling, M. (eds.) ECCV 2016. LNCS, vol. 9911, pp. 135–153. Springer, Cham (2016). https://doi.org/10.1007/978-3-319-46478-7_9
36. Vaswani, A., et al.: Attention is all you need. In: NeurIPS, vol. 30, pp. 5998–6008 (2017)
37. Xiao, T., Li, S., Wang, B., Lin, L., Wang, X.: Joint detection and identification feature learning for person search. In: CVPR, pp. 3376–3385. IEEE (2017)
38. Xu, H., Saenko, K.: Ask, attend and answer: exploring question-guided spatial attention for visual question answering. In: Leibe, B., Matas, J., Sebe, N., Welling, M. (eds.) ECCV 2016. LNCS, vol. 9911, pp. 451–466. Springer, Cham (2016). https://doi.org/10.1007/978-3-319-46478-7_28
39. Xu, J., Zhao, R., Zhu, F., Wang, H., Ouyang, W.: Attention-aware compositional network for person re-identification. In: CVPR, pp. 2119–2128 (2018)
40. Xu, K., et al.: Show, attend and tell: neural image caption generation with visual attention. In: ICML, pp. 2048–2057 (2015)
41. Xu, S., Cheng, Y., Gu, K., Yang, Y., Chang, S., Zhou, P.: Jointly attentive spatial-temporal pooling networks for video-based person re-identification. In: ICCV, pp. 4733–4742 (2017)

42. Yan, Y., Ni, B., Song, Z., Ma, C., Yan, Y., Yang, X.: Person re-identification via recurrent feature aggregation. In: Leibe, B., Matas, J., Sebe, N., Welling, M. (eds.) ECCV 2016. LNCS, vol. 9910, pp. 701–716. Springer, Cham (2016). https://doi.org/10.1007/978-3-319-46466-4_42

43. Yang, J., Zheng, W.S., Yang, Q., Chen, Y.C., Tian, Q.: Spatial-temporal graph convolutional network for video-based person re-identification. In: CVPR, pp. 3289–3299 (2020)

44. Yang, Z., He, X., Gao, J., Deng, L., Smola, A.: Stacked attention networks for image question answering. In: CVPR, pp. 21–29 (2016)

45. Yi, D., Lei, Z., Liao, S., Li, S.Z.: Deep metric learning for person re-identification. In: ICPR, pp. 34–39. IEEE (2014)

46. Yuan, Y., Huang, L., Guo, J., Zhang, C., Chen, X., Wang, J.: OCNet: object context network for scene parsing. In: IJCV (2021)

47. Zeng, K., Ning, M., Wang, Y., Guo, Y.: Hierarchical clustering with hard-batch triplet loss for person re-identification. In: CVPR, pp. 13657–13665 (2020)

48. Zhang, L., et al.: Ordered or orderless: a revisit for video based person re-identification. IEEE Trans. Pattern Anal. Mach. Intell. **43**(4), 1460–1466 (2021)

49. Zhang, Z., Lan, C., Zeng, W., Jin, X., Chen, Z.: Relation-aware global attention for person re-identification. In: CVPR, pp. 3186–3195 (2020)

50. Zheng, L., et al.: MARS: a video benchmark for large-scale person re-identification. In: Leibe, B., Matas, J., Sebe, N., Welling, M. (eds.) ECCV 2016. LNCS, vol. 9910, pp. 868–884. Springer, Cham (2016). https://doi.org/10.1007/978-3-319-46466-4_52

51. Zheng, L., Zhang, H., Sun, S., Chandraker, M., Yang, Y., Tian, Q.: Person re-identification in the wild. In: CVPR (2017)

52. Zheng, Z., Zheng, L., Yang, Y.: A discriminatively learned CNN embedding for person reidentification. ACM Trans. Multimedia Comput. Commun. Appl. (TOMM) **14**(1), 1–20 (2017)

53. Zheng, Z., Zheng, L., Yang, Y.: Unlabeled samples generated by GAN improve the person re-identification baseline in vitro. In: ICCV (2017)

54. Zhong, Z., Zheng, L., Zheng, Z., Li, S., Yang, Y.: Camera style adaptation for person re-identification. In: CVPR (2018)

Artificial Fish Swarm Algorithm for Mining High Utility Itemsets

Wei Song$^{(\boxtimes)}$ ⓘ, Junya Li, and Chaomin Huang

School of Information Science and Technology, North China University of Technology,
Beijing 100144, China
songwei@ncut.edu.cn

Abstract. The discovery of high utility itemsets (HUIs) is an attractive topic in data mining. Because of its high computational cost, using heuristic methods is a promising approach to rapidly discovering sufficient HUIs. The artificial fish swarm algorithm is a heuristic method with many applications. Except the current position, artificial fish do not record additional previous information, as other related methods do. This is consistent with the HUI mining problem: that the results are not always distributed around a few extreme points. Thus, we study HUI mining from the perspective of the artificial fish swarm algorithm, and propose an HUI mining algorithm called HUIM-AF. We model the HUI mining problem with three behaviors of artificial fish: follow, swarm, and prey. We explain the HUIM-AF algorithm and compare it with two related algorithms on four publicly available datasets. The experimental results show that HUIM-AF can discover more HUIs than the existing algorithms, with comparable efficiency.

Keywords: Data mining · Artificial fish swarm algorithm · High utility itemset · Position vector

1 Introduction

High utility itemsets (HUIs) are extensions of frequent itemsets (FIs) that consider both the unit profit and frequency of occurrence. HUI mining (HUIM) [2] is an active research topic in data mining, and various algorithms [6, 11] for it have been proposed. In contrast to the support measure (which is used in FI mining), utility (used in HUIM) does not satisfy the downward closure property. Hence, the computational cost of HUIM is high. Furthermore, for application fields such as recommender systems, it is not necessary to use all HUIs [13].

To reduce the burden of HUIM, heuristic methods—such as genetic algorithm (GA) [3] and particle swarm optimization (PSO) [5]—have been used for HUIM, to discover acceptable itemsets within a reasonable time. For these algorithms, HUIs are discovered iteratively, and results of one iteration affect the HUIs discovered in the next iteration. Thus, the resulting HUIs tend to be clustered around certain itemsets after many iterations, and the number of results is limited if no new individuals are generated randomly.

© Springer Nature Switzerland AG 2021
Y. Tan and Y. Shi (Eds.): ICSI 2021, LNCS 12690, pp. 407–419, 2021.
https://doi.org/10.1007/978-3-030-78811-7_38

It was also verified in [9] that diversity is of great importance for generating a greater number of HUIs in a smaller number of iterations.

In contrast to GA, PSO, and other heuristic methods that are used for HUIM, the artificial fish swarm algorithm (AFSA) only records the current position, but not other previous information, of each artificial fish (AF). This approach is essentially consistent with the problem of HUIM with a large number of diverse results. Therefore, in this paper, we use the AFSA to formulate an HUIM algorithm. The experiments show that the proposed algorithm can discover more HUIs than two other related algorithms.

2 Preliminaries

2.1 HUIM Problem

Let $I = \{i_1, i_2, ..., i_m\}$ be a finite set of items; each set $X \subseteq I$ is called an *itemset*. Let $D = \{T_1, T_2, ..., T_n\}$ be a transaction database. Each transaction $T_d \in D$, with unique identifier d, is a subset of I.

The *internal utility* $q(i_p, T_d)$ represents the quantity of item i_p in transaction T_d. The *external utility* $p(i_p)$ is the unit profit value of i_p. The *utility* of i_p in T_d is defined as $u(i_p, T_d) = p(i_p) \times q(i_p, T_d)$. The utility of itemset X in T_d is defined as $u(X, T_d) = \sum_{i_p \in X \wedge X \subseteq T_d} u(i_p, T_d)$. The utility of X in D is defined as $u(X) = \sum_{X \subseteq T_d \wedge T_d \in D} u(X, T_d)$. The *transaction utility* (TU) of transaction T_d is defined as $TU(T_d) = u(T_d, T_d)$. The *minimum utility threshold* δ, specified by the user, is a percentage of the total TU values of the database, and the *minimum utility value min_util* $= \delta \times \sum_{T_d \in D} TU(T_d)$. An itemset X is called an HUI if $u(X) \geq min\_util$. Given a transaction database D, the task of HUIM is to determine all itemsets whose utility is no less than *min_util*.

The *transaction-weighted utilization* (TWU) of itemset X is the sum of the transaction utilities of all the transactions containing X [6], and is defined as $TWU(X) = \sum_{X \subseteq T_d \wedge T_d \in D} TU(T_d)$. X is a high transaction-weighted utilization itemset (HTWUI) if $TWU(X) \geq min\_util$. An HTWUI containing k items is called a k-HTWUI.

Table 1. Example database.

TID	Transactions	TU
1	(B, 1), (C, 2), (D, 1), (F, 2)	15
2	(A, 4), (B, 1), (C, 3), (D, 1), (E, 1)	18
3	(A, 4), (C, 2), (D,1)	11
4	(C, 2), (D, 1), (E, 1)	11
5	(A, 5), (B, 2), (D, 1), (E, 2)	22
6	(A, 3), (B, 4), (C, 1), (D, 1)	17
7	(D, 1), (E, 1), (F, 1)	12

Table 2. Profit table.

Item	A	B	C	D	E	F
Profit	1	2	1	5	4	3

As a running example, consider the transaction database in Table 1 and the profit table in Table 2. For convenience, an itemset {B, E} is denoted by BE. The utility of E in T_2 is $u(E, T_2) = 4 \times 1 = 4$, the utility of BE in T_2 is $u(BE, T_2) = 2 + 4 = 6$, and the utility of BE in the example database is $u(BE) = u(BE, T_2) + u(BE, T_5) = 18$. Given $min\_util = 35$, because $u(BE) < min\_util$, BE is not an HUI. The TU of T_2 is $TU(T_2) = u(ABCDE, T_2) = 18$; the utility of each transaction is shown in the third column of Table 1. Because BE is contained in transactions T_2 and T_5, $TWU(BE) = TU(T_2) + TU(T_5) = 40$; therefore, BE is an HTWUI.

2.2 Basic Principle of AFSA

The AFSA [4], is inspired by the collective movement of fish and their typical social behaviors. For the AFSA, A_i represents the ith AF. The position of A_i, denoted by $X_i = < x_{i1}, x_{i2}, ..., x_{id} >$, represents a possible solution. In addition, the food concentration at position X is denoted by $Y = f(X)$. Let $d_{i,j} = \|X_i - X_j\|$ be the distance between two positions, X_i and X_j. An AF located at X_i inspects the search space around it, within its visual distance (VD). If there is a new position X_j such that $d_{i,j} \le VD$ and $f(X_j) > f(X_i)$, the AF moves a step toward X_j. To realize this principle, three behaviors of an AF are used iteratively.

Preying Behavior. Preying is the behavior whereby a fish moves to a location with the highest concentration of food. Letting $X_i(t)$ be the current position of the ith AF (A_i) at time t, the position of A_j randomly selected within the visual distance VD) is:

$$X_j = X_i(t) + VD \times rand(). \tag{1}$$

where $rand()$ produces a random number between 0 and 1. If $f(X_j) > f(X_i)$, A_i moves a step toward X_j, as follows:

$$X_i(t+1) = X_i(t) + S \times rand() \times \frac{X_j - X_i(t)}{\|X_j - X_i(t)\|}. \tag{2}$$

where S is the maximum length of a step that an AF can take at each movement. Otherwise, A_i selects a position X_j randomly again, using Eq. 1, and decides whether it satisfies the forward condition. If it cannot satisfy it after try_number times, it moves a step randomly, as follows:

$$X_i(t+1) = X_i(t) + VD \times rand(). \tag{3}$$

Swarming Behavior. In nature, a swarm of fish tends to assemble, so as to be protected from danger while avoiding overcrowded areas. Suppose that there are n_f AFs within distance VD of A_i, if the following conditions are satisfied:

$$\begin{cases} f(X_c) > f(X_i) \\ \frac{n_f}{n} < \delta \end{cases}.$$ (4)

where n is the total number of AFs, $\delta \in (0, 1)$ is the *crowding factor*, and $X_c = <x_{c1}, x_{c2}, ..., x_{cd}>$ is the central position within distance VD of A_i, whose elements are determined by:

$$x_{ck} = \sum_{m=1}^{n_f} x_{mk} / n_f.$$ (5)

where $1 \le k \le d$. This means that there is more food in the center, and the area is not overcrowded. Thus, $X_i(t)$ moves a step toward the companion center using:

$$X_i(t+1) = X_i(t) + S \times rand() \times \frac{X_c - X_i(t)}{\|X_c - X_i(t)\|}.$$ (6)

If the conditions in Eq. 4 are not satisfied, the preying behavior is executed instead.

Following Behavior. When a fish finds a location with a higher concentration of food, other fish follow. Let X_b be the position within distance VD of A_i with the highest food concentration. That is, for all X_j such that $d_{ij} \le VD$, $F(X_j) \le f(X_b)$. If the following conditions are satisfied:

$$\begin{cases} f(X_b) > f(X_i) \\ \frac{n_f}{n} < \delta \end{cases}.$$ (7)

A_i goes forward a step to X_b using:

$$X_i(t+1) = X_i(t) + S \times rand() \times \frac{X_b - X_i(t)}{\|X_b - X_i(t)\|}.$$ (8)

Otherwise, A_i executes a preying behavior.

3   Related Work

Inspired by biological and physical phenomena, heuristic methods are effective for solving combinatorial problems. Based on stochastic methods, heuristic methods can explore very large search spaces to find near-optimal solutions. Because HUIM is a task with high computational cost, heuristic methods are suitable for traversing the very large search spaces of HUIM within an acceptable time.

A GA was the first heuristic method used for HUIM, and two HUIM algorithms, HUPE$_{UMU}$-GARM and HUPE$_{WUMU}$-GARM, were proposed in [3]. The difference between them is that the second algorithm does not require a minimum utility threshold. These two algorithms tend to fall into local optima, leading to low efficiency and fewer mining results. Zhang et al. proposed an HUIM algorithm with four strategies [14]—neighborhood exploration, population diversity improvement, invalid combination avoidance, and HUI loss prevention—to improve the algorithm's performance.

PSO is another heuristic method used for HUIM. Lin et al. proposed an HUIM algorithm based on PSO with a binary coding scheme [5]. Song and Li proposed an HUIM algorithm based on set-based PSO [9]. The main difference is that the concept of cut set is used in the latter algorithm to improve the diversity of the resulting HUIs.

Other heuristic methods have also been used for HUIM, including an artificial bee colony (ABC) algorithm [7] and ant colony optimization (ACO) [12]. Furthermore, heuristic algorithms are also used for mining some other specific HUIs, such as top-k HUIs [10] and high average-utility itemsets [8].

4   Modeling HUIM Using AFSA

We use a *position vector* (PV) to represent the position of an AF. Letting HN be the number of 1-HTWUIs, a PV is represented by an HN-dimensional binary vector, in which each bit corresponds to one 1-HTWUI. Assuming that all 1-HTWUIs are sorted in a total order, if the kth 1-HTWUI appears in a PV, then bit k of the PV is set to 1; otherwise, the bit is set to 0. It is proved in [6] that an item with a TWU value lower than the minimum utility threshold cannot appear in an HUI. Thus, only 1-HTWUIs are considered for representation in a PV.

Letting P be a PV, the jth ($1 \leq j \leq NH$) bit of P is randomly initialized using roulette wheel selection with the probability:

$$Pr(P(j)) = \frac{TWU(i_j)}{\sum_{k=1}^{NH} TWU(i_k)}. \tag{9}$$

In the same manner as other related work [5, 7], we use the utility of the itemset directly as the object for optimization. Letting X be the itemset corresponding to P,

$$f(P) = u(X). \tag{10}$$

5 The Proposed HUIM-AF Algorithm

5.1 Algorithm Description

Algorithm 1 describes our HUIM algorithm, HUIM-AF.

Algorithm 1	HUIM-AF
Input	Transaction database D, minimum utility value min_util, population size N, maximum number of iterations max_iter, maximum number of attempts try_number
Output	HUIs
1	Initialize N PVs using Eq. 9;
2	$SHUI = \varnothing$;
3	$iter = 1$;
4	**while** $iter \leq max\_iter$ **do**
5	**for** $i=1$ to N **do**
6	is_follow = false;
7	is_swarm = false;
8	$P_i = \text{Follow}(P_i)$;
9	**if**(!is_follow) **then**
10	$\text{Swarm}(P_i)$;
11	**end if**
12	**if**(!is_follow AND !is_swarm) **then**
13	$\text{Prey}(P_i)$;
14	**end if**
15	**end for**
16	$iter$ ++;
17	**end while**
18	Output HUIs.

In Algorithm 1, the initial population is generated in Step 1. *SHUI*, the set of discovered HUIs, is initialized as an empty set in Step 2. Step 3 sets the iteration counter to 1. In the loop in Steps 4–17, each AF performs one of the three behaviors, namely follow, swarm, or prey, and then the iteration counter is incremented by 1. This procedure of AFs (performing one action and incrementing the iteration counter) is repeated until the maximum number of iterations is reached. Here, P_i is the PV of A_i in the current population. Finally, Step 18 outputs all discovered HUIs. The procedures for follow, swarm, and prey are described in Algorithms 2, 3, and 4, respectively.

Algorithm 2	Follow(PV P)
1	$Best\_AF = P_1$;
2	**for** i=1 to N **do**
3	Calculate $dis = BitDiff(P, P_i)$ using Eq. 11;
4	**if**($dis \leq VD$ AND $fitness(P_i) > fitness(Best\_AF)$) **then**
5	$Best\_AF = P_i$;
6	**end if**
7	**end for**
8	$dis' = BitDiff(Best\_AF, P)$;
9	**if** ($dis' > 0$) **then**
10	is_follow = true;
11	Randomly generate a positive integer k no higher than dis';
12	Update P by applying bitwise complement operation on k bits of it;
13	**if** ($f(P) \geq min\_util$ AND $IS(P) \notin SHUI$) **then**
14	$IS(P) \rightarrow SHUI$;
15	**end if**
16	**end if**

In Algorithm 2, describing the follow behavior, Steps 1–7 determine the best PV (that with the highest utility value) within distance VD of the enumerating PV. If the best PV discovered is different from the enumerating PV itself, Step 10 sets the follow variable and Steps 11–12 update the enumerating PV by bitwise complement within the bits that are different from the best PV. The operation in Steps 3 and 8, used for calculating the difference between two PVs, is defined as:

$$BitDiff(P_i, P_j) = \left|\left\{n|P_i(n) \oplus P_j(n) - 1\right\}\right|, \tag{11}$$

where $P_i(n)$ is the nth bit of P_i, $\oplus$ is the exclusive disjunction operation, and $|S|$ denotes the number of elements in a set S. According to Eq. 11, the distance between two PVs is represented by the number of their corresponding bits that have different values. If the updated PV has a utility value that is no lower than the minimum threshold, and it is not recorded as an HUI, it is stored in Steps 13–15. The function $IS(P)$ returns itemset X including the items whose corresponding bit in P is one.

Algorithm 3	Swarm(PV P)
1	**for** j=1 to NH **do**
2	count_zero[j] = 0;
3	count_one[j] = 0;
4	**end for**
5	**for** i=1 to N **do**
6	Calculate $dis = BitDiff(P, P_i)$ using Eq. 11;
7	**if**($dis \leq VD$) **then**
8	**for** j=1 to NH **do**
9	count_zero[j]++ if $P_i(j)$ is zero;
10	count_one[j] ++ if $P_i(j)$ is one;
11	**end for**
12	**end if**
13	**end for**
14	**for** j=1 to NH **do**
15	**if**(count_one[j] $\geq$ count_zero[j])
16	$Cen\_AF[j] = 1$;
17	**Else**
18	$Cen\_AF[j] = 0$;
19	**end if**
20	**end for**
21	**if** $(f(Cen\_AF) \geq min\_util$ AND $IS(Cen\_AF) \notin SHUI)$ **then**
22	$IS(Cen\_AF) \rightarrow SHUI$;
23	**end if**
24	**if** $(f(P) < f(Cen\_AF))$ **then**
25	is_swarm = true;
26	$dis' = BitDiff(Cen\_AF, P)$;
27	Randomly generate a positive integer k no higher than dis';
28	Update P by applying bitwise complement operation on k bits of it;
29	**if** $(f(P) \geq min\_util$ AND $IS(P) \notin SHUI)$ **then**
30	$IS(P) \rightarrow SHUI$;
31	**end if**
32	**end if**

In Algorithm 3, describing the swarm behavior, two binary arrays for determining the center PV, within distance VD of the enumerating PV, are initialized in Steps 1–4. The same two arrays are then updated in the loop in Steps 5–13. According to these two arrays, the center PV is determined in the loop in Steps 14–20. If the center PV represents an HUI that has not been discovered before, this HUI is recorded in Steps 21–23. If the center PV has a higher utility than the current PV, the swarm variable is set in Step 25. In addition, the current PV is updated according to the center PV in Steps 26–28. If the updated PV represents an HUI that has not been discovered before, this HUI is recorded in Steps 29–31.

Algorithm 4	Prey(PV P)
1	flag = false;
2	*times* = 1;
3	**while** *times* ≤ *try_number* **do**
4	Randomly generate a positive integer k no higher than VD;
5	Change k bits of P to generate P';
6	**if** ($f(P') ≥ min\_util$ AND $IS(P') ∉ SHUI$) **then**
7	$IS(P') → SHUI$;
8	**end if**
9	**if** ($f(P') > f(P)$) **then**
10	$P = P'$;
11	flag = true;
12	**break**;
13	**end if**
14	*times* ++;
15	**end while**
16	**if** (! flag) **then**
17	Update P by randomly changing several bits of P;
18	**if** ($f(P) ≥ min\_util$ AND $IS(P) ∉ SHUI$) **then**
19	$IS(P) → SHUI$;
20	**end if**
21	**end if**

In Algorithm 4, describing the prey behavior, a flag parameter, representing whether a better position can be found within *try_number*, is initialized to false in Step 1. The number of tries is then initialized to 1 in Step 2. The loop in Steps 3–15 generates new HUIs and updates the current position by trying at most *try_number* times. If no better positions are found in this loop, the current PV is changed by using a bitwise complement operation randomly on several of its bits in Step 17. If the new PV representing an HUI is not discovered before, it is stored in Step 19.

5.2 An Illustrative Example

We use the transaction database in Table 1 and profit table in Table 2 to explain the algorithm. Given $min\_util = 35$, because $TWU(F) < min\_util$, item F is deleted from transactions T_1 and T_7. Thus, the PV of an AF is a five-dimensional binary vector, in which the first five bits represent A, B, C, D, and E, respectively.

Assume that the size of the population N is 5 and the visual distance VD is 3. According to Eq. 9, five PVs are generated randomly: $P_1 = <10110>$, $P_2 = <11010>$, $P_3 = <10111>$, $P_4 = <11001>$, and $P_5 = <00001>$.

Considering P_1, it first performs the follow behavior, in which it is also initialized as *Best_AF*. The PVs within P_1's visual distance are determined to be P_1, P_2, and P_3. According to Eq. 10, $f(P_1) = 32$, $f(P_2) = 41$, and $f(P_3) = 16$, so *Best_AF* = P_2. According to Eq. 11, *dis'* = $BitDiff(Best\_AF, P_1) = 2$, so *is_follow* is set to true. Because the second and third bits of P_1 are different from the corresponding bits of *Best_AF*, one of these two bits (e.g., the third bit) is randomly changed to update the new P_1 to < 10010 >, which represents the itemset AD. Because $u(AD) > min\_util$, $SHUI = \{AD\}$. Because the first PV performs the follow behavior, the other two behaviors (swarm and prey) are not performed in this iteration.

In the same iteration, the other four PVs are processed similarly. The next iteration then starts to process each PV to discover HUIs until the maximal number of iterations is reached.

6   Performance Evaluation

In this section, we evaluate the performance of our HUIM-AF algorithm and compare it with the $HUPE_{UMU}$-GARM [3] and HUIM-BPSO$_{sig}$ [5] algorithms. We downloaded the source code of the two comparison algorithms from the SPMF data mining library [1].

6.1   Test Environment and Datasets

The experiments were performed on a computer with a 4-core 3.20 GHz CPU and 4 GB memory running 64-bit Microsoft Windows 7. Our programs were written in Java. Four real datasets were used to evaluate the performance of the algorithms. The characteristics of the datasets are presented in Table 3.

Table 3. Characteristics of the datasets used for the experimental evaluations.

Datasets	Avg. trans. length	No. of items	No. of trans
Chess	37	76	3,196
Mushroom	23	119	8,124
Accidents_10%	34	469	34,018
Connect	43	130	67,557

The four datasets were also downloaded from the SPMF data mining library [1]. The Chess and Connect datasets originate from game steps. The Mushroom dataset contains various species of mushrooms, and their characteristics. The Accidents dataset is composed of (anonymized) traffic accident data. Similarly to the work of Lin et al. [5], the dataset used for the Accidents_10% experiments contained only 10% of the total dataset.

For all experiments, the population size was set to 20, try_number was set to 3, and the termination criterion was set to 10,000 iterations. Furthermore, VD was set to a value specific to each dataset, by:

$$VD = \lfloor 0.1 \times NH \rfloor + 1, \tag{12}$$

where $\lfloor 0.1 \times NH \rfloor$ denotes the largest integer that is less than or equal to $(0.1 \times NH)$.

6.2 Execution Time

First, we demonstrate the performance of these algorithms. When measuring the execution time, we varied the minimum utility threshold for each dataset.

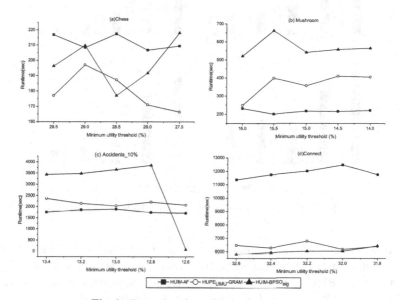

Fig. 1. Execution times for the four datasets.

We can observe from Fig. 1 that the execution time of the proposed HUIM-AF was comparable to that of the other two algorithms. Specifically, it was always faster than the other two algorithms on the Mushroom dataset, whereas its speed was lower than that of the other two algorithms on the Connect dataset. On the Accidents_10% dataset, HUIM-AF was faster than the other two algorithms except when the minimum utility threshold was 12.6%. On the Chess dataset, HUIM-AF was slower than $HUPE_{UMU}$-GARM and comparable to HUIM-BPSO$_{sig}$. The main reason for the speed of HUIM-AF is that the PVs within distance VD of the enumerating PV always need to be determined, which incurs a relatively high computational cost.

6.3 Number of Discovered HUIs

Because HUIM algorithms based on heuristic methods cannot ensure the discovery of all itemsets within a certain number of cycles, we compared the number of HUIs discovered by each of the three algorithms. The results are shown in Fig. 2.

In contrast to the results on efficiency, HUIM-AF always discovered more HUIs than the other two algorithms. In particular, there were cases in which both $HUPE_{UMU}$-GARM and HUIM-BPSO$_{sig}$ could not find any results on the Mushroom or Accidents_10% datasets. For example, in the Mushroom dataset, $HUPE_{UMU}$-GARM could not find any results when the minimum thresholds were 15% and 16%, and the same occurred for

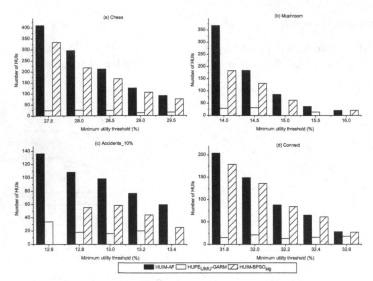

Fig. 2. Number of discovered HUIs for the four datasets.

HUIM-BPSO$_{sig}$ when the minimum threshold was 15.5%. The superiority of HUIM-AF was also demonstrated on Accidents_10% when the threshold was 12.6%. Figure 1 shows that HUIM-BPSO$_{sig}$ was faster than HUIM-AF; however, we can observe from Fig. 2 that HUIM-BPSO$_{sig}$ could not find any results for this threshold.

These experiments show that the AFSA can leap over local optima effectively, so it is more suitable for the HUIM problem because there are multiple targets to optimize.

7 Conclusions

In this paper, we propose an HUIM algorithm following the AFSA paradigm. We formally model the problem in which AFs have three behaviors, and describe the algorithm in detail, with an example. Our experimental results show that the AFSA can discover more HUIs than two other heuristic methods. Our future work includes the design of effective pruning strategies to make this type of algorithm more efficient.

Acknowledgments. This work was partially supported by the National Natural Science Foundation of China (61977001), the Great Wall Scholar Program (CIT&TCD20190305).

References

1. Fournier-Viger, P., et al.: The SPMF open-source data mining library version 2. In: Berendt, B., et al. (eds.) ECML PKDD 2016. LNCS (LNAI), vol. 9853, pp. 36–40. Springer, Cham (2016). https://doi.org/10.1007/978-3-319-46131-1_8
2. Fournier-Viger, P., Lin, J.C.-W., Nkambou, R., Vo, B., Tseng, V.S. (eds.): High-Utility Pattern Mining. SBD, vol. 51. Springer, Cham (2019). https://doi.org/10.1007/978-3-030-04921-8

3. Kannimuthu, S., Premalatha, K.: Discovery of high utility itemsets using genetic algorithm with ranked mutation. Appl. Artif. Intell. **28**(4), 337–359 (2014)
4. Li, X., Shao, Z., Qian, J.: An optimizing method based on autonomous animals: fish-swarm algorithm. Syst. Eng. Theor. Pract. **22**(11), 32–38 (2002). (in Chinese)
5. Lin, J.C.-W., et al.: Mining high-utility itemsets based on particle swarm optimization. Eng. Appl. Artif. Intel. **55**, 320–330 (2016)
6. Liu, Y., Liao, W.-K., Choudhary, A.: A two-phase algorithm for fast discovery of high utility itemsets. In: Ho, T.B., Cheung, D., Liu, H. (eds.) PAKDD 2005. LNCS (LNAI), vol. 3518, pp. 689–695. Springer, Heidelberg (2005). https://doi.org/10.1007/11430919_79
7. Song, W., Huang, C.: Discovering high utility itemsets based on the artificial bee colony algorithm. In: Phung, D., Tseng, V.S., Webb, G.I., Ho, B., Ganji, M., Rashidi, L. (eds.) PAKDD 2018. LNCS (LNAI), vol. 10939, pp. 3–14. Springer, Cham (2018). https://doi.org/10.1007/978-3-319-93040-4_1
8. Song, W., Huang, C.: Mining high average-utility itemsets based on particle swarm optimization. Data Sci. Pattern Recogn. **4**(2), 19–32 (2020)
9. Song, W., Li, J.: Discovering high utility itemsets using set-based particle swarm optimization. In: Yang, X., Wang, C.-D., Islam, M.S., Zhang, Z. (eds.) ADMA 2020. LNCS (LNAI), vol. 12447, pp. 38–53. Springer, Cham (2020). https://doi.org/10.1007/978-3-030-65390-3_4
10. Song, W., Liu, L., Huang, C.: TKU-CE: cross-entropy method for mining top-K high utility itemsets. In: Fujita, H., Fournier-Viger, P., Ali, M., Sasaki, J. (eds.) IEA/AIE 2020. LNCS (LNAI), vol. 12144, pp. 846–857. Springer, Cham (2020). https://doi.org/10.1007/978-3-030-55789-8_72
11. Song, W., Liu, Y., Li, J.: Vertical mining for high utility itemsets. In: Proceedings of the 2012 IEEE International Conference on Granular Computing, pp. 429–434 (2012)
12. Wu, J.M.T., Zhan, J., Lin, J.C.W.: An ACO-based approach to mine high-utility itemsets. Knowl.-Based Syst. **116**, 102–113 (2017)
13. Yang, R., Xu, M., Jones, P., Samatova, N.: Real time utility-based recommendation for revenue optimization via an adaptive online top-k high utility itemsets mining model. In: Proceedings of The 13th International Conference on Natural Computation, Fuzzy Systems and Knowledge Discovery, pp. 1859–1866 (2017)
14. Zhang, Q., Fang, W., Sun, J, Wang, Q.: Improved genetic algorithm for high-utility itemset mining. IEEE Access **7**, 176799–176813 (2019)

Ensemble Recognition Based on the Harmonic Information Gain Ratio for Unsafe Behaviors in Coal Mines

Jian Cheng[1,2](✉), Botao Jiao[3](✉), Yinan Guo[3,4](✉), and Shijie Wang[5]

[1] Research Institute of Mine Big Data, China Coal Research Institute,
Beijing 100013, China
jiancheng@tsinghua.org.cn
[2] State Key Laboratory of Coal Mining and Clean Utilization, Beijing 100013, China
[3] School of Information and Control Engineering, China University of Mining
and Technology, Xuzhou 221116, China
[4] School of Electromechanical and Information Engineering,
China University of Mining and Technology (Beijing), Beijing 100083, China
[5] Shandong Energy Group Co., Ltd., Jinan 250014, China

Abstract. More than 90% accidents occurred in coal mine are caused by unsafe behaviors of human. How to effectively identify unsafe behaviors and decrease the possibility of their occurrence is the fundamental of avoiding accidents. However, the number of unsafe behaviors is far less than that of safe ones in a behavior dataset of coal mine. Serious imbalance has a negative impact on recognition efficiency and accuracy. To address the problem, the harmonic information gain ratio is defined by introducing the degree of imbalance into traditional information gain, and the corresponding feature selection method is presented. By integrating it into Underbagging, a novel ensemble recognition based on the harmonic information gain ratio for unsafe behaviors is presented, with the purpose of avoiding information loss caused by feature reduction and guaranteeing recognition accuracy. Based on a sub-dataset obtained by undersampling, the optimal features subset is selected by the proposed feature selection method, and employed to train a base classifier built by support vector machine. The weighted sum of all base classifiers output forms final recognition result. Each weight is calculated from the corresponding harmonic information gain ratio. Experimental results on UCI dataset and a behavior dataset for a particular coal mine indicate that the proposed ensemble recognition method outperforms the others, especially for a dataset with high imbalance ratio.

Keywords: Feature selection · Imbalanced data · Ensemble learning · Information gain ratio

1 Introduction

In the process of coal production, the accidents seriously threaten the safety of coal mines [8]. Relevant statistics show that 90% of production safety accidents

Y. Tan and Y. Shi (Eds.): ICSI 2021, LNCS 12690, pp. 420–429, 2021.
https://doi.org/10.1007/978-3-030-78811-7_39

in coal mining enterprises are caused by human unsafe behavior [12]. Therefore, determining the factors that cause unsafe behaviors and mastering its laws is one of the fundamental ways to reducing the probability of mines safety accidents. At present, there are many influencing factors related to unsafe behaviors. In addition to human cognitive factors, many related factors have not been recognized and quantified. In order to fully grasp the influencing factors of unsafe behaviors and discover the potential impact rules, a large amount of raw data is collected. However, these data are involved with a huge number of noisy, redundant or unnecessary features which affects the accuracy of behavior recognition and reduces the efficiency of recognition [5]. In addition, because the frequency of unsafe behaviors is lower than safe behaviors, the collected data often has class imbalanced problem. That is, the number of samples that represent one class is significantly less than another class. When a model is trained on an imbalanced data set, it tends to show a strong bias to the majority class (safe behavior) and ignore minority class (unsafe behavior). For these reasons, it is great significance to construct an efficient and accurate method for identify unsafe behaviors of miners.

As the main cause of accidents, the recognition of unsafe behaviors has developed into an important branch of behavior recognition. Most of the research on unsafe behavior recognition is based on data-driven methods [2], which classify unsafe behaviors by extracting behavior characteristics. Kong [11] analyzed the signals of Electro-encephalogram (EEG) to identify the driver's fatigue driving based on the alpha waves in the human brain. Xu [1] extracted the rotation invariant features of the driving image, present a CNN network architecture that deals with driver behavior recognition. Zhu [7] proposed a dangerous behavior detection model based on OpenPose multi-person attitude estimation algorithm, which can be used to improve the intelligence level of the substation video monitoring. In recognition for unsafe behavior in coal mine, Jian, W. [6] used the FPgrowth algorithm to find the recurring and highly dangerous unsafe behaviors of coal mine workers in their daily operations, and proposed a big data storage analysis model based on Hadoops gas behavior security risks.

In machine learning, the information gain can be used to define a preferred sequence of features to rapidly narrow down the data dimensions. Usually, a feature with high information gain should preferred to other features. For high-dimensional sparse data, Dinata [9] performed feature reduction on the original data set based on information gain, effectively reducing the number of iterations of K-mean clustering and improving clustering accuracy. Paramitha [10] introduced information gain to select features to improve the accuracy of user sentiment analysis. However, all these studies ignore the imbalanced problem of unsafe behavior data.

In this paper, a feature selection algorithm based on the harmonic information gain ratio is proposed to tackle the feature redundancy and class imbalanced problems of unsafe behavior data sets. The harmonic factor is used to improve the discriminability of the information gain ratio to the minority class samples, make feature extraction more accurate.

2   Feature Selection Based on Harmonic Information Gain Ratio

As mentioned in Sect. 1, unsafe behavior recognition data sets usually have imbalanced problem. When the traditional information gain ration is applied to the classification of imbalanced data sets, the correlation features of majority class will have a more significant impact. To address the issue, this section proposes the harmonic information gain ratio, which introduces a harmonic factor to correct the influence of the class size on the information gain to achieve a more reasonable feature selection.

2.1   Information Gain Ratio

Note $F = \{F_1, F_2, \ldots, F_D\}$ is the set of features, where $F_i = \{f_{i1}, f_{i2}, \ldots, f_{iK}\}$ represents the ith feature and f_{ik} is the kth feature value. $S \subset F$ represents a feature subset of F. Let $Y = \{y_1, y_2, \ldots, y_N\}$ be the true label of a set of samples, and y_j is the label of the jth sample. Since this paper focuses on the binary classification of unsafe behavior, so $y_j \in \{-1, +1\}$. Normally, $+1$ represents the minority class (positive class), and -1 represents the majority class (negative class). Based on this, the relevant definition of information gain ratio as follows:

Information entropy: Give a feature F_i, $p(f_{ik}) = Pr(F_i = f_{ik})$ is the probability density function of F_i, the information entropy $H(F_i)$ of F_i is formally defined as

$$H(F_i) = - \sum_{f_{ik} \in F_i} p(f_{ik}) \log p(f_{ik}) \tag{1}$$

Conditional entropy: For label set Y, the entropy conditioned on feature F_i is written as $H(Y|F_i)$, and it defined as

$$H(Y|F_i) = - \sum_{f_{ik} \in F_i} p(f_{ik}) H(Y|F_i = f_{ik})$$
$$= \sum_{f_{ik} \in F_i} p(f_{ik}) \sum_{y_j \in Y} p(y_j|f_{ik}) \log p(y_j|f_{ik}) \tag{2}$$

Information gain: Base on the information entropy and conditional entropy, the information gain $I(Y; F_i)$ is defined as

$$I(Y; F_j) = H(Y) - H(Y|F_i) \tag{3}$$

Information gain ratio: It is a ratio of information gain to the intrinsic information, which can be formulated as

$$G(F_j) = \frac{I(Y; F_j)}{H(F_j)} \tag{4}$$

2.2   Harmonic Information Gain Ratio

Let n_{min} and n_{maj} be the number of minority samples and majority samples in data set, respectively. The data set size N satisfies $N = n_{min} + n_{maj}$. Give the definition of the harmonic factor T as

$$T = \frac{n_{maj}}{n_{min}} \tag{5}$$

According to the Eq. (2), conditional information entropy needs to calculate the posterior probability of y_j under the condition of $F_i = f_{ik}$. When $F_i = f_{ik}$ the numbers of minority samples and majority samples are expressed as n_{min}^{ik} and n_{maj}^{ik}, and the harmonic posterior probability is defined as

$$p_t\left(y_j \middle| f_{ik}\right) = \begin{cases} \frac{n_{min}^{ik} T}{n_{min}^{ik} T + n_{maj}^{ik}} & y_j = +1 \\ \frac{n_{maj}^{ik}}{n_{min}^{ik} T + n_{maj}^{ik}} & y_j = -1 \end{cases} \tag{6}$$

It can be seen that by adding the harmonic factor T, the influence of the difference in class size on the information entropy can be weakened without changing the distribution of the samples. Based on this, define the harmonic condition information entropy $H_T(Y|F_i)$ as

$$H_T\left(Y \middle| F_i\right) = - \sum_{f_{ik} \in F_i} p\left(f_{ik}\right) \sum_{y_j \in Y} p_t\left(y_j \middle| f_{ik}\right) \log p_t\left(y_j \middle| f_{ik}\right) \tag{7}$$

The harmonic information gain is defined as $I_T(Y; F_i) = H(Y) - H_T(Y|F_i)$. Then the harmonic information gain ratio $G_T(F_i)$ can be written as

$$G_T\left(F_j\right) = \frac{I_T\left(Y; F_j\right)}{H\left(F_j\right)} \tag{8}$$

It can be seen that when the class size are equal, $G_T(F_i)$ is the same as $G(F_i)$. Consequently, $G(F_i)$ can be regarded as a special case of $G_T(F_i)$. In order to judge the contribution of each feature to behavior recognition, the harmonic information gain ratio of all features is normalized, denoted as

$$G_T^*\left(F_j\right) = \frac{G_T\left(F_j\right) - G_T^{min}}{G_T^{max} - G_T^{min}} \tag{9}$$

G_T^{max} and G_T^{min} represent the maximum value and the minimum value of all features harmonic information gain ratios, respectively. Given a threshold $0 \le \beta \le 1$, then the selected feature subset satisfied $S = \{F_i | G_T^*(F_i) > \beta, i = 1, 2, \ldots, D\}$.

3   Ensemble Recognition for Unsafe Behaviors

The corresponding classifier is trained based on the existing samples after obtaining the feature subset meeting the requirements. This paper combining the feature selection and the UnderBagging (UB) algorithm, proposes an ensemble recognition method of unsafe behavior based on the harmonic information gain ratio, termed as TUB⸱

UnderBagging is an ensemble learning algorithm based on undersampling with replacement. Considering the high imbalance ratio of unsafe behavior data set, the feature selector based on the harmonic information gain ratio is added to the UnderBagging framework to improve the recognition accuracy of each base classifier for the minority class, as shown in Fig. 1.

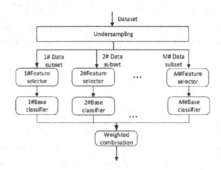

Fig. 1. Ensemble recognition based on the harmonic information gain ratio

TUB first obtains M diversified data subsets through under-sampling. However, being different from traditional UB, the data subset formed by TUB is not completely balanced. The reason is that in the unsafe behavior data set, the number of samples in the minority class (unsafe behavior) is much less than the number of samples in the majority class (safe behavior). Each data subset is forced to balance through under-sampling with replacement, a large number of majority samples will be discarded, which result in information loss. Although this problem can be alleviated by increasing the number of data subsets and training more base classifiers. However, an excessively large ensemble size will seriously affect the training and recognition efficiency. For example, when the imbalance ratio $IR \leq 0.01(IR = n_{min}/n_{maj}$, usually, the real behavior dataset is generally lower), the number of base classifiers is at least greater than 100. Therefore, this paper adopts under-sampling without replacement, which randomly selects n_{max}/M majority samples to form a data subset with all minority samples.

With a relatively balanced data subset is obtained, feature selection method based on harmonic information gain ratio is introduced, and selects unique feature subset for each data subset. In this way, the recognition accuracy of the minority samples is further improved, and the generalization ability of the model is increased. In this paper, support vector machine is used as the base classifier.

The weight of the base classifier $w_m (m = 1, 2, \ldots, M)$ is determined according to the average harmonic information gain ratio of the corresponding data subset features, denoted as

$$w_m = \sum F_i \in S_m \frac{G_T(F_i)}{|S_m|} \tag{10}$$

$S_m(m = 1, 2, \ldots, M)$ represents the feature subset selected by the mth data subset, and $|S_m|$ is the number of features in the feature subset. For the test sample x, the output of each base classifier as $v_m(m = 1, 2, \ldots, M)$ is recorded, and then the prediction of the ensemble classifier v_E is the weighted combination of all base classifiers. The calculation method is as follows:

$$v_E = \sum_{m=1}^{M} w_m v_m \tag{11}$$

4 Experiments and Evaluation

In order to verify the effectiveness of the proposed method, this paper analyzes the recognition performance of the algorithm by adopting three UCI datasets and a behavior dataset of a coal mine. The behavior dataset of a coal mine is obtained from a coal company and contains 68 features such as employee age, health status, work-ing hours, weather conditions, etc. The behavior dataset contains 207373 samples where there are 1131 unsafe behavior samples, 206242 safe behavior samples, and the sample imbalance ratio is 0.005, which belongs to a highly imbalanced dataset. The information of all datasets is shown in Table 1.

Table 1. Imbalanced datasets.

Dataset	Features	Samples	IR
Ablonc	7	731	4/69
Segment	19	1320	11/33
Statlandsat	36	1487	4/11
Coal	68	207373	1/201

In order to evaluate the performance of the proposed algorithm reasonably, this paper adopts performance indicators such as accuracy, recall, G-mean, F1-Score, and AUC value.

Experiment 1: In order to deeply analyze the influence of the different base classifiers on the ensemble recognition model, this paper adopts four common classification algorithms such as Support Vector Machine (SVM), Decision Tree (DT), K Nearest Neighbor (KNN), and Naive Bayes (NB) as the base classification and compares theirs classification performance on four imbalanced datasets. The G-mean values obtained by the different base classifiers TUB are summarized in Table 2, it can be seen that the base classifier composed of SVM and DT has relatively better classification results. In particular, the ensemble recognition model based on SVM has the best unsafe behavior recognition performance in the behavior dataset of a coal mine. Therefore, this paper adopts SVM to form the base classifier of TUB.

Table 2. G-mean value of TUB with different base classifiers.

Dataset	SVM	DT	KNN	B
Abalone	0.9549	**0.9589**	0.9483	0.9259
Segment	**0.9853**	0.9706	0.9669	0.9505
Statlandsat	0.9332	**0.9407**	0.9363	0.9009
Coal	**0.7427**	0.7174	0.7253	0.6931

Experiment 2: To fully verify the effectiveness of the proposed algorithm, the proposed TUB algorithm is compared with the traditional UnderBagging (UB) and support vector machine (SVM). The number of base classifiers for TUB and UB are both set to 10. To ensure fairness, both algorithms use support vector machines as the base classifiers.

Table 3. The performance of different recognition algorithms.

Dataset	Classifier	Acc	F1	Recall	G-mean
Abalone	TUB	0.9180	**0.9341**	**0.9895**	0.9549
	UB	**0.9727**	0.8387	0.8799	**0.9852**
	SVM	0.9617	0.6316	0.4615	0.6794
Segment	TUB	0.9788	0.9626	**0.9500**	**0.9853**
	UB	0.9498	0.9332	0.8000	0.9498
	SVM	**0.9939**	**0.9888**	0.8330	0.9688
Statlandsat	TUB	**0.9323**	**0.9713**	0.8276	0.9332
	UB	**0.9323**	0.9300	**0.8300**	**0.9332**
	SVM	0.9114	0.8986	0.7155	0.9019
Coal	TUB	0.9183	**0.6713**	**0.7107**	**0.7427**
	UB	**0.9528**	0.6527	0.6876	0.6996
	SVM	0.9474	0.4972	0.5525	0.6598

Table 3 shows that the recognition accuracy does not effectively reflect the performance of the classifier for the imbalanced data classification. Regarding F1-Score, G-mean, and recall, it can be seen that the proposed TUB algorithm performs better than the traditional UnderBagging algorithm. The harmonic information gain rate given in this paper can effectively find the features that have a significant contribution to the minority recognition which can improve the recognition accuracy and avoid the influence of redundant features on the efficiency of the algorithm. In addition, the traditional UnderBagging algorithm performs significantly better than the SVM algorithm on all data sets, which indicates that the ensemble learning can significantly improve the recognition performance. Furthermore, the ROC curve and AUC value of different algorithms are compared as shown in Fig. 2. It can be seen that the proposed algorithm has achieved the best AUC value on all datasets.

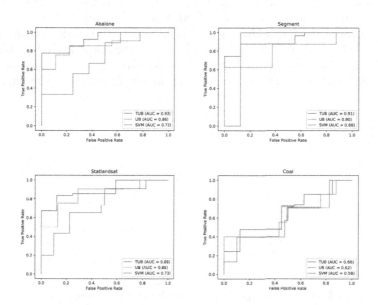

Fig. 2. ROC curve and AUC value of different classifiers

The proposed ensemble recognition model mainly includes two algorithm parameters: the harmonic information gain ratio threshold β and the number of base classifiers M. In order to analyze the influence of algorithm parameters on the performance of unsafe behavior recognition, comparative experiments are carried out when different parameter values are set.

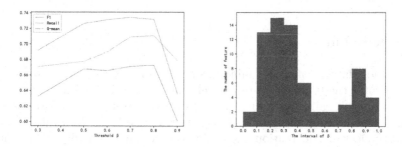

Fig. 3. Left: Recognition performance of behavior dataset in coal mine under different Threshold. Right: The harmonic information gain ratio distribution

First, β is set to $\{0.3, 0.5, 0.6, 0.7, 0.8, 0.9\}$ respectively when the value of M is 10. The changing trend of the proposed algorithm in the three evaluation indicators is shown in Fig. 3. It can be seen that the best recognition performance is achieved when β is equal to 0.7. The recognition performance of the algorithm is significantly reduced when β increases. This is because that some features containing important classification information are eliminated, which worsens the recognition performance. In order to further analyze the influence of different features on behavior recognition, Fig. 3 also shows the normalized distribution of the harmonic information gain rate of all the features in the behavior dataset of a coal mine. It can be seen that there are significant differences in the importance of features.

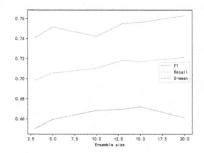

Fig. 4. Recognition performance under different ensemble size

Second, M is set to $\{3, 5, 10, 13, 15, 20\}$ respectively with β is set to 0.7, and the recognition performance is shown in Fig. 4. It can be seen that with the increasing number of base classifiers, the indicator curves of the proposed method have increasing trend which show the improved performance with relevant parameters. This is because that the behavior dataset of a coal mine has a higher imbalance rate, and the increasing number of base classifiers can better learn the data distribution characteristics, thereby obtaining better recognition performance.

5   Conclusion

The behavior dataset of a coal mine has significant imbalanced characteristics. To better solve the problem of unsafe behavior recognition, with the fundamental of information gain, this paper proposes a feature selection method based on the harmonic information gain ratio considering the imbalance ratio. Besides, to avoid the loss of information caused by feature dimensionality reduction and improve the recognition accuracy, the proposed feature selection method is combined with ensemble learning to construct an ensemble recognition framework for unsafe behavior, named TUB. This model reduces the imbalance ratio by adopting undersampling, and the weight of each classifier is calculated from the corresponding harmonic information gain ratio. Experimental results demonstrate that the proposed ensemble recognition model performs well in classification whether it is the UCI dataset or the behavior dataset of coal mine. In the future, evolutionary algorithms [3,4] are introduced to determine the optimal number of base classifiers in the model should be an interesting research direction.

Acknowledgments. This work is supported by the National Natural Science Foundation of China under Grant 61973305, 61573361, Six Talent Peak Project in Jiangsu Province under Grant 2017-DZXX-046, Natural Science Foundation of Liaoning Province for the State Key Laboratory of Robotics under Grant 2020-KF-22-02.

References

1. Dan, X., Yong, D., Junhong, J.: Research on driver behavior recognition method based on convolutional neural network. China Saf. Sci. J. **29**(10), 12–17 (2019)
2. Guo, Y.N., Cheng, J., Luo, S., Gong, D., Xue, Y.: Robust dynamic multi-objective vehicle routing optimization method. IEEE/ACM Trans. Comput. Biol. Bioinf. **15**(6), 1891–1903 (2018). https://doi.org/10.1109/TCBB.2017.2685320
3. Guo, Y.N., Zhang, X., Gong, D.W., Zhang, Z., Yang, J.J.: Novel interactive preference-based multi-objective evolutionary optimization for bolt supporting networks. IEEE Trans. Evol. Comput. **24**(4), 750–764 (2020)
4. Guo, Y., Yang, H., Chen, M., Cheng, J., Gong, D.: Ensemble prediction-based dynamic robust multi-objective optimization methods. Swarm Evol. Comput. **48**, 156–171 (2019)
5. Hamdani, M., et al.: Class association and attribute relevancy based imputation algorithm to reduce twitter data for optimal sentiment analysis. IEEE Access **7**, 136535 136544 (2010). https://doi.org/10.1109/ACCESS.2019.2942112
6. Jian, W., Wanjun, Y.: Hadoop-based behavior safety management model for methane gas. Mod. Electron. Tech. **42**(21), 154–102 (2019)
7. Jianbao, Z., Zhilong, X., Yuwei, S., Qingshan, M.: Detection of dangerous behaviors in power stations based on open pose multi-person attitude recognition. Autom. Instrum. **35**(02), 47–51 (2020)
8. Jiangshi, Z., Gui, F., Qiming, G., Liujun, H., Jiguo, W.: The pre-controlling measures on unsafe behavior. J. China Coal Soc. **037**(A02), 373–377 (2012)
9. Kesuma Dinata, R., Novriando, H., Hasdyna, N., Retno, S.: Reduksi atribut menggunakan information gain untuk optimasi cluster algoritma k-means. Jurnal Edukasi dan Penelitian Informatika (JEPIN) **6**, 48–53 (2020). https://doi.org/10.26418/jp.v6i1.37606
10. Paramitha, A., Indriati, Arum Sari, Y.: Analisis sentimen terhadap ulasan pengguna MRT jakarta menggunakan information gain dan modified k-nearest neighbor. Jurnal Pengembangan Teknologi Informasi dan Ilmu Komputer **4**, 1125–1132 (2020)
11. Wanzeng, K., Weicheng, L., Fabio, B., Sanqing, H., Gianluca, B.: Investigating driver fatigue versus alertness using the granger causality network. Sensors (Basel, Switzerland) **15**(8), 19181–19198 (2015)
12. Yu, W.: Analysis on the improvement of unsafe behavior management. Shaanxi Coal **S2**, 121–123 (2020)

Feature Selection for Image Classification Based on Bacterial Colony Optimization

Hong Wang[1], Zhuo Zhou[1], Yixin Wang[1], and Xiaohui Yan[2(✉)]

[1] College of Management, Shenzhen University, Shenzhen 518060, China
[2] School of Mechanical Engineering, Dongguan University of Techonlogy, Dongguan 523808, China

Abstract. Image classification is an important issue in pattern recognition, the high dimension features is a challenging task since only a few number of them are effective in classification. To improve the classification efficiency, it is necessary to reduce the dimensionality of image features before classification. This study provides a novel image classification application based on Bacterial Colony Optimization, which can decreases the computation burden and improves the classification's efficiency. Specifically, the elimination strategy in original algorithm is removed, and the communication, chemotaxis, migration, and reproduction strategies are kept. Additionally, the communication and chemotaxis step size of the Bacterial Colony Optimization are modified for feature selection in image classification. Several comparision experiments on two public image datasets are conducted to verify the effectiveness of the method. Experimental results prove that the method can greatly improve the classification accuracy and efficiency.

Keywords: Feature selection · Bacterial colony optimization · Image classification

1 Introduction

In recent years, with the deepening of image recognition research, the research objectives have become more complex, and the image feature dimension has become higher. The feature space of many high-dimensional data objects contains redundant features and noise features. These features may reduce the accuracy of classification or clustering on the one hand and increase the time and space complexity of learning and training on the other hand. Therefore, when classifying or clustering high-dimensional data, it is usually necessary to reduce the dimensionality of features to reduce machine learning time and space complexity.

Feature selection is an important feature dimensionality reduction method, and a large number of feature selection algorithms have been proposed. Feature selection uses certain evaluation criteria to select feature subsets from the original feature space. The purpose is to screen out invalid, incomplete and redundant features as much as possible. Besides, a good feature subset can use fewer features to describe most of the original data's attributes and better retain the information that the original data can convey.

Y. Tan and Y. Shi (Eds.): ICSI 2021, LNCS 12690, pp. 430–439, 2021.
https://doi.org/10.1007/978-3-030-78811-7_40

Finding the optimal feature combination is a difficult task. It is almost impossible to consider all the feature combinations to find the optimal feature combination.

Due to the advantages of heuristic algorithms, such as fewer parameters to be adjusted and irrelevant to the optimization target's gradient, more research focuses on using these heuristic algorithms to deal with feature selection problems. At present, many scholars have applied different evolutionary algorithms to the feature selection of image recognition problems. Zhang, X et al. applied the Particle Swarm Optimization [PSO] to the feature selection of hyperspectral image classification [1]. Naeini et al. used PSO for target-based very high spatial resolution satellite image feature selection [2]. Chen, L et al. used Ant Colony Optimization (ACO) or an improved ACO for image feature selection [3, 4]. Rutuparna Panda used the improved Bacterial Foraging Optimization (BFO) for the feature selection of face images [5], Lei L. et al. used Genetic Algorithms (GA) for the feature selection of insect images [6], Uroš Mlakar et al. used differential evolution algorithm for humans Feature selection of facial expression images [7].

By simulating the life cycle behavior of bacterial colonies, a novel heuristic algorithm is proposed in 2012, namely Bacterial Colony Optimization (BCO) [8]. Considering the application of the algorithm, this paper proposes an image classification method based on Bacterial Colony Optimization Feature Selection (BCOFS). In the experiment, two public databases were used to test the effectiveness of the classification method, and a variety of other feature selection methods were used for comparison experiments. Experimental results show that the classification method has shown good performance on both databases.

The rest of the paper is organized as follows: Sect. 2 introduces the image features used in the experiment of this paper. Section 3 provides a brief description of BCO and BCOFS. In Sect. 4, the experiment results and analyses of the experiment are shown and discussed. Finally, the conclusion of this paper is provided in Sect. 5.

2 Image Feature Extraction

The feature is the key to image classification, which directly determines the final classification result as a basis for classification. At present, the three essential features of images are shape feature, texture feature and color feature [9]. The above features describe the image target from different angles to extract effective information to identify the image. Note that different features are often selected for different classification tasks. When selecting features, It is necessary to select features based on specific issues. This paper uses four image features: Average curvature, Histogram of Oriented Gradient (HOG), Grayscale, and Gabor. The following is a detailed description of them.

Average curvature is an external bending measurement standard in differential geometry, which describes the curvature of a curved surface embedded in the surrounding space. The average curvature is the average value of any two orthogonal curvatures at a point on a space surface.

The essence of the HOG feature is the gradient's statistical information, and the gradient mainly exists at the edge. It constructs features by calculating and counting the histogram of the gradient direction of the local area. First, the image needs to be divided into small connected regions, called cells. Then collect the histogram of each direction's gradient or edge direction and each pixel in the cell. Finally, these histograms are combined to form a feature descriptor.

The texture is formed by the repeated occurrence of grayscale distribution in space. A certain grayscale relationship between two pixels is separated by a certain distance in the image space. This relationship is called the spatial correlation of grayscale. The gray-level co-occurrence matrix describes the texture by studying the spatial correlation of gray levels, describing the relationship between adjacent pixels in the image, and reflecting different aspects (uniformity, uniformity, etc.). It gives adjacent information about the intensity conversion between pixels. For a grayscale image I with a size of N $\times$ N, the grayscale co-occurrence matrix P can be defined as:

$$P(i, j) = \sum_{x=1}^{\infty} \sum_{y=1}^{\infty} \{1, \text{if} I(x, y) = i \text{ and } I(x + \Delta x, y + \Delta y) = j\} \tag{1}$$

Among them, $(\Delta x, \Delta y)$ represents the positional relationship between two pixels.

Haralick et al. proposed 14 types of statistics calculated based on the gray-level co-occurrence matrix [10]. Following is only the part used in this paper: Correlation, Uniformity, Contrast, Energy, Entropy.

The Gabor feature is generally obtained by convolving the image and the Gabor filter, which is defined as the Gaussian function's product and the sine function. The Gabor filter's frequency and direction representation are close to the frequency and direction representation of the human visual system, so it is often used for texture representation and description. Gabor filter is a complex exponential signal constructed by Gaussian function. A two-dimensional Gabor filter using Cartesian coordinates in the space domain and polar coordinates in the frequency domain is defined as:

$$g_{x_0, y_0, f_0, \theta_0} = e^{i[2\pi f_0(x_0\cos\theta_0 + y_0\sin\theta_0) + \varphi]} \cdot gauss(x_0, y_0) \tag{2}$$

$$gauss(x_0, y_0) = K \cdot e^{-\pi[a^2(x_0\cos\theta_0 + y_0\sin\theta_0)^2 + b^2(x_0\sin\theta_0 - y_0\cos\theta_0)^2]} \tag{3}$$

a and b describe the shape of the feature, θ_0 indicates the direction, φ indicates phase shift.x_0, y_0, f_0,and θ_0 indicate the position in the space domain and frequency domain respectively.

3   Bacterial Colony Optimization for Feature Selection

Feature selection is a complex combination problem with discrete high-dimensional parameters. Considering the adaptability of image feature selection, the Bacterial Colony Optimization Feature selection, abbreviated as BCOFS, has made some changes to its structure and strategies on the basis of BCO, the following is a detailed description of BCO and BCOFS.

3.1   Bacterial Colony Optimization (BCO)

The BCO [8] is proposed based on simulating the bacterial colony's life cycle behavior. The evolutionary process of bacterial growth model and the growth process of colonies will show the best solution to the problem.

The basic behavior of bacteria in the lifecycle can be simply divided into five parts: chemotaxis, communication, migration, reproduction and elimination. Under the action of these 5 parts, the entire colony system finally obtains the best source of nutrition. The following is a detailed introduction of these 5 parts. For more detailed information, please refer to [8].

3.1.1 Communication and Chemotaxis

Throughout the optimization process, chemotaxis is accompanied by communication. Bacteria will swim and tumble in the environment. However, they must also provide their information to the colony in order to exchange overall information, thereby guiding their direction and methods of action. In the bacterial individual's update process, when the objective function value of the bacterial individual's spatial position is better than the previous position's objective function value, the bacterial individual will adopt a forward movement mode. When the objective function value of the individual's spatial position is When the objective function value is not superior last time, it is simulated that the bacteria's current area environment is not as good as its parent's location. The individual bacteria will not move in this direction, but will tumble around in place, that is, search for nearby spatial locations.

The process of bacterial swim can be formulated as (4), The process of bacterial tumble can be formulated as (5), Their chemotaxis step size is shown in (6).

$$X_{k+1} = X_k + C_k[R_1 \cdot (G_{best} - X_k) + R_2 \cdot (P_{best} - X_k)] \tag{4}$$

$$X_{k+1} = X_k + C_k[R_1 \cdot (G_{best} - X_k) + R_2 \cdot (P_{best} - X_k) + Tu_k] \tag{5}$$

$$C_k - C_{min} + (\frac{iter_{max} - iter_k}{iter_{max}})^n \cdot (C_{max} - C_{min}) \tag{6}$$

In the formula, X_{k+1} indicates where the bacteria will move to in the next step, X_k indicates the current location of the bacteria, G_{best} is the global best location. P_{best} is the best location for individual bacteria. R_1 and R_2 are randomly generated constant, the specified value range belongs to [0,1], $R_1 + R_2 = 1$. C_k is an adaptive chemotaxis step size, $iter_{max}$ indicates the highest number of iterations, $iter_k$ indicates the current number of iterations. Tu_k is tumble direction variance of the bacterium.

3.1.2 Migration

Individual bacteria will migrate according to certain conditions. The purpose is to find a better source of nutrients and reduce the possibility of falling into the best local source of nutrients. The migration strategy in this paper is expressed as random. If the conditions are met, the bacteria will migrate to a random area according to a function, it can be described as:

$$X_{k+1} = rand \cdot (B_u - B_l) + B_l \tag{7}$$

The value range of *rand* is [0, 1], B_l and B_u are the lower and upper boundary.

3.1.3 Reproduction and Elimination

Similar to natural selection in reality, bacteria in locations with low nutrient content will be eliminated. In contrast, bacteria in places with high nutrient content will get more energy for survival and reproduction. In this algorithm, the bacteria with the higher fitness value will split into two identical bacteria, which is advantageous for the subsequent search.

3.2 Bacterial Colony Optimization Feature Selection (BCOFS)

The original BCO is a good method to solve continuous optimization problems, but in this paper, the research goal is to apply it to the feature selection of the image, which is a discrete combination problem with high dimensions. In order to adapt BCO to feature selection, BCOFS has made some changes on the basis of BCO. BCOFS removes the elimination strategy while retaining the communication, chemotaxis, migration and reproduction strategies. Some improvements are made in communication and chemotaxis strategies. It can be formulated as:

$$X_{k+1} = X_k + R_1 \cdot (G_{best} - X_k) + (1 - R_1) \cdot (P_{best} - X_k) \tag{8}$$

$$X_{k+1} = X_k - C_k \cdot Tu_k \tag{9}$$

$$C_k = (\frac{iter_{max} - iter_k}{iter_{max}}) \cdot R \tag{10}$$

$$Tu_k = \Delta_k / \sqrt{\Delta_k^T \cdot \Delta_k} \tag{11}$$

where X_{k+1} and X_k are the location of the bacteria, G_{best} and P_{best} are the global best position and the individual best position. $R_1 \in [0, 1]$. C_k is tumble step size, $iter_{max}$ indicates the highest number of iterations, $iter_k$ indicates the current number of iterations. R is a constant coefficient depending on the total number of iterations. The randomness parameter Tu_k is calculated from Δ_k, $\Delta_k = (2 \cdot rand - 1) \cdot rand$.

The main idea of the algorithm is as follows: The bacteria at the beginning is placed in the solution space and some mechanisms are stipulated. Bacteria update their position through communication and chemotaxis, and the migration mechanism, and move to a place with high nutrient concentration. If the number of bacteria movement reaches a certain set and it still does not reach a higher concentration position, such bacteria will be died, and then the bacteria in the high nutrient concentration will start to reproduce to replace the died bacteria, so as to maintain the population. It uses the number of iterations or accuracy as the end condition like other swarm intelligences.

Figure 1 shows the flow chart of BCOFS for image feature. After preprocessing the image database (multi-feature extraction), it enters the process of feature selection and classification. The whole process can be simply divided into five steps:

Step 1: parameters of population are initialized, the classification accuracy is used as an evaluation function to calculate the fitness of each bacteria, the individual best fitness value and the global best fitness value is recorded.

Step 2: communication and chemotaxis operations are used to update the location of the bacteria, the fitness value of the bacteria is calculated again, the individual best fitness value and the global best fitness value are updated.

Step 3: if the migration conditions are fulfilled, the bacterial individuals will migrate to a random location.

Step 4: after the bacteria positions are updated, the reproduction operation is used to replicate individuals with good positions instead of individuals with poor positions.

Step 5: if the algorithm reaches the specified number of iterations, the best feature combination and best accuracy will be output. If not, repeat the above steps.

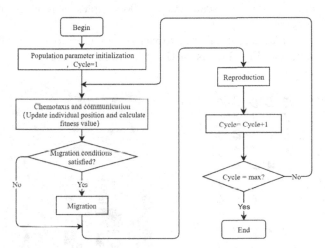

Fig. 1. Flow chart of BCOFS for image feature

4 Experiments

4.1 Experimental Setting

This paper conducts experiments on the MNIST handwritten digits dataset and the NEU surface defect dataset. In the comparison experiment, four other feature selection algorithms were used for comparison, namely the classic Bacterial Foraging Optimization (BFO) [11], the traditional feature selection algorithm Correlation-based Feature Selection (CFS) [12] and the widely used swarm intelligence algorithm Genetic Algorithm (GA) and Binary Particle Swarm Optimization Algorithm (BPSO) [13]. After many experiments, it is found that all algorithms will converge within 50 iterations, so the number of iterations is set to 50, and the number of each algorithm population is set according to the specific data set (The dimension of features extracted from the MNIST database is higher, so the population set to 50, and the dimension of features extracted from the NEU database is lower, so set to 26). KNN algorithm selects K = 1, SVM uses linear kernel function. In addition, all experiments were repeated 30 times.

4.2   Dataset Introduction and Multi-feature Extraction

The MNIST database of handwritten digits comes from the National Institute of Standards and Technology (NIST). There are 70,000 processed two-dimensional grayscale images, including 60,000 training set images and one Ten thousand test set images. Handwritten numbers do the image from 250 different people, 50% of them are high school students, and 50% are staff from the Census Bureau. Due to the excessive number of images in the original data set, this experiment uses 1000 pictures as the data set.

In the Northeastern University (NEU) surface defect database [14], six kinds of typical surface defects of the hot-rolled steel strip are collected, i.e., Rolled-in Scale (RS), Patches (Pa), Crazing (Cr), Pitted surface (PS), Inclusion (In) and Scratches (Sc). The database includes 1,800 grayscale images: 300 samples each of six different kinds of typical surface defects.

Figure 2 shows some examples in the two data sets. Figure 2(a) is the MNIST database of handwritten digits and Fig. 2.(b) is NEU surface defect database.

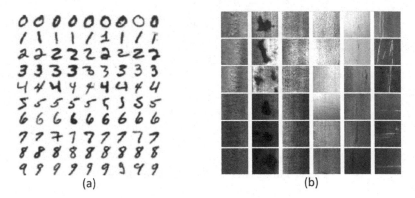
(a)   (b)

Fig. 2. Examples of the two databases

For the MNIST database of handwritten digits, four types of features, namely grayscale, Gabor, average curvature, and HOG, are extracted and integrated together to obtain 247 features. For the NEU surface defect database, two types of features, grayscale and Gabor are extracted and integrated, and the total feature dimension is 69.

4.3   Experiment Results and Analyses

Table 1 shows the classification results of each algorithm. Shown in brackets is the average number of features selected by the algorithm. Only BFO and GA can compare the classification accuracy based on feature dimensions, so Fig. 3 and Fig. 4 only show the relationship between the classification accuracy of these methods and the number of selected features.

The data shows the accuracy of BCOFS is better than the accuracy of no feature selection process. It indicates that the algorithm removes redundant or useless features. Compared with other algorithms, BCOFS has the highest accuracy rate on the two experimental data sets. This proves the effectiveness of BCOFS in image feature selection.

In the actual application process of the classification method, the classification has two phases: training and use. The use phase of the classification method is longer than the training phase, the efficiency of the use phase accounts for a larger weight. The efficiency of the use stage is determined by the number of features selected by the algorithm, so the efficiency of the classification method should be analyzed according to the number of features selected by each method. In the training phase, it is the fastest not to use the feature selection algorithm. However, if the feature selection is not performed in the training phase, it will reduce the efficiency of the use phase, A large number of features will reduce classification efficiency. Therefore, in the long run, the efficiency of not using the feature selection algorithm is the worst. Other classification methods that use feature selection algorithms have correspondingly reduced the number of features. Although the time of the training phase is different, they all show a certain feature selection ability. From the average number of selected features, the efficiency of the classification method based on CFS is the highest. BCOFS does not have a big advantage, only superior to BFO and PSO, and GA shows a big difference between the two databases.

From the perspective of comprehensive classification accuracy and classification efficiency, BCOFS has the highest classification accuracy and the second highest classification efficiency.

Table 1. Comparison of accuracy of each algorithm

Algorithm		Database			
		MNIST database		NEU database	
		Mean(%)	Best(%)	Mean(%)	Best(%)
KNN	ALL	89.21	89.50	59.6	60.2
	BCOFS	92.17(51.67)	92.50	82.6(30)	83
	BFO	91.5(55)	92	80.1(35)	81.5
	CFS	89.60(25)	90.60	76.11(7)	76.6
	GA	90.4(100)	91.2	72.39(5)	72.5
	BPSO	91.6(96)	92	71.06(33)	72.25
SVM	ALL	93.53	93.80	80.8	81.2
	BCOFS	95.72(88.33)	96.2	90.1(25)	90.5
	BFO	95.5(90)	96	86.4(40)	87.0
	CFS	90.70(25)	91.6	86.33(10)	86.5
	GA	95(95)	95.4	87.22(20)	88
	BPSO	94.7(102)	95.2	86.56(35)	87.1

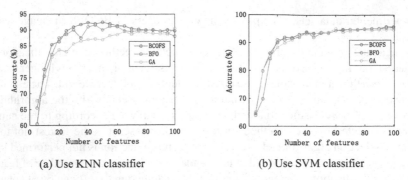

(a) Use KNN classifier   (b) Use SVM classifier

Fig. 3. Classification results of some methods in the MNIST database

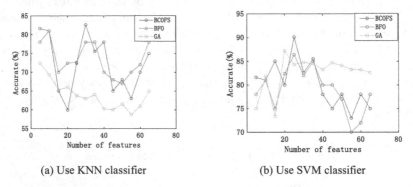

(a) Use KNN classifier   (b) Use SVM classifier

Fig. 4. Classification results of some methods in the NEU database

5   Conclusions

For the feature selection and classification of images, this paper proposes a classification method based on BCOFS. This method aims to improve the classification accuracy as much as possible while selecting as few features as possible. Experiments on the MNIST database of handwritten digits and the NEU surface defect database have proved BCOFS's effectiveness in image feature selection. The experimental results show that the classification method based on BCOFS is effective. There is a lot of application space in the field of image recognition. However, the generalization of this classification method in the field of image recognition requires further research.

Acknowledgement. This work is partially supported by The National Natural Science Foundation of China (Grants Nos. 71901152, 61703102), Natural Science Foundation of Guangdong Province (2020A1515010752), Guangdong Basic and Applied Basic Research Foundation (Project No. 2019A1515011392), and Natural Science Foundation of Shenzhen University (85303/00000155).

References

1. Zhang, X., Wang, W., Li, Y., Jiao, L.C.: PSO-based automatic relevance determination and feature selection system for hyperspectral image classification. Electron. Lett. **48**, 1263–1265 (2012)
2. Alizadeh Naeini, A., Babadi, M., Mirzadeh, S.M.J., Amini, S.: Particle Swarm optimization for object-based feature selection of VHSR satellite images. IEEE Geosci. Remote Sens. Lett. **15**, 379–383 (2018)
3. Chen, L., Chen, B., Chen, Y.: Image feature selection based on ant colonny optimization. Aust. Joint Conf. Artif. Intell. **7106**, 580–589 (2011)
4. Chen, B., Chen, L., Chen, Y.: Efficient ant colony optimization for image feature selection. Signal Process. **93**, 1566–1576 (2013)
5. Panda, R., Kumar, M., Panigrahi, B.K.: Face recognition using bacterial foraging strategy. Swarm Evol. Comput. **1**, 138–146 (2011)
6. Lei, L., Peng, J., Yang, B.: Image feature selection based on genetic algorithm. In: Zhong, Z. (ed.) Proceedings of the International Conference on Information Engineering and Applications (IEA) 2012. LNEE, vol. 219, pp. 825–831. Springer, London (2013). https://doi.org/10.1007/978-1-4471-4853-1_101
7. Mlakar, U., Fister, I., Brest, J., Potočnik, B.: Multi-objective differential evolution for feature selection in facial expression recognition systems. Expert Syst. Appl. **89**, 129–137 (2017)
8. Niu, B., Wang, H.: Bacterial colony optimization. Discret. Dyn. Nat. Soc. **2012**, 1–28 (2012)
9. Wang, Y., Li, S., Mao, J.: Computer image processing and recognition technology. J. China High Education Press, pp. 31–105 (2001)
10. Haralick, R.M., Shanmugam, K.: Textural features for image classification. IEEE Trans. Syst. Man Cybern. **SMC-3**, 610–621 (1973)
11. Passino, K.M.: Biomimicry of bacterial foraging for distributed optimization and control. IEEE Control Syst. Mag. **22**, 52–67 (2002)
12. Hall, M.A.: Correlation-based feature selection of discrete and numeric class machine learning (2000)
13. Kennedy, J., Eberhart, R.C.: A discrete binary version of the particle swarm algorithm. In: IEEE International Conference on Systems, Man, and Cybernetics. Computational Cybernetics and Simulation, vol. 5, pp. 4104–4108 (1997)
14. He, Y., Song, K., Meng, Q., Yan, Y.: An end-to-end steel surface defect detection approach via fusing multiple hierarchical features. IEEE Trans. Instrum. Meas. **69**, 1493–1504 (2020)

Local Binary Pattern Algorithm with Weight Threshold for Image Classification

Zexi Xu, Guangyuan Qiu, Wanying Li, Xiaofu He$^{(\boxtimes)}$, and Shuang Geng

College of Management, Shenzhen University, Shenzhen, China

Abstract. Image classification has attracted the attention in many research field. As an efficient and fast image feature extraction operator, LBP is widely used in the Image classification. The traditional local binary pattern (LBP) algorithm only considers the relationship between the center pixel and the edge pixel in the pixel region, which often leads to the problem of partial important information bias. To solve this problem, this paper proposes an improved LBP with threshold, which can significantly optimize the processing of texture features, and also be used to address the problems of multi-type image classification. The experimental results show that the algorithm can effectively improve the accuracy of image classification.

Keywords: Image classification · Feature extraction · Dimensionality reduction · BP network

1 Introduction

In the modern era of data explosion, people are exposed to a great variety of pictures. The first step to process the data of these pictures is to classify them [1, 2]. With the continuous development of computer technology, artificial intelligence and other fields, the number and types of data are increasingly diverse. Therefore, image classification has attracted people's attention. Image classification usually includes image information acquisition, image preprocessing, image detection, feature extraction and dimensionality reduction and then the image classification is eventually achieved.

LBP (local binary pattern) is a practical and popular texture feature extraction operator. it is widely used for its principle is easy to understand and the extracted texture features are obvious,. Ahonen et al. [3] proposed a face image representation method based on local binary pattern (LBP) texture features. Guoying Zhao et al. [4] improved the extended VLBP(volume local binary patterns) algorithm to deal with dynamic features and applied it in expression recognition. Wei Li et al. [5] proposed a framework of LBP based image local feature extraction and two-level fusion with global Gabor feature and original spectral feature to classify hyperspectral images with high spatial resolution, and achieved good results.

However, the traditional LBP Operator only considers the relationship between the gray value of the center pixel and the neighboring pixel when obtaining the eigenvalues, which is easily affected by the change of gray value and reduces the recognition rate. An

© Springer Nature Switzerland AG 2021
Y. Tan and Y. Shi (Eds.): ICSI 2021, LNCS 12690, pp. 440–451, 2021.
https://doi.org/10.1007/978-3-030-78811-7_41

improved LBP algorithm introduces the concept of threshold. Firstly, the gray values of the neighborhood pixel and the center pixel are obtained by calculation, and the influence on the threshold is further calculated. On this basis, the LBP algorithm of weighted threshold UM proposed in this paper introduces the weight U to the ordinary threshold M, so as to reduce the influence of the maximum and minimum value in the calculation of threshold M, and make the threshold M more distinguishable. The algorithm also optimizes the field of face recognition. When extracting the features of mouth, eyebrow, eye and other parts, we can increase the weight U appropriately to make the features of these parts more obvious.

2 Literature

2.1 LBP Algorithm

LBP (Local Binary Pattern) is first put forward by T. Ojala et al. in 1994 for texture feature extraction, it has the significant advantages in terms of rotation invariance and gray invariance, as well as the extraction of the local texture feature of the image.

Description of LBP Features
The original LBP [6, 7] is based on the Central Pixel of the window as the threshold, comparing the gray values of 8 adjacent pixels in the 3 * 3 window. When the surrounding pixels are larger than the Central Pixel, the position of the central point is marked as 1, otherwise it is marked as 0, eight adjacent pixels can have a string of eight binary digits. Then, the LBP value of the Pixel in the center of the window is calculated and used to reflect the texture information of the region [8].

The image encoding is shown in Eq. (1):

$$\mathit{\Gamma\_ALL} = \sum\nolimits_{I=1}^{N} (P_I) \tag{1}$$

$$s(g_p - g_c) = \begin{cases} 1, g_p \geq g_c \\ 0, g_p < g_c \end{cases} \tag{2}$$

Where: g_c stands for the gray value of the center point; g_p stands for the gray value of each point in the 8 neighborhood; s () is a symbolic function;

Figure 1 shows a 3×3 local binary mode encoding scheme. With 83 as the center pixel, binary processing is carried out in a clockwise direction from the top left. If the gray value is greater than 83, it is marked as 1, otherwise, it is marked as 0, after that the LBP values for the center pixel are achieved.

2.2 KPCA

The Principle Component Analysis (PCA) is widely used in various scenarios which is a mathematical transformation method that interprets most of the information of the original variable by combining a few linear combinations of the original variables and transforming multiple dimensions into a few main components that are not related

88	77	58
109	84	125
99	79	92

1	0	0
1		1
1	0	1

binary system: 10011011   decimal system: 155

Fig. 1. Schematic diagram of the original LBP

to each other. However, the principle component analysis method has its limitations in processing nonlinear data. As an alternative, KPCA method is used to solve the nonlinear problem [9].

Kernel Principal Component Analysis (KPCA) is a nonlinear PCA method that first projects data dimensions into the feature space through nonlinear transformations and analyzes the main components in the feature space. As shown in the following illustration: In a two-tier spherical data, PCA classification is not effective, while KPCA is able to classify data (Fig. 2).

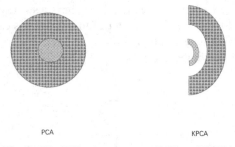

PCA   KPCA

Fig. 2. The difference between PCA and KPCA

2.3 BP Neural Network

BP neural network is a multi layer neural network, it is one of the most widely used neural network, it uses the gradient descent to make the actual output and the expected output error minimum [10].

The BP network adds several layers of neurons between the input and output layers, which become implicit layers without direct connection to the outside world, but their state changes can influence the relationship between input and output. Each layer has several nodes. BP neural network has a strong nonlinear mapping capability and flexible network structure. The performances of different networks with different structures are also distinct [11].

The training of BP neural network is to update its two parameters. The training process of BP neural network includes: forward-looking calculation, input data processing and obtaining actual value;

Reverse calculation: the error signal between the actual output and the expected output is transmitted backwards along the network connection path. The weights and bias of each neuron are corrected to minimize the error signal.

3 Proposed Method

3.1 Improved LBP Algorithm (1): LBP with Threshold Value M

The original LBP solely compares the gray value between the center pixel and the neighborhood, without considering the relationship between them, so it often omits some important feature information.

An improved LBP is proposed to solve this problem. Firstly, a threshold value M [12] is introduced to calculate the difference between the gray values of the neighborhood pixels and the center pixel by adding the absolute values and then taking the average value. The average value is taken as the threshold value. If it is greater than the threshold value, it is 1; if it is less than or equal to the threshold value, it is 0. The LBP value calculated in this way takes into account not only the function of the center pixel, but also the relationship between the neighborhood.

First, the gray value of the center pixel is subtracted from the gray value of the adjacent pixel, and then added to get the threshold value M. the formula (3) is as follows:

$$M = \frac{1}{p} \sum_{i=0}^{p-1} |g_i - g_c| \tag{3}$$

Second, the gray value of GP and GC of the center pixel are subtracted from the absolute value in a certain order to compare with the threshold value M, greater than is 1, less than is 0, as shown in formula (4)

$$s(g_p - g_c) = \begin{cases} 1, |g_p - g_c| > M \\ 0, |g_c - g_c| \le M \end{cases} \tag{4}$$

Then the binary code is converted into decimal number, and the formula (5) is as follows:

$$LBP^*_{P.R} = \sum_{I-1}^{P-1} S(g_p - g_c)2^P \tag{5}$$

Where: p stands for the number neighborhood pixels, where p is 8; gc stands for the gray value of the center point; gp stands for the gray value of each point in the 8 neighborhood; s () is a symbolic function;

An improved LBP coding process is shown in Fig. 3.

88	77	58
109	84	125
99	79	92

0	0	1
1		1
0	0	0

M: 16   binary system: 00110001   decimal system: 49

Fig. 3. Schematic diagram of the LBP with Threshold Value M

3.2 Improved LBP Algorithm (2): LBP with Weight Threshold UM

The simple introduction of a threshold M ignores the importance of the key areas. In the image classification, there are often several core areas of images. Among these core areas, the introduction of a weight threshold called UM can effectively increase the degree of differentiation. The weight refers to the proportion of the neighborhood excluding the maximum value when calculating the threshold value M. The larger the weight is, the less the threshold is affected by the maximum value, which makes it more stable and distinguishable.

Firstly, the absolute value is obtained by subtracting the gray value of the center pixel from the gray value of the adjacent pixel, as is shown in formula (6) (7).

$$Max = Max\,(gi - gc),\ I = 0,\,1....p - 1 \tag{6}$$

$$Min = Min\,(gi - gc),\ I = 0,\,1....p - 1 \tag{7}$$

Then we change the ratio of the minimum to the maximum of the original M to optimize the threshold M, as shown in Eq. (8).

$$M = \frac{U * \sum_{i}^{p-1} |g_i - g_c| - (U - 1) * (Max + Min)}{6U + 2} \tag{8}$$

$$s(g_p - g_c) = \begin{cases} 1, |g_p - g_c| > M \\ 0, |g_c - g_c| \le M \end{cases} \tag{9}$$

Where P is the number of adjacent pixels, gc is the gray value of the central point, gi is the gray value of the current central point, and U is the quality;

The operation is shown in the figure below (Fig. 4):

The following figure shows the LBP feature map after LBP extraction of an image (Fig. 5):

88	77	58
109	84	125
99	79	92

0	0	1
1		0
0	1	1

U=2,UM=15   binary system：00110011   decimal system：51

Fig. 4. Schematic diagram of the LBP with Weight Threshold UM

(a) Original   (b) LBP feature map   (a) Original   (b) LBP feature map

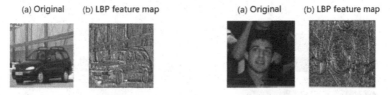

Fig. 5. Feature image of LBP

4 Experiment and Results

4.1 Experiment

This paper is based on the dataset that comes from Imagenet, the largest database of image recognition in the world: colorful pictures of different categories are respectively placed under Car, Dog, Face and Snake. There are about five thousand images in each category. Among them, the dataset is divided into the training set and the test set in a ratio of 7:3 (Fig. 6).

Fig. 6. Pictures of four training sets

The main processes includes:

① Data preprocessing: In this paper, all the color images of the training set are preprocessed and transformed into 256*256 Gy images.
② Feature extraction: In this paper, LBP basic mode, LBP threshold M and LBP weight threshold UM are used for feature extraction successively, and their statistical histogram is used as feature vector [13].

③ Data dimensionality reduction: Since the picture belongs to nonlinear data, KPCA is used to process these data, so that it can be greatly reduced in dimensionality.
④ BP neural network training: The feature vectors obtained from dimension-reduction are stored in trainInputs, and the image categories corresponding to the training set are stored in trainTargets. A BP neural network with four neurons is constructed for training.
⑤ Result evaluation: ACC, Micro-F1, Macro-F1 [14] and other accuracy coefficients are used for evaluation (Fig. 7).

$$P_{I} = \frac{TP_{I}}{TP_{I} + FP_{I}}, (I = 1, \ldots \ldots, N) \tag{10}$$

$$R_{I} = \frac{TP_{I}}{TP_{I} + FN_{I}}, (I = 1, \ldots \ldots, N) \tag{11}$$

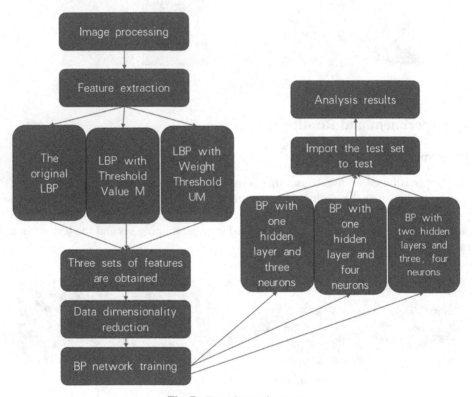

Fig. 7. Experimental process

Where: TP represents true positive, which means the prediction is positive, and it's actually positive; FP represents false positive, which means the prediction is positive, actually negative; FN represents false negative, which means the prediction is negative, actually positive; N represents the number of categories of pictures.

F1 score is an index, which is used to measure the accuracy of binary classification model in statistics. It takes into consideration the accuracy and recall of the classification model. The F1 score can be perceived as a weighted average of the accuracy rate and recall rate of the model, with the maximum value of 1 and the minimum value of 0.

Generally speaking, Micro-F1 and Macro-F1 belong to two types of F1 indicators.These two F1 are used as evaluation indexes in multi-class classification tasks, and they are two different ways to calculate the mean value of F1. There are differences between the calculation methods of Micro-F1 and Macro-F1, and the results are also slightly different.

Calculation method of Micro-F1: first calculate the total precision and recall of all categories, and then calculate the Micro-F1:

$$P\_ALL = \sum_{I=1}^{N}(P_I) \tag{12}$$

$$R\_ALL = \sum_{I=1}^{N}(R_I) \tag{13}$$

$$Micro\_F1 = \frac{2 * P_{ALL} * R\_ALL}{P_{ALL} + R\_ALL} \tag{14}$$

Calculation method of Marco-f1: average precision and recall of all categories, and then calculate the Macro-F1:

$$FL_I = \frac{2 * P_I * R_I}{P_I + R_I} \tag{15}$$

$$Macro\_F1 = \frac{\sum_{I=1}^{N}(FL_I)}{N} \tag{16}$$

4.2 Comparative Analysis of Experimental Results

300 images of each of these four kinds of images are selected as the test set, and the test is conducted according to three kinds of feature extraction methods. The test results are as follows (Tables 1, 2 and 3 and Figs. 8, 9 and 10):

Table 1. The result of BP network test of three neurons

Kind of LBP	ACC	Macro-F1	Micro-F1
1. The original LBP	0.8401	**0.7995**	**0.8401**
2. LBP with Threshold Value M	0.8517	0.8322	0.8517
3. LBP with Weight Threshold UM, U = 1	0.8750	**0.8419**	**0.8750**
4. LBP with Weight Threshold UM, U = 2	0.8625	0.8632	0.8625
5. LBP with Weight Threshold UM, U = 3	0.8642	**0.8655**	0.8642
6. LBP with Weight Threshold UM, U = 4	0.8500	0.8506	0.8500

Table 2. The result of BP network test of four neurons

Kind of LBP	ACC	Macro-F1	Micro-F1
1. The original LBP	0.8514	**0.8519**	**0.8514**
2. LBP with Threshold Value M	0.8762	0.8544	0.8762
3. LBP with Weight Threshold UM, U = 1	0.8912	**0.8777**	**0.8912**
4. LBP with Weight T hreshold UM, U = 2	0.8821	0.8835	0.8821
5. LBP with Weight Threshold UM, U = 3	0.8864	**0.8873**	**0.8864**
6. LBP with Weight Threshold UM, U = 4	0.8650	0.8675	0.8650

Table 3. The result of BP network test of three,four neurons

Kind of LBP	ACC	Macro-F1	Micro-F1
1. The original LBP	0.8382	**0.6892**	**0.8382**
2. LBP with Threshold Value M	0.8673	0.8413	0.8673
3. LBP with Weight Threshold UM, U = 1	0.8874	**0.8345**	**0.8874**
4. LBP with Weight Threshold UM, U = 2	0.8545	0.8588	0.8545
5. LBP with Weight Threshold UM, U = 3	0.8745	**0.8733**	**0.8745**
6. LBP with Weight Threshold UM, U = 4	0.8403	0.8420	0.8403

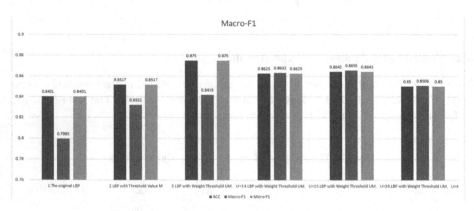

Fig. 8. The result of BP network test of three neurons

As can be seen from the above three histograms, when we use LBP with weight threshold UM, in the same BP network structure, ACC of the original LBP is 0.8401; ACC of LBP with threshold value m is 0.8517, and LBP with weight threshold UM is 0.8517. UM's ACC reaches 0.8750, which increases by approximately 3%. When the total number of pictures climbs, this seemingly small difference can significantly

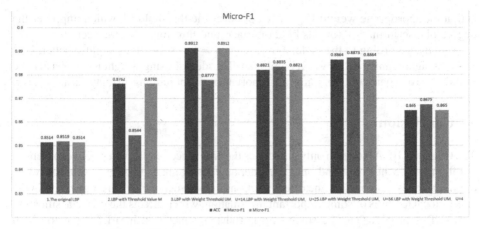

Fig. 9. The result of BP network test of four neurons

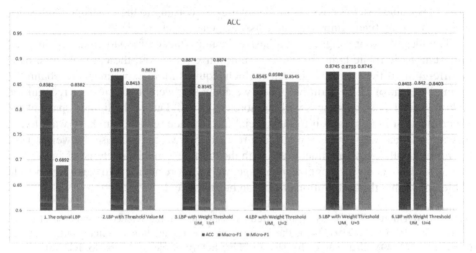

Fig. 10. The result of BP network test of three, four neurons

improves the degree of accuracy of the test. The change of Macro-F1 can also reflect the superiority of the improved UM. Next, we discuss the BP structure and LBP algorithm.

In the structure of BP neural network, in the above test, the increase of the number of hidden layers is not equal to the increase of the accuracy of the experiment, but if the numbers of hidden layers are the same, increasing the number of neurons is conducive to the optimization of the test results and the improvement of the accuracy of the experiment. In three different BP neural networks, it is proved that the LBP model with threshold UM is the best.

In the basic mode of LBP, LBP with threshold M and LBP with weight threshold UM, it can be concluded that the LBP algorithm with weight threshold UM is the best based on the comparison of these three LBP feature extraction methods. The main reason is that the appropriate weight U greatly optimizes the threshold M, which improves the degree of discrimination for this kind of image, and thus improves the accuracy.

Finally, according to the table above, while u = 2 is sometimes slightly better, U = 1 or 3 is significantly better when combined. The idea of using weight to emphasize the importance of a particular region as an important basis for judging an image is innovative.

5   Conclusion

The original LBP Algorithm only calculates the difference of gray value between Central Pixel and adjacent Pixel, and only considers the result analysis of gray value between them, and does not consider the direct relationship between them. This may result in the loss of local features of some important information, which affects the final classification result. This problem can be solved by introducing threshold into LBP, but too high threshold will lead to insufficient differentiation. Therefore, it is suggested to raise the threshold of Body weight (UM). By setting appropriate weight U, the threshold M is more reasonable and the effect is more obvious. According to the experimental results, this method effectively improves the classification efficiency.

The idea of using weight to emphasize the importance of a particular region as an important basis for judging an image is innovative. However, the optimal value of U is different in different research areas, which requires more experiments to obtain the optimal weight of U. In addition, when we apply this weight idea to face recognition and other fields, its advantages will be more confirmed. Due to the limited space of the experiment, I did not apply it to the special field of face recognition, but I want to put forward an idea: when we carry out face recognition, we can increase the weight U of the nose, mouth, eyes and other areas with the most personal characteristics in the face image, so that when we recognize the image, we pay more attention to these high weight U blocks, so that the recognition effect can be more ideal.

Acknowledgement. This study is supported by National Natural Science Foundation of China (71901150, 71702111, 71971143), the Natural Science Foundation of Guangdong Province (2020A151501749), Shenzhen University Teaching Reform Project (Grants No. JG2020119).

References

1. Bin, J., Jia, K., Yang, G.: Research progress of facial expression recognition. Comput. Sci. **38**(4), 25–31 (2011)
2. Yan, O., Nong, S.: Facial expression based on combined features of facial action units recognition. Chin. Stereol. Image Anal. **16**(1), 38–43 (2011)
3. Ahonen, T., Hadid, A., Pietikainen, M.: Face description with local binary patterns: application to face recognition. IEEE Trans. Pattern Anal. Mach. Intell. **28**(12), 2037–2041 (2006). https://doi.org/10.1109/TPAMI.2006.244
4. Zhao, G., Pietikainen, M.: Dynamic texture recognition using local binary patterns with an application to facial expressions. IEEE Trans. Pattern Anal. Mach. Intell. **29**(6), 915–928 (2007). https://doi.org/10.1109/TPAMI.2007.1110

5. Li, W., Chen, C., Hongjun, S., Qian, D.: Local binary patterns and extreme learning machine for hyperspectral imagery classification. IEEE Trans. Geosci. Remote Sens. **53**(7), 3681–3693 (2015). https://doi.org/10.1109/TGRS.2014.2381602
6. Zhou, Y., Wu, Q., Wang, N.: Discriminant complete local binary pattern facial expression recognition. Comput. Eng. Appl. **53**(04), 162–169 (2017)
7. Werghi, N., Tortorici, C., Berretti, S., Del Bimbo, A.: Boosting 3D LBP-based face recognition by fusing shape and texture descriptors on the mesh. IEEE Trans. Inf. Forensics Secur. **11**(5), 964–979 (2016). https://doi.org/10.1109/TIFS.2016.2515505
8. Li, C., Liu, Y., Li, C.: Application of PCA and KPCA in comprehensive evaluation. Yibin College J. **10**(12), 27–30 (2010)
9. Wu, J. (ed.): The Theory and Practice of Integrated Automation System of Water Resources Engineering. China Water and Hydropower Press (2006)
10. Lu, M., Zhou, H.: An improved multi-scale local binary pattern expression recognition method. Hebei Agric. Mach. **10**, 65–69 (2017)
11. Lei, J., Lu, X., Sun, Y.: Expression recognition based on improved local binary pattern algorithm (10), 36–37 (2018)
12. Xin, W., Sheng, Z., Zhu, Y.: Intelligent Troubleshooting Technology: MATLAB Application. Beijing University of Aeronautics and Astronautics Press, Beijing (2015)
13. Li, L.: Expression recognition algorithm based on LBP hierarchical feature **32**, 733–735 (2013)
14. Zhang, H.: An overview and research perspective of local binary pattern. J. Image Signal Process. **05**(3), 121–146 (2016)

Applying Classification Algorithms to Identify Brain Activity Patterns

Marina Murtazina$^{(\boxtimes)}$ and Tatiana Avdeenko

Novosibirsk State Technical University, Karla Marksa Avenue 20, 630073 Novosibirsk, Russia
murtazina@corp.nstu.ru

Abstract. The paper examines the applicability of classification algorithms to identify brain activity patterns according to EEG data using the example of determining the state of eyes. In this area of research, the following main subareas can be distinguished: search for a features' set that can be extracted from EEG data using a minimum number of electrodes, preprocessing techniques development for EEG data, classification techniques comparison and development of new ones, recommendations development for the selection of suitable classification algorithms. Particularly relevant is the question of how accurately classifiers can work with data coming in real time from the EEG neuroheadsets. The paper compares basic classification algorithms implemented in the Weka machine learning tool. For the experiment, six data sets were designed from the "EEG Motor Movement/Imagery" corpus. In the first stage of the experiment, 20 algorithms were investigated on one dataset. The best results were obtained by the IBk, RandomForest and RandomTree algorithms. In the second stage of the experiment, these three algorithms are compared on five additional data sets. The best result on four datasets was obtained for IBk, and one dataset best showed on the RandomForest. The study clearly demonstrated that using a simple data preprocessing procedure it is possible to obtain a classification model that works with an accuracy of 73%–93% even if one applies well-known machine learning algorithms. This preprocessing procedure consists of applying a bandpass filter and excluding outliers from data whose values are greater than three standard deviations from the median.

Keywords: EEG · Classification algorithm · Weka · IBk

1 Introduction

Currently, the problem of recognizing patterns of brain electroencephalography (EEG) data based on machine learning is a rapidly growing field of research [1, 2]. Initially, the attention of scientists was focused on the application of machine learning from EEG data in the field of medical phenomena. The first attempts to classify EEG data were associated with the diagnosis of epilepsy and a number of other diseases. With the emergence and increasing availability of the brain-computer interface technology, the emergence of consumer-grade neurodevices on the market, there have appeared many areas of application of automatic interpretation of EEG data results in the gaming industry

© Springer Nature Switzerland AG 2021
Y. Tan and Y. Shi (Eds.): ICSI 2021, LNCS 12690, pp. 452–461, 2021.
https://doi.org/10.1007/978-3-030-78811-7_42

and education. Currently, machine learning is a promising approach for the classification of EEG signals, which is actively used in identifying body movements [3], recognizing cognitive load [4] and emotions [5], assessing sleep quality [6], analyzing some of mental states [7–9], detection of EEG abnormalities, which may indicate diseases [10]. At the same time, it is extremely important to identify the state of the eyes, since it has a significant impact on the interpretation of EEG data when solving many problems.

In [11], the use of EEG signals for online determination of the eyes state is proposed, k-nearest neighbours and a multilayer perceptron classifiers are investigated. In [12], the issues of classification of the eyes state according to EEG data are investigated using the Weka machine learning algorithm library. In [13], the problem of detecting closed and open eyes using EEG signals from one electrode data is investigated. In [14], it was demonstrated a significant difference between Eye-Closed and Eye-Open in terms of subband frequencies using discrete wavelet transform. In [15], the differences in the functional networks of the brain between closed eyes and open eyes at rest were researched.

Thus, at present, in the research field of brain activity patterns classification according to EEG data, the following main fields of study can be distinguished: (1) search for an informative set of features that can be extracted from EEG data using a minimum number of electrodes, (2) development of preprocessing procedures of EEG data, which will improve the results of classification methods, (3) comparison of existing classification algorithms and the development of new ones, (4) development of recommendations for the selection of suitable classification algorithms. Particularly relevant is the question of how accurately classifiers can work with data received in real time from EEG neuro-headsets, since in this case it is difficult to apply methods that require long-term data processing of EEG data. The purpose of this study is to compare the results of the classification algorithms based on EEG data obtained from 14 sensors. The work is organized as follows. Section 1 describes the research problem. Section 2 discusses theoretical research questions such as processing and analyzing EEG recordings. Section 3 describes the experiment design. Section 4 presents the results of the experiment. Section 5 draws conclusions on the work done.

2 Theoretical Background

2.1 EEG

An electroencephalogram is a recording of electrical biosignals received from the brain. EEG recording carried out using electrodes placed on the head. In order to be able to match the areas of the brain and the location of the electrodes, the 10–20 electrode placement system is used, as well as its extended versions 10–10 and 10–5. Adherence to this family of electrode placement standards is also performed when creating various EEG headsets. Figure 1(a) shows the arrangement of 64 electrodes in a 10–10 system. The electrodes corresponding to the electrodes of the Emotiv Epoc neuro-headset, which is a popular EEG recording instrument for research purposes, are highlighted in gray. Figure 1(b) shows an EEG sample for a resting state with open eyes.

The identifier for each electrode includes the designation of the brain region and side of the cerebral hemisphere. Areas of the brain are designated as follows: prefrontal (Fp),

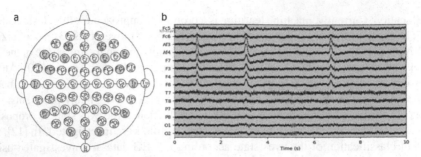

Fig. 1. (a) EEG electrode positions; (b) EEG for a state of rest with open eyes.

frontal (F), temporal (T), parietal (P), occipital (O), central (C). Letter combinations are used for intermediate electrode sites. For example, AF stands for the area between the Fp and F regions. Marker I is used in the names of electrodes placed on the occipital protuberance. Odd numbers of electrodes indicate the left hemisphere of the brain, even ones correspond to the right, for electrodes located on the midline, the letter "z" is used instead of the number.

The main characteristics of an EEG are frequency and amplitude. Frequency is determined by the number of vibrations per second. It is expressed in hertz (Hz). The amplitude is the range of fluctuations in the electric potential on the EEG. It is measured in microvolts (μV). In the EEG, five main frequency ranges are distinguished: δ delta (less than 4 Hz), θ theta (from 4 to less than 8 Hz), α alpha (from 8 to less than 13 Hz), β beta (from 13 to less than 30 Hz), γ gamma (more than 30 Hz).

The raw EEG recordings are usually stored in EDF format. Descriptions of this format for storing multichannel biosignals were published in 1992 in "Electroencephalography and Clinical Neurophysiology" Journal [16]. An EDF file consists of a header record followed by data records. The title record defines the technical characteristics of the recorded signals. EEG data records are multidimensional time series.

2.2 EEG Data Classification

The first step in solving the problem of automatic classification of brain activity patterns based on EEG data is to obtain a data sample suitable for applying machine learning algorithms in accordance with the classification problem. This stage includes EEG data collection, EEG data preprocessing, channel selection and feature extraction. The second step is to build a classifying model from the sample of data (see Fig. 2).

Fig. 2. EEG data classification.

The quality of classification will depend both on how the raw EEG data will be pre-processed and on the algorithm that will be applied at the second stage.

EEG signals are presented as digital numerical values. After collecting the data, it is required to perform preprocessing of the EEG data, during which the detection and removal of various artifacts (unwanted signals), added to the EEG signal during recording sessions, occurs. Artifacts are classified as physiological and non-physiological in origin. The former include, for example, such as normal electrical activity of the heart, muscle activity, blinking of the eyes. The second group includes electrical noise, both in the recording system and outside it [17, 18]. Artifacts appear as outliers in the EEG data. Outlier is defined as an extreme value in the input data that is far beyond the limits of other observations. Outliers can be detected through extreme analysis, approximation methods, or projection methods. One of the most effective approaches to removing certain types of EEG artifacts is bandpass filtering. For example, eye blink artifacts are represented by signals with a frequency below 4 Hz [19], while nonphysiological artifacts caused by electrical noise are observed at frequencies of about 50 Hz or 60 Hz [20].

The channel selection for feature extraction may be due to the following goals: reducing the computational complexity of processing when extracting features from these data, increasing the performance of the classifying model by excluding data unnecessary for predicted activity, and identifying the brain region that generates activity [21]. Feature extraction is a step in which specific features selected and extracted from the data. When classifying according to EEG data, the numerical value of the channel at a certain point in time can be used as a feature. Also, indicators calculated according to channel data for certain periods of time can be used as a feature of EEG data. These features include: standard deviation, mean, median, maximum and minimum values in the frequency range, band power, etc. [22].

Classification and regression are the two main tasks of supervised machine learning. The classification model predicts belonging of objects to one or another class being the discrete value of a certain dependent categorical variable, based on a set of independent variables (features). In order to implement machine learning from EEG data, there exist specialized libraries for the classification of biosignals, for example, pyRiemann [23] Also, in this area, software is actively used that is not specialized in EEG analysis, for example, Weka. For Weka, the input data is usually provided in a *.ARFF (Attribute-Relation File Format) file. The *.ARFF format file consists of two sections. The first section contains the name of the relationship, a list of attributes. The second section contains the rows of data instances. Weka supports four types of attributes: numeric, nominal, string, and data format. When preparing the file for the classification of EEG data, the dependent variable is assigned a nominal type, and the features are assigned a numeric type.

3 Experiment Design

The classification will be carried out using data from 14 electrodes, corresponding to the Emotiv Epoc neuro-headset circuit. The methodology of the research is based on the provisions of researches [11, 12]. The experimental technique includes: (1) defining a strategy of preparing the data for machine learning, (2) determining a list of machine learning algorithms for the research, (3) preparing datasets for an experiment,

(4) machine learning classifiers and assessing accuracy of classifying models in the Weka.

To determine the strategy for obtaining a data sample suitable for applying machine learning algorithms, the initial data set "Eeg-eye-state" was studied, in which measurements from 14 sensors of the Emotiv headset with a duration of 117 s were saved [12]. This kit is actively used in studies of the eyes state. From the normalized initial data, a channels histogram was obtained, shown in Fig. 3.

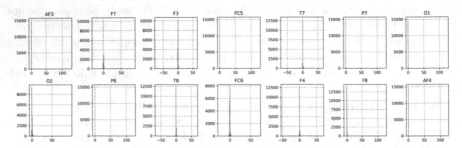

Fig. 3. Channels histogram of raw data after normalization.

The above data was used to study classifiers using the IBk and RandomForest algorithms of the Weka tool. Their choice is explained by the result of the research of the "Eeg-eye-state" dataset author. The classification accuracy for the IBk algorithm was 83.6515%, and for RandomForest – 93.6115%. To assess the accuracy of the classifiers, the cross-validation method with 10 blocks was used. Then, the instances whose values exceed three standard deviations from the median were excluded from the EEG data. The histogram of the channels after cleaning from outliers is shown in Fig. 4. Based on the cleaned data, classifiers were constructed using the IBk and RandomForest algorithms. The accuracy of the first was 97.2252%, the second one was 93.745%. Obviously, the chosen approach to cleaning the data gives a positive result, therefore it will be used for all data sets during the experiment.

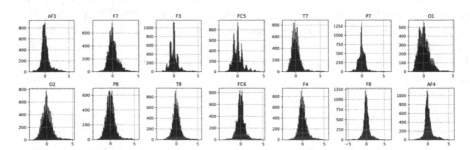

Fig. 4. Channels histogram after cleaning and normalizing data.

The Weka-3-8-4 machine learning algorithm library was used to train the classifying models. 20 algorithms were selected from it (see Table 1).

Table 1. Classification algorithms of the Weka.

Category	Algorithm
Functions	MultilayerPerceptron
Lazy	Kstar, IBk
Meta	AdaBoostM1, AttributeSelectedClassifier, LogitBoost, IterativeClassifierOptimizer, FilteredClassifier, ClassificationViaRegression, RandomizableFilteredClassifier, RandomSubSpace, Bagging, RandomCommittee
Rules	DecisionTable, JRip
Trees	J48, LMT, REPTree, RandomForest, RandomTree

The experiment was carried out using the "EEG Motor Movement/Imagery" data corpus located in the PhysioNet data warehouse. The set consists of one- and two-minute EEG recordings obtained from 109 subjects. EEGs were recorded using a 64-channel electroencephalograph and saved in the EDF format [24]. From the "EEG Motor Movement/Imagery" data set, data were taken from 21 to 51 s for EEG recordings with eyes open and closed at rest. The sample is made for six subjects, starting with the second subject. Let us designate them $S2$, $S3$, $S4$, $S5$, $S6$ and $S7$, respectively. A set of scripts in Python was developed to automate the formation of ARFF data sets from EEG records in EDF format. EEG recordings are filtered in the alpha band in order to exclude some of the artifacts.

4 Experiment Results

Each set was cleaned from the outliers and normalized. Then the histograms of the channels were built to make sure that the data was not distorted, and it is possible to start an experiment. Figure 5 shows a histogram of the channels after cleaning and normalization data for the $S4$ set.

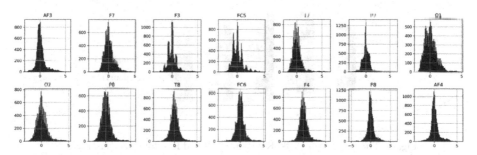

Fig. 5. Channels histogram after clearing and normalizing data for the $S4$ dataset.

At the first stage of the experiment, on the generated data set for the second subject from "EEG Motor Movement/Imagery", machine learning was performed using 20 standard WEKA classifiers. The results are shown in Fig. 6.

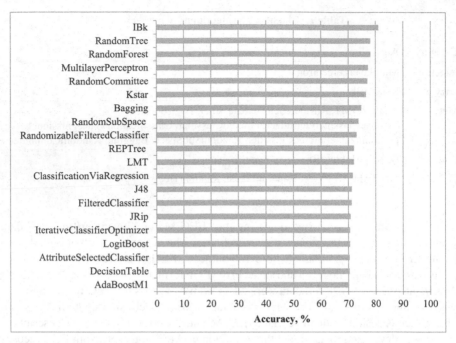

Fig. 6. The accuracy of classifiers obtained in Weka using default settings.

The best results were obtained using the IBk, RandomForest and RandomTree algorithms. At the second stage of the experiment, the models were trained using these three algorithms on five more data sets obtained from the rest of the selected subjects from the "EEG Motor Movement/Imagery" corpus. The best result for four datasets was shown on the IBk algorithm, and for one – on the RandomForest algorithm (see Fig. 7).

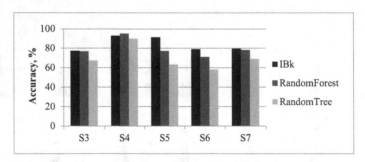

Fig. 7. The accuracy of IBk, RandomForest and RandomTree classifiers.

The best result for four datasets was shown on the IBk algorithm, and for one - on the RandomForest algorithm. Further, the indicators of the classification success were investigated when changing the standard settings for the IBk classifier using the example of the dataset *S5*. The results are shown in Fig. 8 and Table 2.

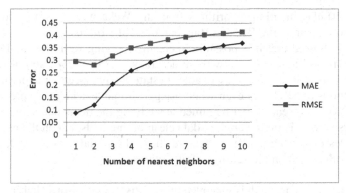

Fig. 8. RMSE and MAE based on the number of nearest neighbors.

Table 2. The accuracy rate and error values of IBk classifier.

Number of nearest neighbors	Accuracy, %	MAE	RMSE
1	91.3355	0.0868	0.2943
2	88.5607	0.1192	0.2799
3	89.3484	0.2035	0.3169
4	83.0290	0.2580	0.3494
5	81.6864	0.2912	0.3681
6	79.4307	0.3151	0.3824
7	78.1418	0.333	0.3928
8	76.8170	0.3482	0.4016
9	76.4053	0.3595	0.4081
10	75.5639	0.3694	0.4146

According to the results shown in Table 2, it can be seen that when using the IBk algorithm, the best classification success for the S5 subject was obtained with the number of nearest neighbors equal to 1. The classification accuracy with this number of nearest neighbors was 91.3355%, mean absolute error (MAE) – 0.0868, and root mean square error (RMSE) – 0.2943. Thus, an increase in the number of nearest neighbors leads to deterioration in the result of the classifier operation.

5   Conclusion

Using the problem of identifying the eyes state as an example, an easy-to-implement approach to solving the automatic classification problem of brain activity patterns based on EEG data was demonstrated. At the stage of preparing the data for classification, band-pass filtering was applied in the alpha band. Instances whose values exceed three standard deviations from the median were excluded from the data sets. As part of the experiment, 20 classification algorithms from the Weka were investigated. The best results were obtained by the IBk and RandomForest algorithms.

The study showed that it is possible to obtain, applying well-known algorithms, a classification model that operates with an accuracy of 73%–93% using a simple data preprocessing procedure. The study enables to state the prospects of applying machine learning methods to EEG data even when using a simple data cleaning procedure. The further direction of research will be aimed at studying the dependence of the result of the classifiers work: (1) on the applied data cleaning methods, (2) on the used features, which includes extracting statistical features from the initial data and calculating the power band values, (3) on the settings of algorithms.

Acknowledgment. The research is supported by Ministry of Science and Higher Education of Russian Federation (project No. FSUN-2020-0009).

References

1. Chan, A., Early, C.E., Subedi, S., Li, Y., Lin, H.: Systematic analysis of machine learning algorithms on EEG data for brain state intelligence. In: 2015 IEEE International Conference on Bioinformatics and Biomedicine (BIBM), pp. 793–799. IEEE, Washington (2015)
2. Doma, V., Pirouz, M.A.: comparative analysis of machine learning methods for emotion recognition using EEG and peripheral physiological signals. J. Big Data **7**, 18 (2020)
3. Sayilgan, E., Kemal, Y., Isler, Y.Y.: Classification of hand movements from EEG signals using machine learning techniques. In: 2019 Innovations in Intelligent Systems and Applications Conference (ASYU), pp. 1–4. IEEE, Izmir (2019)
4. Saha, A., Minz, V., Bonela, S., Sreeja, S.R., Chowdhury, R., Samanta, D.: Classification of EEG signals for cognitive load estimation using deep learning architectures. In: Tiwary, U.S. (ed.) IHCI 2018. LNCS, vol. 11278, pp. 59–68. Springer, Cham (2018). https://doi.org/10.1007/978-3-030-04021-5_6
5. Suhaimi, N.S., Mountstephens, J., Teo, J.: EEG-Based emotion recognition: a state-of-the-art review of current trends and opportunities. Comput. Intell. Neurosci. **2020**, 8875426 (2020)
6. Ravan, M.: Machine learning approach to measure sleep quality using EEG signals. In: 2019 IEEE Signal Processing in Medicine and Biology Symposium (SPMB), pp. 1–6. IEEE, Philadelphia (2019)
7. Jiao, Z., Gao, X., Wang, Y., Li, J., Xu, H.: Deep convolutional neural networks for mental load classification based on EEG data. Pattern Recogn. **76**, 582–595 (2018)
8. Edla, D.R., Mangalorekar, K., Dhavalikar, G., Dodia, S.: Classification of EEG data for human mental state analysis using Random Forest Classifier. Procedia Comput Sci. **132**, 1523–1532 (2018)
9. Antonijevic, M., Zivkovic, M., Arsic, S., Jevremovic, A.: Using AI-based classification techniques to process EEG data collected during the visual short-term memory assessment. J. Sensors **2020**, 8767865 (2020)
10. Gemein, L., et al.: Machine-learning-based diagnostics of EEG pathology. NeuroImage **2020**, 117021 (2020)
11. Sabancı, K., Koklu, M.: The classification of eye state by using kNN and MLP classification models according to the EEG signals. IJISAE **3**(4), 127–130 (2015)
12. Roesler, O., Suendermann, D.: A first step towards eye state prediction using EEG. In: 2013 International Conference on Applied Informatics for Health and Life Sciences, Istanbul, (2013)

13. Öner, M., Hu, G.: Analyzing One-Channel EEG Signals for Detection of Close and Open Eyes Activities. In: 2013 Second IIAI International Conference on Advanced Applied Informatics, pp. 318–323. IEEE, Los Alamitos (2013).
14. Fathillah, M.S., Jaafar, R., Chellappan, K., Remli, R.: A study on EEG signals during eye-closed and eye-open using discrete wavelet transform. In: 2016 IEEE EMBS Conference on Biomedical Engineering and Sciences (IECBES), pp. 674–678. IEEE, Kuala Lumpur (2016)
15. Tan, B., Kong, X., Yang, P., Jin, Z., Li, L.: The difference of brain functional connectivity between eyes-closed and eyes-open using graph theoretical analysis. Comput. Math. Methods Med. **2013**, 976365 (2013)
16. Kemp, B., Värri, A., Rosa, A.C., Nielsen, K.D., Gade, J.: A simple format for exchange of digitized polygraphic recordings. Electroencephalogr. Clin. Neurophysiol. **82**, 391–393 (1992)
17. Jiang, X., Bian, G.B., Tian, Z.: Removal of artifacts from EEG signals: a review. Sensors (Basel, Switzerland) **19**(5), 987 (2019)
18. Urigüen, J.A., Begoña, G.-Z.: EEG artifact removal-state-of-the-art and guidelines. J. Neural Eng. **12**(3), 031001 (2015)
19. Shahbakhti, M., Bavi, M., Eslamizadeh, M.: Elimination of blink from EEG by adaptive filtering without using artifact reference. In: 2013 4th International Conference on Intelligent Systems, Modelling and Simulation, pp. 190–194 (2013)
20. Schembri, P., Anthony, R., Pelc, M.: Detection of electroencephalography artefacts using low fidelity equipment. In: 4th International Conference on Physiological Computing Systems, Madrid, Spain, pp. 65–75 (2017)
21. Alotaiby, T., El-Samie, F.E.A., Alshebeili, S.A., Ahmad, I.: A review of channel selection algorithms for EEG signal processing. EURASIP J. Adv. Signal Process. **2015**(1), 1–21 (2015). https://doi.org/10.1186/s13634-015-0251-9
22. Murtazina, M.S., Avdeenko, T.V.: Classification of brain activity patterns using machine learning based on EEG Data. In: 2020 1st International Conference Problems of Informatics, Electronics, and Radio Engineering (PIERE), pp. 219–224. IEEE, Novosibirsk (2020)
23. Mousavi, M., de Sa, V.R.: Spatio-temporal analysis of error-related brain activity in active and passive brain–computer interfaces. Brain-Comput. Interfaces **6**(4), 118–127 (2019)
24. Schalk, G., McFarland, D.J., Hinterberger, T., Birbaumer, N., Wolpaw, J.R.: BCI2000: a general-purpose brain-computer interface (BCI) system. IEEE Trans. Biomed. Eng. **51**(6), 1034–1043 (2004)

An Improved El Nino Index Forecasting Method Based on Parameters Optimization

Chenxin Shen, Qingjian Ni[✉], Shuai Zhao, Meng Zhang, and Yuhui Wang

School of Computer Science and Engineering, Southeast University, Nanjing, China
nqj@seu.edu.cn

Abstract. El Nino is an important research issue in meteorology. In this paper, we propose a time series model to predict the NINO index. In this model, variational mode decomposition (VMD) is applied to extract multiple sub-signals, the long short term memory network (LSTM) is used to fit these sub-signals. Aiming at the optimization of parameters, we design a K-means neighbor particle swarm optimization (KMPSO) based on comprehensive learning particle swarm optimization (CLPSO), which optimizes the parameters of VMD and LSTM. El Nino data is widely concerned due to its strong relevance to world climate change. We conduct experiments on El Nino data, and put forward a forecast model, which has better forecast skills than other models. Experiments results demonstrate that the proposed method extends forecast time limits, and improves the accuracy prediction.

Keywords: El Nino · Forecast model · LSTM · Signal decomposition · VMD · CLPSO

1 Introduction

The NINO3.4 index is defined as sea surface temperature (SST) anomaly averaged over the equatorial box bounded by 5N, 5S, 170W and 120W, which is an important reference standard for describing El Nino events [1]. The research on El Nino index began decades ago. Meteorologists usually used the method of establishing dynamic climate model to predict El Nino index, such as sintex-f [2] in Japan, HCM and CCA [3] in IRI of Columbia University and cfsv2 [4] in the United States.

With the development of artificial intelligence, deep learning and machine learning methods also enrich the kernel of El Nino index prediction model. In El Nino prediction, single factor forecast models are usually not adopted because El Nino is influenced by many other factors [5,6]. Therefore, selection method for modeling is an important issue. To solve the multi-factor prediction problem, researchers introduce different artificial intelligence methods. Luo [7] from Nanjing University of information engineering introduced CNN model into the prediction of Nino index. They took the sea level temperature and ocean heat

© Springer Nature Switzerland AG 2021
Y. Tan and Y. Shi (Eds.): ICSI 2021, LNCS 12690, pp. 462–471, 2021.
https://doi.org/10.1007/978-3-030-78811-7_43

content of El Nino3.4 area as the input of the model. The model accurately predicted the multi-year El Nino3.4 index, and extended the prediction of El Nino index forecast-lead to 17 months.

In this paper, a NINO3.4 index prediction model based on VMD [8] and LSTM is proposed. VMD is widely used in the field of mechanical automation. In recent years, some researchers have applied this method to the field of wind speed prediction, and achieved better results than other state-of-art models [9]. As a meteorological factor, the uncertainty of El Nino index is similar to that of wind speed, so VMD is also applicable. The model proposed in this paper uses the method of VMD to obtain multiple sub-signals with limited bandwidth, then trains LSTM with these sub-signals.

Since comprehensive learning particle swarm optimization (CLPSO) [10] algorithm doesn't work well sometimes, we design an improved comprehensive learning particle swarm optimization algorithm based on the K-means neighbor, which is proved to have better constringency rate and optimized quality than CLPSO by experiments. We get inspirations from Lee [11], who combined KNN methods with PSO. The optimized prediction model can lengthen the prediction lead-time of El Nino index to 56 months, and is better than the commonly used prediction model.

The structure of this paper is as follows: in the Sect. 2, signal decomposition methods and comprehensive learning particle swarm optimization algorithm are briefly described. In the Sect. 3, we propose the El Nino3.4 index forecast model. In the Sect. 4, we provide analysis about the process and result of our experiments. The Sect. 5 summarizes the work of this paper.

2   Related Methods

2.1   Signal Decomposition

Wavelet decomposition is a signal decomposition method based on Flourier transformation. Flourier transformation [12] is good at processing stationary signals. It obtains components from a signal of multi-frequency, but it cannot determine the time each signal appears. Wavelet decomposition uses the wavelet basis function which can be attenuated to solve the problem. Through the convolution of primary function with signal, Wavelet decomposition gets high frequency and low frequency sub-signals.

Empirical mode decomposition method (EMD) [13] decomposes complex signal into simple intrinsic mode functions (IMF). Each IMF is approximately symmetry to local mean value, a nearly sinusoidal, while the eigenmodes are approximately independent of each other. The variational mode decomposition (VMD) decomposes the signal into several modes with limited bandwidth at the same time, and each mode component is around the corresponding center frequency. Therefore, the algorithm first constructs a variational problem model, and then obtains each mode by solving the model. The variational model can be described as: under the condition that the sum of modal functions is equal to the original input signal, K limited bandwidth modes are found to minimize the sum of modals' bandwidth.

2.2   Comprehensive Learning Particle Swarm Optimization

When the traditional particle swarm optimization (PSO) algorithm updates the particle velocity, it uses the global optimal position of particle swarm and the historical optimal position of particle to update iteratively. As global optimal position is in opposite direction to the historical optimal position of a particle, a large number of invalid motions will be generated [14]. Another effect is that all dimensions of each particle are moving to the same model, making it difficult to jump out of the local optimal position. Therefore, the CLPSO adopts the method of updating particle speed in different dimensions. In this paper, we proposed an improved particle swarm optimization, which references the strategy that particles learn in different dimensions separately.

2.3   Envelope Entropy

Entropy is used to describe the degree of chaos in a system. In the variational mode decomposition, there are two key factors that affect the quality of decomposition: K (the number of layers K) and $alpha$ (penalty parameter). In this paper, the minimum envelope entropy [15] is selected as the optimization objective to optimize these two parameters.

3   El NINO Prediction Model Based on KMPSO

3.1   The Proposed Forecasting Method Process

El Nino3.4 index prediction method proposed in this paper can be divided into two parts. Firstly, we use KMPSO to optimize VMD parameters, and the target is to minimize the minimum envelope entropy of sub-signals decomposed from El Nino3.4 monthly index by VMD. After optimization, the parameters of VMD, mode number k and constraint center frequency $alpha$, are obtained under the condition of minimum envelope entropy. Secondly, after getting sub-signals cut into data set, we create K LSTM models to fit the data set. KMPSO also optimizes the parameters of K LSTM models, while the optimization target is the minimum square error obtained by cross validation. Then prediction model expected in this paper is finally obtained. The process is shown schematically in Fig. 1.

3.2   Improved Particle Swarm Optimization Based on the K-Means Neighbor

In this paper, on the basis of the CLPSO, an improved PSO which counts on the K-means neighbor is proposed. For convenience, we name it KMPSO. Flowchart of KMPSO is shown in Fig. 2.

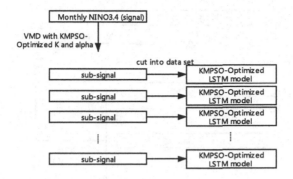

Fig. 1. The process of El Nino3.4 forecasting method

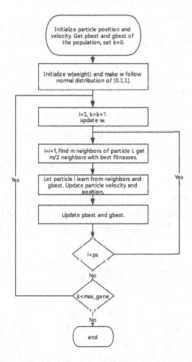

Fig. 2. The flow chart of KMPSO

There are two key parameters in VMD: mode number k and initial center constraint strength ($alpha$) of each mode. The number of modes determines the number of sub-signals after decomposition, the constraint strength controls the bandwidth of sub-signals. In this paper, the minimum value of the envelope entropy in sub-signals is selected as the optimization target, and we use KMPSO to optimize these parameters. The optimization process is shown in Fig. 3.

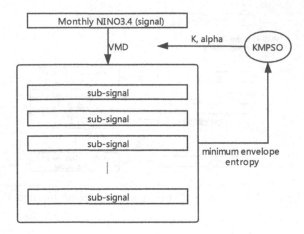

Fig. 3. The optimization process of VMD

The detailed process is shown below:

1. Initialize the particle velocity and position randomly in space, calculate the fitness of each particle, and get the historical and global optimal position of the population.
2. Initialize ω and make it conform to normal distribution in the range of [0.1, 1]. Every particle has its' own ω.
3. For each particle i, find any $2 * m$ particles, and calculate the random-weight Manhattan distance between the particle i and the historical optimal position of $2 * m$ other particles. The distance is calculated with the formula below:

$$dist_{ij} = \sum_{d=1}^{dim} \frac{rand() * abs(pop[i].Pos[d] - pop[j].PbPos[d])}{XMAX[d] - XMIN[d]} \tag{1}$$

In this formule, dim means sequence of particle position, $pop[i]$ means particle i in the population, $pop[i].Pop[d]$ means the position value of particle i in dimension d. $XMAX[d]$ is the max value of particles in dimension d.

4. On the basis of Manhatten distance, find m nearest neighbors from $2 * m$ particles, and find $m/2$ nearest neighbors with higher fitness from m neighbors. Pay attention to that when mention about neighbors' position and fitness, we refer specifically to the historical best position of these particles. The particle i selects different dimensions of $m/2$ best neighbors with a certain probability for learning. If there are multiple nearest-neighbor models in a certain dimension, then particle i would learn from the average value of this dimension of these models. For example, $kmpos_i[d]$ for particle i in dimension d should be obtained below.

$$prob = int(rand(0, 1, m/2) + 0.5)$$

$$r = \frac{1}{\sum_{k=1}^{m/2} prob[k]} \tag{2}$$

$$kmpos_i[d] = r * \sum_{j=1}^{m/2} prob[j] * pop[neibor_j].PbPos[d]$$

In the formula above, $int(rand(m/2) + 0.5)$ means generate a $m/2$-Size array used as possibility,which is about whether particle i learns from the d dimension of $m/2$ neighbors, while $neighbor_j$ is one of the neighbors configured before, history best position (in dimension d) of this particle is $pop[neighbor_j].PbPos[d]$.

5. According to the global optimal position and KMPos position, we update the speed and position of particles, as well as the historical optimal position and the historical optimal fitness of particles. The global optimal position and the global optimal fitness of particle swarm are updated, either. Speed is updated as follows.

$$disp_i = C1 * (pop.GbPos - pop[i].Pos)$$
$$+ C2 * (kmpos_i - pop[i].Pos) \tag{3}$$
$$pop[i].V = \omega[i] * pop[i].V + disp_i$$

In this formula, we update the speed of particle i with global best position and $kmpso$ position.

6. Repeat step 3 to 5, and the weight ω of each generation is attenuated in equal proportion until the number of iterations reaches the target value. Then global optimal position is the optimal solution. The weight attenuation equation is as follows:

$$\omega[i] = \omega[i] * \frac{2 * maxgen - k}{2 * maxgen} \tag{4}$$

k means generation in the formula above, we can know that when k is equal to $maxgen$, each ω changes to half of its' value.

4 Experiments and Results

4.1 Data Set

The data used in this article comes from https://climatedataguide.ucar.edu/. The monthly average NINO3.4 index (°C) of 1900–2019 is obtained with processing the data. The training set and test set are divided by 4:1.

4.2 Experimental Settings

In the experiment, we tried signal decomposition methods on NINO3.4 series firstly. The sub-signals are then cut into datasets. Datasets generated by WT,

VMD and EMD are trained and tested on regression models. Many combinations are made to seek the best one. After that, the best combination is chosen and we optimize the parameters of it by KMPSO.

To compare CLPSO and KMPSO, we use the fitness function introduced in Sect. 3.2, which aims to get appropriate parameters K and $alpha$. Parameters are shown in Table 1.

Table 1. Parameters used in experiment

Parameters	KMPSO-VMD-LSTM	CLPSO-VMD-LSTM	VMD-LSTM
K (VMD)	7	5	3
Alpha (VMD)	1567	1318	1000
Num_layers (LSTM)	3	2	1

4.3 Analysis of Different Combinations Without Optimization

This section introduces the prediction effect of different signal decomposition methods combined with simple regression models. The results of different prediction models based on signal decomposition are drawn in Fig. 4.

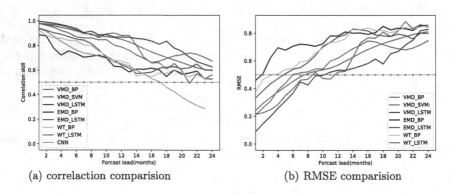

(a) correlaction comparision   (b) RMSE comparision

Fig. 4. Comparision of varied models (correlation skill)

In Fig. 4, VMD, EMD, WT are signal decomposition methods, BP, LSTM, SVM are pure models. We combine different decomposition methods and regression models here to see which combination is better. The blue line (VMD-LSTM) shows the best performance on correlation skill and RMSE among them. From this figure, we can conclude that the LSTM prediction model based on the variational mode decomposition has the best result in the correlation coefficient quota, and the RMSE indicator of this model is also better than other models.

4.4   Analysis of Experiment on KMPSO and CLPSO

In Fig. 5, KMPSO is the improved algorithm proposed in this article, while CLPSO is the comprehensive learning particle swarm optimization algorithm. The optimization parameters in this experiment are K and $alpha$, and optimization goal is minimum value of entropy, which aims to search for global optimum parameters for VMD. It can be seen that KMPSO gets better solution than CLPSO, but the convergence is weaker. In addition, we find that KMPSO runs faster than CLPSO per generation (KMPSO needs 80.45 s per generation, while CLPSO needs 92.24 s per generation, both of them run 100 generations with 200 particles).

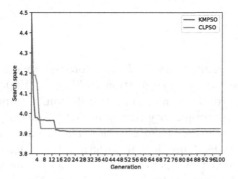

Fig. 5. Result of optimization result

4.5   Analysis of Prediction Model Based on KMPSO

The forecasting method in this paper needs to be optimized in two aspects, parameter selection of VMD and parameter selection of LSTM. After parameter optimization, the prediction model in this paper can lengthen the prediction time-lead to 56 months, and the correlation coefficient comparison with other models is shown in Fig. 6. In the figure, KMPSO-VMD-LSTM model is optimized from VMD-LSTM model by KMPSO. CLPSO-VMD-LSTM model is optimized from VMD-LSTM model by CLPSO. CNN is equal to Luo's model [7], SINTEX-F means Takeshi's model [16], VAR is Harun's model [17]. Some models have shorter bars, because their results have low correlation in the next months.

It can be seen from Fig. 6 that the forecast model optimized by KMPSO performs better than other popular models on correlation skill. From the result, it can be concluded that El Nino index has a high historical similarity. VMD fits El Nino data better when compared with other signal decomposition methods. Based on KMPSO proposed in this paper, the prediction ability of the prediction model is greatly improved.

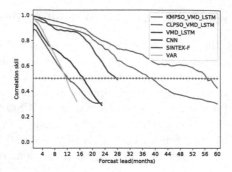

Fig. 6. Result of NINO3.4 forecast models (correlation skill)

5   Conclusion

This paper proposes an improved NINO3.4 forecast model, and puts forward an improved particle swarm optimization (KMPSO). Experiments show that KMPSO is proposed to get more accurate solution. Compared with CLPSO, KMPSO has lower time complexity in population updating process, and it generates more accurate solutions than CLPSO. The forecast model optimized by KMPSO extends time-lead prediction to 56 months, longer than state of art models, and also improves the accuracy of predictions. This work would provide a powerful assistant for meteorologists to predict El Nino events.

Our work shows the potential of machine learning and optimization methods in meteorology. For further work, we will apply these models to other spatial-temporal forecasting scenarios. It is also a direction worth deeper research about the selection of models, design of model architectures and optimization of parameters.

Acknowledgements. This paper is supported by National Key R&D Program of China (2018YFB1004300).

References

1. Saint-Lu, M., Braconnot, P., Leloup, J., Marti, O.: The role of El Nino in the global energy redistribution: a case study in the mid-holocene. Clim. Dyn. **52**(12), 7135–7152 (2019)
2. Malherbe, J., Landman, W.A., Olivier, C., Sakuma, H., Luo, J.: Seasonal forecasts of the SINTEX-F coupled model applied to maize yield and streamflow estimates over north-eastern South Africa. Meteorol. Appl. **21**(3), 733–742 (2014)
3. Goddard, L., Baethgen, W., Bhojwani, H.: The international research institute for climate and society: why, what and how. Earth Perspect. **1**, 1–14 (2014)
4. Pillai, P.A., et al.: How distinct are the two flavors of El Nino in retrospective forecasts of climate forecast system version 2 (CFSv2)? Clim. Dyn. **48**(11–12), 3829–3854 (2017)

5. Lambert, S., Marcus, S., De Viron, O.: Atmospheric torques and earth's rotation: what drove the millisecond-level length-of-day response to the 2015–16 El Nino? Earth Syst. Dyn. Discuss. **8**, 1–14 (2017)
6. Xue, Y., Kumar, A.: Evolution of the 2015/16 El Nino and historical perspective since 1979. Sci. China (Earth Sci.) **60**(09), 1572–1588 (2017)
7. Ham, Y.G., Kim, J.H., Luo, J.J.: Deep learning for multi-year ENSO forecasts. Nature **573**, 568–572 (2019)
8. Wang, D., Luo, H., Grunder, O., Lin, Y.: Multi-step ahead wind speed forecasting using an improved wavelet neural network combining variational mode decomposition and phase space reconstruction. Renew. Energy **113**, 1345–1358 (2017)
9. Han, L., Zhang, R., Wang, X., Bao, A., Jing, H.: Multi-step wind power forecast based on VMD-LSTM. IET Renew. Power Gener. **13**(10), 1690–1700 (2019)
10. Gao, L., Kirby, M., ud Din Ahmad, M., Mainuddin, M., Bryan, B.A.: Automatic calibration of a whole-of-basin water accounting model using a comprehensive learning particle swarm optimiser. J. Hydrol. **581**, 124281 (2020)
11. Lee, C.Y., Huang, K.Y., Shen, Y.X., Lee, Y.C.: Improved weighted k-nearest neighbor based on PSO for wind power system state recognition. Energies **13**(20), 5520 (2020)
12. Griffin, D., Lim, J.: Signal estimation from modified short-time Fourier transform. IEEE Trans. Acoust. Speech Sig. Process. (ASSP) **32**, 804–807 (1983)
13. An, N., Zhao, W., Wang, J., Shang, D., Zhao, E.: Using multi-output feedforward neural network with empirical mode decomposition based signal filtering for electricity demand forecasting. Energy **49**, 279–288 (2013)
14. Liang, J.J., Qin, A.K., Suganthan, P.N., Baskar, S.: Comprehensive learning particle swarm optimizer for global optimization of multimodal functions. IEEE Trans. Evol. Comput. **10**(3), 281–295 (2006)
15. Ong, W., Tan, A., Vengadasalam, V., Tan, C., Ooi, T.: Real-time robust voice activity detection using the upper envelope weighted entropy measure and the dual-rate adaptive nonlinear filter. Entropy **19**, 487 (2017)
16. Doi, T., Storto, A., Behera, S.K., Navarra, A., Yamagata, T.: Improved prediction of the Indian Ocean dipole mode by use of subsurface ocean observations. J. Clim. **30**(19), 7953–7970 (2017)
17. Rashid, H.A.: Factors affecting ENSO predictability in a linear empirical model of tropical air-sea interactions. Sci. Rep. **10**(3), 1740–1745 (2020)

Intrusion Detection System Based on an Updated ANN Model

Yu Xue[1,2](✉), Bernard-marie Onzo[1], and Ferrante Neri[3]

[1] School of Computer and Software,
Nanjing University of Information Science and Technology, Nanjing 210044, China
xueyu@nuist.edu.cn
[2] Jiangsu Key Laboratory of Data Science and Smart Software, Jingling Institute of Technology,
Nanjing 211169, China
[3] COL Laboratory, School of Computer Science, University of Nottingham,
Jubilee Campus, Wollaton Road, Nottingham NG8 1BB, UK
ferrante.neri@nottingham.ac.uk

Abstract. An intrusion detection system (IDS) is a software application or hardware appliance that monitors traffic on networks and systems to search for suspicious activity and known threats, sending up alerts when it finds such items. In these recent years, attention has been focused on artificial neural networks (ANN) techniques, especially Deep Learning approach on anomaly-based detection techniques; because of the huge and unbalanced datasets, IDS encounters real data processing problems. Thus, different techniques have been presented which can handle this problem. In this paper, a deep learning model or technique based on the Convolutional Neural Network (CNN) is proposed to improve the accuracy and precisely detect intrusions. The entire proposed model is divided into four stages: data collection, data pre-processing, the training and testing stage, and performance evaluation.

Keywords: Neural network · Deep learning · Intrusion detection system · CNN

1 Introduction

Like humans, computer systems are also subject to attacks such as viruses, unauthorized access, theft of information, and denial-of-service attacks were the most significant contributors to computer crime. SYMANTEC'S Internet Security Threat Report points out a 56% increase in the number of web attacks in 2010 [1]. Such malicious activities are becoming more numerous, varied, and sophisticated. They pose a very significant risk to nations and organizations such as public health, financial, and government institutions, from a social and economic perspective. Given the complexity of the problem, traditional security tools such as firewalls, anti-spam techniques, and antiviruses cannot recognize new and sophisticated attacks.

Rapid detection of the virus is therefore essential to make quick decisions and to protect networks. This detection in network security is made by monitoring network

© Springer Nature Switzerland AG 2021
Y. Tan and Y. Shi (Eds.): ICSI 2021, LNCS 12690, pp. 472–479, 2021.
https://doi.org/10.1007/978-3-030-78811-7_44

traffic to detect suspicious activity that can be classified into four types of attack (DOS, Probe, U2R, and R2L). However, unfortunately, these methods could malfunction and affect the accuracy of the detection, which is a significant inconvenience. Besides, this testing process is very time-consuming, expensive and the detection rate is also low. Because of these problems, repeated testing must be performed to obtain an accurate detection and low false alert [2].

An intrusion detection system (IDS) [3] watches for abnormalities in traffic and raises the alarm. There are two kinds of IDS, i.e., Host-based IDS and network-based IDS [4, 5]. The basic techniques used for intrusion detection are anomaly detection and misuse or signature-based detection [6–8]. Anomaly detection watches abnormalities in traffic, whereas misuse detection tries to match data with the known attack pattern. One of the significant disadvantages of misuse detection [7] is that new forms of attack are not detected. Therefore most researches focused on anomaly detection techniques [5]. Anomaly-based technique as statistical, neural networks, machine learning [9] and data mining [10], immune system approaches [8].

Indeed several approaches have been used to address these problems. However, the results show that a reasonable margin of progress remains to be made to improve and quality detection to meet the crucial need to avoid intrusions into computer networks. In this perspective, this paper aims to propose a Convolutional Neural Network model to overcome this problem.

This paper is divided into seven sections. Section 2 presents the related works; the methodology used for the proposed framework is discussed in Sect. 3. Section 4 exposes experimental conditions; Sect. 5 shows the experimental results and comparative analysis. Conclusion and future improvements are discussed in Sect. 6.

2   Related Work

Intrusion detection techniques have been actively studied to help the conventional network resist malicious attacks. In artificial neural networks (ANN), various methods and frameworks have been proposed for network intrusion detection. Table 1 shows some of the methods used by researchers and their results.

3   Methodology

We have exploited UNSW-NB15 [23] and CIC-IDS2017 [21] datasets to generate the training and testing samples for this work. We have then selected subsets of samples for training, validation, and test phases; next, we trained our CNN model using the samples of training with validation step following by test our model using the testing samples. We also calculate the performance of our trained model to make a comparison with other models.

Our proposed deep learning-based model comprises several phases summarized in the following four steps:

Step 1: Collect dataset: NSL-KDD, UNSW-NB15, and CIC-IDS2017.
Step 2: Cleaning and uniformization of the different datasets.

Table 1. Related methods.

Authors	Used Algorithms	Method	Results
Hasani et al.[11]	LGP-BA, SVM	The LGP-BA algorithm was used to feature selection and the SVM was used to categorize the acquired features.	Increased accuracy and efficiency.
Gupta and Shrivastava[12]	BC, SVM	SVM was used to classify normal attacks and BC to enhance performance improvements in IDS.	Increased accuracy. 88.46% average accuracy.
Kim et al. [13]	SVM, C4.5	The combination of misuse detection and anomaly detection methods is used to detect intrusions	Reduces the high time complexity of the training and testing processes.
Guo et al. [14]	k-NN, k-means	K-NN and K-means algorithms are used to reduce FPR and FNR.	Detect network anomalies effectively with a low false positive rate.
Hu et al. [15]	AdaBoost	An Intrusion Detection Algorithm is introduced based on the AdaBoost algorithm.	Low computational complexity and error rates.
Mazraeh et al. [16]	AdaBoost, j48, IG	The main learning algorithms, SVM, Bayes Naive, and J48, have been used for feature categorization.	The superiority of the efficiency of the proposed method.
Singh et al. [17]	OS-ELM	A technique based on the Online Sequential Extreme Learning Machine (OS-ELM) is presented.	Outperforms other published techniques in terms of accuracy, false positive rate and detection time. 95.75% average accuracy with a false positive rate of 5.76%
Al-Yaseen et al. [18]	SVM, ELM, k-means	A multi-level hybrid intrusion detection model is presented using a combination of K-means, SVM, and ELM algorithms.	High efficiency in attack detection and its accuracy is the best performance thus far. 95.75% average accuracy
Sujitha and Kavitha [19]	Layered MPSO	A multi-objective particle swarm optimization algorithm is used for feature selection.	The system is highly robust and efficient and can deal with real-time attacks and detect them a fast and quick response.
Horng et al. [20]	SVM, BIRCH	An SVM-based intrusion detection system with BIRCH algorithm is proposed.	Detects DoS and Probe attacks better than previous methods.

Step 3: For each class, divide the data into different sets: training set (81%), validation set (9%), testing set (10%).

Step 4: Evaluate the performance of the model with accuracy, precision, recall, F-measure.

The various layers and parameters applied by our model are shown in Table 2.

Table 2. Model summary.

Layer (type)	Output shape	Param
conv1d (Conv1D)	(None, 76, 64)	256
conv1d_1 (Conv1D)	(None, 74, 64)	12352
conv1d_2 (Conv1D)	(None, 72, 64)	12352
max_pooling1d (MaxPooling1D)	(None, 36, 64)	0
conv1d_3 (Conv1D)	(None, 34, 128)	24704
conv1d_4 (Conv1D)	(None, 32, 128)	49280
max_pooling1d_1 (MaxPooling1)	(None, 16, 128)	0
flatten (Flatten)	(None, 2048)	0
dense (Dense)	(None, 50)	102450
dense_1 (Dense)	(None, 1)	51
		Total: 201,445 Trainable: 201,445 Non-trainable: 0

4 Experiments

4.1 Dataset Description

In this work, a recent dataset includes many modern attacks provided by the Canadian Institute for Cybersecurity called CICIDS2017 [21], and the raw network packets of the UNSW-NB15 [23] dataset have been used. The UNSW-NB 15 dataset has nine types of attacks: Fuzzers, Analysis, Backdoors, DoS, Exploits, Generic, Reconnaissance, Shellcode, and Worms. The Argus, Bro-IDS tools are used, and twelve algorithms are developed to generate 49 features with the class label. It improved the NSL-KDD [22], which became too old and did not reflect the modern attacks anymore. CICIDS2017 dataset contains the most up-to-date common attacks, which resembles actual real-world data (packet capture files, pcap). It also includes the results of the network traffic analysis using CICFlowMeter with labeled flows based on the timestamp, source and destination IPs, source and destination ports, protocols, and attack (as CSV files).

4.2 Data Pre-processing

The CSV files are organized as pcap files, i.e., the columns are the traffic parameters, and the rows represent the packets. Also, we removed the broken lines, made finite infinite elements, and normalized the whole.

4.3 Environment

All the experiments are performed in a python environment running on a Linux Manjaro Mikah workstation with 16 Gb of RAM, an Intel® Core™ i7-8550U CPU @ 1.80 GHz × 8 as a processor, and GeForce MX130 2048 Mb GPU with the following tools: Keras, TensorFlow, Matplotlib, Pandas, Scikit.

5   Results

Experimental results reveal that the proposed model has promising results on training, as shown in Fig. 1 and Fig. 2. Also, the confusion matrix we are getting is in Table 3. Decoding the confusion matrix, out of 30 intrusion predictions, we are getting 0 wrongly classified. Out of 30 normal states, we are getting 28 normal classified rights and 2 as wrongly classified as intrusions. Despite the results shown in Table 4, we believe that improvements can be made to increase accuracy by improving the data pre-cleaning and feature selection.

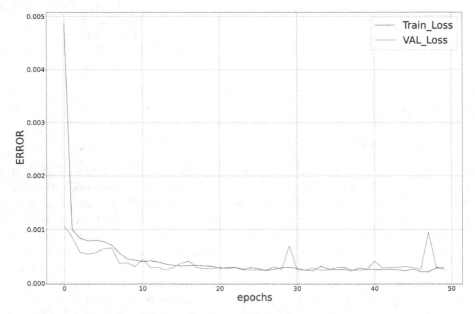

Fig. 1. Loss evolution

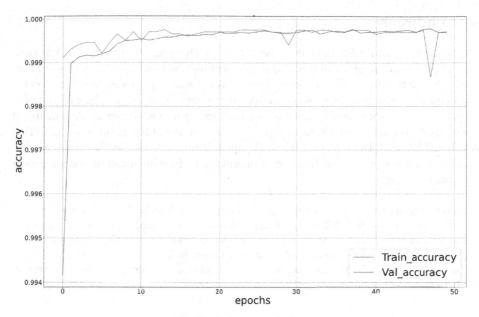

Fig. 2. Accuracy evolution

Table 3. Confusion matrix of the proposed

	Intrusion	Non-intrusion
Intrusion	TP = 9734	FN = 4
Normal	FP = 3	TN = 12834

Table 4. Other metrics

Metric	Formula	Value
Precision	$\frac{TP}{TP+FP}$	0.9996
Recall	$\frac{TP}{TP+FN}$	0.9995
False alarm rate	$\frac{FP}{FP+TN}$	0,0002
True negative rate	$\frac{TN}{TN+FP}$	0.9997
F measure	$\frac{Precision*Recall}{Precison+Recall}$	0.4997
Accuracy	$\frac{TP+TN}{TP+TN+FP+FN}$	0.9996

6 Conclusion

The contribution of this study is to develop is machine-learning model an Intrusion Detection System implementation for high detection rate and low false positives. This will improve the capacity to maintain a high level of security for safe and trusted communication of information between various organizations. Based on the experimental results, it can be concluded that the proposed model is a much better model due to the good results (accuracy of 99.96%) it produces as compared to the other models. This result validates the Deep Neural Networks application for Intrusion Detection System. Our future work will be in the direction of improving the Feature Selection and improving the data cleaning.

Acknowledgement. This work was partially supported by the National Natural Science Foundation of China (61876089, 61876185, 61902281, 61375121), the Opening Project of Jiangsu Key Laboratory of Data Science and Smart Software (No. 2019DS301), the Engineering Research Center of Digital Forensics, Ministry of Education, the Key Research and Development Program of Jiangsu Province (BE2020633), and the Priority Academic Program Development of Jiangsu Higher Education Institutions.

References

1. Symantec: Internet Security Threat Report (ISTR) 2019|Symantec, April 2019. https://www.symantec.com/securitycenter/threat-report. Accessed Jan 2021
2. Rozenblum, D.: Understanding Intrusion Detection Systems, SANS Institute Reading Room site1
3. Rajasekhar, K., Babu, B.S., Prasanna, P.L., Lavanya, D.R., Krishna, T.V.: An overview of intrusion detection system strategies and issues. Int. J. Comput. Sci. Technol. **2**(4) (2011)
4. Sonawane, S., Pardeshi, S., Prasad, G.: A survey on intrusion detection techniques. World J. Sci. Technol. **2**(3), 127–133 (2012)
5. Kumar, V., Sangwan, P.O.: Signature based intrusion detection system using SNORT. Int. J. Comput. Appl. Inf. Technol. **1**(3) (2012) (ISSN: 2278–7720)
6. Uddin, M., Khowaja, K., Rehman, A.: Dynamic multi-layer signature based intrusion detection system using mobile agents. Int. J. Netw. Secur. Appl. (IJNSA) **2**(4) (2010)
7. Ning, P., Jajodia, S.: Intrusion Detection Techniques. http://citeseerx.ist.psu.edu/viewdoc/download?doi=10.1.1.89.2492&rep=rep1&type=pdf
8. Vinchurkar, D., Reshamwala, A.: A review of intrusion detection system using neural network and machine learning technique. Int. J. Eng. Sci. Innov. Technol. (IJESIT) **1**(2) (2012)
9. Lee, W., Stolfo, S.J., Mok, K.: Adaptive intrusion detection: a data mining approach. Artif. Intell. Rev. **14**(6), 533–567, Kluwer Academic Publishers (2000)
10. Kim, J., Bentley, P., Aickelin, U., Greensmith, J., Tedesco, G., Twycross, J.: Immune system approaches to intrusion detection - a review. Nat. Comput. TBA. Springer. https://doi.org/10.1007/s11047-006-9026-4
11. Hasani, S.R., Othman, Z.H., Kahaki, S.M.M.: Hybrid feature selection algorithm for intrusion detection system. J. Comput. Sci. **10**, 1015–1025 (2014)
12. Gupta, M., Shrivastava, S.K.: Intrusion detection system based on SVM and bee colony. Int. J. Comput. Appl. **111**, 27–32 (2015)
13. Kim, G., Lee, S., Kim, S.: A novel hybrid intrusion detection method integrating anomaly detection with misuse detection. Expert Syst. Appl. **41**, 1690–1700 (2014)

14. Guo, C., Ping, Y., Liu, N., Luo, S.S.: A two level hybrid approach for intrusion detection. Neurocomputing **214**, 391–400 (2016)
15. Hu, W., Maybank, S.: AdaBoost-based algorithm for network intrusion detection IEEE Trans. Syst. Man Cybern. B Cybern. **38**, 577–583 (2008)
16. Mazraeh, S., Ghanavati, M., Neysi, S.H.N.: Intrusion detection system with decision tree and combine method algorithm. Int. Acad. J. Sci. Eng. **3**, 21–31 (2016)
17. Singh, R., Kumar, H., Singla, R.K., et al.: An intrusion detection system using network traffic profiling and online sequential extreme learning machine. Expert Syst. Appl. **42**, 8609–8624 (2015)
18. Al-Yaseen, W.L., Othman, Z.A., Nazri, M.Z.A.: Multi-level hybrid support vector machine and extreme learning machine based on modifed K-means for intrusion detection system. Expert Syst. Appl. **67**, 296–303 (2016)
19. Sujitha, B., Kavitha, V.: Layered approach for intrusion detection using multiobjective particle swarm optimization Int. J. Appl. Eng. Res. **10**, 31999–32014 (2015)
20. Horng, S.J., et al.: A novel intrusion detection system based on hierarchical clustering and support vector machines. Expert Syst. Appl. **38**, 306–313 (2011)
21. Intrusion Detection Evaluation Dataset (CIC-IDS2017). https://www.unb.ca/cic/datasets/ids-2017.html
22. NSL-KDD dataset. https://www.unb.ca/cic/datasets/nsl.html
23. The UNSW-NB15 Dataset Description. https://www.unsw.adfa.edu.au/unsw-canberra-cyber/cybersecurity/ADFA-NB15-Datasets/
24. Xue, Y., Tang, T., Pang, W., Liu, A.X.: Self-adaptive parameter and strategy based particle swarm optimization for large-scale feature selection problems with multiple classifiers. Appl. Soft Comput. **88**, 1–12 (2020)
25. Xue, Y., Xue, B., Zhang, M.: Self-adaptive particle swarm optimization for large-scale feature selection in classification. ACM Trans. Knowl. Discov. Data **13**(5), 1–27 (2019)

Bayesian Classifier Based on Discrete Multidimensional Gaussian Distribution

Yihuai Wang and Fei Han[✉]

School of Computer Science and Communication Engineering, Jiangsu University,
Zhenjiang, China
hanfei@ujs.edu.cn

Abstract. Bayesian classifier has become one of the most popular classification methods due to its flexible probability expression and good classification performance. However, in the classification of multidimensional discrete data, the assumption of data independence in naive Bayes classification is unrealistic, the Bayesian network structure is complex and the classification performance is sometimes unstable. In order to more effectively consider the correlation of data in the Bayesian classifier, this paper proposes a Bayesian classifier with multidimensional Gaussian distribution based on discrete data. This classifier uses the moment estimation of the training data to obtain the covariance matrix of the classification model, and reflects the correlation of the data through the covariance matrix. In view of the mathematical model problem caused by the non-positive definite covariance, the largest linearly independent group after feature sorting is selected as a new sample to estimate the covariance matrix, and then the optimized Bayesian classification model is obtained. The classification simulation experiments on several datasets verify the effectiveness and stability of the proposed classifier.

Keywords: Bayesian classifier · multidimensional Gaussian distribution · Maximum linearly independent group

1 Introduction

The Bayesian classification is a simple and effective classification algorithm, which uses the prior distribution of the data to calculate its posterior probability with the Bayesian formula and selects the class with the largest posterior probability as the class to which this sample belongs [1–3]. Bayesian classifiers are mainly divided into two categories the naive Bayesian (NB) classifier and Bayesian network classifier. NB is based on Bayes' theorem and the independent assumption of data characteristics. This classifier is simple to implement and has a good classification accuracy rate [4–7]. Bayesian network classifier uses a directed acyclic graph to describe the dependency relationship of attributes, and uses a conditional probability table to describe the joint probability distribution of attributes [8–10].

The construction of the network structure of the Bayesian network classifier requires a lot of prior knowledge, but a lot of data is actually like a black box. The researcher does

© Springer Nature Switzerland AG 2021
Y. Tan and Y. Shi (Eds.): ICSI 2021, LNCS 12690, pp. 480–490, 2021.
https://doi.org/10.1007/978-3-030-78811-7_45

not know the dependence relationship between the sample features. Using the scoring function to search for Bayesian nets will also fall into the problem of how to construct the scoring function, so the use of Bayesian net classifiers is rare [11]. Although the naive Bayes classifier is widely used, it has the fatal flaw of not considering the correlation of features, so it is not suitable for processing samples with high dimensions and strong correlation [12, 13].

How to avoid falling into the complex problem of constructing the feature-dependent network under the premise of considering the feature relevance is the problem to be solved in this study. We know that the covariance matrix is one of the parameters of the multidimensional Gaussian distribution, and the covariance matrix and the correlation coefficient matrix are the contract matrix. Therefore, the introduction of discrete multidimensional Gaussian distribution in Bayesian classification can solve the above problems. Since the Gaussian distribution is a probability distribution defined on continuous data variables, this paper extends its definition to a discrete space to establish Multidimensional Gaussian Bayes (MGB). The key point of this algorithm is the construction of covariance. To ensure the effectiveness of the algorithm, the largest linearly independent group of samples and the independent assumption of linearly related features are selected to form a new covariance matrix.

2 Preliminaries

2.1 Bayesian Network Classifier

Bayesian nets are also called "belief nets", which use directed acyclic graphs to describe the correlation between attributes [14]. If the correlation between the features is known, only the conditional probability table of the feature needs to be estimated for the training sample to obtain the joint probability density function. However, in real applications, the correlation of features is generally not known. The commonly used solution is "scoring search", which is to define a score function to evaluate the fit between the Bayesian network and the training set, and then based on this scoring function to find the best structured Bayesian network [15, 16].

Artificially setting the attribute dependence in the Bayesian network structure is not objective enough, and using the scoring function to search the network space is more complicated and may get poor classification results. Therefore, the Bayesian network classifier is less practically used.

2.2 Naive Bayes Classifier

Naive Bayes classifier is the most widely used classifier among classifiers formed based on Bayesian theory. Naive Bayes uses the training set samples to calculate the joint probability of all features under the assumption that the features are independent, and then classifies the test set samples after obtaining the classification model. Its algorithm is simple and has a good effect on some sample classification [17].

Naive Bayes is generally divided into Bernoulli Bayes classifier, multinomial Bayes classifier, and Gaussian Bayes classifier according to different distribution functions assumed for data characteristics. Bernoulli distribution is a typical discrete probability distribution, which is mainly used in the classification of binary data. The multinomial distribution is a generalization of the binomial distribution, and the binomial distribution is a generalization of the Bernoulli distribution, which is mainly used for text classification. This Gaussian Bayesian classification is based on a one-dimensional continuous space and is mainly used for continuous sample classification. However, none of the above three types of classifiers consider the correlation between features.

3   Multidimensional Gaussian Bayes Classification Decision

3.1   Gaussian Distribution Mathematical Model

Since the exact distribution of the data is not known, the Bayesian formula cannot be used for classification. The solution is to construct a reasonable hypothesis to approximate the requirements of Bayes' theorem. In the study, the Gaussian distribution density function is used to approximate the Bayesian formula Conditional Probability. When the training set sample N and the number of various samples N_i tend to infinity, according to the law of large numbers, various samples approximately obey a multidimensional normal distribution, and the multidimensional Gaussian distribution density function is denoted as $p(x; \mu, \sum)$:

$$p(x; \mu, \sum) = \frac{1}{(2\pi)^{d/2} det(\sum)^{\frac{1}{2}}} exp\{-\frac{1}{2}(x-\mu)^T \sum^{-1}(x-\mu), x \in R^d \quad (1)$$

where $\mu = [\mu^1, \mu^2 \dots, \mu^d]$, $\sum$ is the $d * d$-dimensional covariance matrix. μ and $\sum$ are the expectation of vector x and matrix $(x-\mu)(x-\mu)^T$. It is not difficult to prove that the covariance matrix is always a symmetric non-negative definite matrix. The relationship between the covariance matrix and the correlation coefficient matrix is as follows:

$$\sum = \begin{bmatrix} \sigma_{11} & 0 & \dots & 0 \\ 0 & \sigma_{22} & \dots & 0 \\ \vdots & \vdots & \ddots & \vdots \\ 0 & 0 & \dots & \sigma_{dd} \end{bmatrix} \begin{bmatrix} R_{11} & R_{12} & \dots & R_{1d} \\ R_{12} & R_{22} & \dots & R_{2d} \\ \vdots & \vdots & \ddots & \vdots \\ R_{1d} & R_{2d} & \dots & R_{dd} \end{bmatrix} \begin{bmatrix} \sigma_{11} & 0 & \dots & 0 \\ 0 & \sigma_{22} & \dots & 0 \\ \vdots & \vdots & \ddots & \vdots \\ 0 & 0 & \dots & \sigma_{dd} \end{bmatrix} \quad (2)$$

The element R_{ij} in the correlation coefficient matrix R represents the correlation coefficient between the i-th feature and the j-th feature. It can be seen from Eq. 2 that the correlation coefficient matrix and the covariance matrix are contractual matrices.

3.2   Bayesian Classification Model Based on Multidimensional Gaussian Distribution

The Bayesian classification decision-making strategy is to use the training set to train each category mean vector μ_i and the parameters in the covariance matrix $\sum_i$, and then the test set samples are classified according to the minimum classification error rate discriminant function. The class sample probability of the Gaussian distribution density function is:

$$P(x|w_i) = \int_{-\infty}^{\infty} \cdots \int_{-\infty}^{\infty} p(x; \mu_i, \sum{}_i) dx_1 dx_2 \ldots dx_{i-1} dx_{i+1} \ldots dx_d \tag{3}$$

The above formula defines the class probability density of Gaussian distribution on continuous variables. When $P(x|w_i)$ is calculated, the amount of calculation is extremely large, and most of the data collected from the natural world are discrete data. In order to extend this model to a discrete space, it is now assumed that the probability density $p(x; \mu_i, \sum_i)$ of the sample point x at tiny neighborhood ΔV remains unchanged. Equation 3 can be simplified to:

$$P(x|w_i) = p(x; \mu_i, \sum{}_i) \Delta V \tag{4}$$

When the probability density function is differentiable, ΔV will not affect the comparison of class conditional probabilities in a very small area, so ΔV can be omitted. Substituting Eq. 4 into the Bayesian decision of the minimum error rate, and then taking the logarithm to remove the polynomial $-\frac{d}{2} ln2\pi$ that is irrelevant to the classification, we can get:

$$\begin{cases} g(w_i|x) = arg\ max\{lnP(w_i) - \frac{1}{2}ln\sum_i - \frac{1}{2}(x - \mu_i)\sum_i^{-1}x - \mu_i^T\},\ x \in w_i \\ w_i = 1, 2, \ldots c \end{cases} \tag{5}$$

Naive Bayes based on Gaussian distribution is a special case where the covariance matrix $\sum$ is a diagonal matrix ($\sigma_{ij} = 0, i \neq j$). At this time, the features are independent of each other. The classification discriminant $g_1(w_i|x)$ for:

$$\begin{cases} g_1(w_i|x) = arg\ max\{lnP(w_i) - \frac{1}{2}\sum_{j=1}^{d} ln(\sigma_{jj}) - \frac{1}{2}\sum_{j=1}^{d} \frac{\left(x^j - \mu_i^j\right)^2}{\sigma_j^2}, x \in w_i \\ w_i = 1, 2, \ldots c \end{cases} \tag{6}$$

Obviously, the key to establishing the Gaussian distribution model of discrete data lies in training various sample mean μ_i and class covariance matrix $\sum_i$. The parameter moments are estimated as follows:

$$\widehat{\mu_i} = M_i, \quad \widehat{\sum}_i = \frac{1}{N_i}\left[(x_i - M_i)^T(x_i - M_i)\right] \tag{7}$$

It should be noted that when the various covariance matrices are not positive definite matrices, the multidimensional Gaussian distribution probability density function is meaningless, so the positive definiteness of the class covariance is a necessary condition to ensure the effectiveness of the classification model.

In order to obtain more training parameters while satisfying the positive definite covariance matrix, a set of largest linearly independent feature groups $[x_i^a, x_i^b, \ldots, x_i^s]$ in the class sample matrix x_i is selected as the new sample data $(a, b, \ldots, s < d)$. The new covariance estimation matrix $\widehat{\sum}_i$ is:

$$\hat{\Sigma}_i = \frac{1}{N_i} \begin{bmatrix} \left(x_i^a - M_i^a\right)^T \left(x_i^a - M_i^a\right) & \left(x_i^a - M_i^a\right)^T \left(x_i^b - M_i^b\right) & \cdots & \left(x_i^a - M_i^a\right)^T \left(x_i^s - M_i^s\right) \\ \left(x_i^b - M_i^b\right)^T \left(x_i^a - M_i^a\right) & \left(x_i^b - M_i^b\right)^T \left(x_i^b - M_i^b\right) & \cdots & \left(x_i^b - M_i^b\right)^T \left(x_i^s - M_i^s\right) \\ \vdots & \vdots & \ddots & \vdots \\ \left(x_i^s - M_i^s\right)^T \left(x_i^a - M_i^a\right) & \left(x_i^s - M_i^s\right)^T \left(x_i^b - M_i^b\right) & \cdots & \left(x_i^s - M_i^s\right)^T \left(x_i^s - M_i^s\right) \end{bmatrix} \tag{8}$$

3.3 Strategies for the Positive Definiteness of Covariance Matrix

In the above scheme, the key is how to select a set of feature data from the sample to make the classification effect good, and how to deal with the remaining features so as not to lose effective feature data is also worth considering. this paper proposes an evaluation criterion function J based on the class spacing. The scoring function of the D-th feature x^D is:

$$J(x^D) = \sum_{i=1}^{C} \sum_{j>i}^{C} \frac{\left(x_i^D - M_j^D\right)^T \left(x_i^D - M_j^D\right)}{\left(M_j^D\right)^2} \tag{9}$$

The larger the value of $J(x^D)$, the lower the expected error rate of the feature classification. After selecting the largest linearly independent group $[x_i^a, x_i^b, \ldots, x_i^s]$ that satisfies various covariances, assume that the remaining features $[(x_i^1, x_i^2, \ldots, x_i^d) \not\subset (x_i^a, x_i^b, \ldots, x_i^s)]$ are independent of each other. Set $[(x_i^1, x_i^2, \ldots, x_i^d) \not\subset (x_i^a, x_i^b, \ldots, x_i^s)]$ as $[x_i^A, x_i^B, \ldots, x_i^T]$, The covariance $\overset{\dots}{\Sigma}_i$ is:

$$\overset{\dots}{\Sigma}_i = \frac{1}{N_i} \begin{bmatrix} \sigma_{aa}^2 & \sigma_{aa}^2 & L & \sigma_{aa}^2 & 0 & 0 & L & 0 \\ \sigma_{aa}^2 & \sigma_{aa}^2 & L & \sigma_{aa}^2 & 0 & 0 & L & 0 \\ M & M & O & M & M & M & O & M \\ \sigma_{aa}^2 & \sigma_{aa}^2 & L & \sigma_{aa}^2 & 0 & 0 & L & 0 \\ 0 & 0 & L & 0 & \sigma_{aa}^2 & 0 & L & 0 \\ 0 & 0 & L & 0 & 0 & \sigma_{aa}^2 & L & 0 \\ M & M & O & M & M & 0 & O & M \\ 0 & 0 & L & 0 & 0 & 0 & L & \sigma_{aa}^2 \end{bmatrix} \tag{10}$$

Where $\sigma_{kl}^2 = \frac{1}{N_i}[(x_i^k - M_i^k)^T (x_i^l - M_i^l)]$, matrix $\hat{\Sigma}_i(1, 1)$ is composed of selected features Covariance matrix subset, $\hat{\Sigma}_i(2, 2)$ is the variance subset obtained after independent assumption of the remaining feature data. Substituting Eq. 10 into Eq. 5 can obtain the improved Gaussian Bayes classification discriminant:

$$\begin{cases} G(w_i|x) = \arg\max\{lnP(w_i) - \frac{1}{2}ln\left\|\overset{\dots}{\Sigma}_i\right\| - \frac{1}{2}\sum_{j=A}^{T}\frac{(x^j-\mu_i^j)^2}{\sigma_j^2} \\ - \frac{1}{2}\left(x^{[a,b,\cdots,s]} - \mu_i^{(x-\mu_i)^T}\right)^T \overset{\dots}{\Sigma}_i(1,1)^{-1}\left(x^{[a,b,\cdots,s]} - \mu_i^{(x-\mu_i)^T}\right)\}, x \in w_i \\ \qquad\qquad\qquad\qquad w_i = 1,2,\ldots c \end{cases} \tag{11}$$

The above formula is applied to the classification of the indefinite covariance matrix, and it is also applicable to the data classification of the positive definite covariance. The process of the MGB classification algorithm is as follows:

Algorithm: Multidimensional Gaussian Bayes classification method

Input: x (sample feature), y (sample label), c (number of categories), M (number of samples), d (feature dimension), s (parameter of s-fold cross-validation);

Output: correct (classification accuracy);

1 $ch \leftarrow$ randperm (M), Sort the features according to Eq.9;

2 **for** i: $= 1$ to s **do**

3      $b \leftarrow ch(1 + (i - 1) * M/s : i * M/s)$;

4   x_test $\leftarrow x (b, :)$, x_train $\leftarrow$ Remove x_test from sample attribute data x;

5   y_test $\leftarrow y (b, :)$, y_train $\leftarrow$ Remove x_test from sample label y;

6   **for** j: $= 1$ to c **do**

7          $\mu_{a_j} \leftarrow$ mean (x_{a_j}); $\Sigma_{a_j} \leftarrow$ cov (x_{a_j});

8   **if** Rank $(M_{a_j}) < d$ **then**

9             $r \leftarrow$ Search for the largest linearly independent group of M_{a_j}, $x \leftarrow x(:, r)$;

10   **end if**

11   **end for**

11      $n \leftarrow$ Compare the test set label with y_train and record the correct number;

12   correct $\leftarrow$ correct $+ n/(M * s)$;

13 **end for**

14 **return** correct.

The algorithm has two core points in the training model: one is the estimation of the sample mean and covariance matrix of discrete data, and the other is to ensure the positive definiteness of various covariances.

4 Experimental Results and Discussion

In order to prove the effectiveness of the multidimensional discrete Gaussian Bayes classifier proposed in this paper and the superiority of the performance of the largest linearly independent group classification model. First, the proposed multidimensional discrete Gaussian Bayes classifier is compared with the KNN classifier and the naive Bayes classifier, and then compare the classification effects of not selecting the maximum linearly independent group and selecting the maximum linearly independent group. The experiment in this paper is run on a PC with 8.0 GB of memory, and the running environment is MATLAB R2018b.

4.1 Datasets

Table 1 is 4 discrete datasets selected from UCI machine learning database. The number of features in these datasets ranges from 19 in Segmentation to 128 in Batch7, the number of samples ranges from 181 in MeterC to 3613 in Batch7, and the categories range from 2 categories in Sonar to 7 categories in Segmentation. Therefore, the selected datasets have sufficient Representative. In this paper, the s cross-validation method is used to divide the test set and the training set.

Table 1. The specification of four data

Datasets	Number of samples	Number of features	Number of classes
Sonar	208	60	2
MeterC	181	43	4
Segmentation	2310	19	7
Batch7	3613	128	6

4.2 Results and Discussion

Performance of Multidimensional Discrete Gaussian Classifier. This part of the experiment mainly analyzes the classification accuracy, stability and rapidity of the classification model in this paper by comparing the three indicators of the classification accuracy average (Avg.), the standard deviation (Std.) and the classification running time (t.). The classification results of the dataset used are shown in the following Table 2:

Table 2. Classification performance of three classifiers on the four data sets

Dataset	Index	KNN	NB	MGB
Sonar	Avg. ± Std	73.59% ± 2.9%	69.28% ± 5.9%	**75.38% ±0.43%**
	t.(s)	1.8019	**1.5182**	2.1318
MeterC	Avg. ± Std	92.77% ± 0.82%	64.86% ± 0.35%	**93.43% ±0.16%**
	t.(s)	3.7281	3.7809	**2.5407**
Sementation	Avg. ± Std	**88.17%** ± 0.35%	79.85% ± 0.40%	87.88% ±0.29%
	t(s)	**1.5611**	3.0534	2.6312
Batch7	Avg. ± Std	98.22% ± 0.07%	70.06% ± 0.10%	**99.89% ±0.01%**
	t.(s)	62.3169	64.1616	**56.5109**

The accuracy of the classification model mainly compares the average accuracy of classification Avg. index. In the comparison of the accuracy of the 4 datasets classifiers, MGB has the best classification accuracy, followed by KNN, and then NB. The main difference between the NB classifier and the other two classifiers is that the classification accuracy of batch7 is lower by more than 20%. This is because the selected dataset features are strongly related, and NB is not effective in classifying these data. MGB not only has a good classification effect on datasets with weak correlation between data features, but also has an excellent classification effect on datasets with strong correlation between features.

If the classification result of the classifier is unstable, the classifier is not suitable for popularization. Therefore, the stability of the classification model is also an important reference index. This paper uses the standard deviation of the classification result to represent the stability of the classifier. In the comparison of the classifier stability of these 8 datasets, the classification stability of MGB is also better than the other two classifiers, and it is a classifier that is worth extending to each data classification.

The running time can directly reflect the complexity of the algorithm. In the comparison of the fastness of the 4 datasets classifiers, the fastness performance of the three classifiers is not much different, but according to the average time comparison of the 4 datasets classification, the MGB classification is the fastest. From the perspective of the amount of calculation, for data classification with a sample size of N, the time complexity of KNN is $O(N^2)$, and the time complexity of MGB is $O(N)$. Therefore, MGB has a better rate when processing data with a high sample size.

The Impact of Selecting the Largest Linearly Independent Group on Classification.
In order to explore the relationship between the amount of sample data, the number of sample features, the rank of sample data (called number of effective features) and the classification accuracy rate, the experiment is carried out with a data set with a large number of sample numbers and sample features in Table 1. The test method is random part of the sample data is selected to participate in the MGB classification in order to observe the relationship between the variables, and then to explore the necessity of selecting the largest linearly independent group for classification.

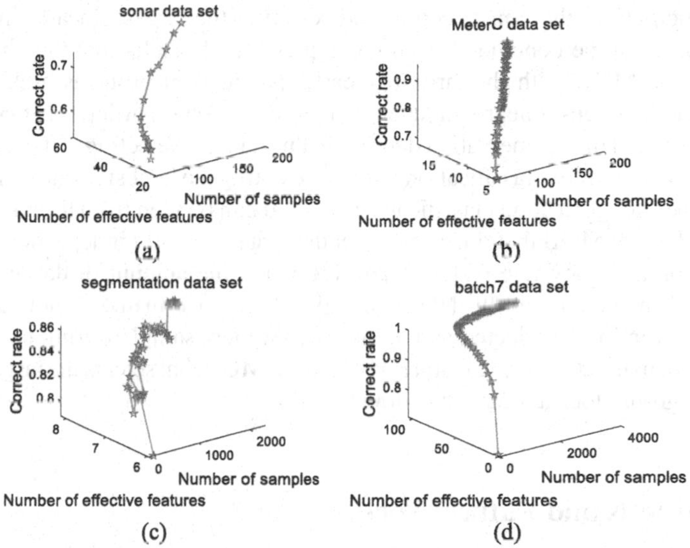

Fig. 1. Three-dimensional graph of sample number, effective features and correct rate

It can be seen from Fig. 1 that as the amount of data increases, the number of effective features and the accuracy of classification are improved. Since the number of elements of the covariance matrix is proportional to the square of the number of effective features, the classification accuracy rate is inferred to be related to the number of elements contained in the fitted covariance matrix and the degree of fitting. Therefore, the more data in the same data set, the better the fitting effect, and the higher the classification accuracy rate. Under the condition of a larger sample size, the more covariance matrix elements will be, and the classification accuracy rate will be higher.

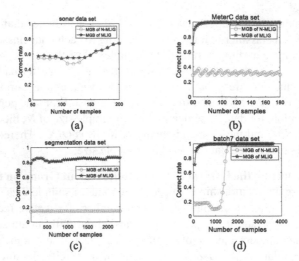

Fig. 2. Classification accuracy of different sample data sizes on the four data sets

Figure 2 show the classification results of different datasets randomly selected M data sizes to participate in the non-selection and selection the largest linearly independent group MGB. It can be concluded from the experimental results that the classification accuracy of the MGB with the largest linearly independent group is higher than the MGB without the largest linearly independent group, which is particularly evident in the MeterC dataset and the segmentation dataset. In this dataset, selecting all the sample data to participate in the MGB classification without selecting the largest linearly independent group still does not get a good classification effect. It can be seen from the batch7 dataset experiment that the MGB that does not select the largest linearly independent group has a classification accuracy rate of less than 20% when the amount of data is low. This is because when the amount of data is low, the covariance matrix is not full of rank, and a class is randomly selected as the class of the new sample during classification, resulting in not using the sample feature data, but the MGB that selects the largest linearly independent group does not have this drawback.

5   Conclusions and Future Work

For the problem that traditional Bayesian classifiers cannot simply and effectively classify data with high feature correlation, this paper proposed a Bayesian classifier based on multidimensional discrete Gaussian distribution. In the new classifier, the correlation between features is considered by estimating the covariance matrix, the largest linearly independent group of the covariance matrix is selected and the linearly related features are independently assumed to ensure the effectiveness of the MGB algorithm. Experimental results have shown that the new classifier is not only superior to NB in data classification with strong feature correlation, but also higher than KNN and NB in data classification with weak feature correlation. In addition, it has less running time and classification the result is more stable. Future research work will use the new model to deal with the classification of extreme data sets (such as the minimal number of samples in the category, the overflow of the covariance matrix determinant in the software).

Acknowledgments. This work was supported by the National Natural Science Foundation of China [Nos. 61976108 and 61572241].

References

1. Boluki, S., Esfahani, M.S., Qian, X., Dougherty, E.R.: Constructing pathway-based priors within a gaussian mixture model for Bayesian regression and classification. IEEE/ACM Trans. Comput. Biol. Bioinf. **16**(2), 524–537 (2019)
2. Yi, W., Park, J., Kim, J.: Classification restricted Boltzmann machine hardware for on-chip semisupervised learning and Bayesian inference. IEEE Trans. Neural Networks Learn. Syst. **31**(1), 53–65 (2020)
3. Luo, L., Yang, J., Zhang, B., Jiang, J., Huang, H.: Nonparametric Bayesian correlated group regression with applications to image classification. IEEE Trans. Neural Networks Learn. Syst. **29**(11), 5330–5344 (2018)
4. Park, S., Chung, W.K., Kim, K.: Training-free Bayesian self-adaptive classification for sEMG pattern recognition including motion transition. IEEE Trans. Biomed. Eng. **67**(6), 1775–1786 (2020)
5. Jung, D.: Data-driven open-set fault classification of residual data using Bayesian filtering. IEEE Trans. Control Syst. Technol. **28**(5), 2045–2052 (2020)
6. Shahtalebi, S., Mohammadi, A.: Bayesian optimized spectral filters coupled with ternary ECOC for single-trial EEG classification. IEEE Trans. Neural Syst. Rehabil. Eng. **26**(12), 2249–2259 (2018)
7. Knight, J.M., Ivanov, I., Triff, K., Chapkin, R.S., Dougherty, E.R.: Detecting multivariate gene interactions in RNA-seq data using optimal Bayesian classification. IEEE/ACM Trans. Comput. Biol. Bioinf. **15**(2), 484–493 (2018)
8. Zhang, H., Wen, B., Liu, J., Zeng, Y.: The prediction and error correction of physiological sign during exercise using Bayesian combined predictor and Naive Bayesian classifier. IEEE Syst. J. **13**(4), 4410–4420 (2019)
9. Mao, X., Li, W., Hu, H., Jin, J., Chen, G.: Improve the classification efficiency of high-frequency phase-tagged SSVEP by a recursive Bayesian-based approach. IEEE Trans. Neural Syst. Rehabil. Eng. **28**(3), 561–572 (2020)
10. Akhtar, N., Mian, A.: Nonparametric coupled Bayesian dictionary and classifier learning for hyperspectral classification. IEEE Trans. Neural Netw. Learn. Syst. **29**(9), 4038–4050 (2018)
11. Dadaneh, S.Z., Dougherty, E.R., Qian, X.: Optimal Bayesian classification with missing values. IEEE Trans. Signal Process. **66**(16), 4182–4192 (2018)
12. Haut, J.M., Paoletti, M.E., Plaza, J., Li, J., Plaza, A.: Active learning with convolutional neural networks for hyperspectral image classification using a new Bayesian approach. IEEE Trans. Geosci. Remote Sens. **56**(11), 6440–6461 (2018)
13. Afshar, P., Mohammadi, A., Plataniotis, K.N.: A Bayesian approach to brain tumor classification using capsule networks. IEEE Signal Process. Lett. **27**, 2024–2028 (2020)
14. Luo, Y., Li, K., Li, Y., Cai, D., Zhao, C., Meng, Q.: Three-layer Bayesian network for classification of complex power quality disturbances. IEEE Trans. Industr. Inf. **14**(9), 3997–4006 (2018)

15. Ke, H.: Improving brain e-health services via high-performance EEG classification with grouping Bayesian optimization. IEEE Trans. Serv. Comput. **13**(4), 696–708 (2020)
16. Zollanvari, A., Dougherty, E.R.: Optimal Bayesian classification with vector autoregressive data dependency. IEEE Trans. Signal Process. **67**(12), 3073–3086 (2019)
17. Yin, H., Xue, M., Xiao, Y., Xia, K., Yu, G.: Intrusion detection classification model on an improved k-dependence Bayesian network. IEEE Access **7**, 157555–157563 (2019)

An Improved Spatial-Temporal Network Based on Residual Correction and Evolutionary Algorithm for Water Quality Prediction

Xin Yu[1], Wenqiang Peng[2], Dongfan Xue[3], and Qingjian Ni[2(✉)]

[1] Nanjing Research Institute of Ecological and Environmental Protection,
Nanjing, China
[2] School of Computer Science and Engineering, Southeast University, Nanjing, China
nqj@seu.edu.cn
[3] Hohai University, Nanjing, China

Abstract. Water quality prediction is of great significance for the supervision of water environment. At present, artificial intelligence method has been tried to be introduced into this field. In this paper, a novel spatial-temporal convolutional attention network based on residual correction and parameter optimization, is proposed for water quality prediction. The model can be divided into three parts. The first part is convolutional attention network in spatial-temporal domain, which uses an one-dimensional convolutional network to extract temporal and spatial information of water quality monitoring station, and adds attention mechanism; the second part is TCN residual correction module, which corrects the residual of the first part; the third part is the parameter optimization module, which introduces PSO algorithm to optimize the model parameters of the first two parts to obtain better results. Based on the real water quality data of a river in South China, this paper carries out relevant comparative experiments, and the experimental results show that the water quality prediction model proposed in this paper is better than other models.

Keywords: Water quality prediction · Spatial-temporal network ·
Residual correction · Evolutionary algorithm

1 Introduction

River water quality directly affects the production and life of human society. With the rapid development of social economy, all kinds of sewage are discharged into the water environment, and water resources are polluted. The pollution of water resources not only leads to the death or even extinction of a large number of aquatic organisms, but also affects the life and health of human beings.

Y. Tan and Y. Shi (Eds.): ICSI 2021, LNCS 12690, pp. 491–499, 2021.
https://doi.org/10.1007/978-3-030-78811-7_46

Therefore, it is very important to reduce and control water pollution, and water quality prediction is an important means of water quality monitoring and management.

At present, water quality prediction methods can be divided into two kinds. One is traditional method, the other is artificial intelligence method. [1] reviewed the current water quality prediction methods, including traditional and artificial intelligence based water quality prediction methods. [2] proposed a new DA−RNN model to predict time series. The experimental results show that the model is effective. [3] improved the DA−RNN model, used multi−layer attention network to predict the geographical perception time series. [4] proposed a temporal pattern attention mechanism. [5] proposed a PSO−DBN−LSSVR model. [6] proposed a SVM−LSTM model. The experimental results show that the prediction effect is significantly improved compared with the single model. [7] combined prediction of CNN and LSTM.[8] proposed an adaptive neuro fuzzy system based on wavelet de−noising technology. [9] combined SWAT model and ANN model to predict water quality. [10] proposed a BHT−ARIMA model to predict short-term time series. [11] proposed a pair-wise bare bones particle swarm optimization algorithm to balance the exploration and exploitation. The experiment results and statistical analysis confirm the performance of PBBPSO with nonlinear functions.

The above results show that most of the model parameters are adjusted manually, which is very troublesome and difficult to ensure whether the parameters have reached the optimal or near the optimal. At the same time, most of the water quality prediction studies only consider the temporal information, but not the spatial information.

In this paper, a new spatial-temporal convolution water quality prediction method is proposed, which makes full use of the data of multiple water quality monitoring points.

The rest of this paper is as follows: the second section mainly elaborates the related methods involved in this paper. The third section mainly introduces the method proposed in this paper. The fourth section is the experimental results and analysis, water quality prediction experiment and analysis of the experimental results. The fifth section summarizes the content of the full text, and prospects the future research direction.

2   Problem Statement and Related Methods

2.1   Problem Statement

In this paper, the prediction target is water quality ammonia nitrogen index. We need to predict the future ammonia nitrogen value according to the known monitoring data. The input data of the model are the monitoring data of the water quality monitoring station that need to be predicted and the monitoring data of other monitoring stations. In this paper, they are combined into $X = \{X_1, X_2, ..., X_{t-1}\}$, X_i stands for the monitoring data of the water quality monitoring station at the i-th moment, and the goal is to predict $y_{t-1+\gamma}$, that is, the future value after γ time. The γ is set according to different needs, and y_t is simply used to represent the predicted target value.

2.2 Related Methods

The related methods used in this paper include one-dimensional convolution neural network, TCN (time convolution network), attention mechanism, GRU (gate recurrent unit) and PSO (Particle swarm optimization) algorithm. This paper uses spatial-temporal attention mechanism to process spatial-temporal data. TCN network is used for residual compensation. PSO is used to optimize the parameters of the model to give full play to the best performance of the model.

3   The Proposed Model

3.1   Framework of the Proposed Model

The model proposed in this paper can be divided into three parts. The first part is the spatial-temporal convolution part, which considers the temporal and spatial information of water quality changes, and introduces the precipitation which has a great impact on water quality. The second part is the residual correction. In this paper, the TCN model is used to correct the residual of the first part. The third part is the optimization part of the first mock exam. The parameter adjustment of the model is one of the main factors that affect the performance of the model. Different parameters of the model are adjusted differently. The parameters of the same model can get different results under different data sets. The parameters of the model are very cumbersome, and the manpower is difficult to make sure that the parameters have reached the optimal or near optimal. In this paper, PSO is used to analyze the parameters of the first two parts of the model, and multiple individuals are randomly initialized to find the optimal parameters. The model framework is shown in Fig. 1.

3.2   Spatial-Temporal Convolution Attention

Feature Selection. There are too many features of spatial-temporal data, some of which have a great impact on the target, and some of which have no impact on the target. Directly inputting all the features into the model will reduce the prediction accuracy and the performance of the model. Therefore, this paper uses Pearson correlation coefficient for feature selection. The range of Pearson correlation coefficient is $(-1,1)$. The larger the absolute value of the correlation coefficient, the greater the correlation between the feature and the target. $GRU's$ hidden layer state $H = \{h_1, h_2, ..., h_{t-1}\}$, h_i contains the hidden layer state of multiple data features at the $i-th$ time, $h_i = \{h_{i1}, h_{i2}, ..., h_{im}\}$, where m is the number of data features.

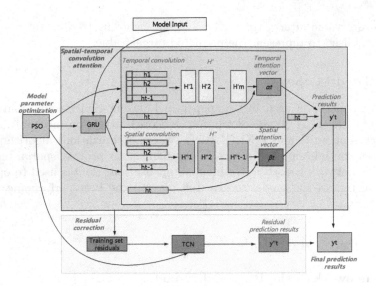

Fig. 1. The framework of the proposed model.

Temporal Convolution Attention. One dimensional convolution network is used to convolute the hidden layer state of each feature from T_1 to T_{t-1}. The convolution kernel is k_1, and the size of each convolution kernel is $t-1$. Therefore, the hidden layer state $H' = \{h_i', h_2', ..., h_m'\}$, $h_i' = \{h_{i1}', h_{i2}', ..., h_{ik1}'\}$. Then, the attention mechanism is adopted for h_i', and the temporal domain attention vector α_t is calculated together with the hidden layer state h_t at time t.

Spatial Convolution Attention. One dimensional convolution network is used to convolute the hidden layer state h_i at the $i-th$ time. The convolution kernel is k_2, and the size of each convolution kernel is m. Therefore, the hidden layer state $H'' = \{h_i'', h_2'', ..., h_{t-1}''\}$, $h_i'' = \{h_{i1}'', h_{i2}'', ..., h_{k2}''\}$. Then, the attention mechanism is adopted for h_i'', and the spatial attention vector β_t is calculated together with the hidden layer state h_t at time t.

Finally, the temporal attention vector α_t, the spatial attention vector β_t and the hidden layer state h_t at time t are combined to predict the target at the next time. The calculation formulas are as follows.

$$y_t = W_y(W_\alpha \alpha_t + W_\beta \beta_t + W_h h_t) \tag{1}$$

W_y represents a full connection layer, and the previous results get the final output through it.

3.3   TCN Residual Correction

No matter what prediction model, there must be errors between the predicted value and the real value. If we can find the variation law of the prediction error and correct the residual error, the prediction accuracy will be improved in theory. In this paper, TCN model is used to modify the residual error of the model proposed in Sect. 3.1. Firstly, the model proposed in Sect. 3.1 is used to fit the training set, and the residual value of the training set is calculated. The residual value is used as a new data set to train the prediction TCN model. The prediction result of TCN model is the residual value of the model proposed in Sect. 3.1. The modified prediction result is obtained by adding the prediction result of TCN model and the prediction result of the model proposed in Sect. 3.1.

3.4 Model Parameter Optimization Based on PSO

In this paper, PSO algorithm is used to optimize the model parameters. The optimization method is to initialize 100 individuals randomly, set the possible value range for each parameter, and then iterate for 100 times. The process of each iteration is the process of model training. After each iteration, the training results of the model are fed back to PSO. PSO adjusts the parameters according to the results, and then advances to the next iteration. Finally, the parameter that makes the model best is selected.

For the parameter optimization of the model proposed in Sect. 3.1, this paper uses PSO to iteratively optimize the time step of GRU model, the number of model nodes and the number of convolution cores of one-dimensional convolution. For TCN model, this paper uses PSO to iteratively optimize the number and size of cores of TCN model.

4 Experiments and Results

4.1 Data Set

In this paper, the real water quality data from an automatic water quality monitoring station of a river in South China is used as the experimental data set. After the preprocessing of data missing, error and other problems, the experimental data is the daily average water quality data of multiple monitoring stations from May 2019 to December 2020. The time span is more than one year, and the prediction target is water quality ammonia nitrogen index. Because the experiment uses real data sets, the experimental results can reflect the actual situation.

4.2 Evaluation Metrics

This paper uses RMSE (root mean square error), MAE (mean absolute error) and MAPE (mean absolute percentage error) to evaluate the accuracy of the model.

4.3   Results and Analysis

In order to verify the effectiveness of the proposed model and the proposed TCN residual correction module and PSO parameter optimization module, the comparative experiments are carried out. The comparative models are LSTM model, GRU model, TCN model, CNN-LSTM model (the combination model of one-dimensional convolution and LSTM) and TPA-LSTM model (a model proposed by [4]). The experimental results show that the proposed model is better than other models in the accuracy of water quality prediction.

In the comparison models, LSTM, GRU and TCN are single models, while CNN-LSTM and TPA-LSTM are hybrid models. Therefore, we compare the prediction results of the two types with the proposed model separately. The comparison of LSTM model, GRU model, TCN model and proposed models is shown in Fig. 2.

It can be seen from Fig. 2 that the predicted values of LSTM, GRU and TCN are quite different from the real values, and there is a problem of prediction delay. Relatively speaking, it is obvious that the difference between the predicted value and the real value of the model proposed in this paper is very small, and the prediction effect is very good. LSTM, GRU and TCN can not predict the changing trend of ammonia nitrogen well, but the model proposed in this paper basically predicts the change trend of ammonia nitrogen.

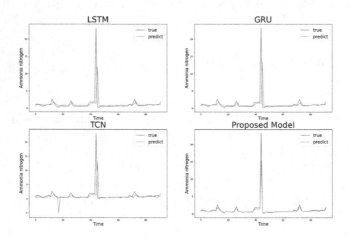

Fig. 2. LSTM model, GRU model, TCN model and proposed model predict results of Ammonia nitrogen.

The comparison of CNN-LSTM model, TPA-LSTM model and proposed model is shown in Fig. 3. As can be seen from Fig. 3, the prediction results of CNN-LSTM and TPA-LSTM are very stable, basically without fluctuation, while the change of real ammonia nitrogen is fluctuating, so the two models can not predict the change rule of ammonia nitrogen.

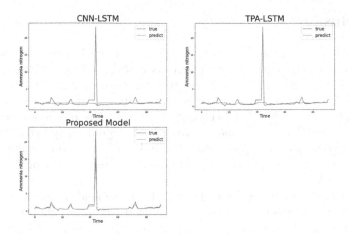

Fig. 3. CNN-LSTM model, TPA-LSTM model and proposed model predict results of Ammonia nitrogen.

In order to better compare the prediction results of the three models, the RMSE, MAE and MAPE index values corresponding to the models are shown in Table 1.

Table 1. Results of experiment

Model	Evaluation metrics		
	RMSE	MAE	MAPE
LSTM	2.3887	0.6597	62.5510%
GRU	2.4094	0.6105	56.4150%
TCN	2.5140	0.6821	71.2000%
CNN-LSTM	2.3615	0.6125	54.8060%
TPA-LSTM	2.3408	0.5557	49.4180%
NoTCN& PSO	2.3369	0.5055	39.5250%
NoPSO	0.4624	0.2490	27.2498%
Proposed model	**0.2488**	**0.1539**	**14.6510%**

Table 1 summarizes the experimental results of all the models in this paper. It can be seen from Table 1 that the prediction accuracy of the water quality prediction model proposed in this paper is significantly better than other prediction models, which is 2.092–2.265 lower in RMSE, 0.402–0.5282 lower in MAE index and 34.767%–56.549% lower in MAPE index, indicating that the model proposed in this paper is very effective in water quality prediction.

5   Conclusion

In this paper, a novel water quality prediction model is proposed, which is a spatial-temporal convolutional attention network based on residual correction and parameter optimization. Based on the real water quality monitoring data, comparative experiments are carried out to verify the effectiveness of the proposed model. The results of experiments show that the prediction accuracy of the proposed model in water quality prediction is significantly better than other models, and verify that the proposed TCN residual correction module and PSO parameter optimization module are of great help to improve the prediction accuracy of the model.

In the future, we will combine the models in the field of water environment and consider the impact of future precipitation on water quality changes to achieve higher prediction accuracy, and try to make a long-term prediction of water quality.

Acknowledgement. This paper is supported by National Key R&D Program of China (2018YFB1004 300).

References

1. Rajaee, T., Khani, S., Ravansalar, M.: Artificial intelligence-based single and hybrid models for prediction of water quality in rivers: a review. Chemometr. Intell. Lab. Syst. **200**, 103978103978 (2020)
2. Qin, Y., Song, D., Cheng, H., Cheng, W., Jiang, G., Cottrell, G.W.: A dual-stage attention-based recurrent neural network for time series prediction. In: Proceedings of the 26th International Joint Conference on Artificial Intelligence, IJCAI 2017, pp. 2627–2633 (2017)
3. Liang, Y., Ke, S., Zhang, J., Yi, X., Zheng, Y.: GeoMAN: multi-level attention networks for geo-sensory time series prediction. In: IJCAI, pp. 3428–3434 (2018)
4. Shih, S.-Y., Sun, F.-K., Lee, H.: Temporal pattern attention for multivariate time series forecasting. Mach. Learn. **108**(8), 1421–1441 (2019). https://doi.org/10.1007/s10994-019-05815-0
5. Yan, J., Gao, Y., Yu, Y., Xu, H., Xu, Z.: A prediction model based on deep belief network and least squares SVR applied to cross-section water quality. Water (Switzerland) **12**(7), 1929 (2020)
6. Peng, W., Ni, Q.: A hybrid SVM-LSTM temperature prediction model based on empirical mode decomposition and residual prediction. In: 2020 IEEE International Conference on Systems, Man, and Cybernetics (SMC), pp. 1616–1621. IEEE (2020)
7. Yao, H., Liu, Y., Wei, Y., Tang, X., Li, Z.: Learning from multiple cities: a meta-learning approach for spatial-temporal prediction. In: The Web Conference 2019 - Proceedings of the World Wide Web Conference, WWW 2019, pp. 2181–2191 (2019)
8. Ahmed, A.N., et al.: Machine learning methods for better water quality prediction. J. Hydrol. **578**, 124084 (2019)
9. Noori, N., Kalin, L., Isik, S.: Water quality prediction using SWAT-ANN coupled approach. J. Hydrol. **590**, 125220 (2020)

10. Shi, Q., et al.: Block Hankel tensor ARIMA for multiple short time series forecasting. In: Proceedings of the AAAI Conference on Artificial Intelligence, vol. 34, pp. 5758–5766 (2020)
11. Guo, J., Sato, Y.: A pair-wise bare bones particle swarm optimization algorithm for nonlinear functions. Int. J. Networked Distrib. Comput. 5(3), 143–151 (2017)

Can Argumentation Help to Forecast Conditional Stock Market Crisis with Multi-agent Sentiment Classification?

Zhi-yong Hao[✉] and Peng-ge Sun

Shenzhen Institute of Information Technology, Shenzhen, China

Abstract. It is well known that investors in stock market making financial decisions are often affected by certain events, like the coronavirus disease pandemic. However, it is very hard to perceive stock market crisis by making use of variety information. In this paper, we investigate whether it is possible to exploit arguments from investor sentiment expressed through financial news and posts, to forecast conditional stock market crisis. Thus, an argumentation enriched multi-agent sentiment classification method is proposed to make full use of variety tone and proliferation under certain events. In particular, the conditional stock market is investigated in our experiment to compare the predictive performance of the argumentation enriched multi-agent sentiment classification system with the existing multiple classifier system for the variance of CSI 300 Index. We find that the proposed argumentation enriched system outperforms the existing popular multiple classifier systems, while giving argumentative explanations through preliminary empirical evaluation.

Keywords: Multi-agent systems · Computational argumentation · Sentiment classification · Conditional stock market

1 Introduction

Multiple classifier systems have been widely used in sentiment classification applications due to its excellent performance [1], such as public opinion monitoring, commodity evaluation, especially stock market forecasting, in which investors making financial decisions often access to plentiful information [2]. Since the beginning of 2020, it is known that there is a significant relationship between the coronavirus disease (COVID-19) pandemic sentiment and stock market [3]. In this circumstance, investor sentiment experienced huge volatility, just like the Reddit-GameStop mania. For no discernible business reason, the stock of GameStop shot up hundreds of percent in a matter of days. These kinds of events indicate that an increase of the specific sentiment has led to crisis of the stock market. Hence, one of the main concerned issues in ensemble learning community, is how to forecast conditional stock market crisis in a way that making full use of variety information like financial news and posts, not just the traditional numerical indicators.

© Springer Nature Switzerland AG 2021
Y. Tan and Y. Shi (Eds.): ICSI 2021, LNCS 12690, pp. 500–510, 2021.
https://doi.org/10.1007/978-3-030-78811-7_47

In fact, one of the popular approaches to combine classifiers is to take the perspective of human deliberation into account and make use of the multi-agent argumentation [4]. It is noted that computational argumentation provides a feasible means of justifying an agent's classification that greatly resemble the way in which humans come to a well-founded consensus. Thus, argumentation based multi-agent classification has aroused great attention in the academic. Our previous work showed that it not only can achieve outstanding predicting performance in ensemble learning systems, but also give exact explanations with easily assimilated reasoning of multi-party argument games [5].

In this paper, we investigate whether argumentation that offers a natural means to commonsense reasoning may help to forecast conditional stock market crisis when used to improve a multi-agent sentiment classification system. The aim of our research is to cope with the problem of inconsistency about risk level encountered by different classifier agents. It is noted that each classifier agent can access its own information for stock market crisis decision. Thus, our key idea is to make use of Multi-Agent System (MAS) for collaborative sentiment classification tasks with the help of computational argumentation. The preliminary empirical evaluation shows that not only better performance, but also intelligent explanation can be obtained by exploiting argumentation in multi-agent sentiment classification to forecast conditional stock market crisis.

The remainder of this paper is organized as follows. Section 2 briefly reviews related work. In Sect. 3, we propose the argumentation enriched multi-agent sentiment classification system. Section 4 presents the comparative evaluation of our method, through preliminary experiments. Finally, this study is concluded in Sect. 5.

2 Related Work

In this section, a summary of work related to our method is presented from the following two aspects. Subsection 2.1 discuss the multiple classifier system for text sentiment analysis, and Subsect. 2.2 outlines the main argumentation based multi agent classification approaches.

2.1 Multiple Classifier Systems for Text Sentiment Analysis

In recent years, multiple classifier systems were introduced to text sentiment analysis, also called sentiment classification [6–8]. For example, an ensemble sentiment analysis model based on three multi-classifier systems is proposed [9]. Their experimental results demonstrate that the multiple classifier system is superior to the traditional sentiment analysis methods. In Araque et al.'s work [10], they present a taxonomy that categorizes multiple classifier systems in sentiment analysis literature. Catal and Nangir [11] developed a multiple classifiers model for Turkish sentiment analysis, and Bagging, Naive Bayes and Support Vector Machines were used to increase the performance of classification. As the opposite of the widely used feature selection method in sentiment analysis, an ensemble approach to stabilize the features is proposed, which can achieve encouraging accuracy when compare to the related research with classifier SMO [12]. To provide an important basis for selecting the seed texts and modifying the training text set for sentiment analysis, Li et al. [13] present a seed selection method based on

Random Swap clustering and a hybrid modification method of the training text set with active learning, using the idea of effectively utilizing unlabeled samples.

However, most multiple classifier systems for text sentiment analysis just combine the base classifiers using majority voting method or weighted averaging method, which do not make full use of the knowledge learned by each basic classifier. Moreover, concerning stock market crisis, the traditional investor sentiment indicators, i.e., quantitative information can also be used as important basis for decision making. Thus, not only text sentiment, but also quantitative sentiment indicators, should be considered in multiple classifier system for stock market crisis forecasting in this paper.

2.2 Argumentation Based Multi-agent Classification

Concerning the issue of how to forecast conditional stock market crisis, using multiple classifier systems in a way that making full use of variety financial information, the idea that not only good classification performance is required, but also the reasons behind the results are very important to human decision makers. Computational argumentation has been regarded an effective technique used by intelligent agents to give explanations for classification results with justification [14–16]. The main idea about this type of approach is that argumentation-based classification could show why arguments for categories are preferred to counterarguments through dialectical reasoning. Earlier, Fan et al. [17] proposed an argumentation-based framework that combine data classifiers with multiagent system, named PISA. The experiments showed that PISA frameworks can not only give excellent classification performance, but also provide argumentative explanations. The advantage of the method of argumentation is that it is not only the category of the object is given, but also the reasons behind it are provided to decision makers in an interpretable manner [18]. Actually, to address the problem of inconsistent in multiple classifier systems, an argumentation based multiagent collaborative method with context-aware layered learning, named CALL [19] is proposed for conflict resolution among multiple agents. The experiments on benchmark datasets showed that the CALL method could increase classification accuracy significantly over state of-the-art, especially in presence of noise.

As far as we know, existing works of argumentation based multiagent classification just provide a common framework for decision making in multiple classifier systems, while ignoring the semantic explanation. In this paper, the rich contextual information of stock market is concerned particularly, aiming to offer argumentative explanations for financial regulators and investors, and further promote the social application of the argumentation approach.

3   Argumentation Enriched Multi-agent Sentiment Classification Method

In this section, we show how to exploit argumentation for multi-agent sentiment classification to forecast conditional stock market crisis. Particularly, it is assumed that each classifier agent has the ability to learn sentiment classification rules for constructing arguments according to its own real time needs in argumentation process. Thus, considering the current conditional stock market presented with a variety of financial information, our multiple classifier system aims to provide early warning decision-making information with justification though multi-party argument game process among agents. As indicated previously, a novel argumentation enriched multi-agent sentiment classification method is proposed for conditional stock market crisis early-warning, which illustrated in Fig. 1. The framework of our method consists of three components, shown in three different colors respectively. The first component, indicated in the purple box, is the argumentation model for conditional stock market crisis early-warning, presented in Subsect. 3.1. Then, in Subsect. 3.2, we describe multiple classifiers for text sentiment analysis from a variety of financial information. Finally, the algorithm for contextual knowledge guided argument constructing is presented in the last subsection. which described in the yellow box, to bridge the gap between the first one and the second one.

3.1 The Argumentation Model

Argumentation has been regarded as an effective technique for conflicts resolution through multi-party argument game. Based on this intuition, we design a multi-agent argumentation model to reach agreement for predicting conditional stock market crisis. As indicated previously, a collection of classifier agents learning from variety financial information try to come to consensus about the future risk level of stock market. To this end, our multiagent argumentation model defined formally as follows.

Definition 1. (Multi-Agent Argumentation Model, MAAM) Given a multiagent system $Ag = \{Ag_1, \cdots, Ag_n\}$, the multi-agent argumentation model is formally defined as MAAM $= < \mathcal{Ref}, \mathcal{Par}, \mathcal{O}, \mathcal{Roles}, \mathcal{ARules}, \mathcal{Q}, \mathcal{R} >$, where: (i) $\mathcal{Ref}$ is the referee agent, managing multiple arguing agents; (ii) $\mathcal{Par} \subseteq Ag$, is the set of all participant agents in argumentation; (iii) $\mathcal{O}$ is the ontology shared by all participant agents; (iv) $\mathcal{Roles} = \{Master, Challenger, Spectator\}$, is the set of roles played by participating agents, including masters, challengers and spectators respectively, (v) $\mathcal{ARules}$ is the set of argumentative rules abided by all participant agents; (vi) $\mathcal{Q}$ is the set of arguments; (vii) $\mathcal{R}$ is the set of attack relations, namely $\mathcal{R} = \mathcal{Q} \times \mathcal{Q}$.

In this paper the argumentation model is used to allow multiple agents to argue about the risk level of stock market during the period under investigation, e.g., the risk level of stock market in the next week. Sentiment classification arguments in favor of or against a particular assertion are defined formally in the following definition.

Definition 2. (sentiment classification argument) A sentiment classification argument $Arg = < Ag, x, ass, s, \vartheta >$, where: (i) $Ag \in \mathcal{Par}$ is the proponent agent; (ii) x is the feature set under investigation; (iii) ass is Ag's assertion; (iv) s is the reasons of ass, a.k.a. prerequisite; (v) $\vartheta \in [0, 1]$ is the strength of Ag's argument, noted as $stren^{Ag}(Arg)$.

It is noted that the strength of a sentiment classification argument is calculated depends on different argument constructing algorithms, which will be described in the next subsection.

In this paper, the sentiment classification arguments exchanged via the speech acts among multiple agents, and the realization of these speech acts is detailed through the argument game dialogue protocol in MAAM. The details of the realization of MAAM will no longer be demonstrated here due to the limited space. Once the dialogue game process has terminated, the prediction of conditional stock market crisis will be presented with its corresponding reasons from the argument game tree.

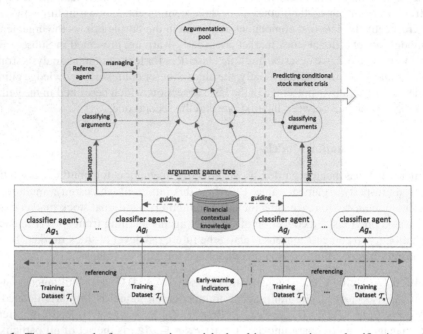

Fig. 1. The framework of argumentation enriched multi-agent sentiment classification method

3.2 Text Conditional Sentiment Mining Within Multiple Classifiers

As already stated above, each agent in MAAM has its own local training data for learning to classify sentiment in stock market. Here each agent could access to its own financial information, like financial news, posts in tieba, etc. In this subsection, we present how these classifier agents can learn from the conditional stock market text, just like the coronavirus disease pandemic sentiment. It is noted that a modular rule induction algorithm, called Prism [19], is chosen in this paper, rather than the "black box" algorithms such as artificial neural network, support vector machine, and so on. We choose Prism for the reason that it can be easily used to construct sentiment classification arguments, which detailed in the next subsection.

The main novelty of our paper is the use of a variety of financial text information as the main source of training data, including financial news on Sina Finance website, posts on EastMoney guba, and Baidu Index platform. As we are interested in the investor sentiment volatility of the coronavirus disease pandemic, which is a conditional stock market, the sample period spans the period January 1, 2020–January 31, 2021. Sina Finance news is considered a reliable source of financial news, which is popular in the text analysis community. Furthermore, EastMoney guba is the largest financial forum in China with a large number of users for individual investment, and Baidu Index is a big data analytics platform based on a large number of Internet user's behavior data. Therefore, it can be easily seen that Sina Fiance news, EastMoney posts and Baidu Index are representatives of institutional investor sentiment, individual investor sentiment and the general public sentiment, respectively.

For the early-warning of conditional stock market crisis, we designed a set of indicators, which can be used to analysis and monitor the stock market crash or the possibility of increased risk. The early-warning indicators also can be used to provide reliable advice for the safe operation of the stock market, thereby minimizing the losses caused by the stock market crisis. Therefore, the purpose of establishing the stock market early-warning indicators is to define and quantify the stock market crisis. The specific steps for constructing early-warning indicators for stock market crisis are described as follows:

Step 1. Calculate the return on the stock market index for each week, namely Market Index Return Rate (MIR):

$$MIR = \frac{\Delta s}{s} = \frac{s_t - s_{t-1}}{s_{t-1}}$$

where s_t is the close price of the market index this week, and s_{t-1} is the close price of the market index in the last week.

Step 2. Calculate the mean μ and standard deviation σ of all MIR in the sample interval.

Step 3. According to the disaster early-warning level, the stock market crisis warning level is divided into four levels, including blue alert, yellow alert, orange alert and red alert. The forecast of a stock market crisis is judged on the basis of each month's MIR and the results of the second step:

When $\mu - 0.5 * \sigma \geq MIR > \mu - 1 * \sigma$, it is means that the price of the stock index fell less than the mean below 0. 5 standard deviations, greater than 1 standard deviation below the average. Referring to the disaster early-warning level, the stock market can be determined to be at level IV crisis status, indicating that the intensity of the crisis is low, with blue alert warning;

When $\mu - 1 * \sigma \geq MIR > \mu - 1.5 * \sigma$, it is means that the price of the stock index fell less than the mean below 1 standard deviations, greater than 1.5 standard deviation below the average. Referring to the disaster early-warning level, the stock market can be determined to be at level III crisis status, indicating that the intensity of the crisis is moderate, with yellow alert warning;

When $\mu - 1.5 * \sigma \geq MIR > \mu - 2 * \sigma$, it is means that the price of the stock index fell less than the mean below 1.5 standard deviations, greater than 2 standard deviation below the average. Referring to the disaster early-warning level, the stock market can be determined to be at level II crisis status, indicating that the intensity of the crisis is intense, with orange alert warning;

When $\mu - 2 * \sigma > MIR$, it is means that the price of the stock index fell less than 2 standard deviation below the average. Referring to the disaster early-warning level, the stock market can be determined to be at level I crisis status, indicating that the intensity of the crisis is high, with red alert warning.

Particularly, we take CSI 300 Index as the research object, considering the MIR of more than 50 weeks in our sample period, from January 1, 2020 to January 31, 2021.

3.3 Contextual Knowledge Guided Arguments Constructing Algorithm

As indicated in Subsect. 3.1, it is known that arguments constructing is the basis of our multi-agent argumentation MAAM. Thus, in this subsection, we detail the construction of sentiment classification arguments used in dialogue games process. Although Prism algorithm can be used to learn modular classification rules, the background knowledge in conditional stock market is still needed to construct classifying arguments. In particular, we assume that each classifier agent applies financial domain ontology with Prism rule learning technique for sentiment classification arguments constructing.

Generally, ontologies are known as a shared and formal understanding of a domain theory, where the term 'shared' means an agreement within a community of experts over the very domain. Financial domain ontology consists of the financial knowledge that a domain expert would exploit in reasoning about a specific situation. For example, a financial domain ontology can be used in reasoning about a specific risk level of stock market crisis.

In fact, the hierarchical relations between concepts in domain ontology, which are also known as class-subClass relations, can be used to make generalization over the values of attributes when learning classification rules for constructing arguments. Therefore, we propose to learn sentiment classification rules for constructing arguments in a new manner. The definition of sentiment classification rule is firstly described as follows.

Definition 3. (sentiment classification rule) A sentiment classification rule scr is defined in the form: *IF* $\mathbb{A}_i\ rel_i\ v_{ix} \wedge \cdots \wedge \mathbb{A}_j rel_j\ v_{jx} \cdots \wedge\ class = c_k\ THEN\ class\ =\ sub\_c_k$, where $\mathbb{A}_i, \ldots, \mathbb{A}_j \in \mathcal{A}$ are attributes, $rel_i, \cdots, rel_j \in \{=, \neq, <, >, \leq, \geq\}$, are relational operators, $v_{ix}, \cdots, v_{jx}$ are attribute values, and sub_c_k is the direct subclass of c_k in the category taxonomy $\mathfrak{Ta}$.

Clearly, the prerequisite of a scr, noted as $prer(scr)$, is a logical conjunction of attribute-value pairs and a categorization. The consequent, noted as $cons(mgcr)$, is a subcategory classification. Here, an attribute-value pair typically has the form $\mathbb{A}_i = v_{ix}$ for nominal attributes, and the other forms for numeric attributes.

Following the description of the foregoing subsection, sentiment classification rules can be learned hierarchically from the training datasets, referring to the category taxonomy $\mathfrak{Ta}$, which obtained from domain Ontology. Our proposed algorithm for learning sentiment classification rules is described in Table 1.

In Algorithm 1, the sentiment classification argument for the situation under investigation is constructed from a training textual dataset $\mathcal{T}$ and category taxonomy $\mathfrak{Ta}$, concerning risk level $\mathcal{RL}$. First, a new training textual dataset $\mathcal{T}'$ is created based on $\mathcal{T}$, leaving the examples labeled by the subclass of $\mathcal{RL}$(line 04 to 11). Then, Prism classification rules are derived from $\mathcal{T}'$ with various attribute sets (line 12 to 15). It is noted that although several rules may have the same consequent, their prerequisites consist of different attribute-value pairs. Next, the given target category $\mathcal{RL}$ is added to sentiment classification rules that are transformed (line 16 to 19). This scr _Learning algorithm executes recursively until the rules for each subclass of $\mathcal{RL}$ are learned (line 21 to 23).

Table 1. The contextual knowledge guided arguments constructing algorithm.

Algorithm 1. *arguments constructing* $(\mathcal{T}, \mathfrak{Ta}, \mathcal{RL})$
input: a training textual dataset $\mathcal{T}$ with attribute set $\mathcal{A}$, category taxonomy $\mathfrak{Ta}$, risk level $\mathcal{RL}$
output: a sentiment classification argument $\mathcal{SCA}$
01     $\mathcal{T}' := \mathcal{T}$
02     $\mathcal{SCA} := \emptyset$
03     $sub\_\mathcal{RL}: = \text{getSubClass}(\mathcal{RL})$
04   **for each** $exa \in \mathcal{T}'$ **do**
05         $cate\_set := sub\_\mathcal{RL} \cap \text{getSuperClass}(\mathcal{T}, \text{cateOf}(exa))$
06   **if** $cate\_set \neq \emptyset$ **then**
07         $exa\_cate := \text{getOneElement}(cate\_set)$
08   **else**
09         $\mathcal{T}' := \mathcal{T}' - \{exa\}$
10   **end if**
11   **end for**
12   **repeat**
13       $attr\_set := attr\_set \cup \text{attributeUsed}(rule\_set)$
14       $rul\_set := \text{Prism Learning}(\mathcal{T}', \mathcal{A} - attr\_set)$
15   **until** $rule\_set == \emptyset$
16   **for each** $rule \in rule\_set$ **do**
17       $mg\_rule := \text{generatedNewRule}(rule, \mathcal{RL})$
18       $\mathcal{SCA} := \mathcal{SCA} \cup \{mg\_rule\}$
19   **end for**
20     $\mathcal{T}'' := \mathcal{T}$
21   **for each** $sub\_cate \in sub\_\mathcal{RL}$ **do**
22       $\mathcal{SCA} := \mathcal{SCA} \cup scr\_\text{Learning}(\mathcal{T}'', \mathfrak{Ta}, \mathcal{RL})$
23   **end for**
24   **return** $\mathcal{SCA}$

4 Experimental Evaluation

The main idea of our research is to assess whether an argumentation enriched multi-agent sentiment classification system outperforms a multiple classifier system in terms of forecasting accuracy in conditional stock market crisis. For empirical evaluation of MAAM, we used CSI 300 Index in China, due to its popularity in Chinese mainland's stock market. In addition, we selected all the components of the CSI 300 Index as the research object of text sentiment analysis.

For textual data, Python packages urllib and requests are used to fetch the financial news data and EastMoney posts, and BeautifulSoup is used to parse the HTML.

The system filters news and posts on a corpus-level using a keyword-based approach, constructing a set of financial terms and the name of the stock of the listed companies.

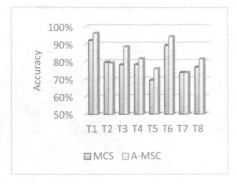

Fig. 2. Prediction accuracy comparison

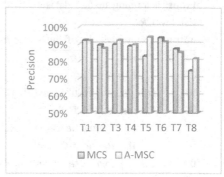

Fig. 3. Prediction precision comparison

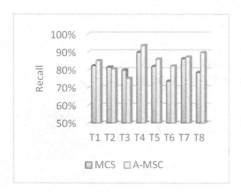

Fig. 4. Prediction recall comparison

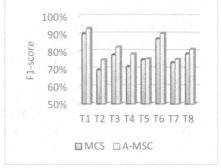

Fig. 5. Prediction F1-score comparison

There are 116 sets of data in this study. Considering the rapid response of the stock market to the economic situation, we take this month's (t) early-warning indicators and next month's (t−1) stock market crisis early-warning signals as a complete set of samples. Hence, we use January 2020 to December 2020 as a training set for the model, a total of 108 samples, and January 2021 as a test set for the model, a total of 8 sets of samples.

The results of predicting conditional stock market crisis are described from Fig. 2 to Fig. 5, by comparing the operation of Argumentation enriched Multi-agent Sentiment Classification (A-MSC), with respect to Multiple Classifier System for sentiment classification (MCS). We use accuracy, precision, and recall as evaluation metrics for early-warning systems. Accuracy refers to the proportion of stock market crises that are actually occurring in the sample that are predicted to be stock market crises. Precision refers to the number of correct predictions divided by the number of all samples. Recall refers to the proportion of the week in which the stock market crisis actually occurred, which is correctly predicted as the stock market crisis.

First, in Fig. 2, it can be seen that, considering accuracy, A-MSC produces the best results in 6 of 8 datasets tested, except for T2 and T7. Second, Fig. 3 and Fig. 4 show that, although A-MCS performs not better than MCS with a few test datasets, it still obtains acceptable performance. Finally, for the recall of prediction, which is showed in Fig. 5, it is noted that the proposed method in this paper performs best in all the 8 datasets compared to multiple classifier system. This is a strong evidence that indicating the superiority of the argumentation enriched multi-agent sentiment classification method.

5 Conclusion

An argumentation enriched multi-agent sentiment classification method has been proposed for conditional stock market crisis early-warning. The method, namely A-MCS, improves classifier agents' sentiment classification performance by exploiting multiple argument games for coming to consensus about the future risk level of stock market. Thus, our key idea is to make use of multi-agent system for collaborative sentiment classification tasks of predicting conditional stock market crisis with the help of argumentation. The preliminary experimental study establishes that the proposed method performs better than state-of-the-art methods, under the current condition of COVID-19 pandemic. With respect to multiagent sentiment classification, a significant advantage provided by the argumentation enriched method is that it not only can improve the prediction performance, but also give consistent explanations with easily assimilated reasons for financial decision makers.

In this paper, we focus on exploiting argumentation for dealing with information fusion issues in multi-agent sentiment classification. Future research will explore domain knowledge transferring among arguing agents further, to increase the efficiency of collaborative sentiment classification. We would like to harness the power of shared financial knowledge for coming to semantic consensus. It is commonly accepted that such explainable artificial intelligence applications would be believed by human users.

Acknowledgements. This work is supported by Guangdong Basic and Applied Basic Research Foundation (Project No. 2019A1515011392), and the basic research project of Shenzhen Institute of Information Technology (Project No. SZIIT2020SK013).

References

1. Ye, X., et al.: Multi-view ensemble learning method for microblog sentiment classification. Expert Syst. Appl. **166**, 113987 (2020)
2. Bellini, V., Guidolin, M., Pedio, M.: Can Big Data Help to Predict Conditional Stock Market Volatility? An Application to Brexit. Machine Learning, Optimization, and Data Science. Springer, Cham, 2020. https://doi.org/10.1007/978-3-030-64583-0_36
3. Costola, M., et al.: Machine learning sentiment analysis, Covid-19 news and stock market reactions. SAFE Working Paper Series (2020)
4. Conţiu, Ş, Groza, A.: Improving remote sensing crop classification by argumentation-based conflict resolution in ensemble learning. Expert Syst. Appl. **64**, 269–286 (2016)

5. Hao, Z.-Y., Liu, T., Yang, C., Chen, X.: Context-aware layered learning for argumentation based multiagent collaborative recognition. In: Tan, Y., Shi, Y., Niu, B. (eds.) ICSI 2019. LNCS, vol. 11656, pp. 23–32. Springer, Cham (2019). https://doi.org/10.1007/978-3-030-26354-6_3

6. Buzea, M.C., Trausan-Matu, S., Rebedea, T.: A three word-level approach used in machine learning for Romanian sentiment analysis. In: 2019 18th RoEduNet Conference: Networking in Education and Research (RoEduNet) IEEE (2019)

7. Hosamani, V., Vimala, H.S.: Data science: prediction and analysis of data using multiple classifier system. Int. J. Comput. Eng. Res. Trends (2019)

8. Huang, F., et al.: Attention-emotion-enhanced convolutional LSTM for sentiment analysis. IEEE Trans. Neural Netw. Learn. Syst. **99**, 1–14 (2021)

9. Yang, K., Liao, C., Zhang, W.: A sentiment classification model based on multiple multi-classifier systems. In: Sun, X., Pan, Z., Bertino, E. (eds.) ICAIS 2019. LNCS, vol. 11633, pp. 287–298. Springer, Cham (2019). https://doi.org/10.1007/978-3-030-24265-7_25

10. Araque, O., et al.: Enhancing deep learning sentiment analysis with ensemble techniques in social applications. Expert Syst. Appl. **77**.C, 236–246 (2017)

11. Catal, C., Nangir, M.: A sentiment classification model based on multiple classifiers. Appl. Soft Comput. **50**, 135–141 (2017)

12. Ghosh, M., Sanyal, G.: An ensemble approach to stabilize the features for multi-domain sentiment analysis using supervised machine learning. J. Big Data **5**(1), 1–25 (2018). https://doi.org/10.1186/s40537-018-0152-5

13. Li, Y., et al.: Cooperative hybrid semi-supervised learning for text sentiment classification. Symmetry **11**.2 (2019)

14. Zeng, Z., et al.: Context-based and explainable decision making with argumentation. In: Proceedings of the 17th International Conference on Autonomous Agents and Multi Agent Systems. International Foundation for Autonomous Agents and Multi-agent Systems (2018)

15. Gaggl, S.A., et al.: Design and results of the second international competition on computational models of argumentation. Artif. Intell. **279**, 103193 (2019)

16. Sirrianni, J.W., Liu, X., Adams, D.: Predicting stance polarity and intensity in cyber argumentation with deep bidirectional transformers. IEEE Trans. Comput. Soc. Syst. **99**, 1–13 (2021)

17. Fan, X., et al.: A first step towards explained activity recognition with computational abstract argumentation. In: 2016 IEEE International Conference on Multisensor Fusion and Integration for Intelligent Systems (MFI). IEEE (2016)

18. Ganzer-Ripoll, J., Criado, N., Lopez-Sanchez, M., Parsons, S., Rodriguez-Aguilar, J.A.: Combining social choice theory and argumentation: enabling collective decision making. Group Decis. Negot. **28**(1), 127–173 (2018). https://doi.org/10.1007/s10726-018-9594-6

19. Hao, Z.-Y., Yang, C., Liu, L., Kustudic, M., Niu, B.: Exploiting skew-adaptive delimitation mechanism for learning expressive classification rules. Appl. Intell. **50**(3), 746–758 (2019). https://doi.org/10.1007/s10489-019-01533-1

Stock Market Movement Prediction by Gated Hierarchical Encoder

Peibin Chen and Ying Tan[✉]

Laboratory of Machine Perception (MOE), Department of Machine Intelligence,
School of Electronics Engineering and Computer Science,
Peking University, Beijing 100871, China
ytan@pku.edu.cn

Abstract. Stock movement prediction is an important but challenging topic in the stock market. Previous methods mainly focus on predicting up or down of one stock, ignoring the significant up and down of the whole stock market that is more related to the final return. In this paper, a novel framework Gated Hierarchical Encoder (GHE) is proposed, which consists of two components: hierarchical feature learning and dynamic gate. Hierarchical feature learning helps the model do prediction from coarse to fine, while dynamic gate dynamically ensembles results from different branches. Experiments show that compared with MLP and LSTM, GHE achieves higher return on multiple stock markets, and predicts more accurately on the significant up and down.

Keywords: Quantitative investment · End-to-end · Classification · Stock market movement prediction · Time series · Deep learning

1 Introduction

Financial activities play a key role in the modern economy, of which stock market is an important part. Stock market movement prediction is conducive to better macroeconomic strategies and monetary policies, and also conducive to more reasonable resource allocation and less investment losses. For the reason that the price of a stock is unpredictable [17,21], researchers mainly focus on predicting the stock price movement.

Stock data is highly chaotic, random and noisy, making it hard to generalize well with traditional machine learning methods, such as SVM [3] and ARIMA [1]. Recently, due to the great success of deep neural network [9,10,19], more and more researchers bring in the technology to stock movement prediction [8,14–16]. Some methods [5] apply adversarial learning to enhance the robustness of the model. Some methods [6] use multiple sources, like news data and stock time series data, to jointly predict the movement. And some methods [11] temp to explore the relationship between stocks. However, these methods are limited in the one-stock-binary-classification framework, failed to warn for the significant up and down of the stock market.

© Springer Nature Switzerland AG 2021
Y. Tan and Y. Shi (Eds.): ICSI 2021, LNCS 12690, pp. 511–521, 2021.
https://doi.org/10.1007/978-3-030-78811-7_48

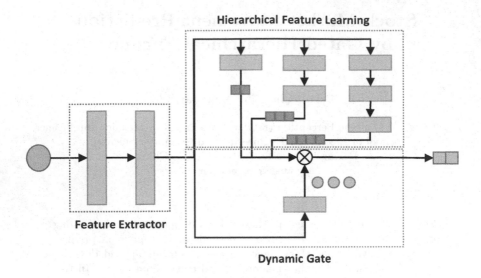

Fig. 1. A brief description of gated hierarchical encoder.

In this paper, the gated hierarchical encoder (GHE) is proposed, with which model is able to predict the up and down as well as the significant up and down. Different from previous methods [5,6], GHE decomposes the task of binary classification into multiple tasks, and splits the top of the model into multiple branches. Each task is still a classification, but the number of classes is positively correlated to the depth of the corresponding branch. Furthermore, a dynamic gate is introduced to ensemble results for different branches according to the stock input. Experiments on China A-shares market and U.S. stock market demonstrate that GHE surpasses baseline methods by an obvious margin, and is able to accurately predict the significant up and down. Ablation study proves each part of GHE contributes to the final strong performance.

The contributions of this paper are summarized as:

- Gated hierarchical encoder is proposed to accurately predict the significant up and down of the market.
- A dynamic gate is proposed to dynamically ensemble results from different branches in GHE according to the stock input.
- Stock indexes are used as input rather than multiple stocks, making it easier to explore the characteristics of the whole stock market.

2   Related Work

Stock movement prediction is a highly challenging topic for academia and industry. Recently, due to the great success of deep learning, it has been possible to extract useful information from stock data automatically for final prediction.

Researches on stock prediction can be divided into two main directions: market-historic-data based analysis and outside market-historic-data based analysis.

Market-historic-data based analysis deals with historical market data, such as opening price and stock volumes. However, this kind of data is highly chaotic, random and noisy. Simply trying to fit the data can easily cause overfitting. Adv-ALSTM [5] introduces adversarial learning in the feature space, forcing the model to be robust against the noise. LSTM-RGCN [11] argues that the movement of one stock has connections with that of others. With this assumption, the authors model the stock relation with a graph neural network, with stocks as nodes and the relation between each other as edges. HMGTF [4] introduce the powerful Transformer [20] that has been widely used in NLP. The method enhances the locality, avoids redundant heads of the Transformer, and utilizes the modified Transformer to learn hierarchical features. With the model, extremely long-term dependencies are mined from financial time series.

Although historical market data is widely used, it is unable to reflect the current market trend in time. Based on this observation, many researchers bring in the outside market-historic-data and build models to predict the emotional signals from the data. HAN [6] uses a multi-level sequence model to extract information, and further uses an attention model to obtain the final signals from the news. DP-LSTM [12] extracts hidden information from the news and integrates multiple news sources through the differential privacy mechanism. CapTE [13] effectively encodes the rich semantics and relation for a certain stock with the powerful Transformer [20] and the capsule network [18].

Different from the methods above, this paper focuses on predicting the significant up and down of the whole stock market, rather than the up and down of a single stock. For this purpose, Gated Hierarchical Encoder is proposed, which significantly outperforms MLP and LSTM on China A-shares market and U.S. stock market.

3   Gated Hierarchical Encoder

Previous methods prefer to consider the stock market movement prediction as a single stock binary classification problem. This brings three problems: (i) Ignoring the change percentage that investors concern about. (ii) Ignoring the connection between China A-shares market and overseas markets. (iii) Single stock prediction is unable to represent the movement of the whole market. Based on these observations, Gated Hierarchical Encoder is proposed, which is formalized by three components: multi indexes data preprocessing, hierarchical feature learning and dynamic gate for ensembling. The details of the model are shown in Fig. 1.

3.1   Multi-indexes Data Preprocessing

Since stock indexes are able to indicate the movement of the stock market, multiple stock indexes are used as input rather than stocks. In order to introduce external information, stock indexes from overseas markets are also used. However, different kind of features and different markets have different scales, making it hard to catch the key clues for prediction. Inspired by the normalization operation in computer vision [7], we normalize the indexes along the data dimension. Assuming the training set $\mathcal{D}$ is consist of N_c China A-shares stock indexes examples x_i^c and N_o overseas markets indexes examples x_i^o, the normalization can be written as

$$\mu_\mathcal{B}^l = \frac{1}{N_l} \sum_{i=1}^{N_j} x_i^l, \tag{1}$$

$$\sigma_\mathcal{B}^l = \sqrt{\frac{1}{N_l} \sum_{i=1}^{N_l} \left(x_i^l - \mu_\mathcal{B}^l\right)^2 + \epsilon}, \tag{2}$$

$$x_i^l = \frac{x_i^l - \mu_\mathcal{B}^l}{\sigma_\mathcal{B}^l}, \tag{3}$$

where $l \in \{c, o\}$.

3.2 Hierarchical Feature Learning

Previous methods ignore the change percentage that investors concern about. But directly predicting the change percentage is impossible because of the insufficient training data [2,17,21]. A novel framework is proposed for hierarchical feature learning, with which model is able to do prediction from coarse to fine.

Formally, given an example x_i, the feature extractor H outputs the embedding of x_i, after which the embedding is passed through K branches $B_1 \ldots B_K$. With deeper the layer, the branches are more complex and are able to encode richer features. Each branch is assigned a classification task whose difficulty depends on the depth of branch. Besides, we assume that the task at lower layer can be solved if the task at higher layer is perfectly solved. For example, B_1 temps to predict the market up or down, and B_K is assigned a more complex task, predicting if market will go significant up, slight up, slight down or significant down. Since different branches share the same feature extractor, and the tasks between branches are highly correlative, the model is expected to learn better with the hierarchical supervise. Defining the classification layer after B_k as F_k, the model forward process can be defined by the following equations,

$$h_i = H(x_i), \tag{4}$$

$$b_i^k = B_k(h_i) \quad \text{for } k \text{ in } 1 \ldots K, \tag{5}$$

$$f_i^k = F_k(b_i^k) \quad \text{for } k \text{ in } 1 \ldots K, \tag{6}$$

$$p_{ij}^k = \frac{exp(f_{ij}^k)}{\sum_{j=1}^{Cls} exp(f_{ij}^k)} \quad \text{for } k \text{ in } 1 \ldots K, \tag{7}$$

where p_{ij}^k represents the output score of class j from branch k for x_i.

Table 1. Comparison on China A-shares market and U.S. stock market. The results are shown as Acc/R.

Market	Methods	MLP	LSTM	Ours
China A-shares market	CSI300	0.592/1.645	0.558/1.752	**0.596/2.037**
	CSI500	0.539/1.387	0.562/1.739	**0.589/2.016**
	CYB	0.552/2.215	0.555/2.686	**0.573/2.917**
	SH	0.544/1.406	0.536/1.461	**0.608/1.691**
	SH50	0.542/1.381	0.546/1.561	**0.613/1.837**
	ZXB	0.546/1.502	0.555/2.248	**0.597/2.485**
U.S. stock market	IXIC	0.563/1.325	**0.587**/1.838	0.561/**2.076**
	SPX	0.523/0.924	**0.578**/1.623	0.558/**1.776**

3.3 Dynamic Gate for Ensembling

Due to the high correlation between tasks of branches, it is possible to ensemble results from different branches to enhance the performance of the model. Different from simply averaging the results, an ensembling gate is proposed to dynamically decide the weight of different branches. The output of the gate is depended on the input data, thus is more flexible and intelligent. Formally, give h_i from the feature extractor, the gate G encodes the feature and outputs K scores, with which the output of K branches are weighted summed to get f_i^e,

$$w_i = G(h_i), \tag{8}$$

$$f_i^e = \sum_{k=1}^{K} M(f_i^k) \times w_i^k, \tag{9}$$

$$p_{ij}^e = \frac{exp(f_{ij}^e)}{\sum_{j=1}^{Cls} exp(f_{ij}^e)}, \tag{10}$$

where p_{ij}^e represents the dynamically ensembling score of class j for x_i.

3.4 Loss Functions

Given multiple stock indexes, Gated Hierarchical Encoder outputs K results for K tasks, and the dynamic gate further ensembles K outputs to get a better prediction. We use cross entropy loss to guide the learning of the model. Furthermore, in order to encourage different branches to be close to the ensembling result, Kullback-Leibler divergence loss is added in all branches. The final loss function is

$$L = w_1 \times L_{cls} + w_2 \times L_{ens} + w_3 \times L_{kl}, \tag{11}$$

where w_1, w_2, w_3 are hyperparameters, and L_{cls}, L_{ens} and L_{kl} are defined as

$$L_{cls} = -\frac{1}{(N_c + N_o) \times K} \sum_{i=1}^{N_c+N_o} \sum_{k=1}^{K} y_i^k \log p_i^k, \tag{12}$$

Table 2. Ablation study on CSI300 and IXIC. The results are shown as Acc/R.

Model	CSI300	IXIC
LSTM	0.558/1.752	0.587/1.838
LSTM+HF	0.573/1.878	0.634/2.039
LSTM +HF + DG (Ours)	0.596/2.037	0.561/2.076

$$L_{ens} = -\frac{1}{(N_c + N_o)} \sum_{i=1}^{N_c+N_o} y_i^e \log p_i^e, \tag{13}$$

$$L_{kl} = \frac{1}{(N_c + N_o) \times K} \sum_{i=1}^{N_c+N_o} \sum_{k=1}^{K} KL(p_i^k || p_i^e), \tag{14}$$

where $KL(\cdot)$ represents the Kullback-Leibler divergence loss.

4   Experiments

In this section, a set of experiments are conducted to verify the ability of Gated Hierarchical Encoder. Firstly, our method is compared with traditional methods on multiple indexes from China A-shares market and U.S. stock market, showing the advantages of our method. Secondly, the dynamic gate is removed from the model for proving the effectiveness of the gate. Finally, the output of the gate is visualized to observe how the gate ensembles branches on different indexes.

4.1   Dataset

We select 6 indexes, including CSI300, CSI500, CYB, SH, SH50 and ZXB, in China A-shares market and 2 indexes, including IXIC and SPX, in U.S. stock market. All data are from 2005-01-01 to 2020-12-14 and split into three parts: training set (from 2005-01-01 to 2017-12-31), validation set (from 2018-01-01 to 2018-12-31) and test set (from 2019-01-01 to 2020-12-14). Accuracy at significant up/down Acc_S and cumulative return R_S of index S are defined as

$$Acc_S = \frac{\sum_{i \in S} \mathcal{I}(y_i = \tilde{y}_i) \times \mathcal{I}(abs(chg_i) > 0.5)}{\sum_{i \in S} \mathcal{I}(abs(chg_i) > 0.5)}, \tag{15}$$

$$R_S = \sum_{i \in S}(1 + \mathcal{I}(y_i = \tilde{y}_i) \times chg_i), \tag{16}$$

where $\mathcal{I}(.)$ is a indicator function that takes on a value of 1 if its augment is true, and 0 otherwise. chg_i is the change percentage of x_i.

4.2   Comparisons

Gated Hierarchical Encoder is compared with MLP and LSTM to show the strength of Gated Hierarchical Encoder. MLP consists of simple interconnected neurons or nodes, learning the nonlinear mapping between input and output vectors. LSTM is a special type of recurrent neural networks which are able to learn long-term dependencies, especially in sequence prediction problems. For fairness, all methods use the same features as input, and are tested using the model whose performance is the best in the validation set. Table 1 shows the results. From Table 1, it is obvious that our method achieves the best results in most of the indexes. The cumulative return earned by different methods in the test set are shown in Fig. 3 and Fig. 4. Gated Hierachical Encoder is able to predict the time of high risk or high returns, thus getting more returns on the stock markets.

4.3   Ablation Study

To clarify the effectiveness of hierarchical feature learning and dynamic gate, we measure the effect of (i) LSTM: Original LSTM model, (ii)LSTM + HF: Adding hierarchical feature learning to LSTM and (iii) LSTM +HF + DG: adding dynamic gate and hierarchical feature learning to LSTM. Experiments are carried out on CSI300 and IXIC. As shown in Table 2, each component improves the origin LSTM and boosts the returns by an obvious margin.

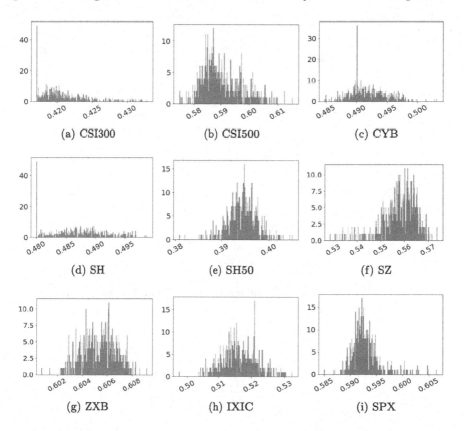

Fig. 2. Visualizations of the dynamic gate for multiple indexes.

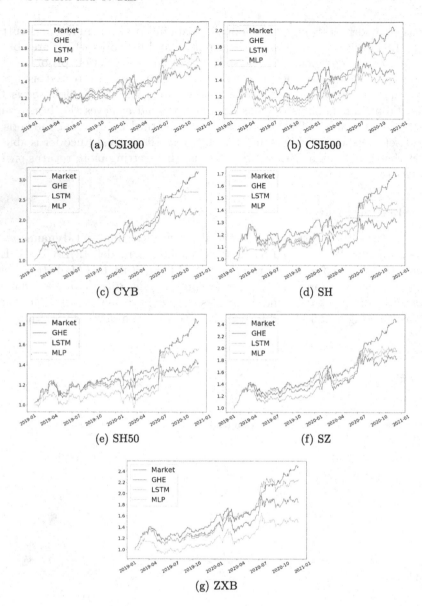

Fig. 3. Comparison of different methods on China A-shares market.

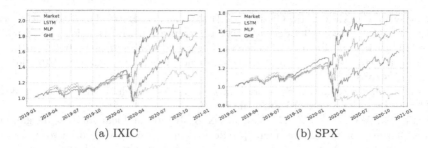

(a) IXIC   (b) SPX

Fig. 4. Comparison of different methods on U.S. stock market.

4.4   Visualizations

Dynamic gate is designed to ensemble multiple branches with dynamic weights. In order to show whether the gate is able to output different scores for different indexes, we count the outputs of dynamic gate and plot the histogram as shown in Fig. 2. For CSI300, CYB, SH and SH50, dynamic gate gives more weights on the multiple classification branch. For the others, dynamic gate gives more weights on the binary classification branch. Therefore, the gate is proved to have the ability to ensemble branches according to different situations.

5   Conclusion

In this paper, Gated Hierarchical Encoder is proposed to predict the significant up and down of the stock market. Stock indexes are used as input for the reason that stock indexes are able to indicate the movement of stock market. Then, the stock movement prediction are divided into multiple highly related classification tasks via hierarchical feature learning. To further boost the performance of the model, a dynamic gate is introduced to ensemble different results from different branches. The ensembling weights are dynamically determined according to the input index. Experiments on China A-shares market and U.S. stock market show that GHE surpasses the baselines by an obvious margin.

Acknowledgements. This work is supported by Science and Technology Innovation 2030 - "New Generation Artificial Intelligence" Major Project (Grant Nos.: 2018AAA0102301, 2018AAA0100302), and the National Natural Science Foundation of China (Grant No. 62076010).

References

1. Ariyo, A.A., Adewumi, A.O., Ayo, C.K.: Stock price prediction using the ARIMA model. In: Al-Dabass, D., Orsoni, A., Cant, R.J., Yunus, J., Ibrahim, Z., Saad, I. (eds.) UKSim-AMSS 16th International Conference on Computer Modelling and Simulation, UKSim 2014, Cambridge, United Kingdom, 26–28 March 2014, pp. 106–112. IEEE (2014). https://doi.org/10.1109/UKSim.2014.67

2. Cogswell, M., Ahmed, F., Girshick, R.B., Zitnick, L., Batra, D.: Reducing overfitting in deep networks by decorrelating representations. In: Bengio, Y., LeCun, Y. (eds.) 4th International Conference on Learning Representations, ICLR 2016, San Juan, Puerto Rico, 2–4 May 2016. Conference Track Proceedings (2016). http://arxiv.org/abs/1511.06068

3. Cortes, C., Vapnik, V.: Support-vector networks. Mach. Learn. **20**(3), 273–297 (1995). https://doi.org/10.1007/BF00994018

4. Ding, Q., Wu, S., Sun, H., Guo, J., Guo, J.: Hierarchical multi-scale gaussian transformer for stock movement prediction. In: Bessiere, C. (ed.) Proceedings of the Twenty-Ninth International Joint Conference on Artificial Intelligence, IJCAI 2020, pp. 4640–4646. ijcai.org (2020). https://doi.org/10.24963/ijcai.2020/640

5. Feng, F., Chen, H., He, X., Ding, J., Sun, M., Chua, T.: Enhancing stock movement prediction with adversarial training. In: Kraus, S. (ed.) Proceedings of the Twenty-Eighth International Joint Conference on Artificial Intelligence, IJCAI 2019, Macao, China, 10–16 August 2019, pp. 5843–5849. ijcai.org (2019). https://doi.org/10.24963/ijcai.2019/810

6. Hu, Z., Liu, W., Bian, J., Liu, X., Liu, T.: Listening to chaotic whispers: a deep learning framework for news-oriented stock trend prediction. In: Chang, Y., Zhai, C., Liu, Y., Maarek, Y. (eds.) Proceedings of the Eleventh ACM International Conference on Web Search and Data Mining, WSDM 2018, Marina Del Rey, CA, USA, 5–9 February 2018, pp. 261–269. ACM (2018). https://doi.org/10.1145/3159652.3159690

7. Ioffe, S., Szegedy, C.: Batch normalization: accelerating deep network training by reducing internal covariate shift. In: Bach, F.R., Blei, D.M. (eds.) Proceedings of the 32nd International Conference on Machine Learning, ICML 2015, Lille, France, 6–11 July 2015. JMLR Workshop and Conference Proceedings, vol. 37, pp. 448–456. JMLR.org (2015). http://proceedings.mlr.press/v37/ioffe15.html

8. Kim, R., So, C.H., Jeong, M., Lee, S., Kim, J., Kang, J.: HATS: a hierarchical graph attention network for stock movement prediction. CoRR abs/1908.07999 (2019). http://arxiv.org/abs/1908.07999

9. Krizhevsky, A., Sutskever, I., Hinton, G.E.: ImageNet classification with deep convolutional neural networks. In: Bartlett, P.L., Pereira, F.C.N., Burges, C.J.C., Bottou, L., Weinberger, K.Q. (eds.) Advances in Neural Information Processing Systems 25: 26th Annual Conference on Neural Information Processing Systems 2012. Proceedings of a Meeting Held Lake Tahoe, Nevada, United States, 3–6 December 2012, pp. 1106–1114 (2012). http://papers.nips.cc/paper/4824-imagenet-classification-with-deep-convolutional-neural-networks

10. LeCun, Y., Bengio, Y., Hinton, G.E.: Deep learning. Nature **521**(7553), 436–444 (2015). https://doi.org/10.1038/nature14539

11. Li, W., Bao, R., Harimoto, K., Chen, D., Xu, J., Su, Q.: Modeling the stock relation with graph network for overnight stock movement prediction. In: Bessiere, C. (ed.) Proceedings of the Twenty-Ninth International Joint Conference on Artificial Intelligence, IJCAI 2020, pp. 4541–4547. ijcai.org (2020). https://doi.org/10.24963/ijcai.2020/626

12. Li, X., Li, Y., Yang, H., Yang, L., Liu, X.: DP-LSTM: differential privacy-inspired LSTM for stock prediction using financial news. CoRR abs/1912.10806 (2019). http://arxiv.org/abs/1912.10806

13. Liu, J., et al.: Transformer-based capsule network for stock movement prediction. In: Proceedings of the First Workshop on Financial Technology and Natural Language Processing, pp. 66–73 (2019)

14. Long, W., Lu, Z., Cui, L.: Deep learning-based feature engineering for stock price movement prediction. Knowl. Based Syst. **164**, 163–173 (2019). https://doi.org/10.1016/j.knosys.2018.10.034

15. Ma, T., Tan, Y.: Multiple stock time series jointly forecasting with multi-task learning. In: 2020 International Joint Conference on Neural Networks, IJCNN 2020, Glasgow, United Kingdom, 19–24 July 2020, pp. 1–8. IEEE (2020). https://doi.org/10.1109/IJCNN48605.2020.9207543

16. Nelson, D.M.Q., Pereira, A.C.M., de Oliveira, R.A.: Stock market's price movement prediction with LSTM neural networks. In: 2017 International Joint Conference on Neural Networks, IJCNN 2017, Anchorage, AK, USA, 14–19 May 2017, pp. 1419–1426. IEEE (2017). https://doi.org/10.1109/IJCNN.2017.7966019

17. Nguyen, T.H., Shirai, K., Velcin, J.: Sentiment analysis on social media for stock movement prediction. Expert Syst. Appl. **42**(24), 9603–9611 (2015). https://doi.org/10.1016/j.eswa.2015.07.052

18. Sabour, S., Frosst, N., Hinton, G.E.: Dynamic routing between capsules. In: Guyon, I., et al. (eds.) Advances in Neural Information Processing Systems 30: Annual Conference on Neural Information Processing Systems 2017, Long Beach, CA, USA, 4–9 December 2017, pp. 3856–3866 (2017). http://papers.nips.cc/paper/6975-dynamic-routing-between-capsules

19. Szegedy, C., et al.: Going deeper with convolutions. In: IEEE Conference on Computer Vision and Pattern Recognition, CVPR 2015, Boston, MA, USA, 7–12 June 2015, pp. 1–9. IEEE Computer Society (2015). https://doi.org/10.1109/CVPR.2015.7298594

20. Vaswani, A., et al.: Attention is all you need. In: Guyon, I., et al. (eds.) Advances in Neural Information Processing Systems 30: Annual Conference on Neural Information Processing Systems 2017, Long Beach, CA, USA, 4–9 December 2017, pp. 5998–6008 (2017). http://papers.nips.cc/paper/7181-attention-is-all-you-need

21. Walczak, S.: An empirical analysis of data requirements for financial forecasting with neural networks. J. Manag. Inf. Syst. **17**(4), 203–222 (2001). http://www.jmis-web.org/articles/997

Other Applications

Spiking Adaptive Dynamic Programming with Poisson Process

Qinglai Wei[1,2(✉)], Liyuan Han[1,3], and Tielin Zhang[4]

[1] The State Key Laboratory for Management and Control of Complex Systems, Institute of Automation, Chinese Academy of Sciences, Beijing 100190, China
qinglai.wei@ia.ac.cn

[2] Institute of Systems Engineering, Macau University of Science and Technology, Macau 999078, China

[3] School of Artificial Intelligence, University of Chinese Academy of Sciences, Beijing 100049, China

[4] Research Center for Brain-Inspired Intelligence, Institute of Automation, Chinese Academy of Sciences, Beijing 100190, China

Abstract. A new iterative spiking adaptive dynamic programming (SADP) algorithm based on the Poisson process for optimal impulsive control problems is investigated with convergence discussion of the iterative process. For a fixed time interval, a 3-tuple can be computed, and then the iterative value functions and control laws can be obtained. Finally, a simulation example verifies the effectiveness of the developed algorithm.

Keywords: Spiking dynamic programming · Poission process · Nonlinear systems

1 Introduction

Impulsive behaviours exist widely in many dynamic systems, such as mathematical biology, engineering control, and information science [1–4]. An impulse is a sudden jump at an instant during the dynamic process. The research of impulsive control system has drawn a lot of attention worldwide. In [5], the stability, robust stabilization and controllability are analyzed for singular-impulsive systems via switching control. In [6], the global stability of switching Hopfield neural networks with state-dependent impulses is described with an equivalent method. It should be mentioned that previous impulsive control methods focus on linear systems [7,8]. However, for nonlinear systems, the hybrid Bellman equation is generally analytically unsolvable.

Adaptive dynamic programming (ADP), proposed by Werbos, is a method of solving optimal control problems, which combines the advantages of dynamic programming, reinforcement learning and function approximation [9–11]. ADP has two branches, value and policy iterations. However, traditional ADP methods [12–14] cannot solve impulsive control problem. To overcome this shortcoming,

Y. Tan and Y. Shi (Eds.): ICSI 2021, LNCS 12690, pp. 525–532, 2021.
https://doi.org/10.1007/978-3-030-78811-7_49

in [15], a new discrete-time impulsive ADP algorithm is proposed to obtain the optimum iteratively, while the impulsive interval is required to constrain in a fixed interval set. Furthermore, the interval set is generally difficult to determine. Until now, to the best of our knowledge, there are no discussions on optimal control problems with the spike train from real biology based on ADP algorithms, and this motivates our research.

2   Problem Statement

We consider the following discrete-time nonlinear control systems

$$x_{k+1} = F(x_k, u_k), k = 0, 1, \ldots \tag{1}$$

where $x_k \in \mathbb{R}^n$ is the state variable and $u_k \in \mathbb{R}^m$ is the spiking control input. Let $F(\cdot)$ be the system function.

Assumption 1. *The system (1) is controllable on a compact set $\Omega_x \subset \mathbb{R}^n$ containing the origin; the system state $x_k = 0$ is an equilibrium state of system (1) under the control $u_k = 0$, i.e., $F(0,0) = 0$; the feedback control law satisfies $u_k(x_k) = \mu\left(\pi_k(x_k), \nu_k(x_k)\right) = 0$ for $x_k = 0$.*

Notations 1. *$\mathbb{R}_+$ and $\mathbb{Z}_+$ are the sets of all non-negative real numbers and integers, respectively. $\mathcal{T} = \{t^s\}$ is the set of spiking instants, where $t^s \in \mathbb{R}_+, s = 1, 2, \ldots$ τ_k is the number of spiking intants in interval $[kT, (k+1)T]$ and λ_k is the firing rate of spike train in $[0, (k+1)T]$, where $T \in \mathbb{R}_+$ and $k = 0, 1, 2, \ldots$. According to $\mathcal{T}$, spiking interval can be expressed as $t_s = t^s - t^{s-1}$, $s = 1, 2, 3\ldots$, where $t^0 = 0$. Let $\Gamma = \{\mathcal{F}_k\}, \mathcal{F}_k \subseteq \mathcal{F}_{k+1} \subseteq \Gamma, k = 0, 1, 2, 3\ldots$, where $\mathcal{F}_k$ includes the information for the computation, such as the state x_k and the number of spiking instants τ_k.*

Let $\mathcal{T}_\theta = \left\{\theta^s | \theta^s = \text{round}(t^{\sum_{i=0}^s \tau_i}), \theta^s \in \mathbb{Z}_+, s = 0, 1, 2, \ldots\right\}$, be the spiking instants, where round$(\cdot)$ is a rounding function. Let $\nu_k = \nu_k(x_k) \in \mathbb{R}^m$ and $\pi_k = \pi_k(x_k) \in \mathcal{Z} = \{0, 1\}$ for $k = 0, 1, 2, \ldots$ When $k = \theta^s$, we have $u_k = \nu_k$ and $\pi_k = 1$. Thus, the spiking control law can be written as $u_k = \mu(\pi_k, \nu_k)$, $\mu(\cdot) : \mathcal{Z} \times \mathbb{R}^m \to \mathbb{R}^m$, where $\mu(\pi_k, \nu_k)$ can be defined as

$$u_k = \mu(\pi_k, \nu_k) = \begin{cases} 0, & \pi_k = 0 \\ \nu_k, & \pi_k = 1. \end{cases} \tag{2}$$

Let $\underline{u}_k = \{u_k, u_{k+1}, \ldots\}$, $\underline{\pi}_k = \{\pi_k, \pi_{k+1}, \ldots\}$ and $\underline{\nu}_k = \{\nu_k, \nu_{k+1}, \ldots\}$ $k = 0, 1, 2, \ldots$, respectively. The given infinite-horizon performace index function for initial state x_0 can be defined as

$$J_0(x_0, \underline{u}_0) = \mathbb{E}\left(\sum_{k=0}^{\infty} U(x_k, u_k) | \mathcal{F}_0\right) = \mathbb{E}\left(\sum_{k=0}^{\infty} U(x_k, \mu(\pi_k, \nu_k)) | \mathcal{F}_0\right) \tag{3}$$

where the utility function $U(x_k, \mu(\pi_k, \nu_k)) \geq 0$ for x_k and $\mu(\cdot)$. We desire to find an optimal spiking control law $u_k^*(x_k) = \mu(\pi_k^*(x_k), \nu_k^*(x_k))$, such that the performace index function is minimum, i.e.,

$$J_k^*(x_k) = \min_{\underline{u}_k} J_k(x_k, \underline{u}_k), \tag{4}$$

satisfying Bellman Equation [16], which is expressed as

$$J_k^*(x_k) = \min_{u_k} \mathbb{E}\{U(x_k, u_k) + J_{k+1}^*(x_{k+1})|\mathcal{F}_k\}. \tag{5}$$

3 SADP Method Based on Poisson Process

In this section, the new iterative SADP algorithm based on Poisson process is described to obtain the optimal spiking control law for a discrete-time nonlinear system (1) with property analysis.

3.1 Transformation of the Utility Function

According to the MLE, the set $\Pi = \{\tau_k\}$ and the set $\Lambda = \{\lambda_k\}$, $k = 0, 1, 2, ...$ can be easily obtained. Let $\bar{\lambda}$ represent the average of $\{\lambda_k\}, k = 0, 1, 2, ...$ For $\mathcal{R} = 0, 1, ...,$ Poisson process [17–19] can be expressed as

$$P(N(t) = \mathcal{R}) = \frac{(\lambda t)^{\mathcal{R}}}{\mathcal{R}!} \exp(-\lambda t). \tag{6}$$

Due to the fixed time interval T, the probability of Poisson distribution in $[kT, (k+1)T], k = 0, 1, 2, ...$ can be calculated as

$$p_{\tau_k} = \frac{(\lambda T)^{\tau_k}}{\tau_k!} \exp(-\bar{\lambda}T). \tag{7}$$

Thus, for each state $x_k \in \Omega_x$, $k = 0, 1, 2, ...,$ we can get a 3-tuple $(x_k, \tau_k, p_{\tau_k})$. Also, the probability p_{τ_k} is added to $\mathcal{F}_k$ for $k = 0, 1, 2, 3...$ Thus, we can obtain a new utility function $\mathcal{U}_{\tau_k}$ expressed as

$$\mathcal{U}_{\tau_k}(x_k, \mu(\pi_{k+\tau_k}, \nu_{k+\tau_k})) = \frac{1 - p_{\tau_k}}{\tau_k} \sum_{j=0}^{\tau_k-1} U(x_{k+j}, 0) + p_{\tau_k} U(x_{k+\tau_k}, \nu_{k+\tau_k}). \tag{8}$$

Thus, the optimal spiking value function $V_k^*(x_k)$ can be defined as

$$V_k^*(x_k) = \min_{\nu_{k+\tau_k}} \left\{ \mathcal{U}_{\tau_k}(x_k, \nu_{k+\tau_k}) + \sum_{j\in\Omega_x} p(j|x_k, \tau_k) J_{k+\tau_k+1}^*(j) \right\}. \tag{9}$$

3.2 Iterative SADP Method Based on Poisson Process

Then, the SADP algorithm based on Poisson process can be derived in Algorithm 1.

Algorithm 1. SADP Algorithm based on Poisson Process

Require:

    Give an initial state x_0 randomly, a computation precision ϵ and an arbitrary positive semi-definite function $\Psi(x)$.

Ensure:

1: Let the iteration index $i = 0$, and the initial iterative value function $V_0(x_k) = \Psi(x_k)$, $k = 0, 1, 2....$

2: Obtain the 3-tuple $(x_k, \tau_k, p_{\tau_k})$, $k = 0, 1, 2, ...$

3: Iterative spiking control law $\nu_i(x_k)$ can be computed as

$$\nu_i(x_k) = \arg\min_{\nu_{k+\tau_k}} \left\{ \mathcal{U}_{\tau_k}(x_k, \nu_{k+\tau_k}) + \sum_{j \in \Omega_x} p(j|x_k, \tau_k) V_i(j) \right\}. \qquad (10)$$

4: Iterative spiking value function $V_{i+1}(x_k)$ can be computed as

$$V_{i+1}(x_k) = \min_{\nu_{k+\tau_k}} \left\{ \mathcal{U}_{\tau_k}(x_k, \nu_{k+\tau_k}) + \sum_{j \in \Omega_x} p(j|x_k, \tau_k) V_i(j) \right\}. \qquad (11)$$

5: If $|V_{i+1}(x_k) - V_i(x_k)| \leq \epsilon, \forall x_k \in \Omega_x$, then the optimal performance index function and optimal spiking control law can be obtained. Goto step 6. Otherwise, let $i = i+1$, and goto step 2.

6: end.

3.3   Property Analysis of the SADP Algorithm Based on Poisson Process

In this section, the property analysis of the SADP algorithm based on Poisson process will be estabilished.

Theorem 1. *Let $J_k^*(x_k)$ and $V_k^*(x_k)$, $k = 0, 1, 2, ...,$ be the optimal performance index function and optimal spiking value function which satisfy (4) and (9), respectively. Then, for each 3-tuple $(x_k, \tau_k, p_{\tau_k})$, $k = 0, 1, 2, ...,$ we have*

$$J_k^*(x_k) = V_k^*(x_k). \qquad (12)$$

Proof. Based on the 3-tuple $(x_k, \tau_k, p_{\tau_k})$ obtained by the real sequence of spike train, for any state $x_k \in \Omega_x$, we can derive that $k + \tau_k$ is a spiking instant, i.e., $\pi_{k+\tau_k} = 1$, with the Poisson probability p_{τ_k}. Thus, according to (4), we can derive the following Bellman equation (13)

$$J_k^*(x_k) = \min_{\underline{u}_k} \left\{ \mathbb{E}\left(\sum_{j=0}^{\infty} U(x_{k+j}, u_{k+j})|\mathcal{F}_k \right) \right\}$$

$$= \min_{\nu_{k+\tau_k}} \left\{ \mathcal{U}_{\tau_k}(x_k, \nu_{k+\tau_k}) + \sum_{j \in \Omega_x} p\left(j|x_k, \tau_k\right) J_{k+\tau_k+1}^*(j) \right\}$$

$$= V_k^*(x_k), \qquad (13)$$

where $p(j|x_k, \tau_k)$ can be expressed as

$$p(j|x_k, \tau_k) = \begin{cases} \dfrac{1 - p_{\tau_k} p_{\tau_{k+\tau_k}}}{N-1}, & j \in \Omega_x, j \neq x_{k+\tau_k} \\ p_{\tau_k} p_{\tau_{k+\tau_k}}, & j = x_{k+\tau_k}. \end{cases} \tag{14}$$

and N represents the number of the states in Ω_x. The Eq. (14) shows that, for state $x_{k+\tau_k}$, the probability is the product of p_{τ_k} and $p_{\tau_{k+\tau_k}}$, while the probability is the same for other states, i.e., $(1 - p_{\tau_k} p_{\tau_{k+\tau_k}})/(N-1)$.

The proof is complete.

According to Theorem 1, for each 3-tuple $(x_k, \tau_k, p_{\tau_k})$, $k = 0, 1, 2, ...$, the Bellman equation (5) can be expressed as

$$J_k^*(x_k) = \frac{1 - p_{\tau_k}}{\tau_k} \sum_{j=0}^{\tau_k - 1} U(x_{k+j}, 0) + \min_{v_{k+\tau_k}} \Big\{ p_{\tau_k} U(x_{k+\tau_k}, u_{k+\tau_k})$$
$$+ \sum_{j \in \Omega_x} p\Big(j|x_k, \tau_k\Big) J_{k+\tau_k+1}^*(j) \Big\}. \tag{15}$$

The Bellman equation (15) can be called "3-tuple Bellman equation".

Lemma 1. *For $i = 0, 1, 2, ...$, and any $(x_k, \tau_k, p_{\tau_k})$ $k = 0, 1, 2, ...$, let $V_{i+1}(x_k)$ and $v_i(x_k)$ be the iterative value function and the iterative control law updated, respectively. According to (1)–(1) in Algorithm 1. Then, the $V_i(x_k)$ converges to the optimal performance index function $J_k^*(x_k)$ as $i \to \infty$, which is defined as Eq. (15), that is*

$$\lim_{i \to \infty} V_i(x_k) = J_k^*(x_k), \tag{16}$$

The conclusion is easily derived and the proof is omitted here.

4   Simulation

We consider the torsional pendulum system to evaluate the performance of our developed algorithm. The dynamic system is expressed as

$$\begin{bmatrix} x_{1,k+1} \\ x_{2,k+1} \end{bmatrix} = \begin{bmatrix} x_{1k} + \Delta t x_{2k} \\ -\dfrac{\Delta t M g l}{J} \sin(x_{1k}) + \left(1 - \dfrac{\Delta t f_d}{J}\right) x_{2k} \end{bmatrix} + \begin{bmatrix} 0 \\ \Delta t \end{bmatrix} u_k \tag{17}$$

where $J = 4/3 \, \text{ml}^2$, $M = 1/3 \, \text{kg}$, $g = 9.8 \, \text{m/s}^2$, $l = 3/2 \, \text{m}$ and $f_d = 0.2$ are the parameters of this system.

The utility function is chosen as $U(x_k, u_k) = x_k^\mathsf{T} P x_k + u_k^\mathsf{T} R u_k$, where $Q = I_1$ and $R = I_2$, denoting the identity matrices with suitable dimensions. Choose the initial value function with the form $\Psi(x_k) = x_k^\mathsf{T} P x_k$, where $P = [10 \ 1; 1 \ 2]$.

In this example, we use the data set shared by Potter Lab [20,21] to establish the 3-tuple. The fixed time is 0.3 s. The spike train is shown in Fig. 1(a)–(c).

We implement Algorithm 1 with $\hat{\Omega}_x$ for 20 iterations in order to urge the iterative value function to be convergent, as shown in Fig. 2(a). where "In" and "Lm" represent first iteration and last iteration, respectively. We can also see that the iterative value function is not smooth in the discretized state space due to the effect of spike train. Thus, the optimal spiking instants may vary with the states. The distribution of the optimal spiking intervals in the discretized state space $\hat{\Omega}_x$ can be seen in Fig. 2(b), existing seven kinds of intervals, from one to seven. In this example, we choose an initial state $x_0^1 = [1.2 \; -0.8]^{\mathsf{T}}$ and we get the corresponding optimal spiking control as shown in Fig. 2(c), respectively.

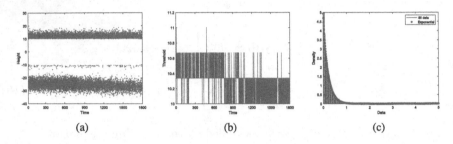

(a)　　　　　　　　(b)　　　　　　　　(c)

Fig. 1. The spike train. (a) Height-time. (b) Threshold-time. (c) Interspike interval.

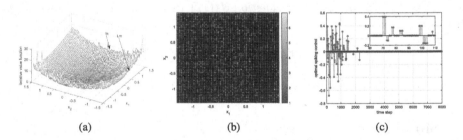

(a)　　　　　　　　(b)　　　　　　　　(c)

Fig. 2. (a) Convergence plots of the iterative value functions. (b) The distribution of the optimal spiking intervals. (c) Optimal spiking control with initial states x_0^1.

5   Conclusion

A new iterative SADP algorithm based on Poisson process is presented to solve optimal control problems for nonlinear systems. By using the model of Poisson process and the method of MLE, we get the 3-tuple. The property analysis is developed to guarantee that the value functions converge iteratively to optimal performance index function. Finally, a simulation example is given to verify the effectiveness of the developed algorithm.

References

1. Wang, X., Yu, J., Huang, Y., Wang, H., Miao, Z.: Adaptive dynamic programming for linear impulse systems. J. Zhejiang Univ. Sci. C **15**(1), 43–50 (2014). https://doi.org/10.1631/jzus.C1300145
2. Li, W., Huang, L., Guo, Z., Ji, J.: Global dynamic behavior of a plant disease model with ratio dependent impulsive control strategy. Math. Comput. Simul. **177**, 120–139 (2020)
3. Haddad, W.M., Chellaboina, V., Kablar, N.A.: Non-linear impulsive dynamical systems. Part II: stability of feedback interconnections and optimality. Int. J. Control **74**, 1659–1677 (2001)
4. Chen, W.-H., Luo, S., Zheng, W.X.: Generating globally stable periodic solutions of delayed neural networks with periodic coefficients via impulsive control. IEEE Trans. Cybern. **47**, 1590–1603 (2016)
5. Yao, J., Guan, Z.-H., Chen, G., et al.: Stability, robust stabilization and H? Control of singular-impulsive systems via switching control. Syst. Control Lett. **55**, 879–886 (2006)
6. Zhang, X., Li, C., Huang, T.: Hybrid impulsive and switching Hopfield neural networks with state-dependent impulses. Neural Netw. **93**, 176 184 (2017)
7. Li, X., Song, S.: Stabilization of delay systems: delay-dependent impulsive control. IEEE Trans. Autom. Control **62**, 406–411 (2016)
8. Zhang, Q., Qiao, L., Zhu, B., et al.: Dissipativity analysis and synthesis for a class of T-S fuzzy descriptor systems. IEEE Trans. Syst. Man Cybern. Syst. **47**, 1774–1784 (2016)
9. Woźniak, S., Pantazi, A., Bohnstingl, T., et al.: Deep learning incorporating biologically inspired neural dynamics and in-memory computing. Nat. Mach. Intell. **2**, 325–336 (2020)
10. Kiumarsi, B., Vamvoudakis, K.G., Modares, H., Lewis, F.L.: Optimal and autonomous control using reinforcement learning: a survey. IEEE Trans. Neural Netw. Learn. Syst. **29**, 2042–2062 (2017)
11. Jiang, Y., Jiang, Z.-P.: Robust Adaptive Dynamic Programming. Wiley, Hoboken (2017)
12. Wen, Y., Si, J., Gao, X., et al.: A new powered lower limb prosthesis control framework based on adaptive dynamic programming. IEEE Trans. Neural Netw. Learn. Syst. **28**, 2215–2220 (2016)
13. Liu, D., Wei, Q., Wang, D., Yang, X., Li, H.: Adaptive Dynamic Programming with Applications in Optimal Control. AIC. Springer, Cham (2017). https://doi.org/10.1007/978-3-319-50815-3
14. Liu, D., Xu, Y., Wei, Q., et al.: Residential energy scheduling for variable weather solar energy based on adaptive dynamic programming. IEEE/CAA J. Automatica Sinica **5**, 36–46 (2017)
15. Wei, Q., Song, R., Liao, Z., et al.: Discrete-time impulsive adaptive dynamic programming. IEEE Trans. Cybern. **50**, 4293–4306 (2019)
16. Puterman, M.L.: Markov Decision Processes: Discrete Stochastic Dynamic Programming. Wiley, Hoboken (2014)
17. Kordovan, M., Rotter, S.: Spike train cumulants for linear-nonlinear Poisson cascade models. arXiv preprint arXiv:2001.05057 (2020)
18. Bux, C.E.R., Pillow, J.W.: Poisson balanced spiking networks. bioRxiv **836601** (2019)

19. Gerhard, F., Deger, M., Truccolo, W.: On the stability and dynamics of stochastic spiking neuron models: nonlinear Hawkes process and point process GLMs. PLoS Comput. Biol. **13**, e1005390 (2017)
20. Newman, J.P., Fong, M.-f., Millard, D.C., et al.: Optogenetic feedback control of neural activity. Elife **4**, e07192 (2015)
21. Fong, M.-F., Newman, J.P., Potter, S.M., et al.: Upward synaptic scaling is dependent on neurotransmission rather than spiking. Nat. Commun. **6**, 1–11 (2015)

Designing a Mathematical Model and Control System for the Makariza Steam Boiler

Taha Ahmadi$^{(\boxtimes)}$ (iD) and Sebastián Soto Gaona$^{(\boxtimes)}$ (iD)

Corporación Unificada Nacional de Educación Superior, Bogotá, Colombia
{taha_ahmadi,sebastian_soto}@cun.edu.co

Abstract. This study provides a method to perform the dynamic analysis of sugarcane bagasse boiler owned by Makariza Company in Colombia. In this method, the values of Makariza industrial boiler has been taken as a reference, which allow to calculate the real values of enthalpy of sub-processes, boiler mass and energy balances. In the proposed dynamic model, the Differential-Algebraic Equations (DAE's) will be used, and the drum steam pressure and flow of the system are controlled using PI controllers. After describing the thermodynamic and mathematical processes, the effect of different operating conditions on system outputs will be examined and compared together considering the existence of the controller and without its use.

Keywords: Boiler · Dynamic model · Industrial boiler · Makariza · PI controller · Steam boiler

1 Introduction

A boiler is an enclosed vessel with special features used to convert water to steam [1–3]. This pressurized steam is then used to transfer heat and energy to one or more processes. Meanwhile, water can be used as a useful and cheap material for energy transfer [4, 5]. When water is converted to steam and given enough energy, its volume can increase up to about 1600 times and for this purpose, it has the ability to transmit a force almost equal to the force produced by the explosion of gunpowder. This makes the steam generation process very dangerous and sensitive and therefore must be properly designed and controlled in high pressures and temperatures [6].

For this reason, mathematical models of boilers are considered into account in studies. In [7], a nonlinear mathematical model of a steam boiler has been introduced. In [8], a linear mathematical modeling of a boiler drum has been presented. The mathematical model of the boiler is of great practical importance. This model can be used to observe the performance of the boiler in different working conditions, as well as for the correct and optimizing the design of different parts of the boiler. For example, in [9], Heat transfer performance of boiler superheaters is optimized using simulation and calculation of fluid dynamics model. Furthermore, the mathematical model of the boiler can be used to design combined heat and power (CHP) plants to operate in a various load demands range in a high efficiency [10].

© Springer Nature Switzerland AG 2021
Y. Tan and Y. Shi (Eds.): ICSI 2021, LNCS 12690, pp. 533–542, 2021.
https://doi.org/10.1007/978-3-030-78811-7_50

Controlling boiler parameters is very complicated due to the multivariate nature of this system and the relationship between these variables (pressure, temperature, flow, level) [11]. This boiler mathematical model can also be used to control and monitor boiler performance online [12–14]. In [15], a dynamic model and controller design of a boiler-turbine is developed.

In this paper, a method for recognizing different operating conditions in the boiler of Makariza Company will be simulated and presented. This simulation shows the actual behavior of the process in the boiler through the resolution of a dynamic model. In this design, the real parameters of the boiler available in Makariza Company have been used. In [16], based on boiler mathematical model of a power plant boiler a network-based model predictive control is presented to adjust the steam temperature of the boiler. Furthermore, dynamic models and control systems are used to adjust drum steam pressure, water level and steam flow [12, 17, 18].

The advantages of using dynamic models can be summarized as follows [19–21]:

- The process of energy production and transfer can be analyzed in depth and determine which parts of the process have a significant impact on the overall behavior of the system.
- They can be used to train operators to gain a general understanding of the system.
- Can be used to optimize operating conditions and improve overall system performance.

In the dynamic model of Makariza boiler, possible changes in system inlet heat are considered as disturbances in the system. In addition, the other values required in the dynamic model of the system are derived from the actual values of the Makariza system. These assessments are impossible without controlling parameters such as the steam flow and pressure, boiler feed water high (BFWH) and inlet heat.

2 Makariza Dynamic Model of the Boiler

Makariza S.A is dedicated to the production and commercialization of products derived from sugarcane [22]. In this context, the company is operating its boiler system with the aim of generating electricity and heat from the combustion of sugarcane bagasse. In order to control the drum steam pressure and flow, improve the efficiency of the production system, identify the required specifications of the electric power generation system, check the compatibility with the boiler system and the power plant steam network, in this study the dynamic model of the system will be reviewed and calculated.

To develop the dynamic system model heat value and BFWH considered as system input parameters and the pressure of dome and steam flow of the super heater (SH) as output parameters. Figure 1 shows a model of a boiler. In a boiler that has reached to steady state, the following parameters are balanced: Mass, Energy and Momentum.

2.1 Boiler Mass Balance

The boiler under study has BFWH feed water inlet and SH saturated steam outlet. These two variables are measured in tons per hour $[\frac{T}{h}]$. Their difference is used to obtain the

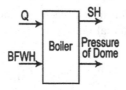

Fig. 1. Boiler model

total mass stored in the boiler (M_C) as follows [15, 23, 24].

$$M_C = \int (BFWH - SH)dt \tag{1}$$

Once the boiler mass (M_C) is obtained, the volume is obtained using the following formula:

$$\frac{d}{dt}(\rho_s.V_{st} + \rho_w.V_{wt}) = BFWH - SH \tag{2}$$

$$(\rho_w - \rho_s)V_{wt} = \int (BFWH - SH)dt \tag{3}$$

$$V_{wt} = \frac{M_C}{\rho_w - \rho_s} \tag{4}$$

Where V_{st} and V_{wt} are the volume of the steam and liquid inside the boiler in m^3. Furthermore, ρ_s and ρ_w are the specific density of steam and water in $\frac{T}{m^3}$.

The total volume (V_t) in the Makariza boiler is 20 m^3 including the dome, boiler and risers. Additionally, the maximum capacity of the dome is 9 m^3. Specific density changes of water and steam are considered as a function of pressure as follows

$$\rho_w = 0.9768 - 9.0803 \times 10^{-3}.P + 1.134 \times 10^{-4}.P^2 \tag{5}$$

$$\rho_s = 7.21 \times 10^{5} + 4.996 \times 10^{-4}.P + 3 \times 10^{-5}.P^2 \tag{6}$$

2.2 Boiler Energy Balance

In a steady state, it is known that the working pressure of the boiler is $21\frac{kg}{cm^2}$ or 300 psi, but this pressure may be disturbed at any moment due to changes in temperature (heat), pressure or water flow. For this purpose, the energy balance of the boiler is written, in which the heat Q is considered as the input of the boiler. The global energy balance is [25–27]:

$$\frac{d}{dt}\left[\rho_s.u_s.V_{st} + \rho_w.u_w.V_{wt} + m_t.C_p.t_m\right] = Q + BFWH.h_f - SH.h_s \tag{7}$$

Where u_s and u_w are specific internal energies of steam and water ($\frac{kJ}{kg}$). Additionally, h_s and h_f are specific enthalpy of steam and feed water ($\frac{kJ}{kg}$). Furthermore, m_t, C_p and

t_m are mass of metal in the boiler, heat capacity of metal, and temperature of metal, respectively. Considering the internal energy $u = h - \frac{p}{\rho}$, the energy balance can be written as:

$$\frac{d}{dt}\left[\rho_s.h_s.V_{st} + \rho_w.h_w.V_{wt} - pV_{wt} + m_t.C_p.t_m\right] = Q + BFWH.h_f - SH.h_s \quad (8)$$

By multiplying the Eq. (2) to h_w (specific enthalpy of saturated liquid water) we have:

$$h_w\left(\frac{d}{dt}(\rho_s.V_{st} + \rho_w.V_{wt})\right) = h_w(BFWH - SH) \quad (9)$$

Subtracting (9) from Eq. (8) we can reach at the following expression:

$$h_c.\frac{d}{dt}(\rho_s.V_{st}) + \rho_s.V_{st}.\frac{dh_s}{dt} + \rho_w.V_{wt}.\frac{dh_w}{dt} - V_t.\frac{dp}{dt} + m_t.C_p.\frac{dt_s}{dt} \quad (10)$$
$$= Q - BFWH(h_w - h_f) - SH.h_c$$

Where $h_c = h_s - h_w$ (specific enthalpy of condensation). Equation (10) shows the relationship of pressure with other terms. If the boiler water level is considered to be well controlled, the volume changes will be small. If these changes are ignored, the following statement can be obtained:

$$K.\frac{dp}{dt} = Q - BFWH(h_w - h_f) - SH.h_c \quad (11)$$

$$K = h_c.V_{st}.\frac{d\rho_s}{dp} + \rho_s.V_{st}.\frac{dh_s}{dp} + \rho_w.V_{wt}.\frac{dh_w}{dp} + m_t.C_p.\frac{dt_s}{dp} - V_t \quad (12)$$

The predominant physical phenomenon is the dynamics of drum steam pressure, water and metal mass in the boiler. Therefore, the appropriate approximation for K would be as follows:

$$K \approx \rho_w.V_{wt}.\frac{dh_w}{dp} + m_t.C_p.\frac{dt_s}{dp} \quad (13)$$

Considering the characteristics of the boiler under study, we can create an approximation of the following terms:

$$\rho_w.V_{wt}.\frac{dh_w}{dp} = 980\,kJ \quad (14)$$

$$m_t.C_p.\frac{dt_s}{dp} = 664\,kJ \quad (15)$$

From the previous equations, all the data are known minus the enthalpies h_s and h_w which will depend on the pressure. The operations have been carried out through tables and a spreadsheet that has given us an equation for each enthalpy as a function of pressure.

$$h_w\left(\frac{kJ}{T}\right) = 0.2769 + 48.949 \times 10^{-3} \times p - 7.054 \times 10^{-4} \times p^2 \quad (16)$$

$$h_s\left(\frac{kJ}{T}\right) = 2.6196 + 15.941 \times 10^{-3} \times p - 2.806 \times 10^{-4} \times p^2 \quad (17)$$

2.3 Enthalpy of BFWH

The enthalpy of feeding water h_f can be calculated by knowing that the approximate value of h in liquid phase states can be calculated using the following expression:

$$h(T, p) \approx u_f(T) + p.v_f(T) \tag{18}$$

$$h(T, p) \approx h_f(T) + v_f(T).\left[p - p_{sat}(T)\right] \tag{19}$$

Knowing that:

– T is the temperature of the liquid (T = 140 °C).
– P_{sat} is the saturation pressure at the given temperature (P_{sat} = 3,614 bar).
– P is the pressure of the liquid (P = 70 bar).
– h_f is the enthalpy of the liquid at 140 °C.
– v_f is the specific volume at 140 °C (v_f = 0.0010798 $\frac{m^3}{kg}$).

$$h_f \approx 589.07 \frac{kJ}{kg} \tag{20}$$

2.4 Drum Temperature

The drum temperature is obtained from its pressure. To do this, constant values of pressure and temperature are obtained on the spreadsheet and therefore its equation is calculated. The equation obtained is as follows:

$$T = 102.45 \times p^{0.22} \tag{21}$$

2.5 Control of the Parameters

Once the dynamic model of the boiler is obtained, different control systems can be considered for it. There are basically three parameters that can be controlled:

- Combustion control
- Feed water control
- Steam temperature or its pressure control

In proper operation, the boiler water level should remain in a band around the control point. Excessive reduction of the surface can empty some of the boiler pipes, which are subjected to overheating, while an excessive increase of the surface can lead to the release of water with the steam generated from the boiler. The water level in the boiler can be obtained by measuring the volume of water in it or other methods without measurement.
   In the studied boiler due to its small size, a simple element controller is used. This control method is used when sever load changes do not occur. In this mode of control, only the water level in the boiler is used to control the feed water through the PI controller. In the following, after implementing the dynamic model of the boiler, the controller is adjusted using the "sisotool" tool and its step response is shown.

3   Simulation of the Boiler Dynamics

In order to verify the developed dynamic model simulation results in Matlab/Simulink will be analyzed. Table 1 shows the values of simulation parameters. Since the system dynamics are slow, the simulation time is estimated at 72,000 s. The PI controllers control the output steam flow and drum pressure of the system by changing the value of heat entering the boiler.

To verify the performance of the controller on the dynamic model, two different type of operating conditions for the boiler are considered. In both of these conditions, it is considered a disturbance in the production of heat with a value of $\pm 1.3 \times 10^5$ J.

3.1   Steam Flow Controller

In this condition, the drum pressure is assumed to be 300 psi. This condition is studied in two cases of controlled and uncontrolled of steam flow.

Table 1. Parameters used in the simulation

Heat Q	3×10^5
Heat disturbances	$\pm 1.3 \times 10^5$
Feed water flow	200
Feed water temperature (T_w)	88 °C
m_t	20000
C_p	452
V_t	20
ρ_w	1000
h_f	$589.0 \; \frac{kJ}{kg}$
$\frac{dt_s}{dp}$	0.073
$V_{wt} \cdot \frac{dh_w}{dp}$	980

The amount of BFWH flow varies due to the changes in turbine demands. The values of BFWH are assumed as the blue line shown in Fig. 2(a).

In the first case, the boiler is considered without the steam flow controller, and the system inlet heat is assumed to be constant and equal to 3×10^5 J. In this case, by decreasing the BFWH value, due to the lack of the output steam flow controller, the boiler temperature increases and as a result, the steam flow of the system increases and vice versa. The steam flow without using controller is shown with an orange line in Fig. 2(a).

In the second case, the desired phase margin and the open-loop cut-off frequency are considered to be 50o and 4712.4 rad/sec. Therefore, the steam flow controller is considered to be P = 1.25 and I = 0.87. The bode diagram will be like Fig. 2(b). In this case, the output steam flow is controlled on 200 psi as shown with green line in Fig. 2(a). As can be seen in this figure, the proposed PI controller can adjust the steam flow at BFWH change moments. The input heat of the system before and after using the controller is shown in the Fig. 3.

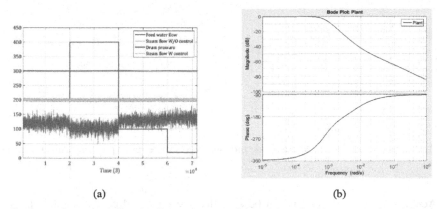

(a)   (b)

Fig. 2. Steam flow controller. (a) Compare steam flow with and without control and (b). Bode plot.

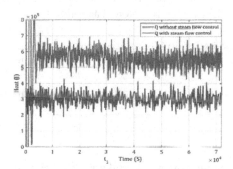

Fig. 3. Input heat of the system before and after using the steam flow controller

3.2 Drum Pressure Controller

The controller can be used also to control the pressure of the system. To verify its function, the reference value of the steam flow changes (as shown in Fig. 4(a)) by increasing the inlet temperature of the system. If the system operates without drum steam pressure control, increasing the inlet temperature of the system in constant steam flow mode will increase the drum steam pressure of the dome (Fig. 4(a)). However, as can be seen in Fig. 4(b) by setting the controller to p = 0.8 and I = 300, the bode diagram will be like Fig. 4(b) and the drum steam pressure follows the reference value, which is 300 psi.

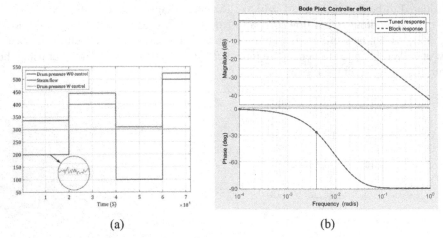

Fig. 4. Steam pressure controller. (a) Bode plot and (b) compare steam pressure with and without control.

According to the results, it can be said that the proposed control model can well achieve the goals of Makariza company. The proposed mathematical model is a suitable tool for observing the behavior of the boiler at the industrial level and can be used for any other boiler. The advantages of the presented dynamic model compared to the dynamic models presented in [10, 28] can be mentioned as follows:

- Ability to measure and access to all variables and parameters
- Observe the behavior of the system in extreme conditions, which is difficult to achieve in reality.
- Simplicity in design and no need to install sensors to measure system parameters

4   Conclusion

This paper presents a dynamic model of a Makariza boiler based on DAEs. This method was used due to the ease of measuring the temperature of the exhaust, pressure and flow of the steam through the analysis of the dome and boiler equations. In this model, the actual values of the Makariza boiler were used.

The performance of the mathematical model was revised in different conditions. In addition, the effect of changes in the input parameters such as temperature and the amount of BFWH was observed on the output parameters.

The mathematical model provided for the boiler also makes it possible to simulate the operation of the bagasse furnace and the heat transfer to the boiler in future studies. Furthermore, this model can be used to optimize the performance of the boiler in different capacities for different types of fuel and additional air inlets. Additionally, this model can be used to select correct boiler operating parameters, for example, fuel mass flow rate or exhaust steam flow to generate electricity in the turbine.

The drum pressure and steam flow of the system were controlled using two PI controllers. The implemented monitoring and control system improved the boiler performance.

The results of calculations and measurements presented in the paper show the accuracy of the mathematical model of the whole boiler. In addition, the presented results showed that by controlling the inlet temperature to the boiler, it is possible to control the pressure and steam flow of the boiler.

References

1. Luo, J., Xue, W., Shao, H.: Thermo-economic comparison of coal-fired boiler-based and groundwater-heat-pump based heating and cooling solution–A Case study on a greenhouse in Hubei China. Energy Buildings **223**, 110214 (2020)
2. Sorn, K., Deethayat, T., Asanakham, A., Vorayos, N., Kiatsiriroat, T.: Subcooling effect in steam heat source on irreversibility reduction during supplying heat to an organic Rankine cycle having a solar-assisted biomass boiler. Energy **194**, 116770 (2020)
3. Akhtari, M.R., Shayegh, I., Karimi, N.: Techno-economic assessment and optimization of a hybrid renewable earth-air heat exchanger coupled with electric boiler, hydrogen, wind and PV configurations. Renewable Energy **148**, 839–851 (2020)
4. Zhang, L., Jiang, Y., Chen, W., Zhou, S., Zhai, P.: Experimental and numerical investigation for hot water boiler with inorganic heat pipes. Int. J. Heat Mass Transf. **114**, 743–747 (2017)
5. Lee, C.-E., Yu, B.-J., Kim, D.-H., Jang, S.-H.: Analysis of the thermodynamic performance of a waste-heat-recovery boiler with additional water spray onto combustion air stream. Appl. Therm. Eng. **135**, 197–205 (2018)
6. Paul, A.R., Alam, F.: Compliance of boiler standards and industrial safety in Indian subcontinent. Int. J. Eng. Mater. Manuf. 3(4), 182–189 (2018)
7. Trojan, M.: Modeling of a steam boiler operation using the boiler nonlinear mathematical model. Energy **175**, 1194–1208 (2019)
8. Sunil, P., Barve, J., Nataraj, P.: Mathematical modeling, simulation and validation of a boiler drum: Some investigations. Energy **126**, 312–325 (2017)
9. Maakala, V., Järvinen, M., Vuorinen, V.: Optimizing the heat transfer performance of the recovery boiler superheaters using simulated annealing, surrogate modeling, and computational fluid dynamics. Energy **160**, 361–377 (2018)
10. Fan, H., Su, Z.-G., Wang, P.-H., Lee, K.Y.: A dynamic mathematical model for once-through boiler-turbine units with superheated steam temperature. Appl. Thermal Eng. **170**, 114912 (2020)
11. Shekhar, S.A., Balaji, R., Jeyanthi, R.: Study of control strategies for a non-linear benchmark boiler. In: 2018 International CET Conference on Control, Communication, and Computing (IC4), pp. 5–10. IEEE (2018)
12. Wang, C., Qiao, Y., Liu, M., Zhao, Y., Yan, J.: Enhancing peak shaving capability by optimizing reheat-steam temperature control of a double-reheat boiler. Appl. Energy **260**, 114341 (2020)
13. Hultgren, M., Ikonen, E., Kovács, J.: Integrated control and process design for improved load changes in fluidized bed boiler steam path. Chem. Eng. Sci. **199**, 164–178 (2019)
14. Chen, C., Zhou, Z., Bollas, G.M.: Dynamic modeling, simulation and optimization of a subcritical steam power plant. part i: plant model and regulatory control. Energy Convers. Manage. **145**, 324–334 (2017)
15. Zhou, Y., Wang, D.: An improved coordinated control technology for coal-fired boiler-turbine plant based on flexible steam extraction system. Appl. Thermal Eng. **125**, 1047–1060 (2017)

16. Tavoosi, J., Mohammadzadeh, A.: A new recurrent radial basis function network-based model predictive control for a power plant boiler temperature control. Int. J. Eng. **34**(3), 667–675 (2021)
17. Jiang, Y., Mao, X., Huang, H., Ma, Z.: Investigation of nonlinear control systems for steam boiler pressure and water level. In: 2020 6th International Conference on Control, Automation and Robotics (ICCAR), pp. 288–293. IEEE (2020)
18. Pástor, M., Lengvarský, P., Trebuňa, F., Čarák, P.: Prediction of failures in steam boiler using quantification of residual stresses. Eng. Fail. Anal. **118**, 4808 (2020)
19. Safdarnejad, S.M., Tuttle, J.F., Powell, K.M.: Dynamic modeling and optimization of a coal-fired utility boiler to forecast and minimize NO_x and CO emissions simultaneously. Comput. Chem. Eng. **124**, 62–79 (2019)
20. Wang, D., Zhou, Y., Li, X.: A dynamic model used for controller design for fast cut back of coal-fired boiler-turbine plant. Energy **144**, 526–534 (2018)
21. Fan, H., Zhang, Y.-F., Su, Z.-G., Wang, B.: A dynamic mathematical model of an ultra-supercritical coal fired once-through boiler-turbine unit. Appl. Energy **189**, 654–666 (2017)
22. Estrada González, M.F.: Factores jurídicos y empresariales que impiden un alza en la exportación panelera en Güepsa Santander (2019)
23. Kaikko, J., Mankonen, A., Vakkilainen, E., Sergeev, V.: Core-annulus model development and simulation of a CFB boiler furnace. Energy Procedia **120**, 572–579 (2017)
24. Zhao, S., et al.: Migration behavior of trace elements at a coal-fired power plant with different boiler loads. Energy Fuels **31**(1), 747–754 (2017)
25. Azami, S., Taheri, M., Pourali, O., Torabi, F.: Energy and exergy analyses of a mass-fired boiler for a proposed waste-to-energy power plant in Tehran. Appl. Therm. Eng. **140**, 520–530 (2018)
26. Chandrasekharan, S., Panda, R.C., Swaminathan, B.N., Panda, A.: Operational control of an integrated drum boiler of a coal fired thermal power plant. Energy **159**, 977–987 (2018)
27. Gao, M., Hong, F., Liu, J.: Investigation on energy storage and quick load change control of subcritical circulating fluidized bed boiler units. Appl. Energy **185**, 463–471 (2017)
28. Lusenko, D.S., Danilushkin, I.A.: Dynamic model of waste heat boiler based on recurrent neural network. Vestnik of Samara State Technical University. Tech. Sci. Ser. **28**(2), 59–72 (2020)

Compositional Object Synthesis in Game of Life Cellular Automata Using SAT Solver

Haruki Nishimura[✉] and Koji Hasebe[✉]

Department of Computer Science, University of Tsukuba, 1-1-1, Tennodai,
Tsukuba 305-8573, Japan
nishimura@mas.cs.tsukuba.ac.jp, hasebe@cs.tsukuba.ac.jp

Abstract. Conway's Game of Life is a two-dimensional cellular automata known for the emergence of objects (i.e., patterns with special properties) from simple transition rules. So far, various interesting objects named still-life, oscillator, and spaceship have been discovered, and many methods to systematically search for such objects have been proposed. Most existing methods for finding objects have comprehensively search all patterns. However, attempting to obtain a large object in this way may cause a state explosion. To tackle this problem and enhance scalability, in this study, we propose a method to generate objects by synthesizing some existing objects. The basic idea is to arrange multiple pieces of existing objects and compose them by complementing the appropriate patterns. The problem of finding complementary patterns is reduced to the propositional satisfiability problem and solved using SAT solver. Our method can reduce the object generation time compared to the case where a large object is generated from the beginning. We also demonstrate the usefulness of our proposed method with an implementation for automatic object generation.

Keywords: Cellular automata · Game of Life · SAT solver · Object synthesis

1 Introduction

Conway's Game of Life [17] is a two-dimensional cellular automata known for the appearance of objects (i.e., patterns with special properties). Objects in Game of Life have attracted significant attention due to their behavior that resembles living things. So far, various interesting objects named still-life, oscillator, and spaceship have been discovered, and many methods to systematically search for such objects have been proposed. Harold [15] proposed a method using De Bruijn's table [8], which was partially implemented as an object generation tool [9]. Knuth mentioned in his book [12] how to reduce the object generation problem to the satisfiability problem then showed that the problem can be solved using a SAT solver. These methods can comprehensively search for all objects

© Springer Nature Switzerland AG 2021
Y. Tan and Y. Shi (Eds.): ICSI 2021, LNCS 12690, pp. 543–552, 2021.
https://doi.org/10.1007/978-3-030-78811-7_51

that fall within a specific rectangular range. However, it is basically a brute force search for patterns that satisfy the conditions for becoming an object, thus there is a problem that the larger the size of the object, the longer it takes to search. To avoid this scalability issue, a possible approach is to reuse the existing objects to reduce the computational cost, that has not been thoroughly investigated.

In this study, we propose a method to generate a large object by synthesizing multiple existing objects in Game of Life. The basic idea is to create a new object (called a chimera) by joining a fragment of existing object with a fragment of another object. The problem of finding complementary patterns is reduced to the propositional satisfiability problem and solved using SAT solver.

We implement an automatic object synthesis tool using Python based on the method described above. Using Z3py [1] as the SAT solver, we obtained that new objects can be created by synthesizing objects belonging to the same category for each of the three categories of still-lifes, oscillators, and spaceships. In our experiments, we observed that new oscillator was obtained from two oscillators with different periods.

The remainder of the paper is organized as follows. Section 2 presents an overview of the Game of Life. Section 3 describes the method for synthesizing objects using the SAT solver. In Sect. 4, we present some examples of synthesizing objects. Section 5 describes related work. Finally, Sect. 6 concludes the paper and presents future work.

2 Overview of the Game of Life

This section briefly presents an overview of the Game of Life and defines some related concepts.

2.1 Rules

A cellular automaton is an automaton in which multiple cells spread over a grid space change their state according to the state of neighboring cells. In Game of Life, cells in the state of 0 or 1 are spread in a two-dimensional lattice space, and transition with time according to the following three rules.

Birth: If the state of a cell is 0 and the state of exactly three cells out of the touching eight cells (called neighbors) is 1, then its state transitions to 1 at the next time.

Survival: If the state of a cell is 1 and the state of two or three cells in the neighbors is 1, then its state transitions to 1 at the next time.

Death: A cell that does not meet either of the above two rules will transition to 0 at the next time.

The formal definition of the Game of Life can be given as follows.

Definition 1 (Game of Life). Let $\mathbb{N}$ be the set of natural numbers. For a two-dimensional lattice space $P \subseteq \mathbb{N} \times \mathbb{N}$, the cell at the intersection point of i-th row (from the top) and j-th column (from the left) is denoted by (i, j). Let the time $t = 0, 1, 2, \ldots (\in T = \mathbb{N})$. The state (0 or 1) of the cell (i, j) at a certain time t is defined as the function $\sigma : P \times T \to \{0, 1\}$, and its value is denoted by $x_{i,j}^t$. A Game of Life with area P is a pair of σ (called the rules of the Game of Life) and the area P_{GoL} satisfying the following formula for any $(i, j) \in P$ and $t \in T$.

$$\sigma(x_{i,j}^t) = \begin{cases} 1 \ (\textit{if } \sum_{m=i-1}^{i+1} \sum_{n=j-1}^{j+1} x_{m,n}^t = 3) \\ 1 \ (\textit{if } x_{i,j}^t = 1 \textit{ and} \sum_{m=i-1}^{i+1} \sum_{n=j-1}^{j+1} x_{m,n}^t = 2) \\ 0 \quad \textit{otherwise.} \end{cases}$$

2.2 Boolean Representation of Rules

By considering the state of a cell as a Boolean value such that 1 is "true" and 0 is "false," function σ representing the rules of Game of Life can be regarded as a Boolean function δ that takes nine states of a cell and its neighbors as arguments and returns its state at the next time step. Based on the above consideration, we define $\delta(V(x_{i,j}^t)) = \sigma(x_{i,j}^t)$, where $V(x_{i,j}^t) = (x_{i-1,j-1}^t, x_{i-1,j}^t, x_{i-,j+1}^t, x_{i,j-1}^t, x_{i,j}^t, x_{i,j+1}^t, x_{i+1,j-1}^t, x_{i+1,j}^t, x_{i+1,j+1}^t)$.

2.3 Objects

There are patterns with special properties in Game of Life. In this study, such a pattern is called an object. Objects can be classified into several categories according to their properties, and the following three are well known.

Still-life: where the shape does not change, no matter how many transitions are made;

Oscillator: that returns to the original shape after p transitions;

Spaceship: that returns to the original shape while moving in a certain direction through p transitions are known.

Formally, these categories of the objects can be defined as follows.

Definition 2 (Object). Rectangular area $R \subseteq P_{GoL}$ is called a spaceship with the direction (a, b) and period p if $x_{i,j}^{t+p} = x_{i-a,j-b}^t$ (i.e., $\sigma^p(x_{i,j}^t) = x_{i-a,j-b}^t$) holds for any $t \in T$ and $(i, j) \in R$. A spaceship with direction $(0, 0)$ is called an oscillator with a period p, and an oscillator with period 0 is called a still-life.

3 Object Synthesis

3.1 Definition

Object synthesis proposed in this paper is based on the method of creating a new object by taking out fragments of two existing objects with the same period and direction and inserting an appropriate pattern between them. Here, the synthesis of two fragments with different periods is possible by taking common multiples for both periods. For the new pattern created by synthesis to become an object with period p, it is required that the states of all cells in the fragments and its surrounding within p do not change after the transition of σ^p (i.e., applying σ p times). Our object synthesis is defined as follows. We consider the case where the two objects are arranged side by side and combined. However, the same definition can be made when two objects are arranged vertically.

Definition 3 (Object Synthesis). Let $m_{max}, m_{min}, n_{max}, n_{min}, mid_u, mid_v, a,$ $b, p \in \mathbb{N}$ be some specific values with $mid_u < mid_v$. Let $R_0, R_1 \subseteq P_{GoL}$ be rectangular areas. For R_0, R_1, we define rectangular areas $R_l(\subseteq R_0)$ and $R_r(\subseteq R_1)$ as follows:

- $R_l = \{(i,j) \mid m_{min} \leq i \leq m_{max}, n_{min} \leq j \leq mid_u\}$.
- $R_r = \{(i,j) \mid m_{min} \leq i \leq m_{max}, mid_v \leq j \leq n_{max}\}$.

A rectangular area $R_2 = \{(i,j) \mid m_{min} \leq i \leq m_{max}, n_{min} \leq j \leq n_{max}\}$ is called a synthesized object from R_l and R_r if the following formula

$$\sigma^p(x_{i,j}) \leftrightarrow x_{(i-a),(j-b)} \tag{1}$$

is true for all $(i,j) \in R_s = \{(i,j) \mid m_{min} - p \leq i \leq m_{max} + p, \ n_{min} - p \leq j \leq n_{max} + p\}$.

Here, R_l and R_r are called left-side and right-side fragments, respectively. Besides, $C_l = \{(i, mid_u) \mid m_{min} \leq i \leq m_{max}\}$ and $C_r = \{(i, mid_v) \mid m_{min} \leq i \leq m_{max}\}$ are called the joint edges of R_l and R_r, respectively. Rectangular area $R_b = \{(i,j) \mid m_{min} \leq i \leq m_{max}, mid_u + 1 \leq j \leq mid_v - 1\}(\nsubseteq R_0, R_1)$ is called a complementary pattern of R_l and R_r.

In Definition 3, the formula (1) must hold for all cells in R_2 and its surrounding within p. To obtain a synthesized object, it is sufficient to choose a pattern of R_b so that the formula (1) holds for only cells in R_b and its surrounding within p. This is because R_l and R_r are parts of the existing objects. Thus, the cells in R_l or R_r and in the area outside R_b and its surrounding within p should satisfy the formula. In this study, we call R_m the area R_b and its surrounding within p. If there exists a pattern such that all cells in R_b satisfy the following formula, then, we obtain a synthesized object.

$$\bigwedge \sigma^p(x_{i,j}) \leftrightarrow x_{i-a,j-b} \quad (x_{i,j} \in R_m). \tag{2}$$

The problem of determining whether there exists an assignment of Boolean values to each variable that satisfies a given propositional formula is known as the Boolean satisfiability problem (SAT) [5].

Although this problem is known to be NP-complete, for the formulas in conjunctive normal form (CNF), Davis-Putnam-Logemann-Loveland (DPLL) algorithm [7] and the conflict-driven clause learning [2,13] have been discovered to reduce the search space. However, various SAT solver based on these algorithms have been developed. The idea of our object synthesis is to derive the complementary pattern by converting the above formula into CNF and solve it using a SAT solver.

3.2   Complementary Pattern

The definition of the complementary pattern in the object synthesis described above is given as follows.

Definition 4 (Complementary Pattern).   Let $e_x, e_y, p, a, b \in \mathbb{N}$ be specific values. Let R_e be a spaceship (with period p and direction (a, b)) defined as follows.

$$R_e = \{(i, j) \mid e_y - p \leq i \leq e_y + m + p - 1,\ e_x - p \leq j \leq e_x + n + p - 1,\ m > 0,\ n > 0\}.$$

A rectangular area $R_c = \{(i, j) \mid e_y \leq i \leq e_y + m - 1,\ e_x \leq j \leq e_x + n - 1,\ m > 0,\ n > 0\}$ ($\subseteq R_e$) is called a complementary pattern of R_e with period p and direction (a, b) if $\sigma^p(x_{i,j}^t) = x_{i-a,j-b}^t(-p \leq a, b \leq p)$ is satisfied for all $(i, j) \in R_e$.

As seen in Definition 2, the complementary patterns for oscillator and still-life can be obtained by considering the case that $(a, b) = (0, 0)$ and the case that both $(a, b) = (0, 0)$ and $p = 0$, respectively.

Since the formula δ^p is equivalent to σ, by Definition 4, the complementary pattern R_c satisfies the following expression.

$$\bigwedge_{i - e_y - p}^{e_y + m + p - 1} \bigwedge_{j = e_x - p}^{e_x + n + p - 1} (\delta^p(V(x_{i,j}^t)) \leftrightarrow x_{i-a,j-b}^t) \tag{3}$$

where $V(x_{i,j}^t) = \{x_{i+u,j+v}^t \mid -p \leq u, v \leq p\}$. By this translation, the problem of object synthesis can be reduced to a satisfiability problem. Formula (3) is called the complementary pattern constraints. The expression $\delta^p(V(x_{i,j}^t)) \leftrightarrow x_{i-a,j-b}^t$ in Formula (3) is called the constraints on cells in the object.

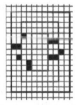

Fig. 1. Example of synthesizing an oscillator with period 2.

Example 1. As a concrete example of the complementary pattern, we present the complementary pattern for synthesizing oscillators. Let us consider the procedure for combining the two oscillators shown in Fig. 1 into one oscillator by filling the blue frame with a complementary pattern. To generate such a complementary pattern, first, cell constraints are created for the 8 × 4 cells shown in the green frame. The complementary pattern constraints are expressed by connecting them with logical product. Since the state of one cell around it is required to create a constraint for one cell, 10 × 6 cells in the red frame are referenced. The state of each cell inside the blue frame is represented as a variable, while that of the outside is represented as a constant. If the cell at the upper left corner of the red area is called $(1, 1)$, We express the constraint of the completion pattern as follows.

$$\bigwedge_{i=2}^{9} \bigwedge_{j=2}^{5} (\delta(V(x_{i,j}^t)) \leftrightarrow x_{i,j}^t). \tag{4}$$

Here, the cell constraints for $(i, j) = (2, 5)$ and $(i, j) = (6, 4)$ are, respectively, represented by the following formulas.

- $0 \leftrightarrow \delta(0, 0, 0, 0, 0, 0, x_{3,4}^t, 1, 1)$.
- $x_{6,4}^t \leftrightarrow \delta(x_{5,3}^t, x_{5,4}^t, 1, x_{6,3}^t, x_{6,4}^t, 0, x_{7,3}^t, x_{7,4}^t, 1)$.

A new object can be obtained by joining the fragments taken out from the two existing objects with a complementary pattern. We call it a chimera. The following proposition guarantees that a chimera obtained from the two objects with period p and direction (a, b) is actually an object with the same period and direction.

Proposition 1. *Let R ($\subseteq P_{GoL}$) be a rectangular area consisting of three rectangular areas R_0, R_1, R_2. Suppose $\sigma^p(x_{i,j}^t) = x_{i-a,j-b}^t (-p \leq a, b \leq p)$ holds for all $(i, j) \in R_0 \cup R_1$, and $\sigma^p(x_{i,j}^t) = x_{i-a,j-b}^t$ holds for all $(i, j) \in R_2$, then R is an object with period p and direction (a, b).*

The two fragments and their complementary patterns correspond to R_0, R_1, and R_2 in the above theorem, respectively. From Definition 4, the complementary pattern does not change the boundary pattern between R_0 and R_2 as well as between R_1 and R_2. Thus, $\sigma^p(x_{i,j}^t) = x_{i-a,j-b}^t(-p \leq a, b \leq p)$ holds for all $(i, j) \in R_0 \cup R_1$. From the same definition, the expression $\sigma^p(x_{i,j}^t) = x_{i-a,j-b}^t(-p \leq a, b \leq p)$ holds for all (i, j) in the complementary pattern. Thus, $\sigma^p(x_{i,j}^t) = x_{i-a,j-b}^t(-p \leq a, b \leq p)$ holds for all $(i, j) \in R_2$. From the above theorem, the chimera is an object with period p and direction (a, b).

In closing this section, we would like to note a limitation on our object synthesis. As mentioned in Definition 3, when synthesizing an object, a fragment taken from an existing object must always share at least three sides with the original object. For example, the fragment of the object shown in the blue frame in Fig. 2 must be cut out so as to share the top, bottom, and left sides with the original object, as in the patterns surrounded by the red frames in (a) and (b).

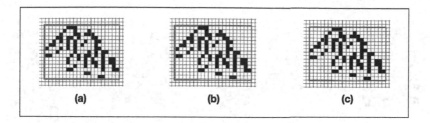

Fig. 2. Example of correct/incorrect fragments of an object.

Therefore, the pattern shown by the red frame in (c) is not a fragment of the original object. The fragment may be the same as the original object. However, in that case, it is allowed that the complementary pattern becomes a pattern composed of all 0 s. In order to avoid such trivial composition, it is necessary to add some new constraints to the complementary pattern.

4 Implementation

We have implemented a simple tool for automating the object synthesis method described above. Here, the Python library Z3py is used to derive the propositional function δp and cell constraints. Besides, this library is used to derive variable assignments, making it possible to satisfy the constraints of the completion pattern.

Below are some examples of objects produced by synthesis using our implementation. Figure 3 shows an example of synthesizing an oscillator. Here, the objects (a), (b), and (c) are oscillator with period of 2, fixed object, and oscillator with a period of 3, respectively. First, we picked a fragment from these three objects (as shown by the red frames in (d) of the figure). Next we synthesized fragments of (a) and (b) by inserting a complementary pattern (as shown by the left blue frame in (d)) and further synthesized the pattern obtained thereby and the fragment of (c) by inserting a complementary pattern (as shown by the right blue frame). As a result, we obtained an oscillator with period of 6, as shown in (d) was finally obtained. The times required for these syntheses were about 60 and 290 s, respectively.

Figure 4 shows an example of synthesizing a spaceship. In this figure, both (a) and (b) are spaceships with period 2 and direction $(0, -1)$, respectively. Similar to the previous example, we first picked a fragment (as shown by the red frames in (c) in the figure) from each of the objects, respectively. By inserting a complementary pattern (as shown by the blue frame) between these patterns, we obtained a spaceship with period 2 and direction $(0, -1)$.

As shown in the above examples, we demonstrated that our method correctly performed the synthesis. In the oscillator's synthesis, it was shown that oscillators with different periods can be synthesized. It is suggested that an oscillator with a larger period can be generated by repeating this operation. From the previous examples, the size of the complementary pattern for synthesis is generally

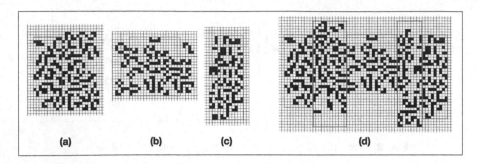

Fig. 3. Example of synthesizing an oscillator.

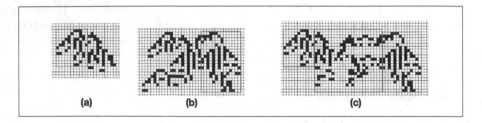

Fig. 4. Example of synthesizing a spaceship.

as large as or larger than each fragment to be synthesized, so it is necessary to keep sufficient space between the fragments for synthesis. If the search for the complementary pattern fails due to insufficient space, it is necessary to redo the search at a wider space. Thus, finding the minimum size of the complementary pattern required for synthesis is useful for shortening the search time. However, this is unclear and further investigation is required. To reduce the time to convert the constraint of each cell to CNF, efficient conversion technique, such as Tseitin conversion would be useful.

5   Related Work

De Bruijn diagram [8] has been widely used as a useful tool in the study of cellular automata. McIntosh [14,16] proposed a method to enumerate still-life by arranging predetermined small patterns. The basic idea was that the patterns of 3×3 square appearing in still-life were classified into 284 types of patterns, then still-lifes of any size and patterns were generated by connecting 3×3 square patterns in the catalog one-by-one according to some connection rules. First, $3 \times n$ patterns were generated by 3×3 square belonging to the catalog is generated as a horizontally long pattern of $3 \times n$ based on the connection rules of two 3×2 rectangles. Next, $m \times n$ pattern is completed by stacking the previously obtained $3 \times n$ patterns while shifting them vertically one cell at a time.

A tool for creating an object using the De Bruijn diagram is Eppstein's gfind [9,10]. This tool used Dr Bruijn diagram to search for a spaceship, which is a kind of gridder.

Bounded model checking [3] is a model checking [4], which limits the transition sequences of a state transition system to a finite length for verification. In [12], Knuth suggested how to apply the logical formula used in the bounded model checking to convert the problem of object generation in Game of Life into satisfiability problem. Let $T(X, X')$ be a logical formula that returns true if the transition from pattern X is reachable to X'. Here, if X_n is reachable from X_0 by n-times state transitions, the formula $\bigwedge_{k=1}^{n} T(X_{k-1}, X_k)$ is true. If $X_0 = X_n$, then this formula returns true when X_0 is an oscillator with period n. Thus, if this equation is converted to CNF and solved using the SAT solver, the concrete pattern of the oscillator can be obtained. By adding some suitable conditions and changing values of the parameters, we can derive still-life, spaceship, and garden of Eden. In [12], it was introduced how to convert the expression $T(X, X')$ to a CNF, which facilitates the solution using the SAT solver. There have been some implementations of this method, such as Cunningham's Logic Life Search [6] and Goucher's ikpx [11].

6 Conclusions and Future Work

In this study, we proposed a method for synthesizing objects in Conway's Game of Life. The basic idea is to join two existing objects by inserting an appropriate pattern to obtain a new object. The constraints that the complementary pattern between the fragments must satisfy is described as a logical formula in which the state of each cell constituting the pattern is a variable. This formula becomes true if and only if a pattern that satisfies the constraint is input. Thus, the object synthesis problem can be reduced to the satisfiability problem, which can be solved using the SAT solver. We implemented the proposed method using Python with Z3py as an SAT solver and demonstrated the synthesis of various objects, such as still-life, oscillator, and spaceship.

Future research will consider finding regularity in shape from a large number of automatically generated objects by synthesis. Objects with the same period and direction of movement partially share some specific patterns while others are composed of different patterns. From this observation, it is expected that various objects can be generated by applying a (set of) certain non-deterministic manipulation rules to a specific pattern. We especially want to clarify such rules by focusing on the relationship with stochastic one-dimensional cellular automata.

References

1. z3py-tutorial. https://github.com/ericpony/z3py-tutorial
2. Bayardo Jr., R.J., Schrag, R.C.: Using CSP look-back techniques to solve real world SAT instances. In: 14th AAAI, pp. 203–208 (1997)

3. Biere, A., Cimatti, A., Clarke, E.M., Strichman, O., Zhu, Y.: Bounded model checking. Adv. Comput. **58**, 118–149 (2003)
4. Clarke, E., Grumberg, O., Kroening, D., Peled, D., Veith, H.: Model Checking, 2nd edn. MIT press, Cambridge (2018)
5. Cook, S.A.: The complexity of theorem-proving procedures. In: 3rd Annual ACM Symposium on Theory of Computing, pp. 151–158 (1971)
6. Cunningham, O.: Logic-life-search. https://github.com/OscarCunningham/logic-life-search
7. Davis, M., Logemann, G., Loveland, D.W.: A machine program for theorem-proving. Commun. ACM **5**(4), 394–397 (1962)
8. De. Bruijn, N.G.: A combinatorial problem. Proc. Koninklijke Nederlandse Academie van Wetenschappen **49**, 758–764 (1946)
9. Eppstein, D.: gfind 4.9 - search for gliders in semitotalistic cellular automata. https://www.ics.uci.edu/eppstein/ca/gfind.c
10. Eppstein, D.: Searching for spaceships. arXiv prepring. cs/0004003 (2000)
11. Goucher, A.P.: Metasat. https://gitlab.com/apgoucher/metasat
12. Knuth, D.E.: The Art of Computer Programming, Volume 4, Fascicle 6: Satisfiability. Addison-Wesley Professional, Boston (2015)
13. Marques-Silva, J.P., Sakallah, K.A.: GRASP: a new search algorithm for satisfiability. In: Digest of ICCAD, pp. 220–227 (1996)
14. McIntosh, H.V.: Linear cellular automata via de Bruijn diagram. Universidad Autónoma de Puebla (1991)
15. McIntosh, H.V.: Life's still lifes. In: Adamatzky, A. (ed.) Game of Life Cellular Automata, pp. 35–50. Springer, London (2010). https://doi.org/10.1007/978-1-84996-217-9_4
16. McIntosh, H.V.: Commentaries on the global dynamics of cellular automata. http://delta.cs.cinvestav.mx/mcintosh/oldweb/wandl/wandl.html
17. Schiff, J.L.: Cellular Automata: A Discrete View of the World. John Wiley and Sons, Inc., Hoboken (2011)

Automatic Detection of Type III Solar Radio Burst

Shicai Liu[1], Guowu Yuan[1,2(✉)], Chengming Tan[2], Hao Zhou[1], and Ruru Cheng[1]

[1] School of Information Science and Engineering, Yunnan University, Kunming 650091, China
yuanguowu@sina.com
[2] CAS Key Laboratory of Solar Activity, National Astronomical Observatories of Chinese Academy of Sciences, Beijing 100012, China

Abstract. With the accuracy improvement of radio telescopes, massive amounts of solar radio spectrum data are received every day. It is inefficient to detect solar radio burst by astronomers, and it is also difficult to meet the real-time requirements of space weather, aerospace and navigation systems and etc. In order to reduce the workload of astronomers and improve the detection accuracy and efficiency, we propose an algorithm for automatic real-time detection of solar radio bursts based on density clustering in this paper. The algorithm firstly uses channel normalization to remove the interference of horizontal stripe in the image. Then, the normal distribution model is used for binarization, and then the DBSCAN clustering algorithm is used to cluster detection of the binarized solar radio burst area. Finally, the Canny operator is used to detect the edge and the time parameter of burst is extracted. Experiments show that the proposed method improves the detection efficiency and accuracy compared with some traditional clustering detection algorithms.

Keywords: Solar radio burst · Channel normalization · DBSACN clustering · Edge detection

1 Introduction

The solar radio burst refers to a sudden radio radiation process that occurs on the sun, and it plays a vital role in understanding the solar atmosphere, solar wind, especially coronal mass ejection (CME). Flares and CME are two of the most intense phenomena of solar activity [1]. According to the radio radiation shape and the frequency drift speed in the dynamic spectrum, astronomers divide solar radio burst events into five main types: I, II, III, IV and V [2].

Type III solar radio bursts contain the most important information about solar activities, so, it has received extensive attention from scientists. Type III radio bursts were initially recognized and named by Wild and McCready [3]. Type III radio bursts were mainly formed by weakly generated relativistic electron beams that move outwards from the solar active area along the open magnetic field lines. The observation feature is that the radiation intensity drifts rapidly shift from high frequency to low frequency, and the

© Springer Nature Switzerland AG 2021
Y. Tan and Y. Shi (Eds.): ICSI 2021, LNCS 12690, pp. 553–562, 2021.
https://doi.org/10.1007/978-3-030-78811-7_52

drift rate is about 100 MHz/s [4] at the range of metric wavelength. The plasma emission mechanism [5] is the most commonly accepted model for triggering type III bursts. The plasma emission theory is a two-step process: firstly, the electron beam excites the Langmuir wave [6]; then part of the energy of the Langmuir wave is converted into the fundamental or second harmonic of the local plasma frequency, also or electromagnetic wave of both. This paper mainly discusses the automatic detection of type III solar radio bursts.

Many scholars at home and abroad have started the detection of solar radio bursts. Xu Long from the National Astronomical Observatory of the Chinese Academy of Sciences proposed algorithm that wavelet transform can remove the noise of solar radio spectrum images [7], they also established a multi-modal deep learning network to detect and identify solar radio bursts. A detection algorithm of solar radio burst based on Hough transform was proposed and improved by Zuo et al. [8]. Hough transform is used for linear feature extraction. It has well detection performance for only type III radio bursts, but for other types of radio bursts, the burst detection effect is ineffective. Then, Dayal Singh·K and Sasikumar Raja et al. proposed a statistical automatic detection algorithm that based on the statistics[9], and the method can detect whether there is a radio burst in the solar radio spectrum image successfully. Cui Zexiao of Yunnan University proposed a clustering algorithm based on outlier detection and K-means [10], but the k-value and the number of categories need to be given in advance, and the clustering effect is poor for the sample set that is not a convex set.

For the observation of the radio spectrum of the sun, manual processing recognition need to spend a lot of cost, timeliness is hard to guarantee, the existing solar radio burst in the traditional methods of detection effect is not obvious. in this paper, the density of DBSCAN clustering algorithm was applied to III type solar radio bursts of detection and identification. Experimental results show that this method can detect and identify III type solar radio bursts well.

2   Data Preprocessing

In this paper, the dynamic spectrogram data come from the National Astronomical Observatory of the Yunnan Observatory. the file formats of solar radio data storage include NUS, NPS, DAT, etc. [11].These file format cannot be intuitive for radio spectrum images displayed, data in these formats need to be transformed into visual images. each dynamic spectral image is composed of 3500*6400 pixel, and each pixel represents the frequency of solar radio bursts corresponding to a particular time point. as shown in Fig. 1(a) the decimeter wave solar radio telescope observations of solar radio spectrum data the outbreak of the edge data fragments, Fig. 1(b) shows the spectrum of the image after conversion data format, transverse represents time, longitudinal represent frequency, spectrum of each pixel in the image is some electricity burst of radiation intensity at a time, each frequency point is independent of each other.At the edge of the solar radio burst, the radio intensity increases gradually with the passage of time.

In the process of forming solar radio spectrum images, the spectral images usually show the phenomenon of horizontal stripes due to the interference of electromagnetic waves in the ground space, the interference of receiving instruments and the channel

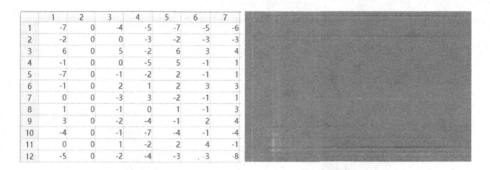

	1	2	3	4	5	6	7
1	-7	0	-4	-5	-7	-5	-6
2	-2	0	0	-3	-2	-3	-3
3	6	0	5	-2	6	3	4
4	-1	0	0	-5	5	-1	1
5	-7	0	-1	-2	2	-1	1
6	-1	0	2	1	2	3	3
7	0	0	-3	3	-2	-1	1
8	1	0	-1	0	1	-1	3
9	3	0	-2	-4	-1	2	4
10	-4	0	-1	-7	-4	-1	-4
11	0	0	1	-2	2	4	-1
12	-5	0	-2	-4	-3	-3	-8

(a) spectral data slice	(b) visual spectral image

Fig. 1. Solar Radio Spectrum Image

effect.This phenomenon has a great interference effect on the determination of image burst area and time detection.In order to eliminate the influence of interference noise and bad channels on detection as much as possible, the channel normalization method [12] is adopted in this paper to deal with the channel effect of horizontal stripes, and the method of subtracting the mean value of each frequency in the spectral image is used. The specific calculation process is shown in Eq. (1).

$$g(x, y) = f(x, y) - M_x \qquad (1)$$

Where $f(x, y)$, $g(x, y)$ respectively represent the original image and the normalized image, and M_x is the mean value of each frequency channel. As shown in Fig. 2(a) is the original image with a large number of horizontal stripes in the background area, and picture Fig. 2(b) is the effect picture after removing the horizontal stripes. By comparison, the channel normalization can perform quite well on removing background noise.

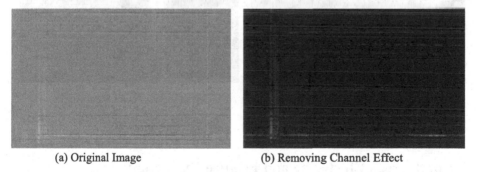

(a) Original Image	(b) Removing Channel Effect

Fig. 2. Channel normalization

3 Detecting Burst Areas by Modeling Quiet Solar Radio

Compared with the calm time of the sun, the frequency intensity of solar radio bursts is accompanied by sharp changes. Solar radio frequency values of two channels are randomly selected for visualization, as shown in Fig. 3, and the visible image presents normal distribution [13]. Therefore, the 3sigma rule can be used to extract 99.74% of the radio frequency data falling in this region.

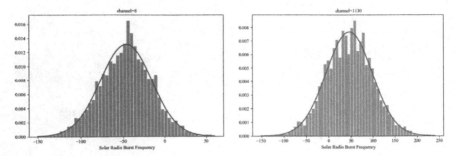

Fig. 3. Corresponding relation between channel frequency histogram and normal distribution

We set the value of pixels within the range of plus or minus three sigma to 1, and those outside the three sigma range to 0. The normalized images of channels are adopted in the normal distribution model, and the effect of binarization of the burst region is shown in Fig. 4, the outline of the burst region in the image becomes clear gradually. However, There are still some white noise points in the non-concentrated areas in the image, as well as discontinuity between the burst region. These interferences will also have an impact on the determination of the burst region.

Fig. 4. The Effect of Separating Burst Areas by Modeling Quiet Solar Radio

4   Burst Areas Clustering and Feature Extraction

4.1   DBSCAN Clustering

After binarization processing, there are still some white noise points scattered in every corner of the image, and the burst region is discontinuous, and some black pixel points will be separated. These problems will affect the determination of the burst region, so it is impossible to accurately locate the bursting time. In order to solve these problems, this paper uses DBSCAN algorithm [14] to cluster the burst region. Since this algorithm can be applied to any shape of cluster and has a good clustering effect for low-dimensional data, it is widely used in scientific research.

In this paper, coordinates of all the white pixel points in the image are extracted as the input of DBSCAN algorithm data. Euclidean distance [15] is used as the clustering metric function between two pixels, as shown in Eq. (2).

$$D_E(x, y) = \sqrt{\sum_{i=1}^{n} (xi - yi)^2} \tag{2}$$

Where (x, y) represents the horizontal and vertical coordinates of a pixel, n represents the dimension of the pixel, and $D_E(x, y)$ represents the set E of Euclidean distances between all pixels. Take the distance between two points from E, and then judge the distance and ε make a comparison. If the distance between a pixel and the core point is less than or equal to ε, The change point is incorporated into the reachable point of the core object.

Various parameters (ε, $minPts$) were used to carry out comparative experiments, The purpose is to find a set of optimal parameters that can effectively separate the noise and solar radio burst region in the image. The selection of various parameters cannot be judged by people's subjective consciousness, and objective evaluation indexes should be used to determine the influence of different experimental parameters on the experimental results. Silhouette Coefficient [16], CH Score (Calinski Harabasz Score) [17] and DBI [18] are commonly used to quality of the cluster result. We use Silhouette Coefficient as a metric. It is defined as Eq. (3).

$$s(i) = \frac{b(i) - a(i)}{\max\{a(i), b(i)\}} \tag{3}$$

Where $s(i)$ is the contour coefficient of the clustering result, and $a(i)$ is the dissimilarity in the cluster between sample i and other samples in the same cluster. The smaller the value of $a(i)$, sample i more close to the cluster, $b(i)$ represents dissimilarity between sample i and different clusters. $s(i)$ is more close to 1, it is reasonable that the sample i is clustered in this class. On the contrary, $s(i)$ is more close to -1, it means that the clustering of sample i is unreasonable, and it should be distributed in another cluster. In summary, the range of the contour coefficient of the clustering effect should be between -1 and 1. The larger the value, the closer the samples of the same kind are, the better the clustering effect, and vice versa.

The experiment uses two images for comparison. The left is the image of the radio burst at 10:05:24 on November 13, 2012, and the right is the image of the solar radio burst at 11:33:46 on July 3, 2012. Three different experimental parameters are used, i.e. a ($\varepsilon = 6$, $minPts = 100$), b($\varepsilon = 8$, $minPts = 100$) and c($\varepsilon = 11$, $minPts = 70$), The noise and the burst points marked with different colors, and they were labeled in the form of scatter plots, as shown in Fig. 5. When using a parameters, the clustering results of the two images could not accurately separate the burst region and the noise points, in which all the pixels belonging to the burst region should be attributed to one color. However, after clustering, this region was divided into 9 clusters (red, yellow, dark blue, etc.). On the contrary, the pixels that should be noisy are divided into the same cluster as the burst region, and the contour coefficients are –0.193 and –0.092, respectively. Parameter a cannot separate the burst region and the noise well, and the clustering results show that the ε is too small, which leads to too many clusters. Therefore, while controlling the *minpts* of the parameter b remain unchanged, expand the value of ε, after clustering, there will still be cases where noise and burst region are confused with each other, but the number of clusters in the burst region is significantly higher than 9 reduced to 5, more continuous and the same kind of sample points, the contour coefficient rises to 0.149 and 0.131. Although the contour coefficient of parameter *b* is improved, it still fails to reach the ideal effect. Therefore, parameter c continues to expand ε to 11, and the number of *minpts* is rduced to 70. As can be seen from Fig. 5(c), the contour coefficient has a qualitative improvement, jumping to 0.568 and 0.547, and this set of parameters will completely separate the burst region and noise of the two images.

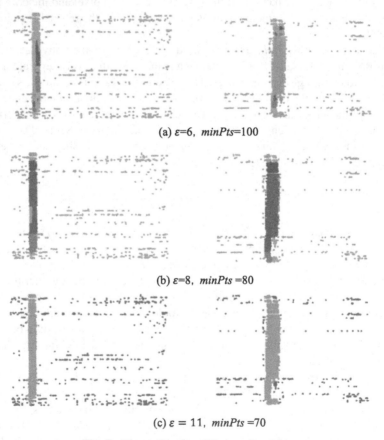

(a) ε=6, *minPts*=100

(b) ε=8, *minPts* =80

(c) ε = 11, *minPts* =70

Fig. 5. Cluster Results of Various Parameters

The image contour coefficients obtained after the two images are clustered with different (ε, *minPts*) are expressed as shown in Table 1. It can be seen from the table that when the ε equals to 11 and *minPts* equals to 70, the contour coefficient values of the two images reach the maximum, meaning that the effect of image clustering can reach the best when using this set of parameters.

Table 1. Contour Coefficients Corresponding to Various Parameters

Parameters	Observation Time	
	2012-11-13 10:05:24	2012-07-03 11:33:46
$\varepsilon = 6, minPts = 100$	− 0.193	− 0.092
$\varepsilon = 8, minPts = 80$	0.149	0.131
$\varepsilon = 11, minPts = 70$	**0.568**	**0.547**

Therefore, parameter c was selected as the target parameter of DBSCAN clustering in this paper to find out the coordinate positions corresponding to all the noise-removed white pixel in the image, and set the pixel value from 255 to 0, so as to achieve the effect of removing noise and improving the accuracy of the solar radio burst region. The result of noise elimination is shown in Fig. 6.

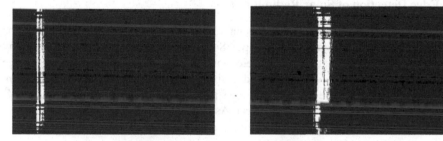

Fig. 6. The Denoise effect of DBSCAN clustering

4.2 Burst Region Continuous

Aiming at the problem of the discontinuity of the burst region in the image after clustering, this paper uses the closed operation [19] to concatenate the discontinuous regin, The closing operation can remove some small holes in the foreground, and a good effect can be obtained in filling some gaps in the burst image. The experiment uses five groups of different kernel sizes to evaluate the effect of continuous filling in the solar radio burst area, which are 3*3, 7*7, 11*11 and 13*13. The filling effect is shown in Fig. 7. When the kernel size of 13*13 is used, the filling effect of the discontinuous area in the burst is the best, and no other redundant edge linear areas are derived, so this paper chooses the 13*13 kernel to close the image.

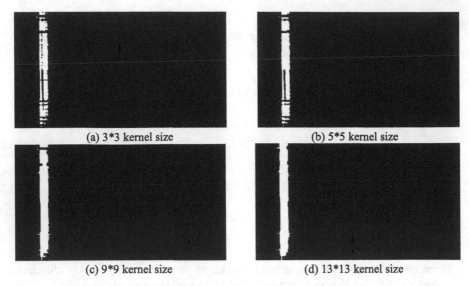

(a) 3*3 kernel size   (b) 5*5 kernel size

(c) 9*9 kernel size   (d) 13*13 kernel size

Fig. 7. Continuous effect of the burst region

5 Experimental Results and Comparison

5.1 Edge Detection of Solar Radio Burst

In order to gain the start and end time of solar radio bursts, edge information needs to be extracted from the burst region. In this paper, Sobel operator, Scharr operator and Canny operator are used to evaluate the edge detection effect of solar radio bursts. The image processed by the closed operation is used as the input of the edge detection algorithm, and the peak signal to noise ratio(PSNR)and structural similarity index measure(SSIM) are used to judge the edge extraction effect. As shown in Fig. 8.

Fig. 8. Canny Edge Detection

5.2 Extraction of Solar Radio Burst Time

The file name of the solar radio observation image is 20121113100524578.data.txt, where 20121113100524578 means that the start time of the image observation is 578 ms at 10:05:24, November 13, 2012. The total observations per image will take 512 s, and then according to the width of the image is 3500. Therefore, the time taken by the width of each pixel in the image can be calculated, and the start time and end time of the solar radio burst can be calculated, as shown in Fig. 9.

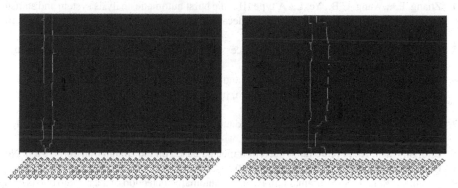

Fig. 9. Start and end time of solar radio burst

6 Conclusion

In this paper, we use DBSCAN clustering algorithm and Canny operator to detect type III solar radio bursts, and the solar radio bursts can be detected successfully and automatically in the spectrum image in real time, which improves the detection efficiency and reduces the workload of scientists. The experimental results show that the method proposed in the paper can detect type III solar radio bursts, and accurately locate the start and end time of eruptions.

Acknowledgement. This work is supported by the Natural Science Foundation of China (Grant No.11790301, 11790305, 11663007, 62061049), the Application and Foundation Project of Yunnan Province (Grant No.202001BB050032, 2018FB100), the Commission for Collaborating Research Program of CAS Key Laboratory of Solar Activity, National Astronomical Observatories (Grant No.KLSA202115) and the Youth Top Talents Project of Yunnan Provincial "Ten Thousands Plan".

References

1. Cane, H.V., Erickson, W.C., Prestage, N.P.: Solar flares, type III radio bursts, coronal mass ejections, and energetic particles. J. Geophys. Res. Space Phys. **107**(A10), SSH-14 (2002)
2. Wild, J.P.: The radioheliograph and the radio astronomy programme of the Culgoora observatory. Publ. Astron. Soc. Austral. **1**(2), 38–39 (1967)
3. Wild, J.P., McCready, L.L.: Observatioas of the spectrum of high-intensity solar radiation at metre wavelengths. i. the apparatus and spectral types of solar burst observed. Aust. J. Chem. **3**(3), 387 (1950)
4. Zhang, P.J., Wang, C.B., Ye, L.: A type III radio burst automatic analysis system and statistic results for a half solar cycle with Nançay decameter array data. Astron. Astrophys. **618**, A165 (2018)
5. Li, C.Y.: Research on the physical process of solar radio bursts (Doctoral dissertation, Shandong university) (2020)
6. Rizzato F.B., et al.: Langmuir turbulence and solar radio bursts. In: Chian A.CL. et al. (eds.) Advances in Space Environment Research, pp 507-514 . Springer, Dordrecht (2003) https://doi.org/10.1007/978-94-007-1069-6_51
7. Xu, L.: Application of wavelet analysis in data processing of solar radio observation (Master's thesis, Xidian university) (2004)
8. Zuo, S., Chen, X.: A radio burst detection method based on the Hough transform. Mon. Not. R. Astron. Soc. **494**(2), 1994–2003 (2020)
9. Singh, D., Raja, K.S., Subramanian, P., et al.: Automated detection of solar radio bursts using a statistical method. Sol. Phys. **294**(8), 1–14 (2019)
10. Cui, Z.: Research on automatic detection method of solar radio spectrum image based on outlier detection and K-means clustering (Master's thesis, Yunnan university) (2019)
11. Chen, S.: Research on classification algorithm of solar radio spectrogram based on convolutional neural network (Master's thesis, Shenzhen university) (2018)
12. Yan, Y., Tan C., Xu L., Ji Hui, R.: Non-linear calibration method and data processing of solar radio bursts. Chin. Sci. (Series A) **31**,73–79 (2001)
13. Mohammad Ahsanullah, B.M., Kibria, G., Shakil, M.: Normal distribution. In: Mohammad Ahsanullah, B.M., Kibria, G., Shakil, M. (eds.) Normal and Student´s t Distributions and Their Applications, pp. 7–50. Atlantis Press, Paris (2014)
14. Ester, M., Kriegel, H.P., Sander, J., Xu, X.: Density-based spatial clustering of applications with noise. In: International Conference Knowledge Discovery and Data Mining, vol. 240, p. 6 (1996)
15. Danielsson, P.E.: Euclidean distance mapping. Comput. Graph. Image Process. **14**(3), 227–248 (1980)
16. Aranganayagi, S., Thangavel, K.: Clustering categorical data using silhouette coefficient as a relocating measure. In: International Conference on Computational Intelligence and Multimedia Applications (ICCIMA 2007), vol. 2, pp. 13–17. IEEE (2007)
17. Van Craenendonck, T., Blockeel, H.: Using internal validity measures to compare clustering algorithms. Benelearn 2015 Poster Presentations (online) 1–8 (2015)
18. Aarts, E., Wichert, R.: Ambient intelligence. In: Bullinger HJ. (eds) Technology Guide, pp 244-249. Springer, Berlin (2009) https://doi.org/10.1007/978-3-540-88546-7_47
19. Castleman, K.R.: Digital image processing (1993)

The Impact of Wechat Red Packet Feature at Achieving Users Satisfaction and Loyalty: Wechat Users in China

Kamal Abubker Abrahim Sleiman[1,2], Lan Juanli[2(✉)], Xiangyu Cai[3], Wang Yubo[2],
Lei Hongzhen[2], and Ru Liu[2]

[1] School of Economics and Management, Yan'an University, Yan'an 716000, China
[2] International School of Business, Shaanxi Normal University, Xi'an 710062, China
[3] Xian University of Finance and Economics, Xi'an 710062, Changan District, China

Abstract. WeChat has become an essential social media platform in China. This research investigates the importance of the WeChat Red packet as a motivator at achieving user's satisfaction and loyalty in China. To investigate its impact, and factors that attribute to its popularity and acceptability, we extended technology acceptance model (TAM), in addition to the main model factors, "perceived usefulness and perceived ease of use" our proposed model include, perceived trust, perceived security, and perceived entertainment. The questionnaire was designed, and SPSS was used for the analysis. The research results provide insight into how WeChat Red Packet can motivate users and build their satisfaction to improve their loyalty which in turn increases WeChat users'. These results have future implication and practice for research.

Keywords: Social media · Red packet · Satisfaction · Technology acceptance model

1 Introduction

Tencent corporation launched Wechat on January 21, 2011, as an instant messaging mobile application software that supports group chatting by transmitting text, speech, video, and images over the network. Wechat has evolved into a promotion tool that makes shopping more convenient as a result of its open platform [1]. WeChat has similar features to WhatsApp to generate both text and voice messages. WeChat provides users an innovative way to communicate and interact with friends through text messaging, hold-to-talk voice messaging, one-to-many messaging, photo, video sharing, location sharing, and contact information exchange [2]. With the advancement of internet and smartphone mobile WeChat become popular with 355 million active users at the end of 2013 and availability in more than 200 countries and 18 different languages [2]. Customer activities within WeChat are range from social events, like entertaining with friends and exchanging information regarding products and services. WeChat has become the most popular and influential social media platform for computer-mediated communication [3]. Psychological motivations are the key in determining social media users' attitudes

© Springer Nature Switzerland AG 2021
Y. Tan and Y. Shi (Eds.): ICSI 2021, LNCS 12690, pp. 563–575, 2021.
https://doi.org/10.1007/978-3-030-78811-7_53

[4]. Inspecting users' psychological needs help marketers to understand what motivates people to use social media.

According to a review of the literature, the majority of recent research concentrated on exploring users' motivations for using Facebook and other social networking sites [4, 5]. There have been very few researches into what motivates Chinese people to use WeChat.

It is important to understand the loyalty drivers in order to understand consumer loyalty. Customer satisfaction is related to customer loyalty, according to marketing literature [6], trust according to [7, 8] and commitment [9], in this study customer satisfaction considered for predicting customer loyalty.

The primary goal of this study is to first investigate the effect of WeChat Red packets as a motivator in achieving user satisfaction and loyalty. Second, how is Red Packet extending Chinese mobile payments as a mobile payment system?

This research is one of the few that empirically investigates the motivation of WeChat users. We used TAM for this study because, to the best of our knowledge, no previous research has looked into Wechat Red Packet adoption and satisfaction.

2   Literature Review

Firms started to use social media as a tool for getting contact with the customer, a channel for direct sales and theatre for social commerce [10]. Social media makes local events global, and this global event will be translated into the local culture. Use of these inexpensive tools by small firms helps achieve their goals and objectives effectively. Instead of all social media success, firms still worry about customer power in autonomous and the abuse of authors right [11]. The big problem with social media is the lack of control over content; anyone can proclaim themselves a specialist in a specific area or subject and influence who cannot differentiate [12]. This doubt makes firms to use social media in an experimental [13]. However, social media is a reality in this age of technology, and it is necessary to firms for achieving goal component.

The most important characteristic of Wechat is the Red Packet "hongbao"; it has long been China's favorite feature. It is based on the Chinese practice of giving gifts to relatives and friends on various occasions such as "birthdays, graduations, and weddings." WeChat was aware of this culture and incorporated this custom into the App [14]. This feature was released in 2014 during the Chinese New Year festival. It can distribute the amount of money randomly or equally. The feature was launched through promotion via China central TV when the audience asked to shake their phones during the broadcast for a chance to win the cash prize (hongbao). One month later, after this feature was launched, WeChat users increased from 30 million to 100 million users, and 20 million (hongbao) were distributed during the Chinese new year holiday. In 2016, during the holiday time, 3.2 billion Red Packets were sent at Chinese New Year midnight [15]. Customer who has provided Bank information may use the platform to pay bills, order goods and services, transfer money to others and pay in the stores [16].

To understand customer loyalty toward using WeChat Red Packet which is very important because it's related to financial transaction, it is important to understand the loyalty drivers. Marketing literature has proved that customer loyalty is linked to customer satisfaction [6], trust according to [7, 8] and commitment [9], in this study customer satisfaction considered for predicting customer loyalty.

3 Theoretical Model and Research Hypotheses

The case study of this research Red Packet, the researcher, is trying to investigate the factors that drive to customer satisfaction and loyalty toward using WeChat platform. In this paper we proposed some additional factors to TAM model to achieve our study goal, from TAM model research selected two factors, perceived ease of use (PEOU), perceived usefulness (PU) [17], since trust the basis of financial transaction perceive of trust was added [18] financial security is the most important factor for mobile payment to users [19]. Since WeChat is a kind of social network it improved some interesting activities to entertain users, therefore, we added perceived sense of enjoyment as a new important factor to the model [18]. Satisfaction is the main customer loyalty driver. In the current study, we are investigating the impact of the red packet on achieving user's satisfaction and loyalty (Fig. 1).

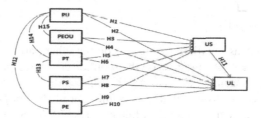

Fig. 1. Research model

3.1 Perceived Usefulness (PU)

Perceived usefulness refers to the degree to which a person believes that using a particular system would enhance his or her job performance [17] in another word when people use specific technology it will help to achieve the goal. Davis ensured that perceived usefulness would affect users to use the new technology. Recently, some relative academic researchers also affirmed the Perceived Usefulness has a positive effect on users' satisfaction and loyalty [20, 21]. User regards the more useful red packet (WeChat), the more satisfaction and loyalty users will have. Thus the following hypotheses were suggested.

H1. Perceived Usefulness has a significant positive impact on Users satisfaction.

H2. Perceived Usefulness has a significant positive impact on Users loyalty.

3.2 Perceived Ease of Use (PEOU)

According to [17] definition, the Perceived Ease of Use is the degree to which a person believes that using a particular system would be free from effort. When people use new information technology, the task can be understood and easy to operate. In this paper perceive ease of use defined as easy for users to use the new technology, Red packet (WeChat) is very easy to operate. [22] Confirmed that perceived ease of use would impact user's online operations. The influence of perceived usefulness (PU), perceived ease of use (PEOU), and trust on satisfaction, arguing that ease of use has a significantly positive effect on satisfaction. This demonstrates that the easier and more effective of the WeChat red packet is to understand and use, the higher the customer satisfaction after using the services of WeChat red packet [23]. Consequently, the following hypotheses are proposed:

H3. Perceived Ease of Use has a significant positive impact on Users satisfaction.

H4. Perceived Ease of Use has a significant positive impact on Users loyalty.

H5. Perceived Ease of Use has a significant positive impact on perceived usefulness.

3.3 Perceived Trust (TF)

Trust has been defined as a willingness to rely on the other partner in whom one confidence [24]. A dishonest of this trust by the supplier or service provider could cause defection. Schurr and Ozanne [25] also defined trust as the belief that will lead to the partner's word or promise to be reliable in fulfillment of the obligation in a relationship. Trust has been defined in so many different ways [7, 26–29]. Trust is a significant factor in a financial transaction especially in e-commerce [30] in e-commerce trust depends on reliability and technology security of the system, the trust degree can also be affected by the relationship and credit records of the customer. When a customer uses the WeChat payment he/she will be asked to add his/her personal information, and credit card information customer may worry about the privacy and financial risk, etc. [18].

H6. Perceived trust has a significant positive impact on Users satisfaction.

H7. Perceived trust has a significant positive impact on Users loyalty.

H8. Perceived trust has a significant positive impact on perceived usefulness.

3.4 Perceived Security (PS)

Security was defined as the system ability to protect users personal data from the unauthorized source, during the online transaction [31]. Security considered as a vital factor for online purchasing [32]. Security and privacy play an essential role in creating trust during online transactions [33]. Customers are willing to buy online if only confident to provides their personal information [34]. According to [31] security can be divided into two part first part is related to data and transactions, the second part directly related to the authenticity of consumers, if the customers are secure, they will tend to be satisfied. Therefore security is essential for e-commerce, and it is a significant factor for a customer to purchase online, based on the above, we come up with proposed hypotheses.

H9. Perceived security has a significant positive impact on Users satisfaction.

H10. Perceived security has a significant positive impact on Users loyalty.

H11. Perceived security has a significant positive impact on perceived trust.

3.5 Perceive Entertainment (PE)

According to "Flow theory" which has been developed by [35] " flow – the state in which people are so involved in an activity that nothing else seems to matter; the experience itself is so enjoyable that people will do it even at great cost, for the sheer sake of doing it." In a study that led [36] about users' adoption behavior of mobile advertisement, the result found that entertainment has a significant effect on the attitude of the user to accept a mobile advertisement. Perceived entertainment should be taken into consideration, especially in designing social media marketing. WeChat red packet is not only a money transfer tool; it has also been developed for fun. For example, it is very common that during certain holidays (e.g., spring festival), people send red packets in a group of friends in WeChat and everyone in the group can receive a random amount of money from that red packet. Normally the person who receives the highest amount of money will continue to send a Red packet in that group, which is for fun and people enjoy such entertainment a lot [37]. From the above, the coming hypotheses were proposed.

H12. Perceive Entertainment has significant positive impact on Users satisfaction.

H13. Perceive Entertainment has a significant positive impact on Users loyalty.

H14. Perceive Entertainment has a significant positive impact on perceived usefulness.

3.6 Customer Satisfaction

It is one of the most critical issues that attracted the attention of the business organization of all types, which is justified by the customer-oriented philosophy and continues improvement principles of modern enterprise [6]. Having a good relationship with the customer will influence the service provider [38], which will result in customer loyalty [39]. It has been proved that the cost of attracting a new customer is high five times the cost of retaining the current customer [40, 41]. Therefore for firms to improve customer satisfaction and loyalty need to research the factors that impacting customer satisfaction and reusing the product/service and achieve loyalty by satisfying the customer. Customer satisfaction considers as a measure for describing how the product/service meets the customer expectations, and it is an important element for ensuring successful business because satisfaction can reflect the firm growth in the future. However, the dimensions of satisfaction and trust were the most common variables in the relationship marketing studies for achieving customer loyalty [7]; these dimensions are also relevant in the context of e-commerce.

H15. Perceived users satisfaction has a significant positive impact on user's loyalty.

4   Research Methodology

4.1   Respondents

The data of the current research were collected from the users of WeChat platform in China. An online questionnaire was designed to gather the data, in this research we mainly targeted on those who have certain experience at using WeChat Red Packet. Researchers have found 804 which were found to be completed and valid for the statistical analysis with a respondent's rate represents (41%) 330 males and (59%) 474 females. The sample size was deemed large enough to generate accurate results for the entire study population.

4.2   Instrument

The questionnaire was written in English then translated to Chinese, and it contents of 40 items each item was measured on 5 points Likert scale. From strongly disagree (1) to strongly agree (5). The questionnaire was based on the previous studies of [17, 18, 42, 43] and it has been reviewed by two experts, and it was modified based on their comments. The first application involves 30 users have been randomly selected to fill the questionnaire for testing validity and reliability for each item specifically, and then some items were improved accordingly.

5   Results

5.1   Reliability and Validity

Cronbach's alpha measurement had been used for testing the reliability and validity display validity coefficient 0.789, 0.821, 0.858, 0.897, 0.852, 0.779, 0.898. According to [44]. The results of the reliability and validity refer to high values for each dimension in the current study questionnaire.

In this study, we use Cronbach's alpha coefficient to test the reliability and validity of the data, we calculated the Cronbach's alpha coefficient of each factor and data, the total Cronbach's alpha coefficient is (0.807) and each of the Cronbach's alpha coefficient dimension is greater than (0.5), according to [45–47] mentioned that prediction for group of (25–50) is acceptable. It means that the collected data meet the general requirement.

The result of (T- value) in this research allows us to realize each variables different effect on user's satisfaction, for user's loyalty (F-value) was used. The variables difference in their influence process and assessment. The T-test supported the 15 original hypotheses, T value was greater than 1.96 (as $p < 0.05$) (see Table 1, 2). That is Users

Table 1. Correlation and regression

Items	R2	R	T-value	F-value	ρ
PU⟶US	0.182	0.426	24.563***	–	0.000
PU⟶UL	0.195	0.442	24.607***	–	0.000
PEOU⟶US	0.307	0.554	10.796***	–	0.000
PEOU⟶UL	0.133	0.365	15.322***	–	0.000
TF⟶US	0.308	0.555	10.799***	–	0.000
TF⟶UL	0.134	0.366	15.324***	–	0.000
SF⟶US	0.373	0.611	14.037***	–	0.000
SF⟶UL	0.239	0.489	16.769***	–	0.000
PE⟶US	0.378	0.616	18.165***	–	0.000
PS⟶UL	0.300	0.548	10.271***	–	0.000
PE⟶PU	0.369	0.607	17.133***	–	0.000
PS⟶PU	0.616	0.379	8.611***	–	0.000
PT⟶PU	0.229	0.480	7.165***	–	0.000
PEOU⟶PU	0.230	0.480	28.178***	–	0.000
US⟶UL	0.315	0.561	14.540***	–	0.000
PU, PEOU, PT, PS, PE⟶US	0.364	0.603	–	458.259***	0.000
PU, PEOU, PT, PS, PE⟶UL	0.309	0.556	–	356.242***	0.000

***$\rho < 0.001$

Table 2. Discriminant validity analysis.

Factors	1	2	3	4	5	6	7
PU							
PEOU	.480**	1					
PT	.480**	1.000**	1				
PS	.457**	.616**	.616**	1			
PE	.426**	.554**	.554**	.611**	1		
US	.442**	.365**	.365**	.489**	.561**	1	
UL	.732**	.798**	.798**	.794**	.786**	.737**	1

**$\rho \leq 0.01$

satisfaction can directly and positively affect uses loyalty; the technology acceptance determinants can directly and positively affect users satisfaction and users loyalty (see Fig. 2). Moreover, the analysis of this study interestingly found that the influence of (PEOU, PT, PS, PE) factors are more on users satisfaction than on users loyalty. From the user's satisfaction side, the perceived entertainment factor (0.161) and security factor (0.611), influence was found to be stronger on user's satisfaction than user's loyalty. From the user's loyalty side, the impact of perceived entertainment was the highest (0.548) it's followed by security factor (0.489). Thus, users, satisfaction plays a vital role as a mediator. The five independent factors influence user's loyalty through user's satisfaction (0.561).

5.2 Correlation and Regression

Technology acceptance model was applied for this study, and it has been extended accordingly. Perceived entertainment, perceived trust, perceived ease of use has a positive relationship with perceived usefulness, interestingly perceived entertainment found to have the strongest correlation with perceived usefulness, and user's satisfaction. Regression, T-value, and F-value, results are shown in Table 1. The correlation of the five dependent variables with US as follow, $R = 0.603$, $R2 = 0.364$, F-value $= 458.259$ and the five dependent variables correlation with UL as follow, $R = 0.556$, $R2 = 0.309$, F-value $= 356.242$, all of the five determinants found to have a significant positive relationship with user's satisfaction as it shown in Fig. 2.

We used Bartlett's test of sphericity and (KMO) measure of sampling adequacy. The results illustrate the suitability of the conducting factor analysis.

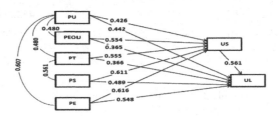

Fig. 2. Research model results

6 Discussion

This study developed a theoretical framework based TAM model, to discuss user satisfaction and loyalty by using WeChat Red Packet as one of the user's loyalty motivator tools.

The study found that, perceived usefulness, has positive direct relationship with users satisfaction (0.426), and indirect contact with users loyalty (0.442), if the platform will help users to improve their job performance, or they realized that its useful for them they will tend to be satisfied and satisfied customer will tend to be loyal customer, many other researchers have proved that perceived usefulness (PU) will have impact on satisfaction and loyalty, this consent with [17, 48, 49]. In consent with [50] 70% of Chinese WeChat users likes to get the red packet incentive. Therefore Red Packet is useful in Chinese daily life and very easy to use.

Perceived ease of use was found to have a strong direct impact on users satisfaction (0.554), and indirect positive impact on users loyalty (0.365), interestingly same with a previous study in the field PEOU has a significant relationship with PU [18]. If the new technology is complicated people will tend to avoid using it, ease of use refers to the degree of free effort at using technology; this means, easy to understand and use [17], easy to learn and easy to operate [51]. Perceived ease of use is the main role of WeChat platform success.

According to previous studies trust has found to be the most important factor in performing an online transaction, in this research we found, trust has a very strong positive relationship with customer satisfaction in the degree of (0.555) directly and indirect relationship with users loyalty (0.366), this matched with [52]. Moreover, trust has found to have an influence on PU more trust more satisfaction. Trust consider being financial transaction foundation [30, 53]. Yue. Q et al. [18] proved that lower trust leads people not to use the service. If the service provider is not a trustworthy customer would not tend to use the service [54].

Security factor has a significant direct relationship with users satisfaction (0.611) in WeChat Red Packet usage, and positive indirect relationship with users loyalty (0.489) this consent with [55–57]. 61% of the survey participant in [33] would be linked with the internet if their privacy and personal data will be protected. In this study, the user considered security as a direct factor for their satisfaction within WeChat Red Packet; however, security is found to have a significant positive relationship with perceived trust (0.561). The finding of this study consent with [18, 32]. Trust is influenced by security, according to [58]. Users who feel comfortable and secured at using WeChat red packet tend to use it more than others [59].

Perceived entertainment have the most significant influence on users satisfaction (0.616), which will reflect on users loyalty (0.548); this finding comes in line with [60]. WeChat payment is different from other online payments, due to social interaction, WeChat introduced a different feature to strengthen interaction between users; the red packet was one of the most important features which attracted millions of peoples to join WeChat. The result of this study consent with [61] that perceived enjoyment enhances users entertainment and increases the use of WeChat payment. People received red packet and coupon they will consider WeChat payment in some situation also instead of others online payment [18]. Finally, strengthen the relationship among the users will guarantee the continuation in using WeChat payment in the future.

Users satisfaction, has significant relationship with users loyalty (R = 0.561), which exactly in line with [49], mobile commerce study, uses loyalty was found that directly and indirectly affected by users satisfaction, also finding of this study consent with the previous studies in the area, which proved that higher customer satisfaction will lead to customer loyalty [7, 28, 32]. A satisfied customer will share his/her experience with others around them; the organization should strive to build and maintain long term lasting relationship with the users through satisfying various users' needs and incentive them to continue using the service that delivered by the organization.

We can notice that in Fig. 2 our five variables showed a significant relationship with customer loyalty, and strongest significant direct relationship with customer satisfaction, this ensured that satisfied customer tends to be loyal [7].

Our proposed model has practical and theoretical implications for researchers and the implementer to create social payment services mobile. This research gives a depth analysis of WeChat Red Packet use in China and how the social network is utilized. Also, the feature of the red packet can be developed in other countries based on their culture, WeChat Company have a chance of developing 25 Red Packet features around the world. This study added new insight into the field of knowledge.

6.1 Limitations and Future Research

Same with other research, there is a limitation in this study which deserves effort for future research. First WeChat is used in 25 countries around the world with the same features, in this study we focused on WeChat Red Packet which has been based on Chinese culture of giving Red Packet to friends and families during the different occasions. This research conducted in China mainland on the WeChat App. The result of this study may not be generalized and inapplicable in other countries. More research is needed in this area to find out what motivate people to use WeChat App in other countries. Also, the role of culture could be considered to give further depth to the research. Second, the selected determinants of customer loyalty may not cover all the reasons that could affect WeChat (Red Packet) user's loyalty in China. Future research can regard other factors such as efficiency, service availability. Moreover, WeChat Company can create the same feature of Red Packet based on the different country culture of giving money.

7 Conclusion

In conclusion, the emergence of the internet on mobile has changed our daily lives. This new social network has attracted billion of users. The finding of this study revealed that perceived usefulness, perceived ease of use, trust factor, security factor, customer, perceived entertainment, satisfaction, are all determinants that effect WeChat (Red Packet) users loyalty in China directly or indirectly. Perceived entertainment was found to have the strongest motive of user satisfaction which indirect influences user loyalty. Also, perceived usefulness has found to have a strong effect on customer loyalty than customer satisfaction. This research sheds new insight into the field of knowledge in the area of mobile payment. It also provides insight into the factors that influence users' loyalty within the mobile payment.

References

1. Xingang, L.: Analysis of criminal activities exploiting social media: with special regards to criminal cases of Wechat fraud in Chinese jurisdiction. J. Legal Stud. "Vasile Goldiş. **26**(40), 19–36 (2020)
2. Hui, C., Cao, Y.: Computers in human behavior examining WeChat users' motivations, trust, attitudes, and positive word-of-mouth: evidence from China. Comput. Human Behav. **41**, 104–111 (2014)
3. Gao, F., Zhang, Y.: Analysis of WeChat on iPhone. In: 2nd International Symposium on Computer, Communication, Control and Automation, pp. 278–281 (2013

4. Chang, Y.P., Zhu, D.H.: Computers in human behavior understanding social networking sites adoption in China: a comparison of pre-adoption and post-adoption. Comput. Human Behav. **27**(5), 1840–1848 (2011)
5. Kim, Y.M.: Gender role and the use of university library website resources: a social cognitive theory perspective. J. Inf. Sci. **36**(5), 603–617 (2010)
6. Arokiasamy, A.R.A.: The impact of customer satisfaction on customer loyalty and intentions to switch in the banking. **5**(1), 14–21
7. Abosag, I., East, B.S., Tynan, C., Lewis, C.: The commitment-trust theory: the British and Saudi Arabian cross-national perspectives. J. Bus. Ind. Mark. **1994**, 1–24 (2006)
8. Veloutsou, C.: Identifying the dimensions of the product-brand and consumer relationship. J. Mark. Manag. **23**(1–2), 7–26 (2007)
9. Ndubisi, N.O.: Relationship marketing and customer loyalty. Marketing intelligence & planning. 13 Feb, (2007)
10. Dong-Hun, L.: Growing popularity of social media and business strategy. SERI Q **3**(4), 112–117 (2010)
11. A. . WILSON, "The internet is destroying the world as we know it," 2019.
12. Constantinides, E., Fountain, S.J.: Web 2.0: conceptual foundations and marketing issues. J. Direct Data Digit. Mark. Pract. **9**(3), 231–244 (2008)
13. Reichheld, F.F., Schefter, P.: E-Loyalty: your secret weapon on the web. Harvard Bus. Rev. **78**(4), 105–113 (2000)
14. China, E.: WeChat e-commerce guide. [Online]. Available: https://www.marketingtochina.com/wechat-e commerce-guide/2019. Accessed 18 Feb 2019
15. Chao, E.: How WeChat became China's app for everything. Innovation Agents. [Online]. Available: https://www.fastcompany.com/3065255/china-wechat-tencent-red-env elopes-and-social-money. Accessed 18 Feb 2019
16. Wechatpay. How to add a foreign credit card to WeChat pay (2018)
17. Davis, F.D., Bagozzi, R.P., Warshaw, P.R.: User acceptance of computer technology: a comparison of two theoretical models. Manag. Sci. **35**(8), 982–1003 (1989)
18. Qu, Y., Rong, W., Chen, H., Ouyang, Y., Xiong, Z.: Influencing factors analysis for a social network web based payment service in China. J. Theor. Appl. Electron. Comer. Res. **13**(3), 99–113 (2018)
19. Zhou, T.: An empirical examination of initial trust in mobile payment. Wireless Pers. Commun. **77**(2), 1519–1531 (2014)
20. Alraimi, K.M., Zo, H., Ciganek, A.P.: Computers & education understanding the MOOCs continuance: the role of openness and reputation. Comput. Educ. **80**, 28–38 (2015)
21. Alonso-Dos-Santos, M., Soto-Fuentes, Y., Valderrama-Palma, V.A.: Determinants of mobile banking users' loyalty. J. Promot. Manag. **26**(5), 615–633 (2020)
22. Katawetawaraks, C., Wang, C.L.: Online shopper behavior: influences of online shopping decision. Asian J. Bus. Res. **1**(2), 66–74 (2011)
23. Hossain, M.S., Russel, A.H., Robidas, L.C.: Analysis of factors affecting the customer's satisfaction with reference to ATM services in Dhaka city. IOSR J. Bus. Manag. **17**(11), 1 (2015)
24. Moorman, C., Deshpande, R., Zaltman, G.: Factors affecting trust in market research relationships. J. Mark. **57**(1), 81–101 (1993)
25. Schurr, P.H., Ozanne, J.L.: Influence on exchange processes: buyers' preconceptions of a seller's trustworthiness and bargaining toughness. J. Consum. Res. **11**(4), 939–953 (1985)
26. Wilson, D.T.: An Integrated Model of Buyer-Seller Relationships. The Pennsylvania State University. Institute for the Study of Business Markets. The Pennsylvania State University, vol. 3004, no. 814. (1995) (ISBM Report lo-1995)
27. Evans, K.R., Crosby, L.A., Cowles, D.: Relationship quality in services selling: an interpersonal influence perspective. J. Mark. **54**(3), 68–81 (1990)

28. James, C.A., Narus, J.A.: A model of the distributor's perspective of a distributor-manufacturer working relationship. J. Mark. Appl. Electron. **48**(1), 62–74 (1984)
29. Bitner, M.: Building service relationships: it's all about promises. J. Acad. Mark. Sci. **23**(4), 246–251 (1995)
30. Gefen, D.: E-commerce: the role of familiarity and trust. Omega **28**(6), 725–737 (2000)
31. Guo, X., Ling, K.C., Liu, M.: Evaluating factors influencing customer satisfaction towards online shopping in China. Asian Soc. Sci. **8**(13), 40–50 (2012)
32. Eid, M.I.: determinants of e-commerce customer satisfaction, trust, and loyalty in Saudi Arabia. J. Electron. Commer. Res. **12**, 78–93 (2007)
33. Chellappa, R.K.: Consumers' trust in electronic commerce transactions: the role of perceived privacy and perceived security. under submission 13 (2008)
34. Whysall, P.: Retailing and the internet: a review of ethical issues. Int. J. Retail Distrib. Manag. **28**(11), 481–489 (2000)
35. Csikszentmihalyi, M.: Flow: The Psychology of Optimal Experience, Harper and Row. Harper and Row, New York (1990)
36. Xiao, Y., Meng, K., Takahashi, D.: Accountability using flow-net: design, implementation, and performance evaluation. Secur. Commun. Netw. **5**(1), 29–49 (2012)
37. Wu, Z., Ma, X.: Money as a social currency to manage group dynamics: red packet gifting in Chinese online communities. In: Proceedings of the 2017 CHI Conference Extended Abstracts on Human Factors in Computing Systems, pp. 2240–2247 (2017)
38. Panda, T.K.: Creating customer lifetime value through effective CRM in the financial services industry. J. Serv. Res. **22**, 157–171 (2003)
39. Jones, M.A., Mothersbaugh, D.L., Beatty, S.E.: Why customers stay: measuring the underlying dimensions of services switching costs and managing their differential strategic outcomes. J. Bus. Res. **55**, 441–450 (2002)
40. Kotler, P., Rackham, N., Krishnaswamy, S.: Ending the war between sales and marketing. Harvard Bus. Rev. **84**(7/8), 68 (2006)
41. Teich, I.: Holding on to customers: the bottom-line benefits of relationship building. Bank Mark. **29**(2), 12–13 (1997)
42. Gerhardt, P., Schilke, O., Wirtz, B.W.: Electronic commerce research and applications understanding consumer acceptance of mobile payment services: An empirical analysis. Electron. Commer. Res. Appl. **9**(3), 209–216 (2010)
43. Lu, Y., Zhou, T., Wang, B.: Computers in human behavior exploring Chinese users' acceptance of instant messaging using the theory of planned behavior, the technology acceptance model, and the flow theory. Comput. Human Behav. **25**(1), 29–39 (2009)
44. Nunnally, J.C., Bernstein, I.: Psychometric Theory, 3rd edn. McGraw-Hill, New York (1994)
45. Peterson, R.A.: A meta-analysis of Cronbach's coefficient alpha. J. Consum. Res. **21**(2), 381–391 (1994)
46. Lance, C.E., Butts, M.M., Michels, L.C.: What did they really say? Organ. Res. Methods **9**(2), 202–220 (2006)
47. Sinclair, K., et al.: Development of a questionnaire to measure the level of reflective thinking. Assess Eval. High. Educ. **25**(4), 381–395 (2007)
48. Rai, A., Lang, S.S., Welker, R.B.: Assessing the validity of IS success models: an empirical test and theoretical analysis. Inf. Syst. Res. **13**(1), 50–69 (2002)
49. Lin, H.H., Wang, Y.S.: An examination of the determinants of customer loyalty in mobile commerce contexts. Inf. Manag. **43**(3), 271–282 (2006)
50. Feng, E.Q.: An exploration of incentives motivating WeChat uses in China. (4) (2016)
51. Ahmed, S.A.A.M., Qiu, T., Xia, F., Jedari, B.: Event-based mobile social networks: services, technologies, and applications. IEEE Access **2**, 500–513 (2014)
52. Cyr, D.: Modeling web site design across cultures: relationships to trust, satisfaction, and e-loyalty. J. Manag. Inf. Syst. **24**(4), 47–72 (2008)

53. Pi, S.-M., Liao, H.-L., Chen, H.-M.: Factors that affect consumers' trust and continuous adoption of online financial services. Int. J. Bus. Manag. **7**(9), 108–119 (2012)
54. Chong, A.Y.-L.: Understanding mobile commerce continuance intentions: an empirical analysis of Chinese consumers. J. Comput. Inf. Syst. **53**(4), 22–30 (2013)
55. Wolfinbarger, M.: eTailQ: Dimensionalizing, measuring and predicting etail quality. J. Retail. **79**, 183–198 (2003)
56. Szymanski, D.M., Hise, R.T.: E-satisfaction: an initial examination. J. Retail. **76**(3), 309–322 (2000)
57. Onn, W., Soon, L.: Determinants of mobile commerce customer loyalty in Malaysia. Procedia - Soc. Behav. Sci. **224**, 60–67 (2016)
58. Flavián, C., Guinalíu, M.: Consumer trust, perceived security and privacy policy: three basic elements of loyalty to a web site. Ind. Manag. Data Syst. **106**(5), 601–620 (2006)
59. Mumin, Y.A., Ustarz, Y., Yakubu, I.: Automated teller machine (atm) operation features and usage in Ghana: implications for managerial decisions. J. Bus. Adm. Educ. **5**(2), 137–157 (2014)
60. Mackey, T.P.: Exploring the relationships between web usability and students perceived learning in web-based multimedia (WBMM) tutorials. Comput. Educ. **50**, 386–409 (2008)
61. Soares, A.M., Pinho, J.C.: Advertising in online social networks: the role of perceived enjoyment and social influence. J. Res. Interact. Mark. **8**(3), 245–263 (2014)

Author Index

Printed in the United States
by Baker & Taylor Publisher Services